Mathematics as a Laboratory Tool

John Milton · Toru Ohira

Mathematics as a Laboratory Tool

Dynamics, Delays and Noise

Second Edition

 Springer

John Milton
W.M. Keck Science Department
The Claremont Colleges
Claremont, CA, USA

Toru Ohira
Graduate School of Mathematics
Nagoya University
Nagoya, Aichi, Japan

ISBN 978-3-030-69581-1   ISBN 978-3-030-69579-8   (eBook)
https://doi.org/10.1007/978-3-030-69579-8

Mathematics Subject Classification: 00A06, 34-01, 39-01, 42-01, 44-01, 60-01, 80-01, 82-01, 92-01, 93-01

1st edition: © Springer Science+Business Media New York 2014
2nd edition: © Springer Nature Switzerland AG 2021

This Springer imprint is published by the registered company Springer Nature Switzerland AG
The registered company address is: Gewerbestrasse 11, 6330 Cham, Switzerland

To our teachers, our students, and the students of our students

Preface to the Second Edition

Increasingly, research into important biological problems is being tackled by interdisciplinary teams of scientists composed of experimentally trained biologists and traditionally trained mathematicians [588]. How can communication between team members be facilitated? The mathematical training of biologists is typically limited to an introductory course in calculus and statistics. Although mathematicians are often able to understand the biology they typically have had little exposure to the types of mathematical problems that arise at the benchtop. This textbook was written to serve as a reference text for teams that investigate the dynamical phenomena generated by real biological systems.

The first 13 chapters form the basis of an undergraduate course in biological dynamics with an emphasis on excitable systems and regulation by feedback control mechanisms. We have used these chapters to successfully teach an undergraduate course on dynamics for biologists in a liberal arts college for over 15 years. We emphasize the study of deterministic dynamical systems using graphical approaches by focusing on the analysis of ordinary differential equations (ODE) and delay differential equations (DDE). We have made every effort to ensure that the material is accessible to those with a first course in calculus supplemented with the use of software packages such as XPPAUT and a few more advanced mathematical concepts which we refer to as Tools. Emphasis is placed on identifying results that can be used at the benchtop, for example, how to distinguish linear from nonlinear dynamical systems, identify bifurcations from experimental observations, calculate a power spectrum, use frequency-domain techniques to analyze feedback control mechanisms, and use phase-resetting approaches to characterize biological oscillators. Given the growing interest of students in computational neuroscience, we have expanded our discussions of excitability, the onset and offset of oscillations in excitable systems (Chapter 11), the design of optimal stimuli to cause switches between attractors (Chapter 12), and multistability in neuronal motifs, in particular the recurrent inhibitory loop (Chapter 13).

The "falling epidemic in the elderly" has motivated applications of DDEs to understand human balance control. Consequently, our analysis of DDEs has been extended to include the analysis of second-order DDEs (Chapter 10) and the occurrence of microchaos in time-delayed digital control of mechanical systems (Chapter 13). In addition, there has been the development of new numerical methods for integrating DDEs, such as semidiscretization. Intelligent use of these techniques requires a more careful discussion of the numerical integration of DDEs (Chapter 10). We have provided templates for computer programs that integrate DDEs in XPPAUT and MATLAB.

The second half of the textbook (Chapters 14–17) focusses on the analysis of dynamics measured at the benchtop. All real biological systems contain time delays, are influenced by noisy perturbations, and are ultimately governed by thermodynamic principles of energy flows. The availability of inexpensive high-speed motion-capture cameras has made it possible to obtain reliable data for a variety of human balancing tasks, including pole balancing at the fingertip, postural sway during quiet standing with eyes closed, and gait. Thus our mathematical discussions focus on the statistical description of noise, the analysis of stochastic and stochastic delay differential equations, random walks, and concepts related to entropy production and power-law behaviors. The few textbooks that are available on these subjects are not typically accessible for an undergraduate readership. Again we have tried to keep our presentation accessible for undergraduates where possible.

Since the first edition of *Mathematics as a laboratory tool*, there has been a growing use of Bayesian inference in biology, particularly in neuroscience (Chapter 14). In our experience, most students have difficulty interpreting the concepts of *a priori* and *a posteriori* probabilities in terms of experimental observations. Therefore we have introduced these concepts by using a number of detective-type examples and in particular, have included a discussion of the Monty Hall problem. Finally, we discuss two neuroscience examples, adaptation in motor learning and audio-visual integration.

We eliminated the laboratory exercises so that those who use this book to teach undergraduates can have fun developing their own laboratory experiences. We have incorporated the material from our own laboratory exercises to generate more examples within the chapters and create more exercises at the end of each chapter.

Perhaps the most embarrassing reason for this second edition was to correct the mistakes we made in the first edition! Although many of these errors arose during the process of preparing the Latex documents, others were simply mistakes. We sincerely hope that the people who use this book enjoy it as much as we have in writing it.

Claremont, CA, USA   John Milton
Nagoya, Japan   Toru Ohira

Preface to the First Edition

It is common in physics and chemistry that the mathematical tools of the discipline are taught by practicing physicists and chemists rather than by mathematicians. Why, then, are biologists not teaching the mathematical tools needed and used by biologists?

The interaction between mathematics and biologists is very much like the relationship between a sheep dog and a sheep: although it is uncomfortable at the beginning, in the end both parties benefit. We believe that by linking mathematics to science through comparisons between prediction and experimental observations rather than by theorems and proofs, it is possible to make a compelling case to biologists that mathematical modeling is something that may actually be useful. Moreover, we find that when the material is presented in this way, biologists are often surprised to discover that mathematics is itself a research-based activity, especially as mathematicians struggle to develop new tools and frameworks to deal with the complexities of the living world.

Perhaps the most important reason for biologists to be teaching their students mathematics arises from the dramatic increase in the number of applications of mathematical and computer models to the study of biological phenomena. On the other hand, rapid advances in laboratory techniques and instrumentation have reached the point where it is no longer feasible for a single researcher to develop an appropriate model and then validate it in the laboratory or real-world setting. Consequently, present-day scientific advances are, more often than not, made through the efforts of multidisciplinary teams composed of biologists, computer scientists, engineers, and mathematicians. One of the challenges facing under-graduate education in the life sciences is identifying strategies to prepare students to work productively as team members.

Successful teamwork critically depends on effective communication among team members. How is it possible for mathematicians and biologists to communicate effectively? On the one hand, the development of the mathematical biology major at undergraduate colleges and universities has prepared students with strong mathe-matical backgrounds who have also taken relevant biology courses with a labora-tory component for further study and research. On the other hand, the mathematical requirements for biology majors are typically limited to an introductory course in differential and integral calculus (or less) and a course in statistics. The availability of computer symbolic manipulation packages, such as Mathematica, Wolfram Alpha, and Sage, makes it possible to teach mathematical concepts without getting bogged down in the technical details of solving a particular example. Armed with computer simulation software packages such as XPPAUT and programming lan-guages such as MATLAB and Scientific Python (SciPy), it is possible for students to see solutions of differential equations evolve before their eyes and to use pow-erful mathematical tools, such as the fast Fourier transform, to analyze data with just a few keystrokes.

However, the indiscriminate use of these computer aids without an understanding of the underlying mathematical concepts is a prescription for disaster. Our approach is to take advantage of computer software tools while teaching the underlying concepts in a manner that is relevant to scientists but at the same time requires only modest mathematical preparation.

The goal of this book is to provide undergraduate biology students an educated overview of the mathematical techniques and concepts useful at the benchtop. We focus on the time evolution, or *dynamics*, of biological systems. Our purpose is not to make benchtop researchers into "chalkboard" modelers of biological phenomena by providing encyclopedic coverage. Rather, our concern is to enable researchers to be flexible enough to change direction quickly, if necessary, in response to new experimental observations. By enabling researchers to read the modeling literature, we hope that they will be motivated to design and perform critical laboratory experiments to test the predictions of existing models and to identify biological phenomena that would benefit from the guidance provided by the development of relevant models by their more mathematically oriented colleagues.

The topics we cover include the following: What can be learned about the nature of the generators of observed dynamics from consideration of the responses (outputs) to inputs (Chapters 2, 4–8, and 11)? What types of inputs are most useful for identifying an underlying dynamical system (Chapters 6, 7, 11, 14)? How are biological systems controlled using time-delayed feedback control mechanisms (Chapter 9)? What are the effects of random perturbations ("noise") (Chapters 14–16)? How does the interplay between deterministic mechanisms and stochastic forces shape the behavior of living systems (Chapters 15 and 16)? How do the dynamics of thermodynamically open systems operating far from equilibrium that characterize life differ from the dynamics of thermodynamically closed systems operating near equilibrium that are typically studied in the laboratory (Chapters 3 and 17)?

In order to accomplish our goals, we have made a number of choices that distinguish this book from standard textbooks in mathematical biology. First, we take advantage of the fact that the simplest cases for a topic can typically be solved exactly and that key insights can often be obtained graphically. However, for our approach to be useful, it is necessary to introduce material that normally would be covered in certain upper-division courses in mathematics. We introduce this material in the form of *tools*—our students prefer to call them tricks. We find that the term "tool" reassures students that they are capable of understanding quite sophisticated concepts when all of the necessary facts are supplied to them.

Second, we use open-source computer software packages to extend the applications learned from the simpler cases to the more complex scenarios encountered at the benchtop. In part, this choice is made pragmatically, since the application of mathematical techniques to real-life experimental situations almost always involves the use of computers. The use of open-source computing tools makes it possible for students to have these tools operating on their own personal computers without concern about excessive expense. Moreover, students involved in the open-source community learn at first hand the power of the Internet for obtaining implementable solutions to real problems in a timely manner.

Finally, we emphasize applications that, at least in principle, can be studied within the time constraint of a 4-h weekly laboratory session. Two types of living dynamical systems are potentially attractive for such a laboratory: (1) the dynamical behaviors of excitable systems, in particular invertebrate neurons and certain green algae, and (2) the dynamics of biomechanical systems in motion, for example, postural sway, stick-balancing at the fingertip, and locomotion. However, most second-year students do not have the skills necessary to record changes in membrane potential generated by excitable cells, such as neurons and muscle cells, as a function of time. Moreover, the cost of the laboratory equipment for such measurements becomes prohibitive as class size increases. On the other hand, the availability of inexpensive motion-capture systems and noninvasive monitoring equipment, some of which use tablet computers and smartphones, makes the second option more attractive. A second advantage is that studying motion taps into student interest in exercise and sporting activities. Applications to problems in ecology, molecular biology, and physiology are included to demonstrate the wide applicability of the laboratory tools.

The laboratory exercises we developed for our classes are listed in *Laboratory Exercises and Projects*. A lab manual that gives the material we provide to our students is provided on a website. It includes information about other websites from which data can be obtained and the materials we use to teach Python programming. Laboratory exercises will be updated annually and new labs added as they are developed. We hope that teachers at other institutions will share their laboratory exercises with others using this repository.

Three recent trends in undergraduate biology education at the Claremont Colleges provided the impetus for this book. First, it was necessary to give students the mathematical and computational tools necessary for participation in a project funded by the National Science Foundation's Undergraduate Biology and Mathematics program entitled Research Experiences at the Biology–Mathematics Interface (REBMI) [588, 590]. The goal of the REBMI project was to enable students to develop skill sets necessary to work in interdisciplinary teams composed of a laboratory-oriented biology student and a quantitatively oriented mathematics student. Together, the REBMI team was expected to obtain an implementable solution in a novel problem-solving setting, often located at an off-campus research facility, within an 8-week period. By the phrase "novel problem-solving setting" was meant topics that were not explicitly covered either in the students' previous coursework or in their research experiences. Most of the REBMI students took a course based on this textbook or the equivalent. The mathematical tools cover the basic skills that the student teams needed to participate in many of these research projects. Second, the dramatic increase in the number of students enrolled in neuroscience majors has made it necessary to prepare students by their junior year to understand topics not typically covered in introductory mathematics courses, including excitability, feedback control, time delays, noise, transfer functions, filters, and power spectra. Finally, funding provided by the Howard Hughes Medical

Institute is being used to make curricular changes at the five Claremont Colleges so that quantitative methods can be more effectively introduced to undergraduate biologists.

There is a fine line between the need for mathematical rigor and making the material accessible to an experimentally oriented audience: too much rigor loses the biological audience, while too little leaves students with insufficient insight into what techniques to apply in what situations and how to judge whether they have been applied correctly. Nonetheless, this is the task that we have tried to accomplish in the hope that more biologists will come to appreciate that mathematics is important and actually makes their work easier, more effective, and more exciting. It is our students who will decide whether we have succeeded.

Claremont, CA, USA   John Milton
Nagoya, Japan   Toru Ohira

Acknowledgements for the Second Edition

We thank the help of Jacques Bélair (Université de Montreal), Sue Ann Campbell (University of Waterloo), Hannah Caris (Pomona College), Richard Cangelosi (Gonzaga College), Joshua Chang (UT Austin), and Leon Glass (McGill University) for pointing our errors out and suggesting ways to correct them. J. M. was supported by the William R. Kenan Jr. Charitable trust and a J. T. Oden visiting faculty fellowship while at the Oden Institute for Computational Engineering and Sciences, UT Austin. T. O. would like to thank the funding support from Ohagi Hospital (Hashimoto, Wakayama, Japan).

Acknowledgments for First Edition

We acknowledge Jennifer Foss for her efforts in teaching the early versions of this material to undergraduates at the University of Chicago (1998–2002) and Roman Natoli, who put the course lectures into the form of a LATEX document (2002). In 2004, John Milton moved to the Claremont Colleges, where he has benefited from many faculty members who have taught him to teach, rather than lecture, undergraduate students. Toru Ohira was added at that time as a coauthor because of the growing importance of random walks and stochastic processes in biology. In addition, he has visited our campuses every summer to work with our students on their research problems.

In 2005, John Hunter and John Milton began talking about how scientific computing could be taught to undergraduate science students in a liberal arts environment. They attended a Scientific Python conference at Caltech, where with Andrew Straw, they began to formulate the idea of a scientific computing workshop for undergraduate students. The first SciPy workshop was held in the autumn of 2007. With the addition of Fernando Perez, the fall-semester SciPy workshops continued annually until 2012. Many of the labs described in the lab manual arose from those workshops.

We thank A. Fucaloro for helpful discussions on equilibrium thermodynamics and J. Bélair, Sue Ann Campbell, A. Landsberg, J. Higdon, and A. Radunskaya for discussions concerning teaching mathematics to biology students. We acknowledge the efforts of Leon Glass, Michael C. Mackey, and the many reviewers who identified errors in our manuscript and the many undergraduates who proofread early versions of the manuscript, in particular Leah Rosenblum. Translating the material in this textbook into a student laboratory experience would not have been possible without the assistance and patience of many people. Specifically, we acknowledge the efforts of Walter J. Cook for electronic hardware development, Boyle Ke for IT support, Bard Ermentrout for his great patience and help with XPPAUT, Arthur H. Lee for his lectures introducing programming in Python, Emily Nordhoff for developing the Arduino lab, and Caleb Mills and Stefan van der

Walt for putting together most of the Python demonstration programs. We thank David Kramer for his careful editing of the first edition and in particular for identifying a number of places that required clarification. Finally, we acknowledge funding from the William R. Kenan Jr. Charitable Trust, the National Institutes of Mental Health, the Brain Research Foundation, and the National Science Foundation.

Contents

Notation

$a := b$	a is by definition equal to b
α	Lévy exponent
β	Power-law exponent
$B(f), B(s)$	Respectively the Fourier and Laplace transforms of the forcing to a differential equation or input to a black box
$c_{xx}(\varDelta)$	Estimated value of the autocorrelation function
$C_{xx}(\varDelta)$	True value of the autocorrelation function
$c_{xy}(\varDelta)$	Estimated value of the cross-correlation function
$C_{xy}(\varDelta)$	True value of the cross-correlation function
$\delta(t - t_0)$	Dirac delta function
$\delta_{n,m}$	Kronecker delta function
$đx$	Change in variable x that depends on the path taken
ϵ	Damping coefficient
ε	Small parameter
λ	Eigenvalue
f	Frequency when used as a parameter; otherwise, a mathematical function
E	Internal energy
$\langle x^n \rangle$	nth moment, or expectation, of a random variable
ξ	Random variable
F	Feedback in the context of control; force in other contexts
G	Gibbs free energy
$C(s)$	Laplace transform of transfer function
γ	Real part of a complex eigenvalue
$g(t), G(f)$	Fourier transform pair. We use different consonants to describe different Fourier transform pairs. Our convention is to use an uppercase consonant to designate the variable in the frequency domain
H	Enthalpy
$H(x, p)$	Hamiltonian
$H(x - c)$	Heaviside, or unit step, function
$I(t, t')$	Impulse function
$\Im(z)$	Imaginary part of the complex number z. If $z = x + jy$, where x, y are real and j is the imaginary unit, then $\Im(z) = y$.
j	The imaginary unit $\sqrt{-1}$
k, K	Parameter (positive unless specifically noted)
Λ	Entropy production
$\mathscr{L}(\cdot)$	Laplace transform of a function
μ	Bifurcation parameter
ω	Angular velocity
Ω	Ohms when used as unit of resistance; number of configurations when used in a thermodynamic context
ϕ	Phase (radians)

$\angle$	Phase (degrees)
p	Momentum
P	Pressure
$p(x)$	Probability density function
$P(x \leq X)$	Probability distribution function
$p(x_1 \mid x_2)$	Conditional probability
$p(x_1; x_2)$	Joint probability
q	Heat
$\Re(z)$	Real part of the complex number z. If $z = x + jy$, where x, y are real and j is the imaginary unit, then $\Re(z) = y$
$r(t)$	Ramp function
S	Entropy
σ^2	Variance
T	Period in the context of oscillations; temperature in context of thermodynamics
τ	Time delay
$\mathcal{T}(x)$	Time scale of variable x
θ	Angle
$U(x, y)$	Lyapunov function
φ	Characteristic function
V	Volume
w	Work
$w(f)$	Estimated power spectrum
$W(f)$	True power spectrum
x, y	Variables
$X(f), X(s)$	Respectively the Fourier and Laplace transforms of the solution of a differential equation or output of a black box

Tools

We assume that readers have completed an introductory course in differential and integral calculus and a course in statistics, but not necessarily courses in linear algebra and differential equations. Our students are encouraged to use symbolic manipulation programs such as Mathematica, WolframAlpha, and Sage to evaluate specific derivatives and integrals as needed. We have identified a few mathematical results that are frequently needed, yet are either not covered at all in introductory calculus courses or if covered, insufficiently emphasized. We have called these methods *tools*. Below, we list each tool together with the section in the text where it is first introduced.

Tool 1 Principle of superposition: Section 1.3.1
Tool 2 Definition of the derivative: Section 2.1
Tool 3 Time scale of change: Section 2.2.2
Tool 4 Linear differential equations with constant coefficients: Section 2.2.3
Tool 5 Law of mass action: Section 3.1
Tool 6 Integrating factor: Section 3.4.2
Tool 7 Linearization using Taylor's theorem: Section 4.2
Tool 8 Euler's formula for complex numbers: Section 7.2.1
Tool 9 Laplace integral transform: Section 7.2.2
Tool 10 Fourier series: Section 8.2.1
Tool 11 Fourier transform: Section 8.2.2
Tool 12 Generating functions: Section 16.8.1

Chapter 1
Science and the Mathematics of Black Boxes

An important activity of biological scientists involves making measurements on living organisms. In the past, biologists focused primarily on the static aspects of life. For example, the recognition that some plants are benign while others are poisonous surely provided the impetus for learning how to tell useful plants from dangerous ones. Progress in the identification and classification of living organisms has been closely tied to the increasing precision of measuring instruments. Indeed, progress in the biological sciences has always been intimately associated with progress in instrumentation: human dissection (315 B.C.E), anatomical illustration (1472), gross anatomy (1858), the cellular structure of tissues (light microscopy, 1668), the structure of cells (electron and scanning microscopy, 1965–1971), the structure of molecules (X-ray diffraction, 1912, and force tunneling microscopy, 1971), including the DNA helix (1963), and the human genome project (2003).

These measurements are of two kinds: those of phenomena that vary little if at all over time (*statics*) and those that change in significant ways over time (*dynamics*). While there is much of interest in static configurations, the prime feature of the biological world is its dynamism. The one thing that this world is not is static! In fact, the only constant is change: the weather changes, organisms age and die, birds fly, hearts beat, and even the stars in the sky execute complex arabesques of motion.

However, there is often an intimate relationship between structure and function. For example, the catalytic activities of enzymes depend on the three-dimensional conformation of the protein, the propulsion of ducks through water on the webbing of their feet, and the performance of athletes on the design and construction of their equipment. Nonetheless, scientists have long realized that the study of dynamics often gives enormous insight into the nature of underlying mechanisms.

Two obstacles have hindered the study of dynamics of living organisms. First, devices to measure time accurately were not developed until the mid-1700s [778]. Second, it took time for mathematics to develop to the point that interesting and meaningful questions about time-varying phenomena could be posed and solved.

© Springer Nature Switzerland AG 2021
J. Milton and T. Ohira, *Mathematics as a Laboratory Tool*,
https://doi.org/10.1007/978-3-030-69579-8_1

Indeed, the concept of the derivative was introduced by Isaac Newton[1] less than 400 years ago, and the impact of nonlinear dynamics in experimental biology began in earnest only about 60 years ago.[2] However, those two obstacles do not sufficiently explain the lag in research on the dynamics of life. The answer lies in the time required to develop our most important instrument for studying time-dependent phenomena, the digital computer: the first "analytical engine" was developed in 1822, the first computer program in 1834, the CRAY supercomputer in 1958, IBM 360s in 1971, the first functional program-controlled digital computer, the Z3, in 1941, and the personal computer in the 1970s.

The purpose of this book is to introduce students to the study of the dynamics of biological systems in a laboratory setting. The area of mathematics that describes the behavior of dynamical systems is referred to as *dynamical systems theory*. However, the use of mathematics in the laboratory is not limited simply to making models. Mathematics is also intimately involved in taking measurements and making comparisons between observation and prediction. The irony is that many students with a scientific bent become biologists in the futile effort to escape mathematics.

There are many very good textbooks that discuss the features of various mathematical models that describe biological dynamical systems [164, 186, 193, 276, 617, 618, 728, 810]. In contrast, there are very few textbooks available to experimental biologists that address issues related to the experimental analysis of dynamical systems (exceptions include [846]). The goal of this book is to fill that gap. Our examples will be largely chosen from physiological systems. This is because many such systems evolve on time scales of seconds to minutes, and thus it is possible to collect the large data sets necessary for making the direct comparisons between observation and prediction that lie at the core of the scientific method. However, all of the methods we discuss can, at least in principle, be applied to any area of biology, provided that the variables under investigation change on suitable length and time scales.

The questions we will answer in this chapter are:

1. What is a system?
2. What are surroundings?
3. What is a variable?
4. What is a parameter?
5. What is a dynamical system?
6. How can we show experimentally that a dynamical system behaves linearly?

---

[1] Isaac Newton (1642–1727), English physicist and mathematician, formulated the laws of motion and universal gravitation and invented the calculus.

[2] We have arbitrarily identified the onset of the mathematical studies of nonlinear biological systems in the laboratory with the development of the Hodgkin–Huxley equations of the neuron [347].

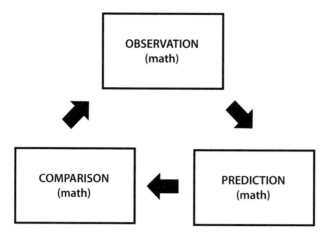

Fig. 1.1 Schematic representation of the scientific method.

1.1 The Scientific Method

Science is the human activity that employs the scientific method to investigate phenomena that can be measured. Applying the scientific method involves three steps (Figure 1.1): collecting experimental observations; formulating a hypothesis, or model, that predicts the mechanism that generates the experimental observations; and carefully comparing predictions to experimental observations. Invariably, there will be discrepancies between predictions and observations. Moreover, new experimental observations will be made that were not available at the time the initial model was formulated. These discrepancies may demand the formulation of new hypotheses, the collection of more observations, and so on.

It is important to realize that there is no guarantee that the scientific method will ever lead to the "perfect hypothesis." All that can be said with certainty is that as the scientific method is employed in refining a model, the range of experimental observations that can be accounted for by the model increases. In other words, proof is not a realizable outcome of the scientific method.

In the context of the scientific method, mathematics is a tool that allows scientists to make better measurements, better comparisons between observation and prediction, and better models. Mathematical analysis can illuminate flaws in proposed models and hence help eliminate such models from further consideration. Estimating the limitations of a model is also established at the benchtop by making direct comparisons between prediction and observation. Mother Nature is the final arbiter!

1.2 Dynamical Systems

The most important questions for a scientist to consider are these: What is the problem that I am interested in studying? What questions do I want to answer?

What is the feasibility of obtaining an answer given the limitations imposed by current technologies? We will use the term *system* to mean the part of the universe in which we are currently interested and the term *surroundings* to include everything else (Figure 1.2). The boundary between the system and its surroundings is not written in stone. We wish to include the most important elements in our system and relegate those that are peripheral to the surroundings. The boundary between system and surroundings changes as new information becomes available, as new technologies enable better measurements, as our understanding of the key mathematical issues increases, and as experimentalists develop better experiments to compare a model's predictions to observed results.

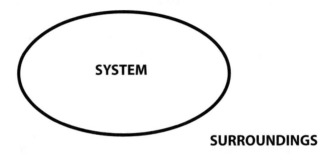

Fig. 1.2 The delineation between a system and its surroundings and the interactions between them is a crucial step in the scientific method.

Equally important is the choice of the appropriate scale at which the problem should be considered. For example, if our interest is the kinematics of animal locomotion, it would likely not be fruitful to consider the interaction between the contractile proteins actin and myosin at the level of single atoms. On the other hand, the dynamics of the molecular motors involved in axonal transport require consideration of molecular motions.

1.2.1 Variables

Our focus is on the interpretation of experiments in which some variable is measured as a function of time. What is a *variable*? A variable is anything that can be measured. Examples of variables include weight, height, concentration, membrane potential, and number of particles. Obviously, the availability of instruments capable of measuring variables of interest with sufficient accuracy and sufficient temporal resolution is a very important concern. An experimentalist may have to abandon temporarily the study of certain variables simply because a suitable instrument does not yet exist or is too expensive to obtain.

The term *time series* is used to describe a table that lists the values of variables together with the times at which each variable was measured. The term *dynamics* describes how variables change as a function of time. For example, do the variables fluctuate rhythmically, do they approach a constant value, or do they vary in a seemingly random manner? Finally, we use the term *dynamical system* to refer to a mathematical model that generates a time series that resembles a system that is measured experimentally.

What is a parameter? *A parameter is a quantity that changes on a time scale that is clearly separable from that of the variables of interest.* In mathematics, the distinction between variables and parameters is self-evident. For example, if we have the equation for a straight line,

$$y = k_1 t + k_2,$$

where y is a function of t, then t and y are variables, and the quantities k_1, k_2 are parameters, since $dk_1/dt = dk_2/dt = 0$. From the mathematical perspective, parameters provide the connection between the independent variable t and the dependent variable y. If we change k_1, we change the slope of the line, whereas changing k_2 changes the intercept of the line, i.e., the value of y when $t = 0$. Thus we can anticipate that there will be relationships between the values of the parameters and the behavior of the dynamical system.

In biology, however, the distinction between a parameter and a variable can be a difficult one to make. There are two problems: (1) there is often not a clear separation in time scales between different variables, and (2) there are limitations in measuring devices. Typically, experimentalists must make compromises in their choice of recording instrument, its response time, and the frequency at which data are digitized by an analog-to-digital (A/D) converter.

As a consequence, differences between variables can be resolved only over a certain range of frequencies and amplitudes: changes that fall outside the capabilities of the measuring device are not resolved.

To illustrate, suppose we wanted to create a model to describe the binding of a drug to a membrane-bound receptor. The binding step takes less than 10^{-9} s, and the number of receptors changes on time scales ranging from hours to months. Thus there is a clear distinction between the time scales for drug binding on the one hand and receptor synthesis and destruction on the other. If we are interested in studying the dynamics of drug binding, then we use an instrument capable of resolving events on very short time scales. On these short time scales, the number of receptors n can be assumed to be constant; that is, we may consider it a parameter for the model, since $dn/dt \approx 0$. If we were interested in studying receptor dynamics, we would choose an instrument that measures the number of receptors on a much slower time scale, perhaps once per hour or once per day. Since such an instrument would not likely be capable of resolving the dynamics of drug binding, the drug would be assumed to be either free or bound to the receptor in the corresponding mathematical model.

More generally, there is not always a clear separation of time scales between biological variables. In such cases, the investigator must make choices concerning how variables not of primary interest are to be handled. First, the investigator must determine what constitutes the system and what constitutes the surroundings. Depending on context, some of these variables might be considered parameters, while others might be lumped together into a time-varying background noisy input signal.

1.2.2 Measurements

Since our emphasis is on how biological systems change as a function of time, we need to consider what types of biological variables can be measured as a function of time, the accuracy of such measurements, and the frequency with which such measurements can and should be made. The answers to these questions are critically important for choosing the type of mathematical model that is meaningful to develop. In general, it is difficult and expensive to measure more than one biological variable simultaneously as a function of time. Thus modeling studies are very closely linked to advances in technology.

Not surprisingly, early mathematical success stories are those in which variables could be measured using electromagnetic (e.g., current flow, potential differences), spectroscopic (e.g., absorption spectra, fluorescent dyes), or cinematographic (e.g., motion-capture cameras) techniques. Examples include excitable cells (heart and other muscle cells, neurons) and certain periodic biochemical reactions. With the growing importance of computers and high-speed imaging technologies, the list of dynamical systems that can be fruitfully investigated is rapidly growing. Nonetheless, limitations of technology remain the major barrier to the application of the scientific method to dynamical systems. These observations suggest that the number of areas of biology that are most amenable to fruitful investigation is presently fairly small, but will grow as advances in technology are made.

A common fallacy is the belief that scientists always make measurements as accurately as possible. In truth, the necessary degree of accuracy depends on, and only on, the question one is trying to answer. Indeed, a good scientist figures out beforehand what level of accuracy is required to answer a particular question and then wastes no effort trying to do substantially better. In terms of testing the validity of mathematical models, consideration of the accuracy needed to verify a model's predictions is important. Practical problems arise when measurement uncertainty is too large to resolve the intrinsic nature of the predicted dynamics. However, many modern-day measuring devices have reached the point that the precision with which a variable can be measured is better than the variability in the value of the variable due to the effects of random perturbations ("noise"). This observation has motivated interest in the properties of stochastic dynamical systems (Chapters 15 and 16).

An important question concerns the evaluation of parameters for mathematical models. Ideally, all parameters should be estimated using experimental methods. However, modelers often combine statistical methods with data to fit parameters to their model (for a nice introduction, see [164]). This procedure provides an estimate of the "best fit" of a model to the observed dynamics. This approach can eliminate a proposed model as a candidate by demonstrating that for the best-fit parameters, the model cannot satisfactorily reproduce the observed dynamics. However, this approach does not circumvent the need to measure parameters. The important question for the validation of a proposed model as a viable candidate is whether it describes the experimental observations for the experimentally determined parameters. Since we cannot avoid estimating parameters experimentally, we believe that it is best to begin with models that are simple enough to be amenable to careful mathematical and experimental investigations. The advantage is that direct comparisons can be made between prediction and observation [232, 238, 520, 584, 652, 736, 906, 907].

1.2.3 Units

Measurements must be accompanied by a unit for them to have physical meaning, and moreover, units are necessary for making decisions and comparisons. For example, one would certainly hesitate before buying a car if its advertised mileage were simply 35. The obvious question is 35 what? There is a big difference between 35 miles per gallon and 35 km per liter, or even 35 yards per gallon!

It is also highly desirable that scientists all over the world use the same units. Historically, different countries used different units for scientific measurements. For example, in the United States of America, a person's height is measured in feet and inches, whereas in most of the rest of the world, height is measured in centimeters. The International System of Units (or SI, from the French *Système International d'Unités*) was established to create an international standard that could be used by scientists worldwide. In the SI system, there are seven fundamental units (Table 1.1): length, mass, time, electric current, temperature, amount of substance, and luminous. Every other unit is considered to be derived from these fundamental units. For example, velocity is length divided by time, force is mass times acceleration, and pressure is force per area. The newest SI rules require in addition (1) that prefixes (e.g., kilo-, centi-) for larger and smaller units progress in steps of three orders of magnitude (e.g., from grams to kilograms) and (2) that the prefixes in complex units be attached only to the numerator, not the denominator (e.g., kg/m^3 is acceptable but kg/mm^3 is not, since the prefix "milli-" appears in the denominator).

There are two points that often cause confusion. The first is to remember that a number of objects has no units. The units of Avogadro's number, namely the number of molecules per mole, is not molecules per mole, but mol^{-1}. Second, angles are measured in radians, and a radian is a dimensionless number. The fact that radian is a dimensionless quantity follows from its definition: the measure of an angle in radians is the length of arc of a circle subtended by that angle divided by the radius of the circle and hence has units of length divided by length (Table 1.1).

In mathematical models, equations are often put into dimensionless form. Although this procedure typically simplifies the form of the equation to be analyzed, there are two important caveats. First, the original equation must be dimensionally correct [81]. Indeed, checking to ensure that the units on both sides of an equation are the same is an important step in developing a model. Moreover, when the units in an equation do not match, consideration of the missing units is often useful

Fundamental unit	Unit symbol	Quantity	Some derived units
Meter	m	Length	Volume (m^3)
			Radians (rad) ($m \cdot m^{-1}$)
			Solid angle (steradian, sr) ($m^2 \cdot m^{-2}$)
Kilogram	kg	Mass	Density ($kg \cdot m^{-3}$)
Second	s	Time	Velocity ($m \cdot s^{-1}$)
			Acceleration ($m \cdot s^{-2}$)
			Force (newtons, N) ($kg \cdot m \cdot s^{-2}$)
			Pressure (pascals) ($N \cdot m^{-2}$)
			Work, energy, heat (joules, J) ($N \cdot m$)
			Power (watts, W) ($J \cdot s^{-1}$)
Ampere	A	Electric current	Voltage (volts, V) ($W \cdot A^{-1}$)
			Electric charge (coulombs, C) ($s \cdot A$)
			Resistance (ohms, Ω) ($V \cdot A^{-1}$)
Kelvin	K	Thermo-dynamic temperature	Heat capacity ($J \cdot K^{-1}$)
Mole	mol	Amount of substance	Molar volume ($m^3 \cdot mol^{-1}$)
			Molar concentration ($mol \cdot m^{-3}$)
Candela	cd	Luminous intensity	Luminous flux (lumens, lm) ($cd \cdot sr$)
			Illuminance (lux, lx) ($lm \cdot m^{-2}$)

Table 1.1 The fundamental SI units and some commonly used derived units.

for identifying what the model has overlooked [81]. Second, once the mathematical analysis for the dimensionless equation has been completed, the parameter values must be checked to see whether they are reasonable. It can happen that the phenomena predicted by the dimensionless model occur for parameter ranges that are not reasonable for a living system. Such a demonstration would suggest that the model is not valid for the system under investigation and should encourage the modeler to seek a more robust model.

1.3 Input–Output Relationships

In the real world, living organisms do not come with the equations that describe their actions imprinted on their bodies for all to see. Therefore, we do not begin our investigations with the assumption that a suitable mathematical model is known.

Indeed, one of the most exciting scientific adventures is the search for plausible mathematical models. From this point of view, experimental research is very much like trying to figure out what is inside a so-called black box (Figure 1.3). The black box in the figure corresponds to the system of interest. The experimentalist is not allowed to look inside to see how things work but only to apply an input and observe the resulting output. In other words, the only way that we can hypothesize how the black box might work is by determining its *input–output relationships*, i.e., the relationship between a measured output and the input given to the system.

Fig. 1.3 From the perspective of an experimentalist, the world looks very much like a black box.

From a practical point of view, experimental research, and hence science, can be thought of as an analysis of input–output relationships. It is this description that most clearly underscores two different approaches for unveiling the contents of the black box. One approach is to model the equations that describe the system (the black box) and then use the proposed equations to ask the question, "what do the outputs look like for all possible inputs?" This is the approach most commonly taken by biomathematicians, and it leads to the analysis of the dynamics of a system as a function of time. The second approach attempts to identify the equations that describe the black box from an analysis of the input–output relationships: "how can we best guess what is inside the black box by looking at the relationship between inputs and outputs?" This is the approach taken by bioengineers, and as we will see, it leads to an analysis of a system's dynamics as a function of frequency f, where $f = 1/t$.

1.3.1 Linear Versus Nonlinear Black Boxes

A fundamental problem is to determine whether the system under investigation behaves like a linear or nonlinear dynamical system. In the laboratory, this determination is made by employing the principle of superposition (see Figure 1.4 and Exercises 1 and 2 at end of this chapter).

Tool 1: Principle of Superposition

Give an input $I_1(t)$ to the black box and measure the output $O_1(t)$. Then give an input $I_2(t)$ to the black box and measure the output $O_2(t)$. Finally, we give the input $I_1(t) + I_2(t)$ and measure the output $O_3(t)$. If $O_3(t) = O_1(t) + O_2(t)$ for a range of values of t, then the black box behaves like a linear dynamical system over that range;

if $O_3(t) \neq O_1(t) + O_2(t)$, then the black box behaves like a nonlinear dynamical system.

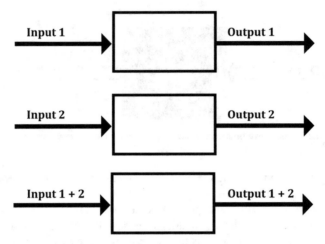

Fig. 1.4 Schematic representation of the principle of superposition for a linear dynamical system.

Tool 1 is extremely useful in the laboratory. If a dynamical system can be shown to behave linearly, then the input–output relations of the system can be completely determined. Surprisingly, it is sometimes much easier to do this in the frequency domain. These approaches are based on the Laplace (Chapter 7) and Fourier (Chapter 8) integral transforms.

Strictly speaking, all real-world dynamical systems are nonlinear. However, it still can be true that to a first approximation, many dynamical systems behave linearly over an appropriate range of parameters and variables. In other words, nonlinear systems can be *almost linear* over a certain range of variables and parameters. Thus an important yet seldom explicitly addressed problem is to determine the ranges of parameters and variables that correspond to a linear system. Moreover, it is possible that although the individual components of a dynamical system are nonlinear, a population of them subjected to random perturbations can behave in a manner that is surprisingly similar to a linear dynamical system (see Chapter 14). Thus before a scientist embarks on the study of a nonlinear dynamical system, it is necessary to define the question to be answered precisely in order to ensure that the questions posed indeed cannot be adequately answered using a linear approximation.

1.3.2 The Neuron as a Dynamical System

We illustrate these concepts with a consideration of the input–output properties of neurons. Neurons are *excitable cells*. An "excitable cell" is one that does not

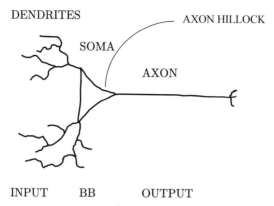

Fig. 1.5 Schematic representation of a neuron, showing the three main compartments—dendrites, soma, and axon—and the location of the axon hillock (top row of labels). The bottom row of labels shows the correspondence to the black box (BB) analogy of a biological dynamical system.

generate an output, or action potential, until it has received sufficient input to cause its membrane potential to exceed a certain threshold, called the *spiking threshold*. In general, a neuron receives both excitatory and inhibitory inputs. The excitatory inputs, or excitatory postsynaptic potentials (EPSPs), increase the membrane potential, while the inhibitory inputs, or inhibitory postsynaptic potentials (IPSPs), decrease the membrane potential. A neuron is organized into three compartments (Figure 1.5): dendrites; a cell body, or soma; and an axon. The basic mechanism of the neuron is that inputs are received at the dendrites and summed at a specialized region on the soma, called the axon hillock. If the spiking threshold is exceeded, an action potential is generated at the hillock, which then travels along the axon. Thus from this simplified point of view, a neuron very much resembles the black box we have used as a metaphor for a biological process. Consequently, measurements of the outputs of neurons as a function of their inputs have been a topic of great interest.

The integrate-and-fire (IF) neuron shown in Figure 1.6 exhibits both linear and nonlinear properties. When the membrane potential is less than the spiking threshold, the neuron sums its inputs linearly (Section 6.4.1). In contrast, dendrites of living neurons exhibit many nonlinear properties and hence do not integrate inputs linearly [110]. An action potential is generated when the membrane potential exceeds the spiking threshold. This violation of the principle of superposition (Tool 1) establishes that action potential formation in the IF neuron is a nonlinear phenomenon. When the action potential occurs, both the membrane potential and the spiking threshold undergo rapid changes (Figure 1.6). The spiking threshold immediately elevates to a very high value for 1 ms and then slowly returns toward its resting value within 1–2 ms. This means that for 1–2 ms after the onset of an action potential, the neuron is *absolutely refractory*, that is, completely unresponsive to any stimulus. During this time, the neuron cannot be made to generate an action potential, no matter how large the input.

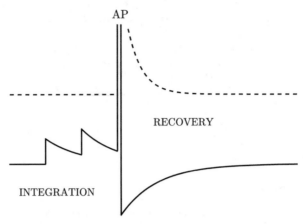

Fig. 1.6 Schematic representation of the changes in membrane potential (solid line) and spiking threshold (dashed line) for an integrate-and-fire (IF) neuron leading up to and following the generation of an action potential (AP). To simplify, we have assumed that prior to AP generation, the IF neuron receives only excitatory inputs and does not receive inputs when the AP is generated. The integration is "leaky" (Section 6.4.1), i.e., the membrane potential decreases between each excitatory input. Note that the recovery of the spiking threshold after the AP occurs faster than the repolarization of the membrane potential to its resting value.

At the same time that these changes are occurring in the spiking threshold, the membrane potential of the neuron quickly decreases to a value that is more negative than its resting potential. In other words, the neuron becomes *hyperpolarized*. In the absence of additional inputs, the membrane potential slowly increases toward its resting potential, typically within about 4–5 ms [59]. Thus there is a period of time after an action potential is generated during which the neuron is said to be in a *relatively refractory state*; namely, a larger input is required to generate an action potential than would be needed if the neuron were at its resting potential.[3] The relative refractory period is quite variable in other excitable cells. For example, it is approximately 70–120 ms in snail peripheral tactile sensory neurons [387], about 200 ms in dog [121] and human [91] cardiac myocytes, about 90 s in the excitable green algae such as *Nitella* [396], and about 150 s in the trabecular reticulum of sponges [480].

There are two limitations of integrate-and-fire approaches to the study of neural dynamics. First, living neurons do not possess a clear-cut membrane potential above which the neuron produces an action potential, but below which it does not. Indeed at the benchtop the spiking threshold might be defined as the membrane potential at which there is a 50 % probability that the neuron will produce an action potential. Second, these models do not account for the observation that under certain circum-

---

[3] It follows from these observations that the spiking threshold depends on the time that the neuron last produced an action potential, and hence on the past history of the neuron [567]. In this way, a neuron can be said to remember its past. Most mathematical and computer modeling studies ignore this effect of the relatively refractory period. For notable exceptions, see [129].

stances a hyperpolarizing inhibitory input can cause the neuron to produce an action potential, for example, anodal break excitation.

The observations in Figure 1.6 correctly emphasize that basic models for the behavior of neurons always include two types of variables [382, 721, 880]: a voltage variable and a recovery variable. Surprisingly it has been shown that two-variable models for neuron spike initiation are sufficient to account for the dynamical behaviors of neuron observed experimentally. Indeed, the biomathematics of neurons and neuronal populations provides one of the very best examples in biology of the power of mathematical and computational modeling guided by careful experimentation. Thus as we develop mathematical tools for laboratory investigators we will typically illustrate their use with applications to neurons and the nervous system. However, we emphasize that the same techniques are being used with great success to unravel the behaviors of other excitable cells including cardiac muscle and pancreatic beta cells.

1.4 Interactions Between System and Surroundings

How does a system interact with its surroundings? Three fundamental forms of isolation of a system from its surroundings are typically considered: (1) *adiabatic systems*: systems enclosed by walls that prevent exchange of both matter and thermal energy between the system and its surroundings; (2) *closed systems*: systems enclosed by walls that allow an exchange of heat or thermal energy with the surroundings but prevent exchanges of mass; (3) *open systems*: systems enclosed by walls that allow both matter and energy exchanges with the surroundings. The properties (e.g., dynamic and thermodynamic) of these three types of interactions between system and surroundings differ from one another (see Chapter 17).

Example 1.1 An example of a closed system is a chemical reaction that takes place within a test tube that is heated from below using the flame produced by a Bunsen burner.   ◇

Since we are distinguishing between a system and its surroundings, it is useful to distinguish between those variables that are internal to the system and those that are external. Of course, the very process of isolating our system from its surroundings is based on specifying some external variables such as those defining geometric dimensions. Additional external variables are the volume of the system, the external pressure exerted by the surroundings onto the system, and the electrical state of the external bodies in contact with the system. In contrast, internal variables describe the local and temporary properties of the system and include the momentary distribution of concentrations in a diffusion process and the distribution of pressures in a rapid-compression process.

Note that for the surroundings to function as true surroundings, the system must not itself influence the surroundings (if it does, then we need to change our definition of the system). In other words, changes in the system do not cause changes in the surroundings, but changes in the surroundings can potentially cause changes in the system. Thus the internal parameters are determined by the external parameters and by the positions, velocities, and other properties of the entities that compose the system, but the external parameters are not influenced by the internal parameters.

Example 1.2 Consider a herd of grazing animals in a pasture. We cannot identify the system as the herd and surroundings as the pasture, because the herd changes the pasture by its grazing activities, and that change, in turn, affects the herd. A better choice for the system would be the herd plus the pasture with the surroundings being that part of the world that lies outside the fence. What about the effects of the fence on the feeding behavior of the herd? Perhaps it would be best to require that the pasture be large enough that it would be unlikely for the herd to come near the fence.   ◇

The study of open systems is of particular importance for the understanding of biological phenomena. Unfortunately, the study of open systems is still very much in its infancy, and few important results have been obtained. However, two types of open system have received some attention. First, we may assume that the surroundings inject a periodic input into the system. Examples include the effects of circadian cycles, tidal cycles, and hormonal cycles (Chapter 11). Second, we may assume that the system receives simultaneously a large number of inputs from independent sources such that the summed input resembles a randomly varying signal, referred to here as *noise* (Chapter 14). On the usual assumption that the central limit theorem holds, these noisy inputs are distributed according to a normal (Gaussian) probability distribution. However, with the advance of better technology, it is becoming increasingly clear that fluctuations in biological systems are not necessarily normally distributed. Sometimes the experimental observations are better described by "distributions with broad shoulders" (see Section 16.7).

1.5 What Have We Learned?

1. What is a system?

   A system is the part of the universe that we are interested in studying.

2. What are surroundings?

   The surroundings comprise everything that is not included in the system. Although the surroundings can influence the system, the boundary between them must be chosen in such a way that the system does not affect the surroundings.

3. What is a variable?

   A variable is something that can be measured.

4. What is a parameter?

A parameter k is a quantity for which $dk/dt \approx 0$ on the time scale of interest.

5. What is a dynamical system?

   A dynamical system is a system in which the variables change as a function of time. We use the time-dependent changes in the variables to construct a model of the mechanisms that caused them to change.

6. How can we show experimentally that a dynamical system behaves linearly?

   We use the principle of superposition (Tool 1).

1.6 Exercises for Practice and Insight

1. Consider a system described by the equation of a straight line, namely

$$y = k_1 x + k_2.$$

Functions of this type are called *affine functions*. In terms of our black-box analogy, what corresponds to the input, the output, and the black box? Does such a dynamical system satisfy the principle of linear superposition (Tool 1)?

2. Show that the function

$$y = kx$$

satisfies the principle of linear superposition (Tool 1). For this reason, functions of the form kx are called *linear functions*.

3. Pick three topics that you are interested in studying in the laboratory. For each, define the system and its surroundings. Do events occurring in the surroundings affect your systems? For each of your choices, design an experiment to determine whether the system behaves linearly. Discuss the time scales you wish to study and then determine, for example, by using the Internet, whether appropriate instruments are available and whether they are affordable. Discuss your conclusions with your classmates and your instructor.

Chapter 2
The Mathematics of Change

How is mathematics used to translate experimental measurements into an understanding of the underlying regulating mechanisms? It was a long time before it was recognized that the values of variables do not in and of themselves provide insight into mechanisms. A key insight was that it is the change in the magnitude of a variable as a function of time that is of importance. Surprisingly, the changes that variables undergo can often be described in simple mathematical terms.

These realizations began with Galileo,[1] who wished to discover the laws that govern motion. He realized that it was not useful to ask about the cause of a state of motion if the motion is occurring at constant speed, any more than it is necessary to determine the reasons for a state of rest. The central problem is the change from rest to motion, from motion to rest, and more generally, all changes in velocity, that is, acceleration. This realization underpinned the subsequent development of Newton's three laws of motion, which make it possible to describe the motion of all objects in the universe traveling at speeds much less than the speed of light. In order to appreciate the importance of these observations, consider that Chinese science discovered the first law of motion and with its help developed gunpowder-powered rockets 2,000 years before Western science managed to do so. However, the Chinese were unable to aim and stabilize the flight of those rockets adequately, because Chinese science did not know the second and third laws of motion.

The questions we will answer in this chapter are:

1. What is a differential equation?
2. What is the order of a differential equation?
3. What is an ordinary differential equation (ODE)?

---

[1] Galileo Galilei (1564–1642), Italian physicist, mathematician, astronomer, and philosopher. He was an important advocate of the Copernican heliocentric theory.

© Springer Nature Switzerland AG 2021
J. Milton and T. Ohira, *Mathematics as a Laboratory Tool*,
https://doi.org/10.1007/978-3-030-69579-8_2

4. What is the time scale of change for a dynamical system?
5. What is the general form of the solution of an nth-order linear ODE with constant coefficients?

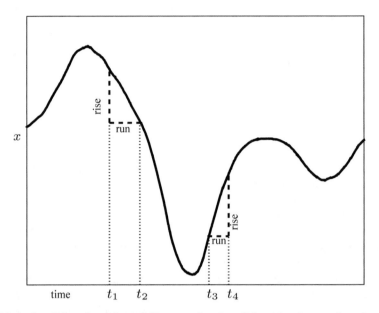

Fig. 2.1 A plot of the values of a variable x as a function of time, showing a region of negative slope (between t_1 and t_2) and a region of positive slope (between t_3 and t_4). The data are for the position of the fingertip in the anterior–posterior plane during stick-balancing at the fingertip.

2.1 Differentiation

Figure 2.1 shows a plot of the variable x as a function of time. Clearly, there are intervals during which x increases and other intervals during which x decreases. We can quantify the "ups and downs" of the changes in the values of x by estimating the slope, i.e., the change in the variable x over a given time interval, say between t_1 and t_2. We denote this slope by $\Delta x/\Delta t$, which can be calculated using the formula

$$\frac{\Delta x}{\Delta t} = \frac{x(t_2) - x(t_1)}{t_2 - t_1} = \frac{x(t + \Delta t) - x(t)}{\Delta t} = \frac{\text{rise}}{\text{run}}, \tag{2.1}$$

where $\Delta t = t_2 - t_1$. The slope is a function of the time interval we choose.

Tool 2: The Derivative

It is natural to ask what happens as the time interval becomes smaller and smaller and approaches zero, i.e., as $\Delta t \to 0$, because that will make it possible to determine the *instantaneous* rate of change. This is the question that Newton posed, and he was led to the following definition of the derivative:

$$\frac{dx}{dt} = \lim_{\Delta t \to 0} \frac{x(t + \Delta t) - x(t)}{\Delta t}. \tag{2.2}$$

In biology, we are often interested in variables that are functions of variables that in turn, are functions of t. For example, the growth of a population depends on its food supply, which varies as a function of time. In mathematics, such relations are referred to as *composite functions*. Specifically, if $x = f(u)$ and $u = g(t)$, then the composite function is $x = f(g(t))$. It is straightforward, but tedious, to use (2.2) to demonstrate that

$$\frac{dx}{dt} = \frac{df}{dg}\frac{dg}{dt}. \tag{2.3}$$

This result is often referred to as the *chain rule*.

2.2 Differential Equations

With the scientific method we seek to make predictions about phenomena that are subject to change. Thus we are particularly interested in equations that tell us how the rate at which a given quantity changes is related to some function of the quantity itself. For example, we would like to know how the rate at which the food supply changes is related to the growth of a given population, which in turn is a function of the food supply. That is, we are interested in equations of the form

$$\frac{dx}{dt} = f(x). \tag{2.4}$$

Equation (2.4) is called a *differential equation*. Since there is only one independent variable, namely time t, (2.4) is further distinguished as an *ordinary differential equation* (ODE). The left-hand side of (2.4) emphasizes that the important concept is the change in the variable per unit time. The right-hand side states the hypothesis that describes the rules proposed to govern the changes in the variable. The solution of this equation corresponds to something that should resemble the experimentally measured time series.

Although Tool 2 is typically introduced during a first course in calculus, its practical uses in the laboratory are seldom emphasized. Tool 2 provides the bridge between discrete-time and continuous-time differential equations. Indeed, it is often easier to derive a model in discrete time and then use Tool 2 to produce the corresponding ODE. Moreover, Tool 2 lies at the basis of a numerical method to integrate differential equations known as *Euler's method*[2] (see also Section 4.4.2).

The usefulness of the model described by (2.4) is determined solely by comparisons between prediction and experimental observation and the ability of this model to suggest new tests of its validity. In other words, there is no more justification for (2.4) than there is for its discrete analog, i.e.,

$$\frac{\Delta x}{\Delta t} = f(x). \tag{2.5}$$

Which of these models is appropriate for a given experiment depends on how frequently the variable can be measured, or sampled, and that is established only by comparison between observation and prediction. Indeed, since many presently available instruments collect data as digital rather than analog signals, it could be argued that (2.5) would be more relevant for laboratory work. Thus an important consideration will be to understand the conditions that ensure that data collected digitally provide a faithful representation of continuous time signals (Chapter 8).

2.2.1 Population Growth

Malthus[3] [532] suggested that the growth of human populations could be described by the linear ordinary differential equation

$$\frac{dx}{dt} = kx, \tag{2.6}$$

where $k > 0$ is the growth rate. This equation states the hypothesis that the growth of a population, i.e., the change in population density x per unit time, at any instant in time, depends solely on the population density at that time, where the population density is the number of organisms per unit area (or perhaps volume for certain aquatic or airborne organisms). Moreover, (2.6) asserts that the rate of population growth is proportional to the size of the population: the bigger the population, the faster it grows.

In order for the equation to be dimensionally correct, the units of k must be time^{-1}. The significance of the constant k is discussed below. By rearranging, this equation becomes

$$\frac{dx}{x} = k\,dt,$$

and we see that we can obtain a solution by integration, i.e.,

$$\int_{x_0}^{x} \frac{du}{u} = \int_{t_0}^{t} k\,dt. \tag{2.7}$$

---

[2] Leonhard Euler (1707–1783), Swiss mathematician and physicist.
[3] Thomas Robert Malthus (1766–1834), English cleric and scholar.

Completing the steps

$$\log x - \log x_0 = k(t - t_0), \quad \log x = \log x_0 + k(t - t_0),$$

where log denotes the natural logarithm, leads to the solution

$$x(t) = x_0 e^{k(t-t_0)}. \tag{2.8}$$

Substituting (2.8) into (2.6) confirms that (2.8) is in fact a solution. Since the choice of initial time t_0 is arbitrary, we can take $t_0 = 0$.

This is one of the few equations of practical significance that we will actually solve in this textbook. Let us look at it closely, so that we understand it very well. In order to solve this equation, we need to specify the time $t = t_0$ when the experiment began and what the population density x_0 was at that time. The value $x(t_0) = x_0$ forms the initial condition for (2.6).

The *order* of a differential equation is equal to the order of the highest derivative (see also Section 2.2.3). Thus (2.6) is a first-order linear ordinary differential equation. The *dimension* of a differential equation is the number of initial conditions required to specify a solution. Thus (2.6) describes a one-dimensional dynamical system. For ordinary differential equations the order equals the dimension.

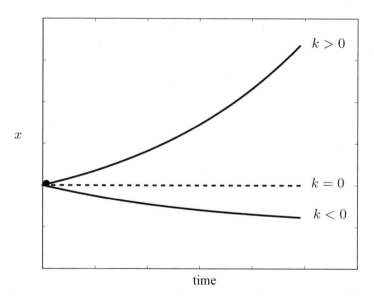

Fig. 2.2 Solution of (2.8) as a function of k. In each case, the initial condition is $x(t_0) = 1$, where $t_0 = 0$. (The point $(0, 1)$ is marked with •.) More details of how this experiment is begun are discussed in Chapter 6.

Figure 2.2 shows qualitatively different predictions (solutions) of (2.6) that depend on the value of k. If $k > 0$, then x grows without bound. This choice describes population growth: indeed, Malthus was the first scientist to point out that populations keep on growing unless there are mechanisms to regulate their size. On the other hand, if $k < 0$, the population decreases toward zero. This situation arises, for example, when a population is moving toward extinction.

How can we determine the value of k? From (2.8), we see that a plot of $\log x$ versus t (often called a semilog plot) should be linear. The slope of this plot provides an estimate of k.

2.2.2 Time Scale of Change

Lee Segel[4] [749] emphasized the concept of *time scale of change* for the analysis of dynamical systems described by a first-order linear differential equation. This time scale refers to how long it takes for a significant part of the change in a variable to occur and is one of an experimentalist's rules of thumb used to estimate the time scales in a dynamical system under investigation.

Tool 3: Time Scale of Change

Figure 2.3a shows how the time scale of change on a given time interval can be estimated as [749]

$$\mathscr{T}(x) := \text{time scale of } x(t) = \frac{x_{\max} - x_{\min}}{\left|\frac{dx}{dt}\right|_{\max}}. \tag{2.9}$$

Example 2.1 Let us calculate $\mathscr{T}(x)$ when $x(t)$ is given by (2.8) for $k < 0$ (Figure 2.3b). Obviously,

$$x_{\max} = x_0, \quad x_{\min} = 0,$$

and hence for fixed x_0, we have

$$\mathscr{T}(x) := \text{time scale of } x(t) \approx \frac{x_{\max}}{k x_{\max}} = k^{-1}. \tag{2.10}$$

This example shows that $\mathscr{T}(x)$ is the time $t = k^{-1}$ that it takes x to decrease to $e^{-k/k} = 1/e$ of its initial value. Scientists and engineers often speak about the $1/e$ *time* to describe the response of the system to a perturbation. The $1/e$ time serves as a useful rule of thumb for comparing the rates of change between different variables.

---

[4] Lee Segel (1932–2005), American applied mathematician and pioneer of modern mathematical and theoretical biology.

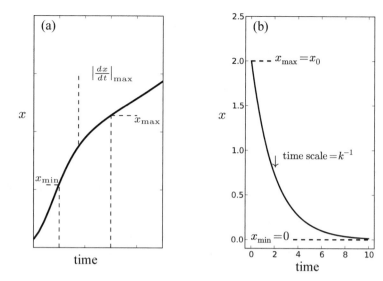

Fig. 2.3 (**a**) Schematic representation of the definition of the time scale given by (2.9). (**b**) Determination of the time scale for $x(t)$ given by (2.8) when $k < 0$ and $t_0 = 0$.

We will use this concept in the next chapter to gain insight into the concept of steady state in enzyme-catalyzed reactions.

2.2.3 Linear ODEs with Constant Coefficients

There are two ways to write an nth-order linear ordinary differential equation: as a single equation whose highest derivative is the nth and as a system of n first-order differential equations, i.e.,

$$
\frac{dx_1}{dt} = f_1(x_1, x_2, \ldots, x_n),
$$

$$
\frac{dx_2}{dt} = f_2(x_1, x_2, \ldots, x_n), \tag{2.11}
$$

$$
\frac{dx_n}{dt} = f_n(x_1, x_2, \ldots, x_n).
$$

If the f_i are linear functions of $x_1, x_2, \ldots, x_n$, that is, if they contain only terms of the form kx (see Exercise 2 in Chapter 1), then (2.11) describes the linear ordinary differential equations

$$\frac{dx_1}{dt} = k_{11}x_1 + k_{12}x_2 + \cdots + k_{1n}x_n,$$

$$\frac{dx_2}{dt} = k_{21}x_1 + k_{22}x_2 + \cdots + k_{2n}x_n, \qquad (2.12)$$

$$\frac{dx_n}{dt} = k_{n1}x_1 + k_{n2}x_2 + \cdots + k_{nn}x_n,$$

where the k_{ij} are parameters to be determined. Note that the order and dimension of (2.11) is equal to n.

The mathematical analysis of linear ordinary differential equations with constant coefficients is a completely solved problem [72, 366]. Thus, it is inexcusable for a scientist not to know about these results, since many experimental systems behave like linear dynamical systems over a certain range of the variables. Moreover, as we will see, many nonlinear systems can behave approximately linearly when subjected to the effects of noisy perturbations from the environment (Chapter 15).

Over the last 300 years, mathematical research has developed a whole arsenal of extremely powerful tools to analyze such systems. We will discuss many of these tools in this text, including topics such as stability theory (Chapter 4), systems identification using the transfer function (Chapter 7), and Fourier analysis (Chapter 8).

Example 2.2 Consider the equation for a damped harmonic oscillator:

$$\frac{d^2x}{dt^2} + k_2 \frac{dx}{dt} + k_1 x = 0, \qquad (2.13)$$

where dx/dt is the velocity and d^2x/dt^2 is the acceleration. We can easily rewrite this equation as the system of first-order equations

$$\frac{dx}{dt} = v, \qquad \frac{dv}{dt} = -k_1 x - k_2 v. \qquad (2.14)$$

The reader should verify that the system described by (2.14) is the same as the differential equation described by (2.13). $\diamond$

There are two reasons why the mathematical descriptions of living organisms are typically written as a system of first-order differential equations. First, it is often much easier to identify each x_i with an experimentally meaningful variable and to determine the mechanism that governs its changes. Second, it is the format typically used by computer algorithms to integrate differential equations.

Tool 4: "The Usual Ansatz"

One useful property of linear ordinary differential equations with constant coefficients is that the solution can often be written as a sum of exponentials, i.e.,

$$x(t) = c_1 e^{\lambda_1 t} + c_2 e^{\lambda_2 t} + \cdots + c_n e^{\lambda_n t}, \qquad (2.15)$$

where the c_i are constants that are determined using the initial conditions, and the λ_i are the rate constants, or as mathematicians like to say, the *eigenvalues*. This means that we can guess that the solution of a linear differential equation will be of the form $x(t) \approx e^{\lambda t}$. We will refer to this procedure as "making the usual ansatz."

The resulting equation in λ is referred to as the *characteristic equation*.[5] By making this ansatz, we change the problem of solving a differential equation into that of finding the roots of an algebraic equation. The eigenvalues can be real or complex. If the eigenvalues are complex, then the real part can be either positive or negative. It is easy to see that if the real part of at least one eigenvalue is positive, then $x(t) \to \infty$ as $t \to \infty$. Systems with eigenvalues having positive real part are said to be *unstable*, and those in which the real part of every eigenvalue is negative are said to be *stable*.

The reader might be tempted to think that unstable systems are irrelevant for experimental study. However, one of the major discoveries of the last century has been the realization that many biological processes, while locally unstable, are stable on a larger scale. On larger scales, nonlinear mechanisms operate to ensure that variables do not grow without bound but are confined to some range (see also the discussion in Chapter 13). The identification of such mechanisms is one of the major contributions of the field of nonlinear dynamics to the study of living organisms.

Example 2.3 Equation (2.6) can be solved by making the usual ansatz. Substitute $x(t) \approx \exp(\lambda t)$ into (2.6) to obtain

$$\lambda e^{\lambda t} + k e^{\lambda t} = 0.$$

Thus

$$\lambda + k = 0, \tag{2.16}$$

and so $\lambda = -k$. Finally, we can write

$$x(t) = c_1 e^{-kt},$$

which is the same answer as (2.8) if we identify c_1 with x_0, the value of $x(t)$ when $t = 0$.   ◇

2.3 Black Boxes

In Section 1.4, we compared the dynamical systems studied by biologists to black boxes whose contents can be inferred only by comparing inputs to outputs. From

---

[5] Eigenvalues are always determined as the roots of a characteristic equation. Since students often first come in contact with the term eigenvalue in a first course on linear algebra, it is tempting to associate eigenvalues with a matrix. However, use of a matrix is simply another way to represent the characteristic equation.

Fig. 2.4 A schematic representation of the relationship between the black-box analogy and (2.17). There is actually a subtle error inside the black box: the two terms in the box do not have the same dimensions and, in particular, do not have the same dimensions as $b(t)$. How should these terms be written so that the dimensions of the terms in the black box and $b(t)$ are the same? The answer is given in Chapter 6.

the point of view of differential equations, this black-box analogy corresponds to the inhomogeneous differential equation

$$\frac{dx}{dt} + kx = b(t). \tag{2.17}$$

If $b(t)$ were identically zero, the differential equation would be homogeneous. We can compare (2.17) with the black box in Figure 2.4 as follows: $b(t)$ is the input, $dx/dt + kx$ describes the contents of the black box, and the output of the black box corresponds to the solution of (2.17).

From an experimental point of view, it is not sufficient simply to write a differential equation such as (2.6). For example, when x describes population density or the concentration of a chemical reactant, such quantities are always nonnegative, and so we need to add the condition $x \geq 0$. And there is another subtlety. Suppose we have a chemical reaction occurring in solution that is described by (2.6) with $k > 0$. Once x gets sufficiently large, the chemical will saturate the solution and precipitate, thus stopping the reaction. Moreover, at intermediate concentrations, it is quite likely that collisions between molecules become significant and invalidate the simple assumptions in (2.6). Thus we expect that (2.6) likely applies only to very dilute solutions or sparse populations (see Section 17.3.1). This observation emphasizes that mathematical models often apply only over restricted ranges of the variables. An experimentalist must always keep this caveat in mind when comparing the solutions of a mathematical model to laboratory observations.

In contrast to biological processes, physical processes are more easily described by differential equations. This is especially true in the study of motion, where it has been established that a single function, the Hamiltonian $H(x,p)$, where x is the position and p is the momentum, is all that is needed to completely describe the movements of a mechanical system. It is by analogy with this success in physics that we apply differential equations to the study of living organisms. But in doing so, we must keep in mind that no theory and no differential equations yet exist that can completely describe a complex biological process.

2.3.1 Nonlinear Differential Equations

The differential equations that arise in models of biological phenomena are almost always nonlinear. Thus the reader may question the relevance of linear differential equations to applications to biology. The important point is that over certain ranges of the variables, a nonlinear differential equation can behave approximately like a linear differential equation (see also Chapter 4 and Section 9.2). Hence a knowledge of linear dynamical systems often provides a good starting point for learning about nonlinear dynamical systems. To illustrate this point, we consider the following example.

Let us use Newton's law of motion

$$F = ma = m\frac{d^2x}{dt^2}, \tag{2.18}$$

where F (units of kilogram meters per second squared) is the force,[6] m is the mass (units of kilograms), and d^2x/dt^2 is the acceleration (units of meters per second squared), to describe the movement of an appendage (limb, fin, tail) of an organism moving in a fluid environment. This equation arises frequently in discussions of the movements of animals and birds, including walking, running, flying, swimming, and bipedal balance control [9, 776, 858].

The forces that influence such movement arise from the muscular activations determined by the central motor program (F_{neural}), the biomechanical properties of the musculoskeletal system (F_{ms}), and the effects of the environment in which the movement occurs (F_{env}) [122, 594].

Suppose that a fish wants to stabilize the position of one of its fins at x_0 in one dimension. Define $x - x_0$ to be the deviation from this position and, without loss of generality, assume that $x_0 = 0$. By the term "stabilize," we mean that if the position of the fin is changed by, for example, a fluctuation in current, then the fish generates a force to return the fin to its original position. Consequently, (2.18) becomes

$$m\frac{d^2x}{dt^2} + F_{env} + F_{ms} = -F_{neural},$$

where the sign convention has been chosen so that the forces act to resist displacement from the $x = 0$ position: when x is positive, the force acts to cause x to decrease, and when x is negative, the force acts to cause x to increase. In order to complete the formulation of our model, it is necessary to identify the forces.

A simple approximation for F_{ms} is

$$F_{ms} = k_1x,$$

where k_1 is a constant. The approximation of $F(x)$ as a linear restoring force is referred to as *Hooke's law*. In a biomechanics context, Hooke's law arises most often

---

[6] The SI unit of force is the newton, N, which is equal to 1 kilogram meter per second squared.

in the description of elastic deformations of connective tissues, such as ligaments and tendons.

In order to determine F_{env}, it is necessary to consider the effects of a liquid environment on fin movement. When a fin moves in the water, the water exerts a *drag* against the movement. If the fin is very small and the movements are very slow, the drag acts to resist the movement by an amount proportional to the velocity.[7] Thus we have

$$F_{env} \approx k_2 \frac{dx}{dt} \qquad \text{(low velocity)},$$

where k_2 is a constant that depends on the shape of the fin and the density of the water. Now suppose that the fin is moved quickly. In this case, it has been observed that the drag is proportional to the square of the velocity, namely

$$F_{env} \approx k_3 \left(\frac{dx}{dt} \right)^2 \qquad \text{(high velocity)},$$

where k_3 is a constant.[8] Since velocity squared is a nonlinear term, we see that the model for the control of the position of the fin will become nonlinear once we consider a range of more reasonable velocities.

The appropriate form of F_{neural} is a subject of current research. However, we can anticipate that models for the feedback control of fin position will have the form

$$m\frac{d^2x}{dt^2} + k_3 \left(\frac{dx}{dt} \right)^2 + k_1 x = -f\left(x(t-\tau), \frac{dx(t-\tau)}{dt} \right), \qquad (2.19)$$

where f is a function that describes the feedback and depends, for example, on the deviation $x(t-\tau)$ of the fin and the speed $dx(t-\tau)/dt$ at which the fin was moving at time $t-\tau$ in the past, where τ is the time delay. Time delays arise in neural control because finite distances separate neurons and axonal conduction velocities are finite as well. Thus τ takes into account the time for the fish's nervous system to detect a deviation in fin position and then respond in order to change it.

A likely first step for the analysis of our model is to analyze the case in which the velocity is very low and the feedback is given by a linear combination of delayed displacement and velocity. Once this situation is well understood, we add back the nonlinearities and use computer simulations to see what differences the nonlinear effects have on the results obtained for the linear model. Differential equations such as (2.19) in which terms of the form $x(t-\tau)$ appear are called *delay differential equations* (DDEs). Since feedback is a universally important biological regulatory mechanism, we can anticipate that it will be important for us as experimentalists to understand models expressed in terms of DDEs (see Chapters 9 and 10).

---

[7] The Reynolds number (Re) for a fluid is the ratio of inertial to viscous forces. The linear approximation of drag will be valid when Re < 0.1, while a quadratic approximation will be valid for higher values of Re, but not so high that the liquid becomes turbulent.

[8] In order to appreciate how complex the study of drag can become, the reader is referred to the literature that surrounds the flight of sports projectiles, including the golf ball [6, 354, 397].

2.4 Existence and Uniqueness

A very important question in the study of linear and nonlinear differential equations concerns whether a unique solution exists for a given choice of initial conditions. In special cases, such as when we are dealing with a linear ordinary differential equation, we can answer this question by solving the differential equation.

Since the solution takes the form of a function that depends on the initial conditions, this solution can be used to establish both the existence and uniqueness of the solution. However, the overwhelming majority of differential equations that arise in the description of biological phenomena are nonlinear, and it is typically not possible to obtain a solution in the form of a closed formula.

Fortunately, it is often possible for mathematicians to demonstrate that unique solutions exist without actually solving the equations. A discussion of the mathematical methods used to establish existence and uniqueness in these cases is well beyond the scope of a textbook for undergraduate biology students (for an accessible review, see [72]). Proofs of existence and uniqueness underlie the use of numerical methods in solving differential equations.

Experimentalists need to keep two points in mind. First of all, the fact that the computer spews out a bunch of numbers as a function of time is not sufficient evidence that a solution exists for the proposed equation. Second, there is no one method for a proof of existence and uniqueness that applies to the wide spectrum of mathematical models proposed to describe biological phenomena. Different approaches are used depending, for example, on whether the differential equation is continuous or piecewise continuous (see below), whether it contains time-delayed variables, and whether it is subject to the effects of random perturbations. It is also important to note that the existence of a solution does not necessarily mean that there exists a unique dependence of the solution on its initial conditions. Indeed, it has been suggested that certain behaviors may reflect the existence of solutions that are not unique [915].

It is difficult to know how these theoretical discussions impact work at the benchtop. Existence seems self-evident, because we have defined a variable as something that can be measured. Thus there appears to be no good reason to deny the existence of an experimental observation unless it can be attributed to an experimental artifact. Moreover, the application of the scientific method excludes all mathematical models whose behaviors are inconsistent with experimental observations.

It is likely that the solutions of models that have been carefully derived from basic scientific principles and that have been vetted by Mother Nature both exist and exhibit a unique dependence on initial conditions. However, we warn the reader against simply adding terms to an existing model in the hope that something interesting might happen. Such a practice can lead to an *ill-posed formulation of the problem*. A problem is ill posed when the form of the differential equation fails to properly take into account its theoretical and experimental foundations.

2.5 What Have We Learned?

1. What is a differential equation?

   A differential equation is a mathematical equation that expresses a relationship between the values of a function of one or more variables and the values of its derivatives of various orders.

2. What is an ordinary differential equation (ODE)?

   An ODE is a differential equation for which there is only one independent variable, typically time.

3. What is the order and dimension of a differential equation?

   The order of the differential equation is equal to the order of the highest derivative. The dimension of a differential equation is the number of initial conditions that must be specified in order to uniquely determine a solution. For an ODE, the order and the dimension are equal. In contrast, a delay differential equation has infinite dimensions (see Section 10.4).

4. What is the time scale of change for a dynamical system?

   The time scale of change provides an estimate of how long it takes for a significant part of the change in a variable to occur [749].

5. What is the general form of the solution for an nth-order linear ODE with constant coefficients?

   The solution will often have the form

   $$x(t) = c_1 e^{\lambda_1 t} + c_2 e^{\lambda_2 t} + \cdots + c_n e^{\lambda_n t},$$

   where the λ_i are the eigenvalues (rate constants). Using Tool 4, we convert the problem of integrating an ODE into that of determining the roots of an algebraic equation in λ. Thus we immediately see an important role for computer programs. Closed-form expressions for the roots of a polynomial are known only for polynomials of degree 4 or less and for certain special polynomials of higher degree. Thus numerical methods must typically be used to compute the eigenvalues.

2.6 Exercises for Practice and Insight

1. The van der Pol equation

   $$\frac{d^2 x}{dt^2} + \varepsilon \left(1 - x^2\right) \frac{dx}{dt} - kx = 0,$$

   where ε is the damping coefficient, plays a central role in the study of excitable systems such as the heart, pancreatic β-cells, and the nervous system [422].

   a. Rewrite this equation as a system of two first-order equations.

b. Explain why the van der Pol equation is not a linear system.

2. The differential equation that describes the movements of a planar pendulum is

$$m\ell^2 \frac{d^2\theta}{dt^2} + \varepsilon\ell\frac{d\theta}{dt} + mg\ell\sin\theta = 0,$$

where m is the mass, θ is the displacement angle, ε is a damping coefficient, ℓ is the length of the pendulum, and g is the acceleration due to gravity.

a. Rewrite this equation as a system of two first-order differential equations.
b. Explain why this version of the pendulum equation is not a linear system.

Chapter 3
Equilibria and Steady States

Mathematical models of dynamical systems often take the general form

$$\frac{dx_1}{dt} = f_1(x_1, x_2, \ldots, x_n),$$
$$\frac{dx_2}{dt} = f_2(x_1, x_2, \ldots, x_n),$$
$$\ldots$$
$$\frac{dx_n}{dt} = f_n(x_1, x_2, \ldots, x_n),$$

(3.1)

where the x_i are the variables and the f_i describe the interactions between variables. From the perspective of (3.1), the goal of science is to identify the x_i and the form of the functions f_i. A typical starting point at the benchtop is to consider a dynamical system initially at rest, perturb it, and watch what happens. A rest state is one in which all of the variables other than time have constant values. Consequently, the rest state is time-independent. In other words, every time we measure the variables we get the same answer. Mathematically, this observation corresponds to

$$\frac{dx_i}{dt} = 0, \quad i = 1, \ldots, n.$$

(3.2)

We shall use the term *fixed point* to indicate the values of the x_i, denoted by x_i^*, that satisfy

$$f_i(x_1^*, x_2^*, \ldots, x_n^*) = 0$$

for $i = 1, \ldots, n$ [291, 810].

There are two types of time-independent states (fixed points) of interest to biologists: *equilibrium points* and *steady states*. Mathematically oriented biologists tend to refer to these fixed points collectively as *equilibria* [74, 186, 193] (for a notable exception, see [617]). However, at the benchtop, the distinction between an equilibrium and a steady state is not based simply on (3.2), but also takes

© Springer Nature Switzerland AG 2021
J. Milton and T. Ohira, *Mathematics as a Laboratory Tool*,
https://doi.org/10.1007/978-3-030-69579-8_3

into account thermodynamic considerations regarding the nature of the interaction between the system and its surroundings. Indeed, many of the phenomena of most interest to biologists, such as the emergence of self-organized behaviors and spatially coherent structures, are properties of open dynamical systems.

In contrast to closed dynamical systems, for example, an experiment performed in a test tube, the behaviors of open systems are maintained by flows of energy and mass between the system and its surroundings. Thus thermodynamic considerations can be very important for developing the appropriate mathematical models. Consequently, we recommend using the term "fixed point" generically and reserving the terms "equilibrium" and "steady state" for situations in which the nature of the fixed point is well understood both dynamically and thermodynamically.

The goal of this chapter is to illustrate the differences between equilibria and steady states from a dynamic point of view. In Chapter 17, we discuss these differences from a thermodynamic viewpoint. We introduce two tools to facilitate the discussion: the law of mass action (Tool 5) and the integrating factor (Tool 6). First, we examine the nature of the time-independent behaviors that arise in a closed dynamical system and then contrast them with those observed in an open dynamical system. Finally, we examine the conditions for which a steady-state approximation is valid.

The questions we will answer in this chapter are:

1. What is the law of mass action and when is it valid?
2. What are the conditions for the existence of a steady state for a Michaelis–Menten enzymatic reaction and for consecutive chemical reactions?
3. Why must the steady-state approximation be used with caution?
4. Is an equilibrium state the same as a steady state?

3.1 Law of Mass Action

Many models developed in biomathematics assume the validity of the law of mass action [541, 756]. In this chapter, we use models based on this law to examine the nature of equilibria and steady states using examples taken primarily from the biochemical literature. In chemical applications, this empirical law states that the rate of an elementary chemical reaction is proportional to the product of the concentration of the participating reactants. An elementary chemical reaction is a reaction that proceeds through just one transition, or mechanistic step. For example, consider the reaction

$$A \xrightarrow{k} B.$$

The law of mass action implies that if we double the initial concentration $[A]$ of A (where the notation $[\cdot]$ denotes concentration), we observe a doubling of the rate of increase of $[B]$. This is the result typically observed experimentally. The differential equation that corresponds to this observation is

$$\frac{d[B]}{dt} = k[A].$$

Similarly, for the chemical reaction

$$A + B \xrightarrow{k} C,$$

we have

$$\frac{d[C]}{dt} = k[A][B],$$

and for the chemical reaction

$$2A + B \xrightarrow{k} C,$$

we have

$$\frac{d[C]}{dt} = k[A]^2[B].$$

Tool 5: Law of Mass Action

For the general chemical reaction

$$nA + mB \xrightarrow{k} C,$$

the *law of mass action* implies that we have

$$\frac{d[C]}{dt} = k[A]^n[B]^m. \tag{3.3}$$

Tool 5 provides a convenient starting point for the development of a mathematical model. However, the law of mass action is not always valid. This law implies random encounters between A and B and hence will be valid only for sufficiently dilute concentrations. Difficulties arise when the law of mass action is applied to situations outside the realm of chemical kinetics (see, for example, [289, 350]). Thus the only practical way to justify the use of this law is by making careful comparisons between prediction and observation. In other words, we must apply the scientific method. Pragmatically, it is always "worth a try" to see what a mass-action-law model predicts and use subsequent experimental observations to guide extensions of the model.

3.2 Closed Dynamical Systems

As discussed in Chapter 1, a closed system is one that exchanges heat, but not mass, with its surroundings. The simplest laboratory example of a closed system is a chemical reaction that takes place within a test tube. Two types of fixed points can arise in a closed dynamical system. On long time scales, there must always be an equilibrium point.[1] However, on shorter time scales there may be the transient appearance of steady states and possibly transient time-dependent states, such as oscillations.

3.2.1 Equilibria: Drug Binding

Students of chemistry are familiar with the concept of chemical equilibrium. Put some chemical reactants into a test tube, stopper it, and wait a sufficiently long time. Once time-dependent changes can no longer be detected, the system is at equilibrium. In biology, measurements of equilibrium states are very important for studying the binding of molecules to molecules, such as the binding of drugs to drug receptors. Here we illustrate the use of the law of mass action by determining the fraction of bound receptors as a function of drug concentration (Figure 3.1).

Our model is

$$M + nC \underset{k_{-1}}{\overset{k_1}{\rightleftharpoons}} L,$$

where M denotes the free receptor, L the bound receptor, C the drug molecule, and n the number of drug molecules bound per receptor. We define $[T] := [M] + [L]$ as, respectively, the total concentration of receptor, the concentration of free receptor, and the concentration of bound receptor. In choosing the subscripts for the k_i, we use the convention that positive integers refer to reactions that proceed from left to right, and negative integers refer to reactions that proceed from right to left. Applying the law of mass action, we obtain

$$\frac{d[L]}{dt} = k_1 [M][C]^n - k_{-1}[L].$$

At equilibrium, we have

$$\frac{d[L]}{dt} = 0.$$

Therefore,

$$[M][C]^n = \frac{k_{-1}}{k_1}[L] = \hat{K}[L], \tag{3.4}$$

where $\hat{K} := k_{-1}/k_1$. Since

$$[L] = [T] - [M],$$

---

[1] This conclusion is a consequence of the second law of thermodynamics, discussed in Chapter 17.

on rearranging, we have

$$G = \frac{[M]}{[T]} = \frac{\hat{K}}{\hat{K} + [C]^n},$$ (3.5)

where G is the fraction of free receptors, and the fraction $1 - G$ of bound receptors is

$$1 - G = \frac{[C]^n}{\hat{K} + [C]^n}.$$ (3.6)

The right-hand side of (3.6) is an example of a *Hill function*. The Hill function was first introduced by A.V. Hill[2] in his study of the binding of oxygen to hemoglobin [341, 870]. This function has a number of useful properties including the following: it is a monotone increasing function of $[C]$; its value goes from 0 to 1 as $[C]$ goes from 0 to ∞; and finally, when $\hat{K} = [C]^n$, the Hill function is equal to 0.5. In Chapter 9, we will use a variety of Hill-type functions to examine the properties of biological feedback control mechanisms.

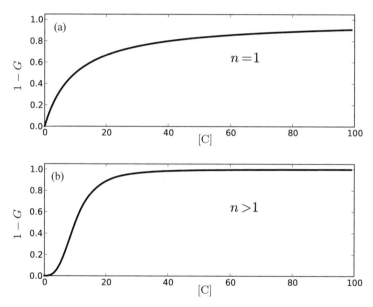

Fig. 3.1 Fraction of bound receptors, $1 - G$, given by (3.6) as a function of the concentration $[C]$ of drug molecules for different values of n: (a) $n = 1$; (b) $n > 1$.

There are different formulations of these results depending on what we are able to measure for a given system and the goal of the experiment. For example, Figure 3.1 shows a plot of the fraction of bound receptors $1 - G$ versus $[C]$ for different values of n. The requirement that a receptor be saturable requires that $n \geq 1$. Certain

---

[2] Archibald Vivian Hill (1886–1977), English physiologist, one of the founders of biophysics and operations research.

proteins, e.g., allosteric proteins, have multiple binding sites. When $n > 1$, $1 - G$ exhibits a sigmoidal dependence on $[C]$. Sigmoidal relationships are likely the most common nonlinearities that arise in biology (see [695, 763, 871] and Section 9.4). An example of a mathematical model that incorporates this binding nonlinearity arises in the setting of the dynamics of the recurrent inhibitory loop in the hippocampus [521]. It has been suggested that the fact that more than one molecule of the antibiotic penicillin can bind to the inhibitory GABA receptor may lie at the basis of the complexity of the dynamics generated by these loops.

It should be noted that as n becomes larger, the binding curve becomes more like an on–off switch; namely, for large n there are only two states: either all binding sites are occupied, or none of them are. Hence recent attention has focused on the possibility that molecular switches of this type can be used to construct molecular computers (see, for example, [258]).

An alternative formulation can be obtained by writing (3.4) as

$$\hat{K}^{-1} = \frac{[L]}{[M][C]^n},$$

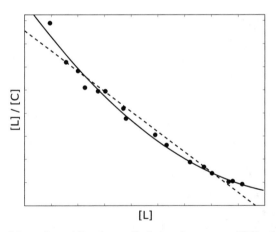

Fig. 3.2 Binding of the aminoacridine dye proflavin to salmon sperm DNA when the total ionic strength is 0.01. Here $[L]$ is the concentration of bound receptor, and $[C]$ is the concentration of proflavin. The dashed line is the best-fit linear relation $[L]/[C] \approx -0.87[L] + 186.2$, and the solid line is the best-fit quadratic relation $[L]/[C] \approx -0.0037[L]^2 - 1.866[L] + 244.3$. Data from [674] with permission.

which on rearranging becomes

$$\frac{[L]}{[C]^n} = \hat{K}^{-1}([T] - [L]). \tag{3.7}$$

This formulation is used to characterize the binding of drugs to biologically important molecules. Figure 3.2 shows an experiment in which the binding of proflavin, an acridine dye, to DNA was investigated [674]. The experimentally observed curvature of this type of plot implies that $n > 1$ (compare solid and dashed lines). This was the first evidence of the existence of more than one type of binding site on DNA for acridine dyes.

3.2.2 Transient Steady States: Enzyme Kinetics

Two of the earliest examples of biotechnology are the use of yeast to ferment alcohol and its use to leaven bread. Consequently, it should not be surprising that one of the first applications of mathematics to molecular biology was to study the rates of these processes and, in particular, the kinetics of the underlying enzymatic reactions. A typical result is shown in Figure 3.3.

A striking observation is that the rate of the reaction did not increase indefinitely with substrate concentration, but approached a finite constant velocity once the substrate concentration became sufficiently high. For our discussion, it is important to note that although both the enzyme and its substrate are contained within the same vessel, the measurements are not made under equilibrium conditions, since there is a continual increase in the product concentration.[3] How are the observations in Figure 3.3 to be explained?

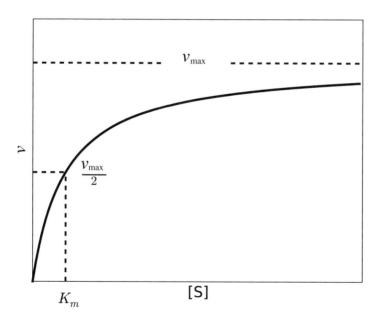

Fig. 3.3 The dependence of the velocity v of an enzymatic reaction as a function of the substrate concentration $[S]$ as predicted by (3.10).

---

[3] Of course, if we waited a sufficiently long time, we would eventually arrive at an equilibrium state.

In 1913, Michaelis and Menten[4] proposed the following model for enzyme kinetics [562]:

$$E + S \underset{k_{-1}}{\overset{k_1}{\rightleftharpoons}} ES \overset{k_2}{\longrightarrow} E + P, \tag{3.8}$$

where $[E]$, $[S]$, $[ES]$, and $[P]$ are respectively the concentrations of enzyme, substrate, enzyme–substrate complex, and product. Their novel idea was that the rate of change of $[ES]$ should satisfy the steady-state approximation, namely

$$\frac{d[ES]}{dt} \approx 0. \tag{3.9}$$

Although the nature of the mathematical conditions for (3.9) was subsequently refined in 1925 by Briggs and Haldane[5] [307], convincing experimental evidence to support these claims was not obtained until 1943, by Chance[6] [111]. Chance studied the enzyme reaction in which hydrogen peroxide is hydrolyzed by the enzyme horseradish peroxidase:

$$\text{peroxidase} + H_2O_2 \overset{k_1}{\longrightarrow} \text{peroxidase} \bullet H_2O_2.$$

There are two reasons why he chose to study this reaction. First, the enzyme–substrate (ES) complex can be spectroscopically identified, so the dynamics could be measured accurately on fast (ms) time scales using a spectrometer. Second, this ES complex rapidly decomposes in the presence of an oxygen acceptor, such as the dye leucomalachite green (A). Thus by putting peroxidase (E), hydrogen peroxide (S), and this dye (A) in the same test tube, we have the reaction

$$\text{peroxidase} + H_2O_2 \overset{k_1}{\longrightarrow} \text{peroxidase} \bullet H_2O_2 + A \overset{k_2}{\longrightarrow} \text{peroxidase} + H_2O + AO.$$

Figure 3.4 shows the results of a typical experiment that measures the time course for $[ES]$. There is a small interval during which (3.9) is approximately satisfied. The length of this interval increases as the substrate concentration increases (data not shown).

The Michaelis–Menten equations for enzyme kinetics can be derived readily using the following three-step recipe, which affords an easy way to derive the changes that occur in enzyme kinetics in the presence of various types of inhibitors (see Exercises 2, 3, and 5 at the end of this chapter).

Step 1: Steady-state approximation

Using the law of mass action, we can write

---

[4] Leonor Michaelis (1875–1949), German physician and biochemist. Maud Leonora Menten (1879–1960), Canadian physician and biochemist.

[5] George Edward Briggs (1893–1985), British botanist. John Burdon Sanderson Haldane (1892–1964), British-born Indian biologist and mathematician.

[6] Britton Chance (1913–2010), American biochemist.

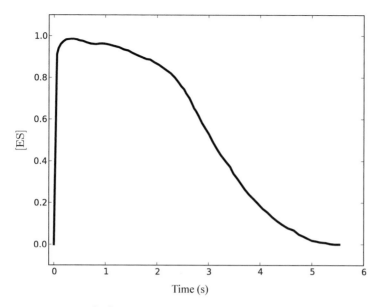

Fig. 3.4 Time course for $[ES]$ formed by the enzyme horseradish peroxidase and the substrate hydrogen peroxide. The initial concentration of hydrogen peroxide was 8 micromoles per liter. Data from [111] with permission.

$$\frac{d[ES]}{dt} = k_1[E][S] - (k_2 + k_{-1})[ES].$$

The steady-state approximation (3.9) results in

$$k_1[E][S] = (k_2 + k_{-1})[ES]$$

(see, however, Section 3.4). We define the Michaelis–Menten constant K_{MM} as

$$K_{MM} = \frac{k_2 + k_{-1}}{k_1} = \frac{[E][S]}{[ES]}.$$

Step 2: Enzyme conservation equation

The next step is to formulate the enzyme conservation equation, i.e.,

$$[E]_T = [E] + [ES],$$

where $[E]_T$ is the total enzyme concentration, $[E]$ is the free enzyme concentration, and $[ES]$ is the concentration of the enzyme in the form of the $[ES]$ complex. In other words, enzyme is neither created nor destroyed, but exists in one of two states: E and ES.

Step 3: Determine the maximum velocity

The velocity v of the enzyme reaction is the rate at which product is formed, i.e.,

$$v = k_2[ES].$$

The maximum velocity occurs when all of the enzyme is involved in the *ES* complex, i.e.,

$$v_{max} = k_2[E]_T.$$

To complete this derivation, we proceed as follows:

$$\frac{v_{max}}{v} = \frac{k_2[E]_T}{k_2[ES]} = \frac{[E]}{[ES]} + 1,$$

but

$$\frac{[E]}{[ES]} = \frac{K_{MM}}{[S]},$$

and hence

$$\frac{v_{max}}{v} = \frac{K_{MM}}{[S]} + 1.$$

On rearranging, we obtain the Michaelis–Menten equation

$$v = \frac{v_{max}[S]}{K_{MM} + [S]}. \tag{3.10}$$

The evolution of an enzyme reaction that follows the Michaelis–Menten equation is shown in Figure 3.3. By setting $K_{MM} = [S]$, it can be seen that K_{MM} is the substrate concentration at which $v = v_{max}/2$.

3.3 Open Dynamical Systems

An open system exchanges both heat and mass with its surroundings. The only time-independent state that can exist in an open system is a steady state; however, every open system is not necessarily associated with a steady state. In fact, open systems are more typically associated with complex time-dependent states, referred to as *dissipative states*, such as limit-cycle (Section 4.5.1) and chaotic (Section 13.1) attractors and mesoscopic states (Section 6.6.1). Moreover, in spatially extended open dynamical systems, we can have the formation of spatial patterns and other self-organized behaviors [617]. A precise description of the dynamics of open systems is beyond the scope of this introductory book and is, in fact, a work in progress in the scientific community. It suffices for our purpose to use the following example to illustrate the behavior of an open system that possesses a steady state.

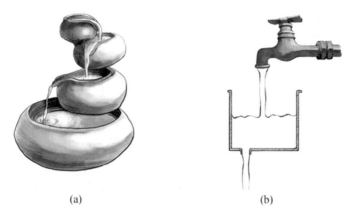

(a)   (b)

Fig. 3.5 (a) Water fountain with four reservoirs, (b) A simple water fountain constructed using a water faucet and a paper cup with a hole in the bottom.

3.3.1 Water Fountains

The water fountain shown in Figure 3.5a is an example of an open dynamical system. It consists of four reservoirs arranged sequentially above each other in such a way that water flows from the highest reservoir to the next highest, and the next, and finally to the bottom reservoir. Water is pumped back from the lowest reservoir to the highest, so that the process repeats. How do we expect the water level to behave in the reservoirs as a function of time?

Intuitively, we would expect that if we adjusted the water pump appropriately, then the water level in each reservoir would remain constant. This constant-level state occurs when the rate of water flowing into a reservoir is exactly balanced by the rate of water flowing out. However, if the water pump is turned off, the water quickly empties into the lowest reservoir.

Figure 3.5b shows a simplified version of a water fountain constructed using a water faucet and a paper cup that has a hole in the bottom (diameter equal to that of a pencil, $\approx 7\,\text{mm}$). The cup is positioned under the faucet so that the top of the cup is located a distance away from the bottom of the faucet. Adjust the flow of water from the faucet so that the level of fluid in the cup remains constant. This condition corresponds to a steady state (see below). Now turn the faucet off. Obviously, the cup empties. In this situation, the open system becomes a closed system, and hence we know that after a sufficiently long time, the system will possess an equilibrium, namely that all of the water is in the basin below the cup. The current terminology in the literature is that the steady state observed in the open system is *far from equilibrium*.

The dynamics of this water-cup fountain are described by

$$\frac{dV_{\text{cup}}}{dt} = \frac{dV_{\text{in}}}{dt} - \frac{dV_{\text{out}}}{dt}, \tag{3.11}$$

where dV_{cup}/dt is the rate of change in the volume of water in the cup, dV_{in}/dt is the rate of volume of water flowing from the tap into the cup, and dV_{out}/dt is the rate of volume of water flowing out of the cup. At steady state,

$$\frac{dV_{cup}}{dt} = 0,$$

or

$$\frac{dV_{in}}{dt} = \frac{dV_{out}}{dt}.$$

We observe that the rate of flow of water out of the cup is greater when the fluid level h is high than when it is low. Thus we make the simplifying assumption[7]

$$\frac{dV_{out}}{dt} \approx kV_{cup}.$$

Thus for a constant faucet flow rate I, (3.11) becomes

$$\frac{dV_{cup}}{dt} + kV_{cup} = I. \tag{3.12}$$

The solution of differential equations of this form is discussed in Section 3.4.2.

Now readjust the faucet flow rate so that the water level in the cup remains constant. Suddenly input an extra amount of water into the cup (for example, pour some water into the cup from another cup). Figure 3.6 (first transient) shows changes in h as a function of time following the addition of this extra quantity of water.

It can be seen that the fluid level gradually returns to the original steady-state fluid level. Similarly, if we suddenly remove some fluid from the cup, the fluid level initially decreases but then slowly increases to the original steady-state fluid level.

This experiment demonstrates that a steady state in an open system can be resistant to change (this is also a property of equilibria and steady states in closed systems). This resistance to change is referred to as *stability* and is the subject of Chapter 4.

3.4 The "Steady-State Approximation"

It is a common misconception that dynamical systems in biology always possess a steady state. Here we discuss two examples that demonstrate that steady states are possible only if certain conditions are satisfied.

---

[7] Bernoulli's equations for fluid flow predict that the rate of fluid flow out of the cup will be, to a first approximation, proportional to $\sqrt{2gh}$, where g is the acceleration due to gravity. However, we hypothesize that because of irregularities in the edge of the hole and the nonlaminar nature of water flow through the hole and from the faucet, the flow rate is more likely to be proportional to h [916]. The interested reader should perform this experiment and see whether our hypothesis is correct.

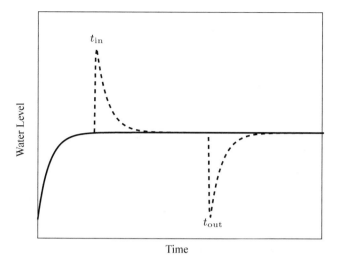

Fig. 3.6 The water level in the paper cup shown in Figure 3.5b as a function of time. At time t_{in}, a small amount of water was quickly added to the cup. At time t_{out}, a small amount of water was quickly removed from the cup. After each maneuver, the water level returned to its steady-state level. These solutions were obtained using (3.12) and constructing I from a combination of a Heaviside function and appropriately timed delta functions (see Chapter 6).

3.4.1 Steady State: Enzyme–Substrate Reactions

In Section 3.2.2, we used the steady-state approximation given by (3.9) to derive the Michaelis–Menten equation.

We can examine the validity of this assumption using the law of mass action (Tool 5) to rewrite (3.8) as a system of four differential equations [749, 750, 752]:

$$\frac{d[E]}{dt} = (k_{-1} + k_2)[ES] - k_1[E][S], \tag{3.13a}$$

$$\frac{d[S]}{dt} = k_{-1}[ES] - k_1[E][S], \tag{3.13b}$$

$$\frac{d[ES]}{dt} = k_1[E][S] - (k_{-1} + k_2)[ES], \tag{3.13c}$$

$$\frac{d[P]}{dt} = k_2[ES]. \tag{3.13d}$$

We can reduce this model to two equations with two unknowns by noting that

$$\frac{d[E]}{dt} + \frac{d[ES]}{dt} = 0.$$

This is just another way of restating the enzyme conservation equation, i.e.,

$$[E]_T = [E] + [ES].$$

Thus we can rewrite (3.13b) and (3.13c) as

$$\frac{d[S]}{dt} = k_{-1}[ES] - k_1([E_T] - [ES])[S], \tag{3.14a}$$

$$\frac{d[ES]}{dt} = k_1([E]_T - [ES])[S] - (k_{-1} + k_2)[ES], \tag{3.14b}$$

with the initial conditions $[S](0) = [S]_0$ and $[ES](0) = [ES]_0$. Once the solution for $[ES](t)$ is obtained, we can use (3.13d) to obtain $[P](t)$.

In order to determine the time scale for the changes in $[ES]$, let us assume that the concentration of S is much larger than that of E, namely $[S]_0 \gg [E]_T$. Experimentally, this is almost always true (or at the very least can be easily arranged). In particular, we assume that $[S](t)$ is constant for all $t \geq 0$. This means that (3.14a) becomes irrelevant and (3.14b) can be replaced by

$$\frac{d[ES]}{dt} = k_1[E]_T[S]_0 - (k_1[S]_0 + k_{-1} + k_2)[ES],$$

$$= \text{constant} - \text{constant}([ES]) = \bar{k} - \hat{k}([ES]),$$

where $\bar{k} := k_1[E]_T[S]_0$, $\hat{k} := k_1[S]_0 + k_{-1} + k_2$, and we have chosen $[ES]_0 = 0$. The solution can be obtained by solving the integral equation

$$\int_0^{[ES]} \frac{d[ES]}{\bar{k} - \hat{k}[ES]} = \int_0^t dt,$$

which yields

$$[ES](t) = \frac{\bar{k}}{\hat{k}}\left(1 - e^{-\hat{k}t}\right) = \frac{[E]_T[S]_0}{[S]_0 + K_{MM}}\left(1 - e^{-\hat{k}t}\right). \tag{3.15}$$

If we repeat this calculation and assume that $[ES]_0 > 0$, then (3.15) becomes

$$[ES](t) = \frac{[E]_T[S]_0}{[S]_0 + K_{MM}}\left(1 - e^{-\hat{k}t}\right) + [ES]_0 e^{-\hat{k}t}. \tag{3.16}$$

There are two similarities between this solution and the results of the water-fountain experiment. First, there is a positive steady-state value of $[ES]$, and in particular,

$$[ES] \rightarrow \frac{[E]_T[S]_0}{[S]_0 + K_{MM}}$$

as $t \rightarrow \infty$. Second, this steady-state value exhibits the same resistance to change; from (3.16), we see that no matter what we choose for $[ES]_0$, we always get the same limiting value of $[ES]$.

The most important concept to understand is that the steady-state approximation given by (3.9) does not imply that $[ES]$ is constant. All (3.9) implies is that $[S]$

changes slowly enough, relative to $[ES]$, that $[ES]$ behaves as if $[S]$ were "temporarily constant." Thus we can use the concept of a time scale (Tool 3) to determine under what conditions the time scale for the changes in $[S]$ is long compared to the time scale for the changes in $[ES]$ [749, 750, 752]. Since the time scale for functions of the form $\exp(-kt)$ is k^{-1}, it follows that

$$\mathscr{T}([ES]) \approx \hat{k}^{-1}. \tag{3.17}$$

The time scale for changes in $[S]$ is

$$\mathscr{T}([S]) = \frac{[S]_{\max} - [S]_{\min}}{\left|\frac{d[S]}{dt}\right|_{\max}}.$$

What does the numerator equal? We know that $[S]$ eventually changes from an initial value $[S]_0$ to a much smaller value. We assume that $[S]_{\max} - [S]_{\min} \approx [S]_0$. Thus the numerator is just $[S]_0$. What does the denominator equal? We know the answer to this, because we have just shown that

$$v = \frac{v_{\max}[S]_0}{K_{\mathrm{MM}} + [S]_0} = \frac{k_2[E_{\mathrm{T}}][S]_0}{K_{\mathrm{MM}} + [S]_0},$$

and thus the time scale for $[S]$ is

$$\mathscr{T}([S]) \approx \frac{[S]_0}{(k_2[E]_{\mathrm{T}}[S]_0/(K_{\mathrm{MM}} + [S]_0))} = \frac{(K_{\mathrm{MM}} + [S]_0)[S]_0}{k_2[E]_{\mathrm{T}}[S]_0},$$

that is,

$$\mathscr{T}([S]) = \frac{K_{\mathrm{MM}} + [S]_0}{k_2[E]_{\mathrm{T}}}.$$

By comparing $\mathscr{T}([S])$ with $\mathscr{T}([ES])$, we find that the condition for the validity of the steady-state approximation given by (3.9) is

$$\hat{k}^{-1} \ll \frac{K_{\mathrm{MM}} + [S]_0}{k_2[E]_{\mathrm{T}}}, \tag{3.18}$$

or (see Exercise 5 at end of chapter)

$$\frac{k_2[E]_{\mathrm{T}}}{k_1} \ll (K_{\mathrm{MM}} + [S]_0)^2. \tag{3.19}$$

Relations (3.18) and (3.19) have recently become known as the *Segel–Slemrod*[8] conditions [644, 752], and they can be satisfied in two ways. One possibility is that $[S]_0 \gg [E]_{\mathrm{T}}$. This is the assumption that Briggs and Haldane made in 1925 [307].

---

[8] Named for Lee Segel (1932–2005) and Marshall Slemrod (b. 1944), American applied mathematicians.

The other possibility is that k_2 is small. This is the original assumption made by Michaelis and Menten in 1913 [562].

3.4.2 Steady State: Consecutive Reactions

In consecutive first-order reactions such as

$$A \xrightarrow{k_1} B \xrightarrow{k_2} C, \tag{3.20}$$

the steady-state approximation can be justified when the intermediates are very reactive and are therefore present in very low concentrations [245]. In order to understand how this is possible, we use the law of mass action to obtain

$$\frac{d[A]}{dt} = -k_1[A], \tag{3.21a}$$

$$\frac{d[B]}{dt} = k_1[A] - k_2[B], \tag{3.21b}$$

$$\frac{d[C]}{dt} = k_2[B]. \tag{3.21c}$$

Since these equations are linear, it is possible to obtain a solution [315]. From (3.21a), we obtain

$$[A](t) = [A]_0 \exp(-k_1 t), \tag{3.22}$$

where $[A]_0$ is the initial concentration of A. We also assume that the initial concentrations of B and C are zero. By substituting (3.22) for $[A](t)$ into (3.21b), we obtain

$$\frac{d[B]}{dt} + k_2[B] = k_1[A]_0 \exp(-k_1 t). \tag{3.23}$$

In order to solve (3.23), we will first introduce the concept of an *integrating factor*. This concept turns out to be quite useful, and in particular, we will use integrating factors to study the effects of forcing functions (Chapter 6) and noise (Chapter 15) on the dynamics of systems that possess a stable fixed point.

Tool 6: The Integrating Factor

Equation (3.23) has the general form

$$\frac{dx}{dt} + kx = b(t), \tag{3.24}$$

with the initial condition $x(t_0) = X_0$. Here we have identified k_2 with k and $k_1[A]_0 e^{-k_1 t}$ with $b(t)$. The term $b(t) \neq 0$ is referred to as an inhomogeneity and hence (3.24) is called an inhomogeneous, or non-homogeneous, ordinary differential equation. Non-homogeneous differential equations arise frequently in descriptions of the response of a dynamical system to an input forcing function (Chapter 6). Here we will first solve (3.24) and then use the general solution to solve (3.23) in order to complete our analysis of the consecutive reactions.

We obtain the solution of (3.24) in two steps. First, in Section 2.2.1, we showed that when $b(t) = 0$, the solution is

$$x(t) = x(t_0)e^{kt_0}e^{-kt} = x(0)e^{-kt},$$

where we have taken $t_0 = 0$. The term e^{-kt} is an example of an integrating factor. Now we consider the case $b(t) \neq 0$. A useful ansatz is to assume that the solution of (3.24) has the form [72, 756]

$$x(t) = e^{-kt}y(t). \tag{3.25}$$

In other words, we assume that the solution of the inhomogeneous differential equation is the product of the solution we know (namely, the solution of the related homogeneous equation) and a function to be determined. We remind the reader that this ansatz is justified if we can determine $y(t)$. If we substitute (3.25) into (3.24), we obtain

$$-ke^{-kt}y(t) + e^{-kt}\frac{dy}{dt} + ke^{-kt}y(t) = b(t),$$

and hence $y(t)$ must satisfy the differential equation

$$\frac{dy}{dt} = b(t)e^{kt},$$

and thus

$$y(t) = \int_{t_0}^{t} b(s)e^{ks}ds + y(t_0).$$

The solution of (3.24) when $t_0 = 0$ is

$$x(t) = e^{-kt}\left[\int_0^t b(s)e^{ks}ds + c\right]. \tag{3.26}$$

The value of c can be determined using the initial condition $x(t_0) = X_0$.

Returning to our problem concerning the dynamics of a system composed of three consecutive reactions, we can use (3.26) to obtain the solution of (3.23) as

$$[B](t) = \frac{1}{\exp(k_2 t)}\left[\int_0^t k_1[A]_0 e^{(k_2 - k_1)s}ds + c\right]$$

and hence

$$[B](t) = \frac{k_1[A]_0}{k_2 - k_1} \exp(-k_1 t) - \frac{k_1[A]_0}{k_2 - k_1} \exp(-k_2 t) + c \exp(-k_2 t). \qquad (3.27)$$

When $t = 0$, we have $[B](0) = 0$, and hence $c = 0$. Thus

$$[B](t) = \frac{k_1[A]_0}{k_2 - k_1} [\exp(-k_1 t) - \exp(-k_2 t)]. \qquad (3.28)$$

All that remains is to determine an expression for $[C](t)$. It must be true that

$$[A]_0 = [A](t) + [B](t) + [C](t),$$

or equivalently,

$$[C](t) = [A]_0 - [A](t) - [B](t).$$

Using this relationship, we obtain

$$[C](t) = [A]_0 \left[1 + \left(1 - \frac{k_1}{k_1 - k_2} e^{-k_2 t} \right) - \left(1 - \frac{k_2}{k_1 - k_2} e^{-k_1 t} \right) \right]. \qquad (3.29)$$

We can use these results to determine whether a steady state exists for (3.20). Thus

$$\frac{d[B]}{dt} = 0 = k_1[A] - k_2[B],$$

and hence

$$[B] = \frac{k_1}{k_2}[A] = [A]_0 \frac{k_1}{k_2} e^{-k_1 t}. \qquad (3.30)$$

In order to assess the validity of this steady-state approximation, it is necessary to compare (3.28) with (3.30). It is easy to see that these equations will be similar if $k_2 \gg k_1$ and $t \gg 1/k_2$ (see below). The first condition means that B is very reactive compared to A. The result is that $[B] \approx 0$ at steady state (and hence $d[B]/dt \approx 0$). This is the observation that we cited at the beginning of this section.

Figure 3.7a compares the changes in $[A], [B], [C]$ as a function of time (see also Exercise 7 at the end of the chapter). The induction period is arbitrarily taken as the time to reach the point of inflection of the $[C]$ versus time curve (the ↓ in Figure 3.7a). The second condition, namely $t \gg 1/k_2$, ensures that the induction period has passed. Figure 3.7b shows the time course of $[B]$ as a function of the ratio k_2/k_1. As we can see, the steady-state assumption is not always true, but it becomes more valid as k_2/k_1 becomes larger. In particular, for large k_2/k_1, the value of $[B]$ is low. A similar observation occurs in the enzyme–substrate reaction we have just discussed. When $[S]_0 \gg K_{MM}$, the steady state $[ES]$ is approximately $[E]_T$, which is typically a small number.

The steady-state approximation must be used with great caution. For example, oscillatory behaviors are observed when a Michaelis–Menten enzyme is coupled to other biochemical pathways, particularly in an open reaction scheme (e.g.,

[637, 734]) and when the Van Slyke–Cullen condition[9] $k_2 \gg k_{-1}$ is satisfied [229, 230]. In these cases, the analysis of the full set of equations correctly predicts the observed oscillation. However, when the same system of equations is simplified using the steady-state approximation, an oscillatory solution is not predicted to occur. The reader may quite rightly ask whether given modern computer capabilities, there is any point in using the steady-state approximation at all. Again we appeal to laboratory observation: if the approximation can be validated experimentally, then it is acceptable. The bonus of making the steady-state approximation is that it is often possible to gain greater insight into the dynamical system, which, in turn, often simplifies the design of experiments.

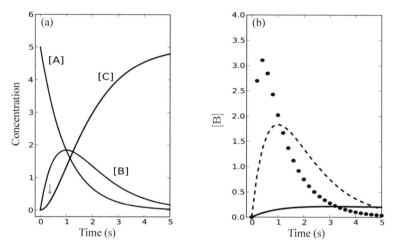

Fig. 3.7 (a) The time courses of $[A]$, $[B]$, $[C]$ for the consecutive chemical reactions described by (3.20). The values of the rate constants were $k_1 = k_2 = 1\ \text{s}^{-1}$, and the initial conditions were $[A]_0 = 5$ and $[B]_0 = [C]_0 = 0$. (b) The time course for $[B]$ as a function of the ratio k_2/k_1. The values of k_2/k_1 are 0.1 (dotted line), 1.0 (dashed line), and 20.0 (solid line), and the initial conditions are the same as in (a).

3.5 Existence of Fixed Points

Given our emphasis on the analysis of fixed points, it is fair to ask whether it is possible to imagine a biologically plausible dynamical system for which no fixed point exists. The answer to this question has long attracted the interest of mathematicians [72] and biologists [866, 867]. The answer intimately depends on whether the f_i in (3.1) are continuous. A requirement of biologically plausible mathematical models is that the value of each variable x_i must always be nonnegative and at most 1 ($0 \le x_i \le 1$). The upper bound of 1 arises because we can divide each x_i by its largest possible value to obtain a number between 0 and 1.

---

[9] Donald Dexter Van Slyke (1883–1971) and Glenn Ernest Cullen (1890–1940), American biochemists.

The consequence of this observation is that for a second-order dynamical system, namely a dynamical system with two variables, the solution trajectories must be confined to the first quadrant. If we pick an arbitrary point in the first quadrant, the action of the f_i is to direct the trajectory to a point again located in the first quadrant. Under these conditions, mathematicians have shown that there is at least one point in the first quadrant that is unchanged by the actions of the f_i. In other words, the requirement that the f_i be continuous functions guarantees that there must be at least one fixed point (there can be more than one).

However, it is quite possible for a biological dynamical system to have no fixed point when the f_i are not continuous but are only piecewise continuous. The most common examples arise in the context of feedback control, such as the model we discussed for the control of the position of a fish's fin in Section 2.3.1. Feedback control refers to a controlling mechanism in which the output of a dynamical system is fed back to influence the system itself. To illustrate this concept, consider

$$\frac{dx}{dt} + k_1 x = f(x(t - \tau)), \tag{3.31}$$

where $x = x(t)$ is the variable to be controlled at time t, while $x(t - \tau)$ is the variable at time $t - \tau$, where τ is the time delay and f is the feedback. The fixed point x^* is determined by setting $dx/dt = 0$ and assuming that $x = x^* = x(t - \tau)$. The *error signal* is equal to $x - x^*$. The action of f is either to decrease (negative feedback) or increase (positive feedback) the error signal. Time delays are essential components of feedback control and represent the time taken to detect the error signal and then act on it. Thus the dynamics of feedback control are discussed in the context of delay differential equations (DDE). A detailed discussion of feedback is postponed until Chapter 9. Here our focus is on the fixed point of (3.31).

There are two situations in feedback control in which a fixed point does not exist. The first possibility occurs when a continuous f is replaced by a piecewise continuous f. To illustrate, assume that $f(x(t - \tau))$ is given by the Hill-type function

$$f(x(t - \tau)) = \frac{k_2 K^n}{K^n + x^n(t - \tau)},$$

where k_2, K, n are positive constants.

The fixed point of (3.31) is given by the solution of

$$kx^* = \frac{k_2 K^n}{K^n + x^{*n}}.$$

For arbitrary $n > 0$, this equation cannot be solved analytically. However, as shown in Figure 3.8a, x^* can be determined graphically. Specifically, if we plot kx versus x and $k_2 K^n / (K^n + x^n)$ versus x using the same axes, then x^* corresponds to the intersection(s) between these two curves. It can be readily seen that for all $n > 0$, only one intersection, and hence one x^*, is possible.

As $n \to \infty$, $k_2 K^n / (K^n + x^n)$ becomes more and more like an on–off switch. When (3.31) describes the pupil light reflex, it is possible to replace f experimentally with $1 - H(x - c)$, where $H(x - c)$ is the unit step (Heaviside)[10] function (shown in Figure 3.8b). The unit step function is a piecewise continuous function in which c is a constant chosen such that $H(x - c)$ is 0 when $x - c < 0$ and 1 when $x - c > 0$ (see Section 9.6 for more details). It should be noted that it is impossible to demonstrate formally that the limit of $k_2 K^n / (K^n + x^n)$ as $n \rightarrow \infty$ is $1 - H(x - k_2/2)$; however, this type of feedback can be readily constructed in the laboratory. For this choice of feedback, (3.31) does not have a fixed point.

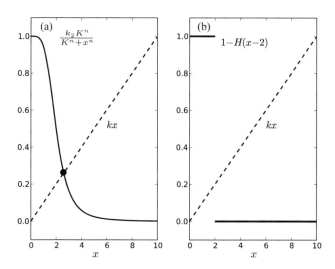

Fig. 3.8 Graphical determination of the fixed point of (3.31). (a) When f is a continuous function of x, the fixed point is the intersection (indicated by •) of the line $y = kx$ (dashed line) and the curve $y = k_2 K^n / (K^n + x^n)$ (solid line). The parameter values are $k_1 = 1$, $k_2 = 1$, $c = 2$ and $n = 4$. (b) No fixed point exists when f is given by $1 - H(x - 2)$, where $H(x - 2)$ is the unit step (Heaviside) function.

The second reason that a fixed point may not exist in the setting of feedback control arises because of the practical limitations in the measurement of the error signal. In general, error signals are measured by sensors that have finite sensitivity. Thus as the error signal becomes smaller and smaller, there comes a point at which the sensor cannot distinguish between a small error and an error equal to zero. Once the error signal enters this sensory dead zone, there is no active feedback control, and the system simply "drifts." Corrective actions are taken once the error becomes large enough so that the sensor can detect it [209, 589, 592].

In order to account for the effects of a sensory dead zone, mathematical models take the form [371, 595, 596]

$$\frac{dx}{dt} + kx(t) = \begin{cases} f(x(t - \tau)) & \text{if } x \text{ is greater than some threshold,} \\ 0 & \text{otherwise.} \end{cases}$$

---

[10] Oliver Heaviside (1850–1925), English electrical engineer, mathematician, and physicist.

As in the previous example, this equation also does not have a fixed point. The solutions of equations of this type include complex periodic and aperiodic functions, and even chaotic behaviors depending on the choice of the feedback function f.

3.6 What Have We learned?

1. What is the law of mass action and when is it valid?

   The law of mass action is an empirical law that states that the rate of a process is proportional to the concentration of the participating reactants. Strictly speaking, the law of mass action holds only for reactions that proceed through one transition state and that occur in dilute solution. However, it is observed that the law of mass action sometimes holds in situations in which these conditions are not met. Perhaps a more generous statement is that the law of mass action is valid when a prediction of a model based on this hypothesis agrees with experimental observation.

2. What are the conditions for the existence of a steady state for a Michaelis–Menten enzymatic reaction and for consecutive chemical reactions?

   Essentially the same condition applies to both situations. With respect to (3.20), the steady-state condition does not imply that $[B]$ is constant. Instead, this conditions requires that $[A]$ change so slowly that $[B]$ behaves as though $[A]$ was constant.

3. Why must the steady-state approximation be used with caution?

   The steady-state approximation must be used with caution because it is not always valid. Sometimes, the same mathematical model yields different predictions depending on whether or not the steady-state approximation is assumed.

4. Is an equilibrium state the same as a steady state?

   No. Although both of these time-independent states are defined dynamically by (3.2), they are different from a thermodynamic point of view. As we discuss in Chapter 17, the important thermodynamic variable for distinguishing between an equilibrium and a steady state is the entropy production, denoted by Λ. Briefly, for an equilibrium, one has $\Lambda = 0$, and for a steady state, $\Lambda > 0$.

3.7 Exercises for Practice and Insight

1. An enzyme and its substrate are placed together in a test tube, and the enzymatic reaction is allowed to proceed for a very long time. At the end of this time, you measure the substrate concentration and observe that it is not equal to zero. Explain why this result is not surprising.
2. Consider the enzyme–substrate reaction described by (3.8). We can measure v and $[S]$. To obtain an estimate of K_{MM} and v_{max}, we can construct a straight-line

plot, often referred to as a *Lineweaver–Burk plot*,[11] described by the equation

$$\frac{1}{v} = \frac{K_{MM}}{v_{max}} \frac{1}{[S]} + \frac{1}{v_{max}}.$$

a. Show that the Lineweaver–Burk equation can be derived from (3.10). How are K_{MM} and v_{max} estimated? Hint: what are the slope and the intercept?

b. Show that alternative ways of obtaining estimates of K_{MM} and v_{max} are

$$\frac{[S]}{v} = \frac{1}{v_{max}}[S] + \frac{K_{MM}}{v_{max}}$$

and

$$v + K_{MM}\frac{v}{[S]} = v_{max}.$$

For each case, determine the slope and the intercepts and explain how you would determine v_{max} and K_{MM}.

c. Notice how the above three equations rely on v and $[S]$, each in a slightly different way. Considering measurement as a source of error, what are the relative advantages among these three types of plots for determining K_{MM} and v_{max}? For example, suppose that it was harder to measure fast v and small $[S]$, or vice versa.

3. Consider the following scheme for an enzyme–substrate reaction and the effect of an inhibitor I:

$$E + S \underset{k_{-1}}{\overset{k_1}{\rightleftharpoons}} ES \overset{k_2}{\longrightarrow} E + P,$$

$$E + I \underset{k_{-3}}{\overset{k_3}{\rightleftharpoons}} EI.$$

This reaction scheme describes the effects of competitive inhibition on enzyme kinetics.

a. What are the definitions of v and v_{max}?

b. What is the enzyme conservation equation?

c. What is the definition of the steady state?

d. In this model, there will be two steady-state constants, K_{MM} and K_I. What are these constants in terms of the k's?

e. Show that

$$v = \frac{v_{max}[S]}{K_{MM}\left(1 + \frac{[I]}{K_I}\right) + [S]}.$$

---

[11] Hans Lineweaver (1907–2009), American physical chemist, and Dean Burk (1904–1988), American biochemist.

    f. Rearrange this relationship to see how $1/v$ behaves with respect to $1/[S]$. Compare this result to the results obtained from the previous exercise.

4. Verify (3.19) noting that $K_{MM} + [S]_0 = (k_{-1} + k_2 + k_1[S]_0)/k_1$.

5. The Mackey–Glass equation[12] arises in the description of white blood cell populations [522]:

$$\frac{dx}{dt} + k_1 x = k_2 x(t - \tau) \frac{K^n}{K^n + x^n(t - \tau)},$$

where $x = x(t)$ is the density of the white blood cell population at time t and $x(t - \tau)$ is the density at time $t - \tau$, where τ is the time delay and K, n are constants.

    a. What is the condition for the steady state?
    b. What is the value of x^* at the steady state?

6. The Longtin–Milton equation[13] arises in the description of the pupil light reflex [498–501],

$$\frac{dx}{dt} + k_1 x = \frac{k_2 K^n}{K^n + x^n(t - \tau)},$$

where $x = x(t)$ is the pupil area at time t, while $x(t - \tau)$ is the area at time $t - \tau$, where τ is the time delay and K, n are constants.

    a. What is the condition for the steady state?
    b. Take $k_1 = 3.21 \text{ s}^{-1}$, $k_2 = 200 \text{ mm}^2 \text{ s}^{-1}$, $K = 50$, $n = 3$, and $\tau = 0.3 \text{ s}$. What is the value of x^* at the steady state?

7. Write a computer program to solve (3.21a)–(3.21c) and verify the results shown in Figure 3.7.

---

[12] Michael C. Mackey (b. 1942), Canadian–American biomathematician and physiologist; Leon Glass (b. 1943), American mathematical biologist.

[13] André Longtin, Canadian neurophysicist (b. 1961); John Milton, (b. 1950) American–Canadian physician–scientist and mathematical biologist.

Chapter 4
Stability

A reasonable starting point for the study of input–output relationships is to assume that the black box shown in Figure 4.1 contains a dynamical system operating at its fixed point. In a laboratory setting, this is equivalent to a system that is either at equilibrium or in a steady state.

Fig. 4.1 A black box with input $b(t)$ and output $x(t)$.

What happens if we suddenly displace the system away from its fixed point and then, just as quickly, release the displacement force? In the previous chapter, we showed that a fixed point can exhibit a certain resistance to change. In mathematics, the study of this resistance to change falls under the topic of *stability theory*. If over time, the system returns to the fixed point, then we say that the fixed point is *stable*; if not, we say that the fixed point is *unstable*. This definition of stability is consistent with everyday experience and coincides with the mathematical concept of *asymptotic stability*. However, other types of stability are recognized (see Sections 4.3.2 and 6.6.1). Since all real dynamical systems (black boxes) are continually subjected to random perturbations, intuition would suggest that the vast majority of systems studied in the laboratory must be stable in some sense. However, one of the most amazing discoveries of the last 120 years has been the realization that a dynamical system can be stable "in some sense" on a large scale even though the fixed points may be locally unstable. An example of this behavior is a limit-cycle oscillation (see Section 4.5.1 and Chapter 11).

© Springer Nature Switzerland AG 2021
J. Milton and T. Ohira, *Mathematics as a Laboratory Tool*,
https://doi.org/10.1007/978-3-030-69579-8_4

In this chapter, we introduce three approaches to determine stability. First, we use a graphical approach to develop intuitions that can help us to interpret and design experiments. Second, we show how stability can be determined if the differential equations that describe the dynamics are known. Strictly speaking, this method applies to the consideration of stability when the magnitude of the perturbation is not too large.

We use these mathematical insights to begin to catalog the possible behaviors that dynamical systems can exhibit. We build on this catalog throughout this book. Recognizing that one or another of these behaviors arises during an experiment can provide important clues as to the nature of the mathematical model that might explain the observation.

Finally, we extend our graphical techniques to introduce a powerful method for examining stability, referred to as *Lyapunov's direct method*.

The questions we will answer in this chapter are:

1. How is the stability of a dynamical system determined?
2. Why do we use the term "local stability" to refer to the classification of fixed points for a nonlinear dynamical system?
3. What condition ensures the local stability of a fixed point?
4. Why is a saddle point important?
5. When the eigenvalues are a pair of complex numbers, what properties of the solution are determined by their real and imaginary parts?
6. Can a linear dynamical system generate a stable and sustained oscillation?
7. Suppose that a dynamical system has one unstable fixed point. Is it possible for the system to be stable in some sense? Give an example.
8. What is the characteristic equation for a linear second-order ordinary differential equation?
9. What is a conservative dynamical system?
10. What is a dissipative dynamical system?
11. What is a Lyapunov function and why is it useful?

4.1 Landscapes in Stability

Many of the important results of physics and engineering are expressed in terms of a difference in *potential*. In such applications, the potential is equal to the work done when an object is moved between two positions in the presence of a force field, for example, particles in a gravitational field or charges in an electric field. If the potential calculated in this way does not depend on the path taken (see also

Sections 4.6.1 and 17.1), then we can define a potential function[1] $U(x)$ that depends typically on the distance between the two positions of the object that was moved. Moreover, we have

$$F(x) = -\frac{d}{dx}U(x),\qquad(4.1)$$

where $F(x)$ is the force.

The concept of a potential function can be extended to modeling dynamics as a flow on a potential surface constructed by plotting $U(x)$ versus x [308]. The advantage of this approach is that it provides a graphical demonstration of the meaning of stability and instability. To illustrate, consider the one-dimensional system

$$\frac{dx}{dt} = f(x).\qquad(4.2)$$

The key point for constructing a dynamic potential $U(x)$ is to recognize that the right-hand side of (4.2) gives the value of the derivative, or slope, at a given point on the potential surface $U(x)$. Thus

$$f(x) = -\frac{d}{dx}U(x),$$

so that (4.2) becomes

$$\frac{dx}{dt} = f(x) = -\frac{d}{dx}U(x),\qquad(4.3)$$

where

$$U(x) = -\int_0^{x(t)} f(s)\,ds.$$

Suppose that $x(t)$ is a solution of (4.3). Then, using the chain rule for differentiation (Tool 2), we have

$$\frac{d}{dt}U(x(t)) = \frac{d}{dx}U(x(t))\cdot\frac{d}{dt}x(t) = -|f(x(t))|^2.$$

Thus $U(x)$ is always decreasing along the solution curves ("water flows downhill"). This approach to estimating stability is a special case of the gradient systems approach to the study of dynamical systems and their stability [308].

The following two examples illustrate the applications of this dynamic potential to questions related to stability. First, let us assume that

$$f(x) = -kx.\qquad(4.4)$$

In this case, we see that the potential function $U(x)$ is the harmonic (parabolic) function (Figure 4.2)

---

[1] Our use of the term "potential" is in keeping with the concept of the Lyapunov function introduced in Section 4.6 [308]. Thus we use the same symbol, $U(x)$, to denote the potential function and the Lyapunov function.

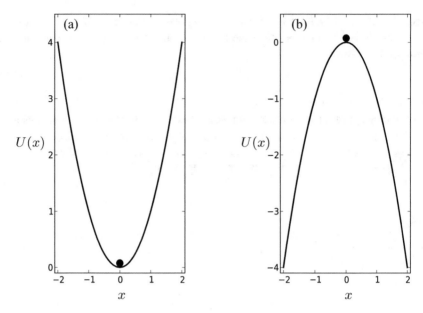

Fig. 4.2 The potential function $U(x)$ for (4.2) when (a) $k > 0$ ("valley") and (b) $k < 0$ ("hill"). The initial condition is indicated by the position of •.

$$U(x) = -\int_0^x -ku \, du = \frac{kx^2}{2}.$$

By making an analogy to the movements of a ball (the • in the figure) over a surface when displaced, we expect that following a small displacement, the ball in the valley (Figure 4.2a) returns (flows) back to its initial position at the bottom of the valley. This result will hold for all initial displacements or perturbations. Hence the fixed point, which corresponds to the bottom of the potential well, is stable. If we change the sign of k from positive to negative, the potential well becomes a potential hilltop (Figure 4.2b). In this case, a small displacement causes the ball to roll away from the hilltop. That is, we are in the presence of an unstable fixed point.

Now consider the case

$$f(x) = x - x^3.$$

This choice of $f(x)$ corresponds to a *cubic nonlinearity* (Figure 4.3a), a nonlinearity important for excitable systems such as those that describe the dynamics of neurons and cardiac cells [422] (see also Sections 4.5.2, 6.6, and 11.1). In this case, we have

$$U(x) = -\int_0^x \left(u - u^3\right) du = -\frac{x^2}{2} + \frac{x^4}{4},$$

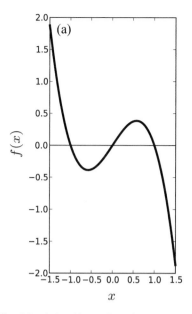
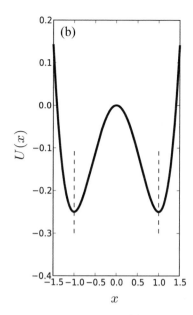

Fig. 4.3 (a) A cubic nonlinearity and (b) the corresponding potential function $U(x)$. The dashed vertical lines indicate the positions of the two stable fixed points.

and we see that $U(x)$ corresponds to a double-well potential (Figure 4.3b). There are three fixed points: $x = 0$, $x = \pm 1$. The fixed point at zero is unstable at the "hilltop" (more generally called a *repeller*), whereas the two fixed points at -1 and $+1$ are stable "valleys" (more generally called *attractors*). The coexistence of two attractors is referred to as *bistability*. Which attractor will be observed experimentally depends on the initial conditions. The set of initial conditions for which $x = -1$ is the attractor (all $x < 0$ in this case) is called the *basin of attraction* for the fixed point $x = -1$. Similarly, the set of all $x > 0$ corresponds to the basin of attraction for the attractor $x = 1$.

Coexistent basins of attraction are separated by a boundary. A familiar example is the continental divide in North America. If we stand on the continental divide facing north, then to our left, all water flows westward, and to our right, all water flows eastward. In the case of the one-dimensional double-well potential shown in Figure 4.3b, the boundary is the unstable fixed point at $x = 0$. In higher-dimensional systems, the boundary is formed by a *separatrix*. Often, such a separatrix is associated with a *saddle*, or hyperbolic fixed point, described in Section 4.3.3. Other possibilities exist as well. The important point for our discussion is that the boundary (separatrix) cannot be crossed unless the system is perturbed by, for example, electrical or mechanical impulses. Even noisy perturbations can cause switches between basins of attraction (see Section 15.2.3).

The concepts of a basin of attraction and bistability (or more generally, multistability) arise very frequently in discussions of the dynamics of biological systems. We anticipate that qualitatively similar potential functions $U(x)$ will serve to describe all situations in which there are two or more coexisting attractors. Thus it is possible for an experimentalist to propose and perform interesting experiments based only on these sketches. Can perturbations be introduced to determine the shape and size of the basin of attraction? If two or more basins of attraction coexist, can stimuli be used to cause switches between attractors? Can switches between attractors be the result of random perturbations? Below, we briefly describe three examples in which questions related to basins of attraction are important for understanding the underlying biology.

4.1.1 Postural Stability

Falls are a leading cause of mortality and morbidity in the elderly [610]. Thus considerable effort is being devoted to understanding how the upright posture is maintained in response to perturbations. From the biomechanical point of view, the condition for maintaining balance is that the body's center of mass lies within the base of support. In the clinic, the center of gravity is estimated to be at the level of the second sacral vertebra [75] ($\approx 55\%$ of height [337]), and the base of support is the area under and between the feet.

In the language of dynamical systems, we say that a stable upright position lies within a basin of attraction that it is roughly of the dimensions of the base of support. It is possible to construct the basin of attraction for an individual's postural stability through the use of carefully honed perturbations [808, 911]. Bipedal locomotion, namely walking and running on two feet, requires that the center of mass be displaced outside the basin of attraction. Hence the impetus for locomotion is the body's effort to catch up with the displaced center of mass. Since many falls occur during the transition between standing and walking [505], it is of interest to study the dynamics at the edge of the basin of attraction.

4.1.2 Perception of Ambiguous Figures

Historically, the perception of ambiguous figures has been considered an example of bistability in neural perception [68, 447]. The most famous example is Necker's cube[2] [627], shown in Figure 4.4. If you stare at Necker's cube long enough, you will soon observe two interesting phenomena: (1) Is the cube oriented down and to the right or up and to the left? (2) Even if you try to keep one interpretation in mind, your perception of the orientation of the cube continually changes; first you see the cube oriented one way, then the other.

Why does our perception of the ambiguous figures change? One possibility is that the changes are related to involuntary eye movements, such as saccades.

---

[2] After Louis Albert Necker (1786–1861), Swiss crystallographer, zoologist, and geographer.

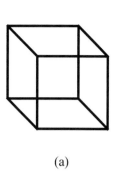

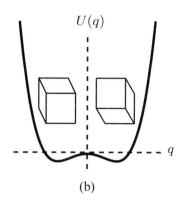

$U(q)$

q

(a)   (b)

Fig. 4.4 (a) Necker's cube. (b) Two different perceptions of Necker's cube corresponding to two different attractors in a dynamical system possessing a double-well potential $U(q)$, where q is an appropriately identified variable. Figure reproduced from [575] with permission.

Indeed, recent studies suggest that eye movements can be trained to cause perceptual switches [189, 318]. However, the switches in perception occur even when the image is stabilized onto the retina [699]. The reader can verify this for themselves by conducting the following experiment. Draw a Necker's cube with white lines on a black background and then illuminate it with a brief pulse of very bright light while looking at the cube, you will "see" a Necker's cube even if you subsequently close your eyes. This is because the image of the Necker's cube has been "temporarily burned onto your retina," and the photoreceptors in your retina take time to recover. The consequence of this experimental trick is that the effects of eye movements on perception have been temporarily eliminated. Since subjects continue to experience spontaneous changes in perception of the Necker's cube, these changes cannot be due to eye movement. This demonstration suggests that there may be a dynamic explanation for this phenomenon. Indeed, analysis of the switching times in perception provides evidence that the human brain is subject to random perturbations that serve to switch the perception of Necker's cube back and forth [68].

4.1.3 Stopping Epileptic Seizures

In the case of a double well, we might wonder whether we could introduce stimuli to cause switching between two attractors [238, 303]. It is sometimes possible to abort an epileptic seizure by introducing a brief electrical [611, 661] or sensory [576] stimulus (Figure 4.5). This observation is very suggestive of an underlying multistable dynamical system [158, 514], and it raises the possibility of developing a brain defibrillator for the nonpharmacological treatment of epilepsy [574].

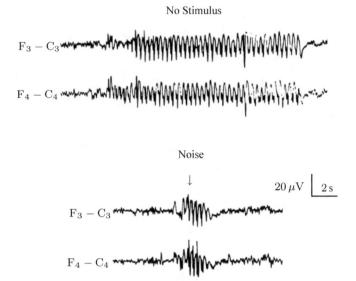

Fig. 4.5 Shortening the length of a seizure using a brief auditory stimulus in an adolescent with a generalized epilepsy. The EEG channels were recorded corresponding to the frontal–central regions bilaterally: $F_3 - C_3$ is from the left hemisphere and $F_4 - C_4$ is from the right hemisphere. The top two traces represent the average length of 71 absence seizures without auditory stimulus, and the bottom two traces represent the average of 69 absence seizures when an auditory stimulus was given near the beginning of the seizure ($\downarrow$). The auditory stimulus was produced by dropping a metal trash can on the floor. Figure reproduced from [576] with permission.

4.2 Fixed-Point Stability

In Section 2.3.1, we introduced the idea that a nonlinear differential equation can behave like a linear differential equation over a certain range of the variables. It is reasonable to ask whether it is possible to linearize such a nonlinear differential equation in a neighborhood of a fixed point. The answer is yes. The importance of this observation is that we can use the roots of the characteristic equation (see Section 2.2.3) to learn about the stability of fixed points with respect to sufficiently small perturbations. Since this approach to stability relies on the linearization of a nonlinear differential equation in a neighborhood of a fixed point, it is often referred to as *local stability analysis*.

The most straightforward way to determine the stability of (4.2) is to plot dx/dt versus x. Figure 4.6 shows an example of such a plot for the differential equation

$$\frac{dx}{dt} = x - x^3 .$$

The fixed points correspond to the values of x for which $dx/dt = 0$. The local stability of each of the fixed points can readily be determined by looking at the slope of $f(x)$ close to the fixed point as follows. Provided that we are sufficiently close to the fixed point, we can approximate the nonlinear differential equation by a linear one. The dashed line in Figure 4.6 indicates the linear approximation to $f(x)$ about the fixed point $x = 0$. The approximate equation that describes the dashed line is

$$\frac{dx}{dt} \approx kx.$$

We know that the fixed point will be stable if $k < 0$, that is, if the slope of the dashed line is negative. The observed slope of this approximation is positive for the $x = 0$ fixed point; hence $k > 0$, and the $x = 0$ fixed point is unstable. Repeating this argument for the $x = \pm 1$ fixed points, we see that both correspond to stable fixed points. Another, perhaps simpler, approach is to interpret Figure 4.6 in terms of flows (the right and left arrows): when $f(x) > 0$, the flow is to the right (x increasing), and when $f(x) < 0$, the flow is to the left (x decreasing).

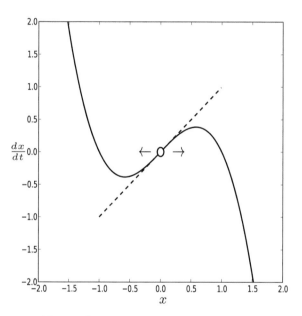

Fig. 4.6 The function $f(x) = x - x^3$ linearized about the point $x = 0$ ($\circ$). The linear approximation is represented by the dashed line. Since the slope of the dashed line is positive, the fixed point at $x = 0$ is unstable. By convention, $\circ$ indicates an unstable fixed point, and the arrows indicate the direction of the flow.

We can also obtain the same conclusions analytically. Here we introduce Tool 7, a very useful tool in approximating the dynamics of (4.2) in a neighborhood of its fixed point by a linear dynamical system. This tool is based on an important theorem in mathematics known as *Taylor's theorem*,[3] which makes it possible to

---

[3] Brook Taylor (1685–1731), English mathematician.

write a function for which $n+1$ derivatives exist on an interval as a polynomial of the form

$$f(x) = f(0) + x\frac{df}{dx}\Big|_{x=0} + \frac{x^2}{2!}\frac{d^2f}{dx^2}\Big|_{x=0} + \cdots + \frac{x^n}{n!}\frac{d^nf}{dx^n}\Big|_{x=0} + R_n(x), \qquad (4.5)$$

where $R_n(x)$ is a remainder, $\frac{d^nf}{dx^n}\big|_{x=0}$ means that the derivative is evaluated at $x = 0$, and $n!$ denotes the factorial function, $n! = 1 \cdot 2 \cdots (n-1) \cdot n$. As the following examples show, the series expansions of a number of well-known functions represent an application of Taylor's theorem.

Example 4.1. Let $f(x) = e^x$. Since the derivatives of e^x of all orders are all simply e^x, and $e^0 = 1$, we see from (4.5) that

$$e^x = 1 + x + \frac{x^2}{2!} + \frac{x^3}{3!} + \cdots.$$

Similarly, we can show that

$$\sin x = x - \frac{x^3}{3!} + \frac{x^5}{5!} + \cdots, \qquad \cos x = 1 - \frac{x^2}{2!} + \frac{x^4}{4!} + \cdots.$$

Taylor series expansions such as these will arise frequently in our work. The reader may consult a mathematical handbook to obtain the appropriate series expansion of a given function; such information is also easily obtained via the Internet. $\diamondsuit$

In order to apply Tool 7 to the determination of the local stability of the fixed points of (4.2), we need two additional steps. First, we define a new variable $\hat{x} := x - x^*$, where x^* is the fixed point. When $x = x^*$, we have $\hat{x} = 0$. Second, we ignore the terms $\hat{x}^n$ for $n \geq 2$. Thus (4.2) becomes

$$\frac{d\hat{x}}{dt} \approx k\hat{x},$$

where $k = \frac{df}{dx}\big|_{x=x^*}$. In addition, we see that the term "local" means that $\hat{x}$ is small enough that the linear approximation to $f(x)$ is valid for our purposes.

4.3 Stability of Second-Order ODEs

We can readily extend the observations in the previous section to the determination of fixed-point stability in a system of two nonlinear first-order differential equations

$$\frac{dx}{dt} = f(x,y), \qquad \frac{dy}{dt} = g(x,y), \qquad (4.6)$$

where $f(x,y)$ and $g(x,y)$ are nonlinear functions of two variables x,y. The fixed points (x_i^*, y_i^*) are the solutions of the equations

$$f(x_i^*, y_i^*) = 0, \quad g(x_i^*, y_i^*) = 0,$$

where the subscript i indicates that there may be more than one fixed point. The linearized equations are

$$\frac{d\hat{x}}{dt} = \left(\frac{\partial f}{\partial x}\right)_{x^*, y^*} \hat{x} + \left(\frac{\partial f}{\partial y}\right)_{x^*, y^*} \hat{y} = k_{11}\hat{x} + k_{12}\hat{y}, \tag{4.7}$$

$$\frac{d\hat{y}}{dt} = \left(\frac{\partial g}{\partial x}\right)_{x^*, y^*} \hat{x} + \left(\frac{\partial g}{\partial y}\right)_{x^*, y^*} \hat{y} = k_{21}\hat{x} + k_{22}\hat{y},$$

where $\hat{x} = x - x^*$ and $\hat{y} = y - y^*$, and the notation

$$\left(\frac{\partial f}{\partial x}\right)_{x^*, y^*}$$

means that the partial derivative[4] of $f(x,y)$ with respect to x is evaluated at (x^*, y^*). Equation 4.7 can be rewritten as

$$\frac{d^2\hat{x}}{dt^2} - k_1 \frac{d\hat{x}}{dt} + k_2\hat{x} = 0, \tag{4.8}$$

where $k_1 := k_{11} + k_{22}$ and $k_2 := (k_{11}k_{22} - k_{12}k_{21})$. Using Tool 4, we obtain the characteristic equation

$$\lambda^2 - k_1\lambda + k_2 = 0. \tag{4.9}$$

This is a quadratic equation in λ, and hence there are two roots. The roots are given by

$$\lambda_{1,2} = \frac{k_1 \pm \sqrt{k_1^2 - 4k_2}}{2}. \tag{4.10}$$

Depending on the values of k_1 and k_2, these roots can be two real numbers or a pair of conjugate complex numbers. The solution of (4.7) has the form

$$\hat{x}(t) = A_1 e^{\lambda_1 t} + A_2 e^{\lambda_2 t} \tag{4.11}$$

in terms of $\hat{x}$, or

$$\hat{y}(t) = B_1 e^{\lambda_1 t} + B_2 e^{\lambda_2 t}$$

in terms of $\hat{y}$, where A_1, A_2, B_1, B_2 are constants determined from the initial conditions. These observations indicate the intimate connections between the parameters (k_1, k_2), the roots of (4.9), and the predicted dynamics in a neighborhood of the fixed point.

---

[4] The partial derivative of $f(x,y)$ with respect to x is calculated by assuming that y is constant. For example, if $f(x,y) = 6xy$, then $\partial f/\partial x = 6y$. Similarly, $\partial f/\partial y = 6x$.

Our next step is to determine the dynamics that are associated with each pair of λ's. We strongly recommend that the student use a computer program to examine the dynamical behavior of (4.7) for different choices of the parameters. This is by far the most efficient way to understand the terminology and the mathematics.

It should be noted that the computer program estimates $x(t), y(t)$ using suitably chosen numerical algorithms. For a linear ordinary differential equation, we have $x(t) = \hat{x}(t)$ and $y(t) = \hat{y}(t)$; however, for a nonlinear ordinary differential equation, $x(t) \approx \hat{x}(t)$ and $y(t) \approx \hat{y}(t)$ only in a neighborhood of the fixed point. In the figures that follow, we adopt the widely used convention of plotting only the numerically determined solutions $x(t), y(t)$. We leave it to the interested reader to determine the extent of the neighborhood about the fixed point for which the solutions of the linearized equations are valid.

4.3.1 Real Eigenvalues

When $k_1^2 \geq 4k_2$, we have a pair of real eigenvalues. The condition for the local stability of the fixed point is

$$\lambda_1 < \lambda_2 < 0,$$

which requires that $k_1 < 0$. In this case, we see from (4.11) that $\hat{x}(t) \to 0$ as $t \to \infty$ for all initial conditions (namely, for all choices of A_1 and A_2).

Figure 4.7a shows the behavior for three different choices of the initial conditions. In all cases, there is a nonoscillatory decrease of $\hat{x}$ to 0. These behaviors are collectively referred to as *overdamping*. The qualitative nature of the approach to the fixed point depends very much on the choice of initial conditions [162, 826]. If we choose $\hat{x}(t_0) > 0$, then there will be three different qualitative patterns in the time-dependent changes in amplitude as a function of $d\hat{x}/dt$. When $d\hat{x}/dt > 0$, the solution will pass through a maximum before finally approaching the fixed point (Case I in Figure 4.7a). A more complex picture arises when $d\hat{x}/dt < 0$. If $d\hat{x}/dt < 0$ is small, the solution approaches the fixed point monotonically (Case II in Figure 4.7a), but if $d\hat{x}/dt$ is sufficiently large, it is possible for the solution to change sign and pass through a minimum before asymptotically approaching the fixed point (Case III in Figure 4.7a). An important property of these solutions is that they cross the line $x = 0$ at most once. Considerations of these transient behaviors have attracted the interest of biomathematicians studying, for example, the effects of toxins and the immune response involved in the acute inflammatory response [162] (see also Chapter 6).

When $\lambda_1 = \lambda_2 = \lambda$, the general solution (4.11) becomes (see Exercise 4)

$$\hat{x}(t) = C_1 e^{\lambda_1 t} + C_2 t e^{\lambda_2 t}.$$

This occurs when $k_1^2 - 4k_2 = 0$, so that $\lambda_{1,2} = k_1/2$. As before, the sign of the real λ determines whether the fixed point is stable ($\lambda < 0$) or unstable ($\lambda > 0$). This case is referred to as *critical damping*. The importance of critical damping is that for

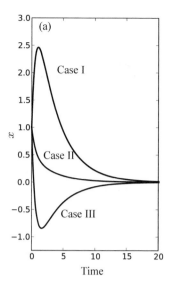
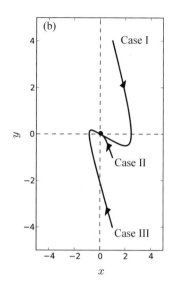

Fig. 4.7 (a) The changes in $\hat{x}$ as a function of time for (4.8) when $k_1 = 2.0$ and $k_2 = 0.5$ for three different choices of initial conditions. In all three cases, $\hat{x}(0) = 1.0$. For Case I, $d\hat{x}/dt(0) = 4.0$; Case II, $d\hat{x}/dt(0) = -1.0$; and Case III, $d\hat{x}/dt(0) = -4.0$. (b) Phase-plane representation for the three cases shown in (a). The phase plane is constructed by plotting $y(t)$ versus $x(t)$.

any perturbation, the approach to the fixed point will be more rapid for a critically damped system than for an overdamped one [826].

It is possible that the $\lambda_{1,2}$ are real but have opposite signs, i.e., $\lambda_1 < 0 < \lambda_2$. An example arises in the dynamics of the inverted planar pendulum (see Section 10.6.1) described by

$$\frac{d^2\theta}{dt^2} - \omega_n^2\theta = 0$$

where θ is the vertical displacement angle and ω_n is the natural angular frequency of small oscillations when the pendulum is hung downwards [801]. In all cases except one, the fixed point is unstable since the dynamics are always eventually dominated by the $A_2e^{\lambda_2 t}$ term in (4.11). The lone exception occurs when the initial conditions are chosen such that $A_2 = 0$. This type of fixed point is referred to as a *saddle point* (see below).

4.3.2 Complex Eigenvalues

If $4k_2 > k_1^2$, then it becomes necessary to take the square root of a negative number. Since a negative number does not have a real square root, we define the *imaginary unit*

$$j = \sqrt{-1}$$

using the notation of engineers and computer scientists; other scientists and mathematicians use i instead of j. In this case, the eigenvalues are complex numbers, and they take the form

$$\lambda = \frac{k_1 \pm j\sqrt{4k_2 - k_1^2}}{2} := \Re(\lambda) \pm j\Im(\lambda),$$

where $\Re(\lambda)$ and $\Im(\lambda)$ are the real and imaginary parts of the eigenvalues. In Section 7.2.1, we show that we can write the solution of (4.8) as

$$x(t) = Ae^{\Re(\lambda)t}\sin(\Im(\lambda)t + \phi), \tag{4.12}$$

where ϕ is the phase shift and A is the amplitude.

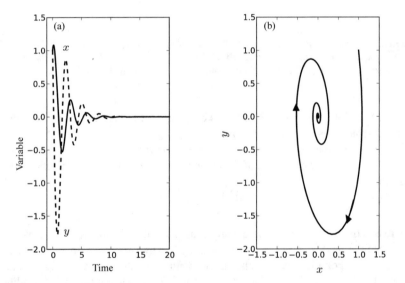

Fig. 4.8 (a) The changes in x and dx/dt as a function of time following a brief perturbation away from the fixed point at $(0,0)$ when the λ are a pair of complex conjugate eigenvalues with negative real parts. (b) Phase-plane representation of a spiral point shown in (a).

Equation (4.12) is very useful, since it shows that when the λ are complex numbers, the solutions of (4.8) are oscillatory. Moreover, the stability of the oscillatory solutions depends on the sign of $\Re(\lambda)$, and the frequency of the oscillatory part depends on $\Im(\lambda)$. Thus if $\Re(\lambda) < 0$, we have an oscillatory solution whose amplitude decreases exponentially, as shown in Figure 4.8a. These responses are referred to as *underdamping*.

On the other hand, if $k_1 = 0$, we have a pair of pure imaginary eigenvalues. Since $\Re(\lambda) = 0$, it follows that (4.12) becomes a pure sinusoidal oscillation (Figure 4.9a). An example of such a system is the harmonic oscillator

$$\frac{d^2x}{dt^2} + k_2 x = 0$$

(see also Section 4.6.1). In this case, a perturbation from the fixed point neither grows nor decreases as a function of time. Another way of thinking about this is that the response to a perturbation occurs infinitely rapidly, and the recovery occurs infinitely slowly. A consequence is that the period and amplitude of the oscillation depend on the initial condition (for an example see Figure 4.11). Systems that exhibit this type of stability are referred to as *neutrally*, or *marginally, stable dynamical systems*.

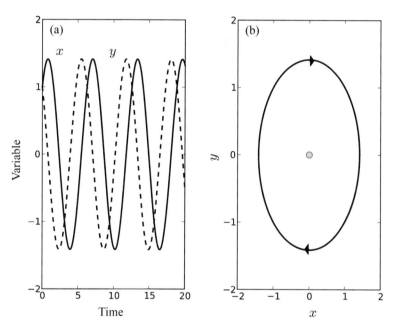

Fig. 4.9 (a) The changes in x and dx/dt as a function of time following a brief perturbation away from the fixed point at $(0,0)$ when $\lambda_{1,2}$ are pure imaginary eigenvalues (i.e., the real part is 0). Note that the oscillations in x and y are phase shifted with respect to each other. (b) Phase-plane representation of a center shown in (a).

4.3.3 Phase-Plane Representation

It is cumbersome to discuss the stability of fixed points using the approach we have outlined. A widely used terminology for the stability of fixed points is based on a graphical approach known as the *phase plane*. The phase plane is constructed from the solutions of a differential equation by plotting $y(t)$ versus $x(t)$ for various initial conditions. In general, these solutions are estimated numerically using appropriate computer software packages. Phase-plane analysis plays a prominent role in the investigation of the properties of nonlinear second-order differential equations, in particular those that arise in the description of excitable systems (see also Section 4.5.2).

There are three important properties of the phase plane. First, time does not appear explicitly. Second, solutions corresponding to different choices of the initial conditions appear as different *trajectories* in the phase plane. The trajectories never intersect. Suppose that two trajectories intersected at some point. Then choose the initial condition to be that point. There would then be two possible solutions for the same initial condition, which violates the requirement that a differential equation exhibit uniqueness (Section 2.4). Third, since x and y are constructed from the solution of the differential equation, we have a complete description of the stability of each fixed point. In the case of a stable fixed point for a nonlinear dynamical system, the basin of attraction will, in general, be larger than the region where the linearization provided by Tool 7 is valid. Nonetheless, the fixed point is classified based on its characteristics in the linear regime.

Table 4.1 Classification of stability of fixed points of (4.8).

Roots of characteristic equation	Phase plane descriptor	Solution of linearized equation
$\lambda_1 > \lambda_2 > 0$	Unstable node	$\hat{x}(t) = A_1 e^{\lambda_1 t} + A_2 e^{\lambda_2 t}$
$\lambda_1 < \lambda_2 < 0$	Stable node	
$\lambda_2 < 0 < \lambda_1$	Saddle point	$\hat{x}(t) = A_1 e^{\lambda_1 t} + A_2 e^{\lambda_2 t}$
$\lambda_1 = \lambda_2 > 0$	Unstable node	$\hat{x}(t) = (A_1 + A_2 t)e^{\lambda t}$
$\lambda_1 = \lambda_2 < 0$	Stable node	
$\lambda_{1,2} = \gamma \pm j2\pi f$		
$\gamma > 0$	Unstable spiral point	$\hat{x}(t) = e^{\gamma t}[A_1 \cos 2\pi f t + A_2 \sin 2\pi f t]$
$\gamma < 0$	Stable spiral point	
$\lambda_{1,2} = \pm j2\pi f$	Center	As above with $\gamma = 0$

The advantage of the phase-plane representation is that it is easy to associate fixed-point stability with pictures constructed in the phase plane from trajectories obtained from different choices of initial conditions. Four terms are used to describe the geometry of the flows in a neighborhood of the fixed point: *node, spiral point*,[5]

---

[5] Historically, mathematicians favored the term focal point. However, the recent trend is to use the term spiral point [72, 164, 810, 868, 880]. The advantage of the latter term is that it describes what one sees in the phase plane. Most undergraduate students do not intuitively associate this phase-plane portrait with the word focus.

saddle point, and *center*. Node and spiral points are further classified as stable or unstable. Table 4.1 compares the stability of the fixed point determined from the roots of the characteristic equation to the terminology that describes the corresponding phase-plane portrait. Let us see how to use this table.

Figure 4.7b shows the phase-plane portrait of a node. From Table 4.1, we see that a node occurs when both roots of the characteristic equation are real numbers and have the same sign. The defining feature of a node in the phase plane is that the trajectories never completely encircle the fixed point. Nodes can be stable or unstable; this figure shows an example of a stable node.

Figure 4.8b shows the phase-plane portrait when the roots of the characteristic equations are a pair of complex numbers. Not surprisingly, this fixed point is called a spiral point. In contrast to a node, the trajectories of a spiral point can completely encircle the fixed point. Spiral points can be unstable or stable; the figure shows a stable spiral point. Finally, Figure 4.9b shows the phase-plane portrait, referred to as a center, that occurs when the roots of the characteristic equation are a pair of pure imaginary numbers.

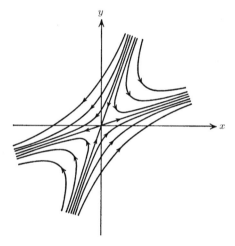

Fig. 4.10 Phase-plane representation for a saddle point. It should be noted that there is only one trajectory that approaches the fixed point. Can you spot it?

Figure 4.10 shows the phase-plane portrait when the roots of the characteristic equation are real numbers with opposite signs. A saddle point is unstable, since all trajectories except one evolve toward $\pm\infty$. The lone stable trajectory would not survive in the real world, since even a very small perturbation would cause the solution to evolve toward $\pm\infty$. Despite their instability, saddle points are very important. For example, the upright position of an inverted pendulum is a saddle point (see Exercise 10). Consideration of how the upright position of a pendulum can be stabilized serves as an important benchmark for comparing different control strategies [606], and such considerations have recently become important in studies of human balance control [13], including stick-balancing at the fingertip [94–97].

This fixed-point classification forms the basis of a common dialogue that spans many disciplines. Consequently, biologists can often obtain important insights by comparing their experimental results to those of, for example, a physical system that has the same types of fixed points. This descriptive way of thinking about the stability of fixed points allows investigators to exchange ideas across disciplines.

4.4 Illustrative Examples

Up to this point there has been a considerable amount of notation. However, the procedure to determine fixed-point stability is actually quite easy once you get the hang of it. Here we illustrate these techniques by discussing four examples.

4.4.1 The Lotka–Volterra Equation

The Lotka–Volterra model[6] describes the interactions between a predator and its prey [507]. Historically, this model represented one of the first uses of mathematics in an active debate involving field biologists and biostatisticians. At stake was the explanation for the fluctuations in the number of hare and lynx pelts collected by the Hudson Bay Company (Figure 4.11). A number of theories were proposed, including the possibility that the cycles represented the effect of sunspot cycles. The Lotka–Volterra model was one of the first models to demonstrate that oscillations in variables could arise autonomously, i.e., could occur solely because of interactions between the variables within the dynamical model. Although this model is not currently favored as providing an accurate description of population dynamics (see below), it provides a nice example of the application of local stability analysis to a nonlinear differential equation.

Let x be the density of the prey (hare) population,[7] and let y be the density of the predator (lynx) population, and assume that we can use the law of mass action to describe the growth, death, and interactions between predators and prey. Consequently, the prey multiply and grow at a rate proportional to their density, while their numbers decrease by predation at a rate proportional to both predator and prey populations, that is, at a rate proportional to xy. On the other hand, the predators decrease at a rate proportional to their density (the greater their number, the less food per predator) and increase at a rate proportional to predation, that is, to xy. These observations can be summarized mathematically as

---

[6] Developed independently by Alfred James Lotka (1880–1949), American biophysicist and biostatistician, and Vito Volterra (1860–1940), Italian mathematician and physicist.

[7] Although it seems intuitive that density should be number of animals per unit area, the SI unit is in fact area^{-1}, since, as the reader will recall from Section 1.2.3, a number of objects has no units in the SI system.

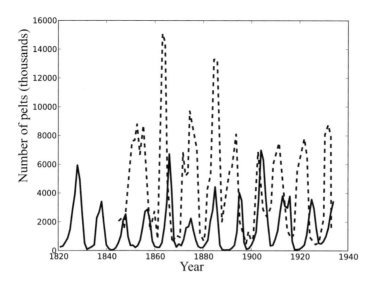

Fig. 4.11 The lynx (solid line) and hare (dashed line) population cycles. Population densities estimated from the number of pelts bought by the Hudson Bay Company each year. Data for lynx obtained from Hyndman, R.J., Time Series Data Library, http://datais/TSDLdemo, accessed on December 8, 2013, from DataMatket.com. Data for hare were obtained from [650].

$$\frac{dx}{dt} = x - xy = x(1-y), \quad \frac{dy}{dt} = xy - y = y(x-1). \tag{4.13}$$

For simplicity, we have assumed that all of the parameters are equal to 1. Figure 4.12 shows the output of a computer program that integrates these equations. There is a qualitative similarity between the observations in Figure 4.11 and the solution of (4.13) shown in Figure 4.12a. In both cases, there is an oscillatory relationship between predator and prey populations in which the predator population density lags behind that of the prey.

There are three steps involved in determining the stability of a nonlinear differential equation such as that described by (4.13). The first is to determine the fixed points (x^*, y^*), which are the values of (x, y) for which $dx/dt = dy/dt = 0$, i.e.,

$$x(1-y) = 0, \quad y(x-1) = 0.$$

We see that there are two fixed points: $(x^*, y^*) = (0,0)$ and $(x^*, y^*) = (1,1)$. We evaluate the stability of each fixed point separately.

The fixed point $(x^*, y^*) = (0,0)$: In this case, the second step is to linearize (4.13) at each fixed point using Tool 7. The linearized (4.13) at the fixed point $(0,0)$ is

$$\frac{d\hat{x}}{dt} = \hat{x}, \quad \frac{d\hat{y}}{dt} = -\hat{y}.$$

The third step is to determine the eigenvalues and then to classify the nature of the fixed point. Using (4.8) with $k_{11} = 1$, $k_{12} = 0$, $k_{21} = 0$, and $k_{22} = -1$ we obtain the characteristic equation

$$\lambda^2 - 1 = 0,$$

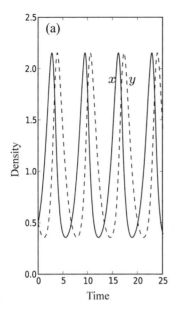
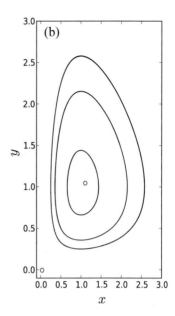

Fig. 4.12 (a) Population cycles in predator and prey produced by the Lotka–Volterra equation (4.13). Note that as expected, the predator population density lags behind that of the prey. (b) Phase-plane representation of (4.13). The unstable fixed points are denoted by ∘.

and hence the eigenvalues are ± 1.

The fixed point $(x^*, y^*) = (1,1)$: Similarly, we can linearize about the fixed point $(1,1)$ to obtain

$$\frac{d\hat{x}}{dt} = -\hat{y}, \quad \frac{d\hat{y}}{dt} = \hat{x}.$$

Using (4.8) with $k_{11} = 0$, $k_{12} = -1$, $k_{21} = 1$, and $k_{22} = 0$ we obtain the characteristic equation

$$\lambda^2 + 1 = 0.$$

Thus the eigenvalues are $\pm j$.

We can use Table 4.1 to complete the description of the dynamics of the Lotka–Volterra equation. The eigenvalues for the $(x^*, y^*) = (0,0)$ fixed point are $\lambda = \pm 1$. Thus this fixed point is a saddle point. The eigenvalues for $(x^*, y^*) = (1,1)$ are $\lambda = \pm j$. Thus this fixed point is a center. As is shown in Figure 4.12b, we can use this information to construct the phase-plane portrait. The interpretation is that the predator and prey population densities will oscillate periodically. This proves the point proposed by Lotka and Volterra that such oscillations could arise autonomously due to interactions between the variables.

There are several well-known problems associated with the Lotka–Volterra model as a description of the dynamics of predator–prey populations. A detailed critique of this model from a population biology point of view is beyond the scope of the present discussion; however, we can draw attention to some mathematical aspects of the model that are problematic. First, the observation that the $(0,0)$ fixed point is a saddle point implies that no matter how small the predator and prey populations might become, they would always recover and, in fact, would attain a population cycle with a very large amplitude. Such, however, is not observed experimentally. Second, the oscillations are neutrally stable, which means that the amplitude and period of the oscillation are not constant, as observed for a limit cycle, but are functions of the initial conditions.

Problems arise when neutrally stable dynamical systems are subject to the effects of random perturbations ("noise"). In particular, noisy Lotka–Volterra systems are unstable [472] and under certain conditions can exhibit a variety of complex and counterintuitive properties [787]. Finally, the Lotka–Volterra equations do not take into account the effect of finite gestation and maturation times. The responses of the predator and prey populations to changes in density do not occur instantaneously, but are influenced by gestation times and the time required for infants to become functioning adults. Thus continuous-time models of population dynamics should be in the form of delay differential equations (see Chapter 9). Although time-delayed Lotka–Volterra equations have received little attention in the ecological literature, they are actively investigated in the context of the dynamics of populations of neurons [22, 904]. Despite these limitations, the Lotka–Volterra model is often used as a convenient starting point for investigations of dynamical systems. (See Exercises 9, 13, and 16 at the end of this chapter.)

4.4.2 Computer: Friend or Foe?

Two major advances in the study of nonlinear dynamics have been the development of fast, cheap computers and the development of efficient methods for the numerical integration of differential equations. So many computer packages are now available that the reader may conclude that there is no longer any point in knowing mathematics at all. After all, why bother with math when you can simply view the dynamics on the computer screen? However, there are constraints on the application of computers and algorithms because of the fact that there are limitations on the representation of real numbers in a digital machine. Furthermore, often overlooked is the fact that the algorithms used to integrate differential equations numerically are themselves dynamical systems. Thus the validity of the computerized simulation requires that the algorithm performs in a stable manner.

To illustrate the problem, let us compute the solution of

$$\frac{dx}{dt} = -kx \tag{4.14}$$

using a numerical method known as *Euler's method*. This is the simplest numerical method for integrating a differential equation. It relies on Tool 2. Specifically, the definition of a derivative is used to transform a continuous differential equation into an equation in which time changes in discrete units Δt of the time step. Hence (4.14) becomes

$$x(t + \Delta t) = x(t) - k\Delta t x(t) = (1 - k\Delta t)x(t), \qquad (4.15)$$

or

$$x(t + \Delta t) = \beta x(t), \qquad (4.16)$$

where $\beta := 1 - k\Delta t$.

Equations (4.15) and (4.16) are examples of a one-dimensional map. The stability analysis of such maps is identical to the methods used for differential equations; however, the condition for stability is different (see Section 13.1):

$$|\beta| < 1, \quad \text{or equivalently,} \quad -1 < \beta < 1. \qquad (4.17)$$

Using this rule, we see that Euler's method for solving (4.14) will yield a stable solution, provided that Δt is chosen to satisfy the stability condition

$$-1 < \beta < 1, \quad \text{or equivalently,} \quad 0 < \Delta t < \frac{2}{k}.$$

If Δt does not satisfy this condition, then the numerical solution of (4.14) will exhibit growing oscillations, even though we know that the solution of (4.14) decays exponentially to zero!

Problems introduced by truncation error and algorithm instability plague the lives of investigators who rely solely on computer simulations. On the one hand, prudent scientists are justified in being wary; on the other hand, pragmatic scientists realize that there is often no choice but to rely on the computer, though they make certain that they have a sufficient understanding of what is going on. For example, current models of neural networks contain $\approx 10^5$ nonidentical neurons [383] (a computer model of the human brain would involve $\approx 10^{11}$ neurons and $\approx 10^{15}$ connections). Although certain statistical generalizations can be obtained without the use of a computer [89], detailed consideration of the dynamics requires numerical simulation. Thus we are left in the very uncomfortable position of not knowing for sure whether the results we see on the computer screen represent artifact or reality.

Although we cannot avoid recourse to computer simulations, we can take steps to try to increase our faith that the results of the computer simulation do not simply represent numerical artifacts. First, we can determine what happens when we decrease the step size Δt. If the computed solution remains qualitatively unchanged, then this favors the possibility that the computed solution is not a numerical artifact. However, as step size decreases, computational time dramatically increases. Second, we can check the results of the computer program for choices of the parameters for which we can calculate the behavior in a neighborhood of a fixed point. Do the dynamics of the computer solution behave as expected from a local stability analysis? Finally, our faith in the computer and mathematical models increases if predictions are consistent with experimental observation.

4.5 Cubic nonlinearity: excitable cells

A defining property of excitable cells is their "all or none" response to inputs. For example, a neuron does not generate an action potential unless its input becomes sufficiently high. It is this behavior that is captured in the integrate-and-fire model for a neuron that we discussed in Section 1.3.2. However, excitability is not restricted to neurons but is a property exhibited by cardiac, skeletal, and intestinal smooth muscle cells; insulin-producing pancreatic beta cells; and even certain plant cells (see Figure 12.13). Thus it is very possible that life on this planet would not be possible without *excitability* [552]. The molecular engine for excitability is the membrane ion channel (see Section 11.1). At the heart of mathematical models for excitable systems is a cubic nonlinearity. Here we illustrate this concept by examining two simple mathematical models for an excitable system, namely the van der Pol oscillator and the Fitzhugh–Nagumo equation for a neuron.

4.5.1 The van der Pol Oscillator

Up to this point, the analysis of stability has been based on characteristic equations and hence is a local analysis. In other words, the results are valid only when a linear approximation is valid. However, a locally unstable dynamical system can exhibit behaviors that are stable in some sense. The first insights into the nature of these more global forms of stability were obtained by the French mathematician Henri Poincaré (1854–1912). One of the first experimental demonstrations that a dynamical system could exhibit a sustained periodic solution whose amplitude and period were independent of the choice of initial condition was provided by van der Pol[8] in 1927–1928 [844, 845]. These dynamical behaviors are called *limit-cycle oscillations* or *limit-cycle attractors*. The van der Pol oscillator was first realized in an electronic circuit containing a triode resistor whose resistance depends on the applied current. The initial motivation was to understand the development of certain cardiac arrhythmias [845]. However, it very quickly became apparent that this model offered a basis for many important biological phenomena including the heartbeat and the generation of action potentials by neurons.

The van der Pol equation is

$$\frac{d^2x}{dt^2} - \varepsilon(1 - x^2)\frac{dx}{dt} + x = 0. \tag{4.18}$$

In order to make the connection between the van der Pol equation and the second-order differential equations used to describe the dynamics of a neuron, it is useful to rewrite (4.18) as

---

[8] Balthasar van der Pol (1889–1959), Dutch physicist.

$$\frac{dx}{dt} = \varepsilon \left[x - \frac{x^3}{3} - y \right], \tag{4.19}$$

$$\frac{dy}{dt} = -\frac{1}{\varepsilon} x, \tag{4.20}$$

where $\varepsilon > 0$ is the damping coefficient and the right-hand side of (4.20) contains a cubic nonlinearity. The fixed point is $(x^*, y^*) = (0,0)$, and the van der Pol oscillator linearized about this fixed point is

$$\frac{d\hat{x}}{dt} = \varepsilon\hat{x} + \varepsilon\hat{y},$$

$$\frac{d\hat{y}}{dt} = -\frac{1}{\varepsilon}\hat{x}.$$

Using (4.8) with $k_{11} = \varepsilon$, $k_{12} = \varepsilon$, $k_{21} = -\frac{1}{\varepsilon}$, and $k_{22} = 0$ we obtain the characteristic equation

$$\lambda^2 - \varepsilon\lambda + 1 = 0 \tag{4.21}$$

Thus the eigenvalues are

$$\lambda_{1,2} = \frac{\varepsilon \pm \sqrt{\varepsilon^2 - 4}}{2}, \tag{4.22}$$

Hence the fixed point is either an unstable node ($\varepsilon > 2$) or an unstable spiral point ($\varepsilon < 2$).

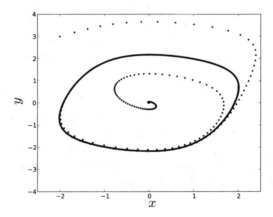

Fig. 4.13 Phase representation of a van der Pol limit-cycle oscillator (solid curve). The dotted curves show two different choices of initial conditions: one near the unstable fixed point $(0,0)$, and the other for a choice of (x,y) outside the limit cycle.

If we examine the van der Pol equations more closely, we see that if $x \gg 0$, then $dx/dt < 0$. This means that x cannot grow without bound. If x cannot grow without bound, then neither can y (since $\dot{y} = -\varepsilon x$). Thus we have the picture that for large x and y, the dynamical system moves toward the fixed point, whereas for x,y close to the critical point, the system moves away (recall that the critical point is unstable).

What is the behavior that results from these two opposing tendencies? The Poincaré–Bendixson theorem[9] states that there must exist a closed trajectory around the fixed point. The concept is simple, but the proof of the formal theorem is actually quite technical. Suppose that a limit cycle did not exist. Then at some point in time, the trajectory would have to cross the trajectory it made earlier. At the crossing point, we would have two different values for the pair $(dx/dt, dy/dt)$, and that would violate the uniqueness of the solution, i.e., for each initial point, there is only one solution (Section 2.4). The only way to avoid this problem is to assume that there is a closed curve that encircles the fixed point. This closed trajectory is called a *limit cycle* (Figure 4.13).

In contrast to the closed trajectories that arise when the fixed point is a center (e.g., the Lotka–Volterra model), a limit cycle in the van der Pol oscillator is stable (Figure 4.13): if we start at a point inside, we evolve outward to the limit cycle; if we start at a point outside, we evolve inward toward the limit cycle. Thus the limit cycle is a time-dependent solution of the dynamical system that is independent of the choice of initial conditions. However, the fixed point is unstable. The van der Pol oscillator was the first example of a dynamical system that is "stable in some sense" even though the fixed point is unstable. In thermodynamic jargon (Chapter 17), the van der Pol oscillator represents a phenomenon that evolves *far from equilibrium*. Dynamical systems also exist that exhibit unstable limit-cycle oscillations (Section 13.6).

4.5.2 Fitzhugh–Nagumo equation

The first mathematical model that associated a cubic nonlinearity with excitability in a neuron was the Fitzhugh–Nagumo equation[10] (FHN) for a neuron. This model takes the form of two ordinary differential equations

$$\frac{dv}{dt} = f(v) - w + I_{ext}, \qquad \frac{dw}{dt} = bv - \gamma w, \qquad (4.23)$$

where v plays the role of membrane potential, w is a recovery variable, I_{ext} is an externally applied current, $f(v)$ is a cubic nonlinearity given by

$$f(v) = v(a-v)(v-1),$$

and the parameters are $0 < a < 1$, $b > 0$, and $\gamma > 0$ [207, 226, 227, 617, 621]. The FHN equations can be derived from the Hodgkin–Huxley equations for a neuron by

---

[9] Formulated in a weaker version by Poincaré and later strengthened and proved rigorously by Ivar Otto Bendixson (1861–1935), Swedish mathematician.

[10] This equation is also sometimes referred to as the Bonhoeffer–van der Pol–Fitzhugh–Nagumo equation: Karl Friedrich Bonhoeffer (1899–1957), German chemist; Richard FitzHugh (1922–2007), American biophysicist; Jin-Ichi Nagumo (1926–1999), Japanese bioengineer and mathematical biologist.

taking into account the time scales for the different ion channel dynamics across the neuronal membrane [207, 422, 617].

Our purpose here is to show that phase-plane analysis can provide insights into excitability [226, 227]. Phase-plane analysis makes use of nullclines to infer the nature of the dynamics generated by second-order ordinary differential equations [186, 721]. In order to demonstrate the "all or none" nature of excitability we take $I_{ext} = 0$ and chose a, b, γ such that only a single fixed point exists. The importance of phase-plane analysis is that it illustrates that the transient behaviors of a dynamical system occur on the scaffolding formed by the deterministic aspects of the dynamical system, namely the fixed points and nullclines.

Figure 4.14 shows the phase plane for the Fitzhugh–Nagumo (FHN) . The v-nullcline, obtained by setting $dv/dt = 0$, is

$$w = f(v) = v(a - v)(v - 1).$$

Since $dv/dt = f(v) - w$, we see that above the v-nullcline, we have $dv/dt < 0$, and below the v-nullcline, we have $dv/dt > 0$. The w-nullcline, obtained by setting $dw/dt = 0$, is

$$w = \frac{b}{\gamma} v,$$

i.e., a straight line through the origin of the phase plane with slope b/γ. The fixed point occurs at the point where the v and w nullclines intersect. Since $dw/dt = bv - \gamma w$, we have $dw/dt < 0$ above the w-nullcline, and $dw/dt > 0$ below. Thus the v- and w-nullclines divide the phase plane into four regions according to the signs of dv/dt and dw/dt.

The meaning of the term "excitable" can be understood by choosing two initial points on the $w = 0$ axis, one on either side of the v-nullcline. We denote these initial points by A and B: the elevation of the membrane potential represented by B is larger than that represented by A. At point A, we are in the region $dv/dt < 0, dw/dt > 0$. The changes in v occur faster than those in w [306]. The net effect is that the trajectory (dashed line) takes a short path to return from A toward the fixed point $(v, w) = (0, 0)$, crossing the w-nullcline near $(0, 0)$, before arriving at the fixed point (since $dv/dt < 0$ and $dw/dt < 0$). In contrast, the path from B to $(0, 0)$ is much longer and crosses both nullclines. Since $dv/dt > 0$ and $dw/dt > 0$, there is an initial increase in v and w. However, when the trajectory crosses the v-nullcline, it moves toward $(0, 0)$, since $dv/dt < 0$. It crosses the w-nullcline at a larger distance from $(0, 0)$ before finally arriving at the fixed point.

The short trajectory from point A to $(0, 0)$ corresponds to the changes in membrane potential and the recovery variable in response to an EPSP that is not large enough to increase the membrane potential above the spiking threshold. The long trajectory from point B to $(0, 0)$ corresponds to the changes in v, w associated with an action potential. The spiking threshold is related to the v-nullcline and hence to the presence of a cubic nonlinearity.

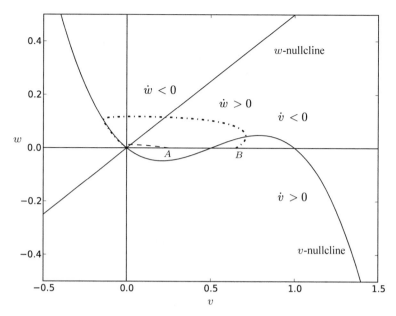

Fig. 4.14 Phase plane for a Fitzhugh–Nagumo neuron whose parameters have been tuned to the excitable regime. For initial point A, the trajectory returns monotonically to the stable fixed point (dashed line). However, for initial point B, the trajectory takes a longer path (dash-dotted line) to the fixed point. These are the typical time courses for an *excitable system*.

4.6 Lyapunov's Insight

We can combine the concepts we have developed up to this point to refine the concept of the "dynamic potential" $U(x)$, introduced in Section 4.1. This approach is called *Lyapunov's direct method* [513].[11] This new approach allows us, at least in principle, to describe the basin of attraction associated with a stable fixed point. In order to understand how this method works, it is useful to introduce the concepts of conservative and dissipative dynamical systems.

4.6.1 Conservative Dynamical Systems

A conservative dynamical system is one for which the sum of the potential energy and the kinetic energy is a constant. An example is the harmonic oscillator

$$\frac{dx}{dt} = p, \quad \frac{dp}{dt} = -kx, \tag{4.24}$$

---

[11] Aleksandr Mikhailovich Lyapunov (1857–1918), Russian mathematician and physicist.

where the momentum is $p := m\frac{dx}{dt}$, and the mass, m, is assumed to be constant.

Using what we have learned, we can readily see that the fixed point is a center (see below). Since the trajectories in the phase plane are closed curves, we can guess that there exists a continuous function $H(x, p)$ that is constant along each of the trajectories:

$$H(x, p) = \text{constant.} \tag{4.25}$$

The function $H(x, p)$ is called the *Hamiltonian*.[12] The constant is determined from the initial conditions.

For mechanical systems such as the harmonic oscillator, there is a systematic way to calculate $H(x, p)$. From (4.25), we know that

$$\frac{dH}{dt} = 0.$$

However, since H is a function of variables that themselves are time-dependent, we must use the chain rule (Tool 2) to obtain the derivative of H with respect to time, i.e.,

$$\frac{dH}{dt} = \frac{\partial H}{\partial x}\frac{dx}{dt} + \frac{\partial H}{\partial p}\frac{dp}{dt} = 0. \tag{4.26}$$

A necessary condition for (4.26) to be valid is that

$$\frac{\partial H}{\partial p} = \frac{dx}{dt}, \quad \frac{\partial H}{\partial x} = -\frac{dp}{dt}. \tag{4.27}$$

These equations are called *Hamilton's equations*.

The function

$$H(x, p) = \frac{p^2}{2} + \frac{kx^2}{2} \tag{4.28}$$

is called the *Hamiltonian* for (4.24). It can be interpreted as the total energy of the dynamical system in state (x, p). The fact that $H(x, p)$ is constant means that energy is conserved; such systems are called *conservative*. For a conservative system, the change in $H(x, p)$ in moving from one position to another is independent of the path taken between the two positions (see also Section 17.1). Moreover, we see that the total energy is divided into two components: a component that depends only on displacement $kx^2/2$, identified as the potential energy, and a component that depends only on the velocity $p^2/2$, identified as the kinetic energy.

The equation for the harmonic oscillator can also be investigated from the point of view of a dynamical system. In this case, it is common to rewrite (4.24) as

$$\frac{dx}{dt} = v, \quad \frac{dv}{dt} = -\frac{k}{m}x = -\omega_n^2 x, \tag{4.29}$$

where $\omega_n^2 = k/m$ is the natural angular frequency of the oscillator.

---

[12] After William Rowan Hamilton (1805–1865), Irish physicist, astronomer, and mathematician.

Using what we have learned, we can see that the eigenvalues are pure imaginary, $\pm j\omega_n$, and hence the solution is of the form

$$x(t) = A\sin(\omega_n t + \phi).$$ (4.30)

Finally, if we rewrite (4.24) as

$$m\frac{d^2 x}{dt^2} = -kx,$$

then we see that the force is the negative derivative of the potential energy:

$$\frac{d}{dx}\left(\frac{kx^2}{2}\right) = kx.$$

This relationship between force and potential energy is always true for a conservative mechanical system. However, it must be emphasized that nonlinear dynamical systems that possess a center can be conservative yet not satisfy Hamilton's equations (see Exercises 16 and 17). In other words, it is possible to have a dynamical system that is conservative but does not possess a Hamiltonian which satisfies (4.27).

4.6.2 Lyapunov's Direct Method

Many dynamical systems are *dissipative*.[13] In other words, the total energy is not constant but decreases as a consequence of additional processes, such as frictional resistance, drag, and viscosity, which dissipate the energy. An example of a dissipative dynamical system is the damped harmonic oscillator

$$\frac{dx}{dt} = p, \quad \frac{dp}{dt} = -k_1 x - k_2 p.$$ (4.31)

It is easy to show that when $k_1 > 0$ and $k_2 > 0$, the fixed point is a stable spiral point. Using (4.27) and (4.28), we have

$$\frac{dH}{dt} = p(k_1 x) + p(-k_1 x - k_2 p) = -k_2 p^2 \le 0.$$ (4.32)

---

[13] In biology, the term *dissipative system* is used to describe a thermodynamic open system that is operating out of, and often far from, thermodynamic equilibrium and freely exchanges both energy and mass with its surroundings. From this point of view, the damped harmonic oscillator describes a transient response of a closed dynamical system. A simple example of a dissipative system is that of the water fountain we considered in Section 3.3.1. A remarkable property of dissipative systems is their ability to form the large-scale spatially coherent structures that characterize biology, including cells, tissues, and organs [99, 638]. In Chapter 17, we more carefully examine the thermodynamic differences between systems at equilibrium and those not at equilibrium.

Thus H is not constant on the trajectory. In other words, as t increases, $H(x,p)$ decreases monotonically until it eventually becomes zero. This observation is a manifestation of the fact that the dissipative system continuously loses energy. In contrast, if we had a positively damped oscillator ($k_2 < 0$), then $dH/dt \geq 0$, and $H(x,p)$ would increase monotonically.

Lyapunov's direct method can be regarded as a generalization of these observations concerning the behavior of damped harmonic oscillators. This approach attempts to identify a function $U(x,y)$, referred to as a *Lyapunov function*, that satisfies the following conditions:

- $U(x,y) = 0$ for $(x,y) = (0,0)$,
- $U(x,y) > 0$ for $(x,y) \neq (0,0)$,
- $dU/dt \leq 0$

where x and y are variables that describe the behavior of a two-dimensional dynamical system. When $dU/dt < 0$, the origin is asymptotically stable.[14] We note in passing that the Hamiltonian of a conservative mechanical system is a Lyapunov function if the potential energy assumes an isolated minimum at the origin in phase space [728].

Example 4.2. We shall show that $U(x) = kx^2/2$ is a Lyapunov function for (4.4). Clearly, the first two conditions are satisfied. The third condition also holds, since

$$\frac{dU}{dt} = \frac{dU}{dx}\frac{dx}{dt} = -k^2x^2 \leq 0.$$

Thus $U(x)$ is a Lyapunov function for this system, and the fixed point is stable. The association of a valley with a stable fixed point and a hill with an unstable fixed point remains valid in general as long as we remember that we are actually describing the Lyapunov function and not the actual potential. The curious reader will note that in fact, the fixed point for (4.4) is asymptotically stable. Exercise 15 shows how to construct a Lyapunov function that demonstrates the asymptotic stability of this fixed point.   ◇

Lyapunov functions are extremely powerful tools for assessing the stability of fixed points. It can happen that the conditions for $U(x,y)$ are satisfied for all x and y. In this case, the fixed point is said to be globally stable. Globally stable Lyapunov functions play a central role in the study of the feedback control of dynamical systems [131, 782]. In other situations, the conditions for $U(x,y)$ are satisfied only over a certain range of x and y. Thus $U(x,y)$ can be used to describe the basin of attraction of the fixed point [150, 266].

There are two limitations of this method: (1) A general recipe for determining $U(x,y)$ is not known, and thus $U(x,y)$ must be "guessed." (2) For a given dynamical system, there may be more than one possible choice of $U(x,y)$. Fortunately, extensive experience has been accumulated by the mathematics community for constructing Lyapunov functions and even for delay differential equations [448].

---

[14] Asymptotic stability means that as $t \to \infty$, the trajectory approaches the fixed point itself. In contrast, stability means that as $t \to \infty$, the trajectory approaches a neighborhood of the fixed point.

4.7 What Have We Learned?

1. How is the stability of a dynamical system determined?

   If the equations are known, we can determine stability of the fixed points either by determining the eigenvalues or using a Lyapunov function (if one is known). In the laboratory, we can see what happens when we can perturb the system using a brief mechanical or electrical stimulus. Of course, most phenomena observed in the laboratory are stable in some sense. An exception would be an explosive compound, such as nitroglycerin, that could be triggered to explode using a small perturbation!

2. Why do we use the term "local stability" to refer to the classification of fixed points for a nonlinear dynamical system?

   Determining the stability of fixed points for nonlinear dynamical systems requires that we linearize the system in a neighborhood of the fixed point (Tool 7). The linearized approximation is valid only in a neighborhood of the fixed point, and hence this is a local stability result.

3. What condition ensures the local stability of a fixed point?

   A fixed point is stable if all of the eigenvalues have negative real part.

4. Why is a saddle point important?

   Saddle points separate basins of attraction in multistable dynamical systems. They also arise in studies of balance control.

5. When the eigenvalues are a pair of complex numbers, what properties of the solution are determined by the real and imaginary parts?

   The real parts determine stability, and the imaginary parts determine the frequency of the solution.

6. Can a linear dynamical system generate a stable and sustained oscillation?

   No. It is true that a sustained oscillation can occur in a linear dynamical system when the eigenvalues are pure imaginary. An example is the harmonic oscillator. However, the harmonic oscillator is not stable under perturbation.

7. Suppose that a dynamical system has one unstable fixed point. Is it possible for the system to be stable in some sense? Give an example.

   Yes, such a system can be stable. A variety of attractors can arise including a limit-cycle attractor and a chaotic attractor. An example is the van der Pol oscillator.

8. What is the characteristic equation for the linear second-order ordinary differential equations described by (4.8)?

   It is given by (4.9).

9. What is a conservative dynamical system?

A conservative dynamical system is one in which dynamical quantities are conserved. For example, in classical physics the sum of the potential and kinetic energy for a conservative dynamical system is constant implying that the system neither loses nor gains energy. Such systems satisfy Hamilton's equations and there exists a quantity $H(x, p)$ which is constant along the trajectories. In certain mathematical models in biology it is also possible to obtain a function $\Psi(x, y)$ which is constant along a trajectory. These situations typically arise when the fixed point is a center. However, these equations may (Exercise 16) or may not (Exercise 17) satisfy Hamilton's equations. Although some authors refer to these biological systems as Hamiltonian systems [649, 685], the conclusion that the biological quantities are conserved in these applications is inconsistent with experimental observations. Thus we leave the investigation of such systems to the more mathematically inclined reader.

10. **What is a dissipative dynamical system?**

A dissipative dynamical system is one in which dynamical quantities are not conserved.

11. **What is a Lyapunov function and why is it useful?**

A Lyapunov function is a generalization of the concepts of conservative and dissipative dynamical systems. Lyapunov functions are useful because they provide a geometric picture of stability (flow over an undulating surface) and, in principle, a method for determining the basin of attraction of a fixed point. Sometimes, for example, when the feedback is piecewise continuous, it is impossible to determine the eigenvalues, and hence stability must be determined using a Lyapunov-function-type approach. Unfortunately, it is often difficult to construct an appropriate Lyapunov function.

4.8 Exercises for Practice and Insight

1. Consider the following two differential equations:

$$\frac{d^2x}{dt^2} - 3\frac{dx}{dt} + 6x = 0, \quad \frac{d^2x}{dt^2} + 3\frac{dx}{dt} - 6x = 0.$$

   a. Make the usual ansatz and determine the characteristic equation for each of these differential equations.
   b. Which equation will have a stable solution and which an unstable solution? Why?

2. Consider the system of two differential equations

$$\frac{dx}{dt} = k_{12}y, \quad \frac{dy}{dt} = k_{21}x + k_{22}y.$$

   Show that the characteristic equation is

$$\lambda^2 - k_{22}\lambda - k_{12}k_{21} = 0.$$

   (Hint: Rewrite the system of differential equations as a single differential equation and then make the "usual ansatz.")

3. Consider the system of two differential equations

$$\frac{dx}{dt} = k_{11}x + k_{12}y, \quad \frac{dy}{dt} = k_{21}x + k_{22}y.$$

Show that the characteristic equation is

$$\lambda^2 - (k_{11} + k_{22})\lambda + (k_{11}k_{22} - k_{12}k_{21}) = 0.$$

4. Suppose that the general solution of

$$\frac{dx}{dt} + k_1 \frac{dx}{dt} + k_2 x = 0$$

is

$$x = c_1 e^{\lambda_1 t} + c_2 e^{\lambda_2 t}$$

and that $\lambda_1 = \lambda_2 = \overline{\lambda}$. Show that

$$x = c_1 e^{\overline{\lambda} t} + c_2 t e^{\overline{\lambda} t}$$

is also a solution.

5. Determine values of k_1, k_2 such that the solution of (4.8) is:

   a. A spiral point.
   b. A center.
   c. A node.
   d. A saddle.

   Write a computer program to verify your choices.

6. Consider the second-order, linear ordinary differential equation.

$$\frac{d^2x}{dt^2} + a\frac{dx}{dt} + cx = 0.$$

   Determine values of a and b so that this equation has:

   a. Two real λ that are both negative.
   b. Two real λ that are both positive.
   c. Two real λ: one of these is negative and one is positive.
   d. Two complex λ for which the real part is negative.
   e. Two complex λ for which the real part is positive.
   f. Two complex λ with real part equal to zero.

7. Show that

$$\frac{d^2x}{dt^2} - \frac{1}{x}\left(\frac{dx}{dt}\right)^2 - (x-1)\frac{dx}{dt} - (x - x^2) = 0$$

   can be rewritten as the system of two first-order differential equations

$$\frac{dx}{dt} = p, \quad \frac{dp}{dt} = \frac{1}{x}p^2 + (x-1)p + (x-x^2).$$

Now define $p = x - xy$ and show that this system of two differential equations is equivalent to the system

$$\frac{dx}{dt} = x - xy, \quad \frac{dy}{dt} = xy - y.$$

This equation is the typical way that the famous Lotka–Volterra equation for a population composed of one type of predator and one type of prey is written (discussed in Section 4.4.1). These two systems of equations are equivalent [728]. The point of this exercise is to show that there is nothing unique about our choice of state variables. It is possible that certain aspects of system behavior may be far more conveniently displayed in some other set of state variables than in those that initially might seem to be most "natural."

8. Show that (4.18) and (4.19) are equivalent.
9. A variant of the Lotka–Volterra equation is

$$\frac{dx}{dt} = k_{11}x - k_{12}x^2 - xy, \quad \frac{dy}{dt} = k_{21}xy - k_{22}y,$$

where $k_{11}, k_{12}, k_{21}, k_{22}$ are positive constants.

a. Can you guess which variable represents the prey and which the predator?
b. What are the fixed points?
c. What are the linearized equations about each fixed point?
d. What is the stability of each fixed point?
e. Assume that $k_{11}/k_{12} > k_{22}/k_{21}$. Using the nullcline approach described in Section 4.5.2, determine whether this model will generate a limit cycle.

10. The differential equation that describes the movements of the plane pendulum is

$$m\ell^2 \frac{d^2\theta}{dt^2} + \varepsilon\ell\frac{d\theta}{dt} + mg\ell\sin\theta = 0,$$

where m is the mass, θ is the displacement angle, ε is a damping coefficient, ℓ is the length of the pendulum, and g is the acceleration due to gravity.

a. Rewrite this equation as a system of two first-order differential equations.
b. What are the fixed points? (Hint: There are two.)
c. Evaluate the stability of each fixed point.

11. The Duffing equation[15] describes the motion of a particle in a double-well potential described by a cubic nonlinearity. In the simplest situation it takes the form

---

[15] Georg Duffing (1861–1944), German engineer and inventor.

$$\frac{dx}{dt} = y, \quad \frac{dy}{dt} = x - x^3.$$

a. What are the fixed points?
b. What are the linearized equations about each fixed point?
c. What is the stability of each fixed point?
d. Write a computer program to numerically integrate these differential equations and verify your conclusions.

12. In their textbook *Understanding Nonlinear Dynamics*, Kaplan and Glass [414] pose the following model for the control of blood cells (x) by a hormonal agent (y):

$$\frac{dx}{dt} = \frac{2y}{1+y^2} - x, \quad \frac{dy}{dt} = x - y.$$

a. What are the fixed points?
b. What are the linearized equations about each fixed point?
c. What is the stability of each fixed point?
d. Write a computer program to numerically integrate these differential equations and verify your conclusions.

13. Mathematical models have long been used to study the transmission of malaria by insect bites. The first such model was developed by Sir Ronald Ross[16] [730]. It takes the form (for extensions to this model, see [20, 534])

$$\frac{dx}{dt} = k_{11}y(1-x) - k_{12}y(1-x), \quad \frac{dy}{dt} = k_{21}x(1-y) - k_{22}y.$$

a. What are the fixed points?
b. What are the linearized equations about each fixed point?
c. What is the stability of each fixed point?
d. Write a computer program to integrate these differential equations numerically and verify your conclusions.

14. The harmonic oscillator described by (4.24) is a conservative mechanical system whose equations of motion can be solved.

a. Calculate the total energy from the sum of the kinetic and potential energies. Two hints: (1) It is easiest to write the solution in the form of (4.30). (2) The potential energy can be obtained by calculating the work required to displace a particle through a distance x.
b. Show that the force is the negative of the derivative of the potential energy.
c. Show that the total energy is independent of time.
d. Show that the total energy is proportional to the square of the amplitude.

---

[16] Ronald Ross (1857–1932), Indian-born British physician.

15. Consider the damped harmonic oscillator

$$\frac{dx}{dt} = y, \quad \frac{dy}{dt} = -k_2 y - k_1 x,$$

which, as we know, has an asymptotically stable fixed point (which is a stable spiral point). A seemingly obvious choice for the Lyapunov function $U(x,y)$ is the total energy

$$U(x,y) = \frac{1}{2} \left[x^2 + k_2 y^2 \right].$$

Show that for this choice of $U(x,y)$, it is impossible to conclude that the fixed point is asymptotically stable. Another possible choice of $U(x,y)$ is [72, 619]

$$U(x,y) = \frac{y^2}{2} + \frac{k_1 x^2}{2} + \varepsilon xy.$$

Verify that for ε sufficiently small and positive, we have $U(x,y) < 0$, so the fixed point is asymptotically stable.

16. Brian Goodwin[17] was the first biomathematician to consider the dynamics of gene regulatory networks [283]. Under certain conditions, the Goodwin model takes the form (see Section 9.3.1 for more details)

$$\frac{dx}{dt} = \frac{k_1}{k_2 y} - k_3, \quad \frac{dy}{dt} = k_4 x - k_5,$$

where x is the RNA concentration and y is the concentration of protein.

a. There is a single fixed point. Show that this fixed point is a center.
b. Show that the given system of two linear differential equations can be written as

$$\frac{d^2 y}{dt^2} - A k_4 \left(\frac{k_1}{k_2 y} - k_3 \right) = 0.$$

c. Define $p = dy/dt$ and rewrite this second-order differential equation as

$$\frac{dy}{dt} = p, \quad \frac{dp}{dt} = k_4 \left(\frac{k_1}{k_2 y} - k_3 \right).$$

Show that this form of the Goodwin model satisfies Hamilton's equations.
d. Show that

$$H(y,p) = \frac{p^2}{2} - \frac{AK}{B} \log y + \alpha y.$$

Brian Goodwin used the fact that this model has a Hamiltonian to apply statistical mechanics to model the temporal organization of cells [283]. This was quite an undertaking, given that this project was begun in 1963! However, this approach ultimately failed, because it does not adequately incorporate the dissipative, far-from-equilibrium nature of living systems.

---

[17] Brian Carey Goodwin (1931–2009), Canadian biomathematician.

17. Consider the Lotka–Volterra equation given by (4.13).

    a. Show that the Lotka–Volterra model does not satisfy Hamilton's equations.

    b. By inspection, it can be seen that [728]

$$(x - xy)\left(1 - \frac{1}{x}\right) + (xy - y)\left(1 - \frac{1}{y}\right) = 0.$$

Show that $d\Psi/dt = 0$ where

$$\Psi(x,y) = (x+y) - (\log x + \log y).$$

It is not surprising that trajectories that surround a center can be described by a continuous function $\Psi(x,y)$ that is constant on the trajectory. This is because the trajectories about a center neither approach nor diverge from another. Thus we can anticipate that Lotka–Volterra systems would share properties with those exhibited by Hamiltonian dynamical systems [649, 685].

Chapter 5
Fixed Points: Creation and Destruction

The fixed point of a dynamical system describes a time-independent state. Depending on the nature of the interactions between a system and its surroundings, the fixed point can be either an equilibrium or a steady state. In the last chapter, we saw that the responses of the system following a perturbation away from the fixed point can be used to assess the stability of the fixed point, namely its resistance to change. Stability depends on the values of the parameters. This statement follows from the fact that the eigenvalues are determined by the parameters. This observation suggests another way to explore dynamical systems: we vary the parameters and observe how the dynamics change. There are two types of changes that we can expect to observe as parameters are changed. First, there can be changes in the stability of a fixed point: it may be stable for one range of parameter values and unstable for another range. Second, it is possible that as parameter values change, new fixed points are created or existing fixed points are destroyed.

A *bifurcation diagram* summarizes the number of fixed points and their stability as parameters are changed.[1] In mathematical models, the occurrence of a qualitative change in dynamics when a parameter changes is called a *bifurcation*. An example is a stable fixed point that becomes unstable, or vice versa, as a parameter is changed. A bifurcation diagram is potentially one of the most exciting pieces of information that a modeler can provide to an experimentalist. The key experiments to be done are laid out for all to see: What happens if a parameter is changed? What happens at stability boundaries? And so on.

The classification of bifurcations in mathematical models of biological systems is a dominant theme of textbooks that deal with nonlinear differential equations [291, 810] and biomathematics [186, 193, 414, 617, 728]. However, the mathematics can be quite difficult, and often great emphasis is placed on specialized topics, such as

---

[1] Of course, we can also include other types of attractors and repellers in the bifurcation diagram, such as limit cycles, tori, and chaotic attractors. However, in this chapter we consider only fixed points.

© Springer Nature Switzerland AG 2021
J. Milton and T. Ohira, *Mathematics as a Laboratory Tool*,
https://doi.org/10.1007/978-3-030-69579-8_5

bursting neurons [382], which most biologists will seldom encounter. Fortunately, the common bifurcations that involve a fixed point can be readily understood using graphical approaches [810] and the mathematical techniques introduced up to this point. These are the *saddle-node*, *transcritical*, and *pitchfork bifurcations*. In this chapter, we discuss the properties of these bifurcations and how they arise in the laboratory.

The "trick" is to associate a given type of bifurcation with a prototypical, or generic, first-order differential equation [291, 294, 810].[2] As we showed in Section 4.2, the stability of the fixed points for such equations can be readily determined by eye by plotting dx/dt versus x. Thus it becomes straightforward to determine the number of fixed points and their stability as critical parameters are changed.

The importance of understanding the properties of these bifurcations for benchtop scientists stems from two observations: (1) complex dynamical systems tend to organize near a stability boundary [291, 848]; (2) the number of known different bifurcations that involve fixed points is small [291, 294, 810]. The first observation implies that we can expect to observe frequently the types of behavior typical of systems tuned close to stability boundaries in biological systems, while the second suggests that experimental observations can be used to identify, at least in principle, the likely nature of the underlying bifurcation.

The questions we will answer in this chapter are:

1. What is a bifurcation?
2. What is a bifurcation diagram?
3. Why is it important for an experimentalist to have a working knowledge of the different types of bifurcations?
4. Why is it impossible to determine a bifurcation point by slowly increasing (or decreasing) the bifurcation parameter? What is the resulting error?
5. Which type of bifurcation is associated with the creation or destruction of a fixed point?
6. What is the bottleneck phenomenon and with which bifurcation it is associated?
7. What does the term "symmetry" mean in the context of a pitchfork bifurcation?
8. What is the slowing-down phenomenon and how does it influence experimental and computational approaches to the study of dynamical systems?
9. What are critical phenomena and why are they important in biology?
10. What type of bifurcation corresponds to a first-order phase transition and what type corresponds to a critical phase transition?
11. What is the critical slowing-down phenomenon?

---

[2] Mathematicians refer to these generic equations as *normal forms*.

5.1 Saddle-Node Bifurcation

The *saddle-node bifurcation* is a basic mechanism by which fixed points are created and destroyed. These bifurcations are particularly important for understanding the generation of oscillations in excitable systems (see Chapter 11). The generic equation for a saddle-node bifurcation is [291, 294, 810]

$$\frac{dx}{dt} = \mu + x^2, \tag{5.1}$$

where μ is the bifurcation parameter. Figure 5.1 shows a plot of dx/dt versus x for different choices of μ. The fixed points correspond to those values of x for which $dx/dt = 0$. There are three cases:

1. $\mu > 0$: no real fixed points.
2. $\mu = 0$: one fixed point.
3. $\mu < 0$: two fixed points, $x = \pm\sqrt{\mu}$.

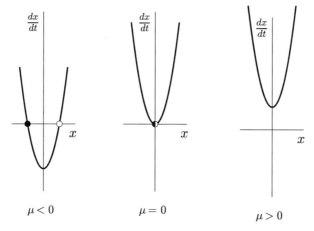

Fig. 5.1 The behavior of (5.1) as the bifurcation parameter μ changes.

Our convention is to indicate fixed-point stability by the shading of the small circle that encloses the fixed point: a stable fixed point corresponds to a solid circle ($\bullet$) and an unstable one to a hollow circle ($\circ$). When $\mu < 0$, there are two fixed points. The fixed point at $x^* = -\sqrt{\mu}$ is stable (negative slope), and the fixed point at $x^* = \sqrt{\mu}$ is unstable (positive slope). When $\mu = 0$, there is only one fixed point, at $x^* = 0$. The stability of the fixed point at $x = 0$ depends on the direction in which x approaches the origin. This fixed point is represented by a circle whose left half is solid, since it is stable when approached from below (negative slope), and whose right half is open, since it is unstable when approached from above (positive slope).

In view of the dual character of this fixed point it is called a saddle node[3]. Finally, when $\mu > 0$, there are no fixed points.

Figure 5.2 summarizes our observations for all real-valued choices of μ in the form of a *bifurcation diagram*.[4] By convention, unstable fixed points are represented by dashed lines and stable ones by solid lines. It should be noted that Figure 5.2 condenses the information obtained from graphs such as those shown in Figure 5.1 into a single graph. The saddle-node bifurcation occurs at $\mu = 0$. Suppose we approach the bifurcation point from the right. Then at $\mu = 0$, a pair of fixed points (one stable, the other unstable) appear "out of the clear blue sky" as μ crosses the stability boundary. For this reason, a saddle-node bifurcation is sometimes referred to as a "blue-sky bifurcation" [2, 765]. Historically, mathematical topologists referred to this type of bifurcation as a fold bifurcation and associated it with one of the elementary catastrophes, namely the *fold catastrophe* [825]. However, this terminology is no longer in vogue.

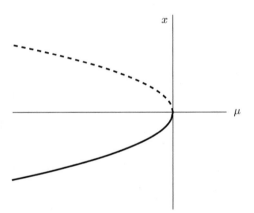

Fig. 5.2 Bifurcation diagram for a saddle-node bifurcation.

A curious and important property of a saddle-node bifurcation is that the fixed point has an effect on the dynamics even when it ceases to exist! This is called the *bottleneck phenomenon* and the remnants of the saddle node are referred to as its *ghost* [235, 810]. The bottleneck phenomena can readily be demonstrated by integrating (5.1) (see Exercise 1). Choose an initial value of $x < 0$. As $\mu > 0$ gets

---

[3] In higher dimensions it is also possible that a saddle-spiral (or saddle-focus) bifurcations can occur [455, 764]. Saddle-spiral bifurcations have recently been emphasized in association with pathological cardiac potential dynamics [450].

[4] In general, bifurcation diagrams in which the μ are complex-valued lead to considerations of Hopf bifurcations. Such bifurcations are sometimes referred to as Poincaré–Andronov–Hopf bifurcations in honor of the mathematicians who first studied their properties: Henri Poincaré; Aleksandr Aleksandrovich Andronov (1901–1952), Soviet physicist and mathematician; Eberhard Frederich Ferdinand Hopf (1902–1983), Austrian-born German and American mathematician and astronomer.

smaller and smaller it takes longer and longer for the trajectory to escape from the neighborhood of $x = 0$ before it escapes toward infinity. We can use (5.1) to estimate the amount of time spent in the bottleneck, $T_{\text{bottleneck}}$, going from from $x = -\infty$ to $x = +\infty$ as [810]

$$T_{\text{bottleneck}} = \int_{-\infty}^{\infty} \frac{dx}{\mu + x^2} = \frac{\pi}{\sqrt{\mu}}.$$

In other words the frequency $f_{\text{bottleneck}}$ scales as $\approx \sqrt{\mu - \mu_c}$.

5.1.1 Neuron Bistability

In 1959, Tasaki[5] [821] demonstrated that a squid giant axon could exhibit bistability, namely the coexistence of two locally stable resting membrane potentials. Within 3 years, biomathematicians had confirmed the existence of bistability in a simplified model neuron [226, 621], and in 1983, it was shown that this behavior could be reproduced by the Hodgkin–Huxley[6] (HH) equation for a neuron [8].

The first mathematical model of a neuron that predicted bistability in the resting membrane potential was the Fitzhugh–Nagumo (FHN) equation [7]

$$\frac{dv}{dt} = f(v) - w + I_{\text{ext}}, \quad \frac{dw}{dt} = bv - \gamma w, \tag{5.2}$$

where v plays the role of membrane potential, w is a recovery variable, I_{ext} is an externally applied current, $f(v)$ is a cubic nonlinearity given by

$$f(v) = v(a - v)(v - 1),$$

and the parameters are $0 < a < 1$, $b > 0$, and $\gamma > 0$ [207, 226, 227, 617, 621]. The FHN equations arise from a consideration of the time scales for the variables in the HH model for a neuron. Since we are interested in the resting membrane potential, we take $I_{\text{ext}} = 0$. Periodic solutions can arise when $I_{\text{ext}} \neq 0$.

A graphical approach to determining the number (and stability) of fixed points of a second-order differential equation is based on the construction of the *nullclines*. The v-nullcline is obtained by setting $dv/dt = 0$,

$$w = f(v) = v(a - v)(v - 1),$$

and the w-nullcline is obtained by setting $dw/dt = 0$,

---

[5] Ichiji Tasaki (1910–2009), Japanese-born American biophysicist and physician.

[6] Alan Lloyd Hodgkin (1914–1998), English physiologist and biophysicist; Andrew Fielding Huxley (1917–2012), English physiologist and biophysicist.

[7] This equation is also sometimes referred to as the Bonhoeffer–van der Pol–Fitzhugh–Nagumo equation: Karl Friedrich Bonhoeffer (1899–1957), German chemist; Richard FitzHugh (1922–2007), American biophysicist; Jin-Ichi Nagumo (1926–1999), Japanese bioengineer and mathematical biologist.

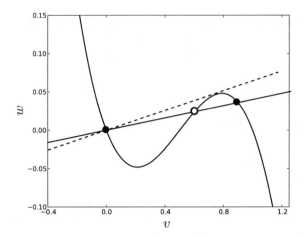

Fig. 5.3 Phase-plane plot of (5.2) for two values of the bifurcation parameter defined as the slope of the w-nullcline. For the lower slope (solid line), there are three fixed points: two stable (•) and one unstable (○). The dashed line shows the critical slope of the w-nullcline that separates the bistable regime from the excitable regime (one stable fixed point). Thus as the slope of the w-nullcline increases with all other parameters held constant, there will be a saddle-node bifurcation [441].

$$w = \frac{b}{\gamma}v.$$

As shown in Figure 5.3, the v-nullcline is a cubic curve, and the w-nullcline is a straight line. The fixed points correspond to the intersections of the v- and w-nullclines. Clearly, the number of fixed points can be one or three, depending on the slope of the w-nullcline. The time constant for the changes in v is about 12.5 times the constant for w [306]. This observation can be used to show that the local stability is determined by the slope of the v-nullcline evaluated at the fixed point. When three fixed points coexist, one of them is unstable, and hence we have bistability. Thus we see that the existence of bistability, when it occurs, is intimately connected with the cubic shape of the v-nullcline, as anticipated in Section 4.1.

We can also use this graphical approach to confirm that a saddle-node bifurcation can occur in the FHN equation for an appropriate choice of the bifurcation parameter [441]. First, we note that the fixed point $(v^*, w^*) = (0,0)$ always exists and is stable when $I_{ext} = 0$. Consider that the bifurcation parameter is the slope of the w-nullcline. The dashed line in Figure 5.3 gives the critical slope of the w-nullcline: small deviations away from this critical slope result in either the creation or destruction of two fixed points. If we start with a very high slope of the w-nullcline and slowly decrease the slope toward the critical slope, we see that two fixed points (one unstable, the other stable) appear "out of the clear blue sky" when the slope equals the critical slope. Similarly, if we start with a w-nullcline that has a small slope and then increase this slope, we see that at the critical slope, the two fixed points collide and are destroyed. In other words, we can have a saddle-node bifurcation.

However, by changing other parameters, or combinations of parameters, other types of bifurcations can occur in the FHN [441] and related models (see Section 11.6 and [380–382]). Thus the FHN neuron is an example of a mathematical model that can exhibit a variety of bifurcations depending on what parameter or combination of parameters is considered to be the bifurcation parameter.

5.2 Transcritical Bifurcation

More commonly, the fixed point persists on both sides of the bifurcation point, but it changes its stability. In particular, there is an *exchange of stability* between the fixed point and another attractor (or attractors). The "other attractor" may be a limit cycle (an Andronov–Hopf bifurcation, discussed further in Section 9.4) or two fixed points (a pitchfork bifurcation, discussed in Section 5.3). Here we discuss the *transcritical bifurcation*, which involves an exchange of stability between two fixed points.

The generic equation (normal form) for a transcritical bifurcation is [291, 294, 810]

$$\frac{dx}{dt} = \mu x - x^2 . \tag{5.3}$$

The bifurcation diagram can be obtained by following the same procedures we used before. It is shown in Figure 5.4. The reader should verify that this is the correct bifurcation diagram (see Exercise 2).

A biologically relevant model that possesses a transcritical bifurcation is the Verhulst,[8] or logistic, model for population growth [186]

$$\frac{dx}{dt} = k \left(1 - \frac{x}{K} \right) x = kx - \frac{x^2}{K} ,$$

where k is the growth rate (bifurcation parameter) of a spatially homogeneous population with density x living in an environment with limited resources. The parameter K is called the *carrying capacity* of the environment. It represents the critical density above which the population growth rate is negative. However, $k < 0$ is not a biologically reasonable choice, since it leads to a negative fixed point. Thus, there is little to be gained in recognizing that a transcritical bifurcation occurs at $k = \mu_c = 0$. Models with transcritical bifurcations also arise in descriptions of biological processes in which there is a succession of different states. Examples include the maturation of plant ecosystems [672], the cell cycle [817], and spread of ideas (such as rumors) and infections in populations [173]. The following example illustrates a technical issue that arises in the identification of bifurcation points characterized by an exchange of stability.

---

[8] Pierre François Verhulst (1804–1849), Belgian mathematician.

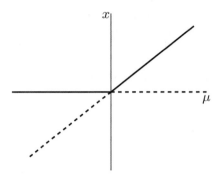

Fig. 5.4 Bifurcation diagram for a transcritical bifurcation.

5.2.1 Postponement of Instability

In the laboratory setting, it is tempting to try to measure a bifurcation point by slowly varying μ. For example, we could design an experiment that begins with $\mu < \mu_c$ and then slowly increase μ until $\mu > \mu_c$. Intuitively, it should be possible to estimate μ_c in this manner. Surprisingly, this method does not work [25, 720, 832]. In fact, the estimate of μ_c thus obtained is always displaced from its true value no matter how slowly μ is increased. One consequence of this phenomenon in the study of neurons is that the onset of neural spiking depends on the rate at which the membrane potential increases [384, 905].

To illustrate, consider the transcritical bifurcation described by (5.3). If we are sufficiently close to μ_c, then $x^2 \approx 0$, and (5.3) becomes

$$\frac{dx}{dt} = \mu(t)x, \tag{5.4}$$

where $\mu(t)$ describes the monotone increase in μ as a function of time. When this approximation is no longer valid, the solution $x(t)$ diverges to the nonzero fixed point. Two definitions of this bifurcation are possible [832]: μ_c, which is the bifurcation point determined Analytically, and μ_d, which is the bifurcation point determined dynamically by the divergence of the solution from the nonzero fixed point. Are the bifurcations points μ_c and μ_d the same?

In order to determine the time t_c that corresponds to μ_c, we need to choose a functional form for $\mu(t)$. In experimental applications, $\mu(t)$ is typically a monotonically increasing function of time, starting at $\mu(t) < \mu_c$ and ending at $\mu(t') > \mu_c$, where $t' > t$. Without loss of generality, we can assume that $\mu(t)$ increases linearly with time [832]:

$$\mu(t) = -C + kt, \tag{5.5}$$

where $C > 0$ and $k > 0$. The condition for μ_c is

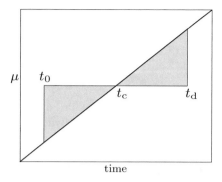

Fig. 5.5 Bifurcation diagram for a transcritical bifurcation showing the definitions for t_c and t_d. The geometric interpretation of (5.9) is that the areas of the two gray triangles are equal.

$$\mu(t_c) = 0, \quad \text{or} \quad t_c = \frac{C}{k}.$$

In contrast, in order to determine the time t_d that corresponds to μ_d, we need to solve (5.4). The solution of (5.4) can be readily obtained as

$$x(t) = x(0)\exp\left[\int_0^t \mu(s)ds\right].\tag{5.6}$$

For this choice of $\mu(t)$, we have

$$x(t) = x(0)\exp\left[-Ct_d + \frac{kt_d^2}{2}\right].\tag{5.7}$$

In this case, the condition for t_d is

$$-Ct_d + \frac{kt_d^2}{2} = 0.\tag{5.8}$$

Referring to Figure 5.5, we see that when $t = t_c$, we have

$$\int_0^{t_c} \mu(s)\,ds < 0.$$

Hence the zero fixed point is stable, even though at $t = t_c$, we have $\mu = \mu_c$. This is because the exponent in (5.6) depends on an integral evaluated over a time interval. Now define $t = t_d$ as the time it takes for this integral to become equal to zero. Referring to Figure 5.5, we see that t_d is determined by the condition

$$\left|\int_0^{t_c} \mu(s)\,ds\right| = \left|\int_{t_c}^{t_d} \mu(s)\,ds\right|.\tag{5.9}$$

In other words, the area of the triangle in Figure 5.5 below the time axis must be equal to the area of the triangle above the time axis. Thus $t_d > t_c$, or equivalently, $\mu_d > \mu_c$. Furthermore, from (5.5) and (5.8), we see that

$$t_d = 2t_c.$$

Hence the time to reach the bifurcation determined dynamically is twice the time that it takes to reach the bifurcation determined analytically. This conclusion is valid regardless of the value of $k > 0$. Thus it is impossible, for example, to determine μ_c by sweeping through a range of μ in the laboratory [25, 720]. Indeed, the only way to avoid this error in determining μ_c is to choose $k = 0$. This is what is done for the mathematical determination of the bifurcation diagrams discussed in Section 13.1. However, as we will soon see in Section 5.4.1, even this procedure has its limitations.

There is a silver lining to this story. Namely, we can anticipate that it may be possible to postpone the instability of the fixed point. Indeed, this intuition turns out to be correct, and in particular, it lies at the basis of the fact that noisy fluctuations in μ can be stabilizing in this sense. These issues are discussed further in Section 15.3.

5.3 Pitchfork Bifurcation

In a pitchfork bifurcation there is an exchange of stability between a single fixed point and a pair of fixed points. Pitchfork bifurcations are characteristic of dynamical systems that exhibit *symmetry* in the sense that changing the sign of the variable x (replacing x by $-x$) does not change the equation. Symmetric differential equations often arise in the description of mechanical systems. Pitchfork bifurcations come in two forms [810]: supercritical and subcritical. The generic equation for a supercritical pitchfork bifurcation is

$$\frac{dx}{dt} = \mu x - x^3, \tag{5.10}$$

and that for a subcritical bifurcation is

$$\frac{dx}{dt} = \mu x + x^3. \tag{5.11}$$

Figure 5.6 shows the bifurcation diagrams (see Exercises 3 and 4 at the end of this chapter). By comparing the bifurcation diagrams for (5.10) and (5.11), we can understand the difference between the terms *supercritical* and *subcritical*. In the bifurcation diagram for (5.10) (Figure 5.6a), the two fixed points occur above the bifurcation, i.e., for $\mu > 0$, hence the term "supercritical." However, for systems described by (5.11) (Figure 5.6b), the two fixed points occur below the bifurcation, i.e., for $\mu < 0$, and hence the term "subcritical."

(a)   (b)

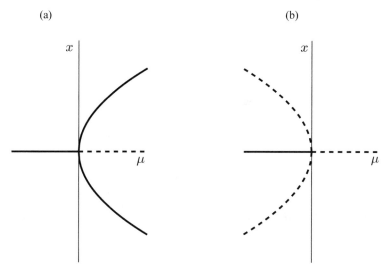

Fig. 5.6 (a) Bifurcation diagram for a supercritical bifurcation for (5.10). (b) Bifurcation diagram for a subcritical bifurcation for (5.11).

5.3.1 Finger-Spring Compressions

Dynamic sensorimotor behaviors are quintessentially complex, nonlinear, and high-dimensional. For example, we handle objects easily without looking at them, but must rely on vision when our fingers are numb. A novel experimental paradigm enables the dynamics of manipulation to be examined [848]: the finger is used to compress a slender spring as far as possible without letting the spring slip or buckle (Figure 5.7). This experiment takes advantage of the hypothesis that complex dynamical systems tend to self-organize near a stability boundary, so to a first approximation, the dynamics are governed by the generic equation for a subcritical pitchfork bifurcation [291, 848]. Amazingly, this assumption is sufficient for understanding some aspects of this experiment.

The application of a downward force to a slender spring causes it to buckle through a pitchfork bifurcation, which occurs when a specific configuration of the spring (typically, the straight configuration) becomes unstable beyond a critical load, and the spring bends or deflects away from the previously stable configuration. However, once unstable, the spring does not attain a new fixed point. This means that spring-buckling represents a subcritical pitchfork bifurcation in which the downward compression of the spring serves as the control parameter.

In real dynamical systems, "blowup instabilities" are usually opposed by the stabilizing influence of higher-order terms. If we assume that the higher-order terms must preserve symmetry, then we anticipate that the generic equation for the thumb-spring compression bifurcation would be (see also Exercise 5).

(a)   (b)

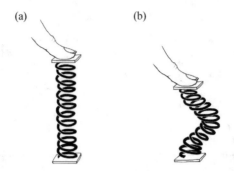

Fig. 5.7 Schematic representation of experimental design. The end caps are represented by the small squares attached to each end of the spring. (a) The subject is required to compress the spring slowly so as to minimize the volume of the audio feedback (not shown) without letting the spring bend. (b) The trial is over when the spring bends. For more details, see [848]. Figure prepared by Rachel Lee.

$$\frac{dx}{dt} = \mu x + x^3 - x^5. \tag{5.12}$$

As before, we can determine the critical points and their stability and thereby obtain the bifurcation diagram (Figure 5.8).

This bifurcation diagram has a number of interesting properties:

1. There is bistability if $\mu_s < \mu < 0$. What is μ_s? We can determine this value by solving for the fixed points. The condition $dx/dt = 0$ means that

$$\mu x + x^3 - x^5 = 0,$$

and hence there are five fixed points, one of which is $x_1^* = 0$. The remaining four can be determined by noting that

$$x^4 - x^2 - \mu = 0.$$

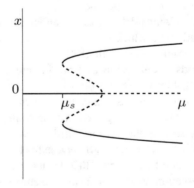

Fig. 5.8 Prototypical example for a "real-world" subcritical pitchfork bifurcation.

If we define $z = x^2$, then we have a quadratic polynomial in z, i.e.,

$$z^2 - z - \mu = 0,$$

for which the solutions are

$$z_{1,2} = \frac{1 \pm \sqrt{1 + 4\mu}}{2},$$

and hence the five fixed points are

$$x_1^* = 0, \quad x_{2,3}^* = \pm\sqrt{\frac{1 + \sqrt{1 + 4\mu}}{2}}, \quad x_{4,5}^* = \pm\sqrt{\frac{1 - \sqrt{1 + 4\mu}}{2}}.$$

The condition for the existence of five real roots is $0 > \mu > -\frac{1}{4}$, so we see that $\mu_s = -1/4$.
2. There is hysteresis as μ is varied.
3. There are two saddle-node bifurcations. Where are they? See also Exercise 4.

There are two types of spring-buckling that can occur when a spring is subjected to a "dead load": lateral and rotational. During experimental trials with human subjects, it was found that lateral buckling occurred before rotational buckling [848]. However, it was observed that the rotational movements better captured the dynamics of active control. In particular, high-speed motion capture in three dimensions showed that the end cap remained horizontal for successful compression trials, but rotated when trials were unsuccessful. These experimental observations were interpreted in the context of the bifurcation diagram shown in Figure 5.9 by measuring the rotation of the end cap. Figure 5.9a shows the rotating end cap's time history for a "successful" trial. The end cap's rotation angle stays well within the domain of attraction predicted by the subcritical bifurcation normal form. Figure 5.9b shows an "unsuccessful" trial. In this case, the spring slipped when the end cap's rotation angle exceeded the domain of attraction. It can be seen that the load at the moment of slippage was lower than it was when the spring did not slip.

5.4 Near the Bifurcation Point

Dynamical systems tuned near a supercritical pitchfork bifurcation exhibit a number of counterintuitive properties. These properties, collectively referred to as *critical phenomena*, include slower recovery from perturbations ("slowing-down phenomenon"), increased variance of fluctuations, increased autocorrelation, intermittency, and the presence of power laws [351, 496]. It has been suggested that the observation of such phenomena implies that the dynamical system is tuned close to the *edge of stability* [94, 501, 741] and that these phenomena may provide useful clues for the prediction of impending crises [741, 784, 785].

The reader must be forewarned that the use of the phrase "critical phenomena" in the current literature has become quite confusing. Statistical physicists use this

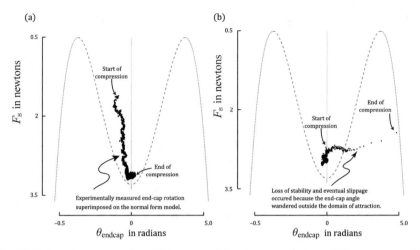

Fig. 5.9 Schematic representation of (a) a successful spring compression and (b) an unsuccessful one. The vertical axis represents the mean compressive spring force; the horizontal axis, the end-cap's rotation angle. Figure reproduced from [848] with permission.

terminology in the context of systems undergoing a phase transition, such as ferro-magnetism [128] and Rayleigh–Bénard convection[9] [107]. Near the critical (bifur-cation) point, these dynamical systems are characterized by collective behaviors for which it is not possible to define a specific correlation length. The existence of power-law behaviors is the cardinal feature of these scale-free systems [784]. However, from the dynamical systems point of view, the term "critical phenomena" typically refers to behaviors occurring in association with a supercritical bifurcation [785].

The confusion arises because certain critical phenomena, such as the slowing-down phenomenon, can also be associated with other types of bifurcations that involve a fixed point (see below). The importance of this observation is that the existence of one of the critical phenomena is insufficient evidence to conclude that all of the critical phenomena must also be present. In other words, the existence of each critical phenomenon must be demonstrated in the relevant mathematical model and/or at the benchtop.

5.4.1 The Slowing-Down Phenomenon

The most obvious dynamical properties of a bifurcation point are determined by the fact that the real part of the associated eigenvalue is zero. The characteristic time

---

[9] John William Strutt, Lord Rayleigh (1842–1919), English physicist; Henri Claude Bénard (1874–1939), French physicist.

scale $\mathscr{T}(x)$ for change, or relaxation time, associated with a given eigenvalue is equal to the reciprocal of its real part (see Section 2.2.2). Consequently, at the bifurcation point, the relaxation time becomes infinite. This observation means that if we perturbed a dynamical system tuned at its bifurcation point away from the fixed point, then it would take an infinitely long time for the system to relax toward the fixed point. In other words, as μ approaches μ_c from below, $\Re(\lambda)$ approaches 0 (where $\Re$ denotes the real part of a complex number), and perturbations are damped less and less rapidly. This is the *slowing-down phenomenon*. Figure 5.10 demonstrates the existence of the slowing-down phenomenon in a simple dynamical system that does not exhibit a supercritical pitchfork bifurcation, namely $dx/dt = kx$. Slowing-down behaviors have been observed in optical systems [62], semiconductor lasers [832], ecological and climate models [741], squid giant axon [546], and human coordinated motions [746]. The observation of the slowing-down phenomenon has been interpreted as providing a warning that a critical transition, such as one leading to the extinction of a species, is about to occur [123, 254, 458, 741]. Of course, the slowing-down phenomenon makes it difficult to determine the exact values of μ, μ_c for which a bifurcation occurs using numerical simulation.[10]

Matsumoto and Kunisawa[11] [546] were among the first investigators to demonstrate that the slowing-down phenomenon could occur in a biological preparation. They studied the onset of self-sustained repetitive spiking in squid giant axon. This onset is expected to occur via a subcritical Hopf bifurcation. In order to induce a slow change in the bifurcation parameter for the squid giant axon, they changed the

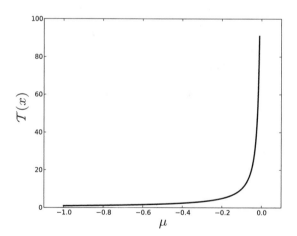

Fig. 5.10 Demonstration of the critical slowing-down phenomenon in the first-order differential equation $dx/dt = \mu x$. The time scale for change $\mathscr{T}(x)$ is equal to μ^{-1}, and at $\mu := \mu_c = 0$, there is a transcritical bifurcation.

---

[10] By this point, the reader should be convinced that methods based on mathematical analyses are the only practical way to determine the bifurcation point!.

[11] Gen Matsumoto (1940–2003), British-born Japanese biophysicist; Takashi Kunisawa (PhD 1978), Japanese bioinformatician.

immersion medium from normal seawater to a 1:4 mixture of seawater and 550 mM NaCl. This idea was based on three observations: (1) The squid axon is quiescent in normal seawater, and it produces a single action potential in response to a sufficiently large elevation in resting potential. (2) In the modified seawater medium, the axon spikes repetitively. (3) The repetitive spiking does not begin until 13–15 min following the change in the immersion medium[12]. They suggested that when the medium is changed an unidentified bifurcation parameter begins changing and at some point this parameter crosses a boundary resulting in a change in stability. The resulting exchange in stability between a stable fixed point (upper left panel in Figure 5.11) and a stable limit cycle (lower right panel in Figure 5.11) is consistent with these expectations.

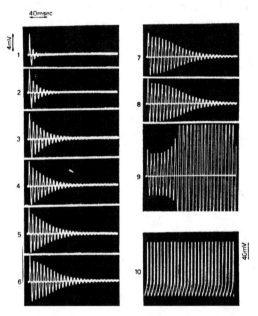

Fig. 5.11 Changes in membrane potential induced by applying a single 1-μ, 1,200-μs pulse in a squid axon following a change in immersion medium. Reproduced from [546] with permission.

Matsumoto and Kunisawa used the slowing-down phenomenon to confirm the fact that an exchange in stability occurred in this preparation. Figure 5.11 shows the effects of applying a brief electrical pulse to the squid axon as a function of time following the change in the immersion medium. Initially (top left-hand panel), the perturbation in the membrane potential dies out quickly. However, as time goes on (top to bottom on the left, then top to bottom on the right), the membrane potential perturbation takes longer and longer to die out. In addition, the recovery appears

---

[12] To our knowledge this mechanism for this bifurcation has not been investigated. However, we speculate that it might be related to a lowering in the calcium ion concentration.

to lose its exponential nature. At a certain point, the perturbation does not die out, but produces repetitive spiking.

5.4.2 Critical Phenomena

Phenomena that occur near bifurcation points also provide insight into one of the most vexing questions in biology, namely, how do large-scale patterns emerge from microscopic and even molecular processes? We suspect that many readers will be surprised to discover that the rules that govern large-scale behavior depend less on the detailed properties of the component parts and more on the nature of the dynamical behaviors that arise near bifurcation points. Knowledge of the properties of a single water molecule is not sufficient to understand the formation of ice any more than understanding a single liver cell provides much insight into the formation and functions of the liver.

In physics, a *phase transition* describes a generally sudden change between two states of a system that exhibit a marked difference in the degree of small- and large-scale organization of its components. The first thing to note is that a phase transition corresponds to a bifurcation point. In the physics literature, two types of phase transitions are recognized.[13] Most commonly, the phase transition occurs abruptly. This is called a *first-order phase transition*, and it corresponds dynamically to a subcritical pitchfork bifurcation. The experiment shown in Figure 5.11 is an example of a first-order phase transition in a dynamical system that exhibits a subcritical Hopf bifurcation (see Chapter 11). Less commonly, phase transition can occur in a continuous manner. This type of phase transition is called a *critical phase transition*, and it corresponds dynamically to a supercritical pitchfork bifurcation. Both types of phase transitions occur in biological systems. However, current interest in biology is focused on the possible role of critical phase transitions in the formation of large-scale coherent dynamical structures.

---

[13] The terminology used by physicists to categorize phase transitions is based on thermodynamic considerations related to a system's *entropy* [128, 784]. In Chapter 17, we show that at equilibrium, the entropy is related to the first derivative of the free energy G with respect to temperature T at constant pressure P via

$$\left(\frac{\partial G}{\partial T}\right)_P = -S.$$

For a first-order phase transition, $(\partial G/\partial T)_P$ is discontinuous at the bifurcation point. The physical quantity that corresponds to this discontinuity is the release or absorption of heat. In contrast, at a critical phase transition, the entropy changes continuously. Singularities exist in the second- and higher-order derivatives of entropy with respect to the control parameter. These singularities most often take the form of a power-law divergence at the critical temperature.

5.5 Bifurcations at the Benchtop

The hallmark of a bifurcation in mathematical models is a sudden and qualitative change in dynamics when a parameter is changed by a small amount. The hypothesis is that qualitative changes in dynamics observed at the benchtop correspond to bifurcations in the relevant mathematical models. Therefore, establishing the validity of a hypothesis typically requires both very careful experimentation and the development of mathematical models that incorporate the essential dynamical details.

How can parameter values be changed in an experimental preparation? There is no simple recipe. Since the rate of a chemical reaction depends on temperature, a simple way to change parameter values is to change the temperature (see Section 17.3). However, this approach has rarely been used to study the dynamics of biological systems because it lacks specificity. In other words, temperature affects all of the rate constants in the dynamical system (though not necessarily in the same way), making it difficult to pinpoint the change in parameter values that resulted in the observed change in dynamics. Moreover, the integrity and viability of biological systems are typically limited to a small temperature range. Consequently, experiments in which parameters have been carefully changed to reveal the nature of underlying bifurcations have up to now relied on the cleverness and ingenuity of the experimentalist. Throughout this book, we will highlight experiments in which control over important parameters has been possible.

5.6 What Have We Learned?

1. What is a bifurcation?

   A bifurcation is a large qualitative change in dynamics produced by a small change in a parameter. Examples of a qualitative change in behavior are a stable fixed point becoming unstable and the appearance of an oscillation.

2. What is a bifurcation diagram?

   A bifurcation diagram is a summary of the dynamics of a system as a parameter is varied.

3. Why is it important for an experimentalist to have a working knowledge of the different types of bifurcations?

   The number of known bifurcations is surprisingly small. Suppose that it is possible to vary a parameter in a dynamical system experimentally, and in doing so, a qualitative change in dynamics is observed. Identifying this change with a bifurcation that exhibits the same changes provides insight into the nature of the dynamical system. This insight, in turn, can sometimes be used to suggest new experiments. Moreover, it has been suggested that complex biological systems tend to self-organize near, or at, the edge of stability, where the generic forms of the bifurcation describe well the expected dynamics [291, 848].

4. Why is it impossible to determine a bifurcation point by slowly increasing (or decreasing) the bifurcation parameter? What is the resulting error?

This approach will always result in an error in the estimation of μ_c. If we design an experimental paradigm that slowly increases μ as a function of time, the estimated μ_c' will be greater than μ_c. If we slowly decrease μ as a function of time, the estimated μ_c' will be less than μ_c.

5. Which type of bifurcation is associated with the creation or destruction of a fixed point?

A saddle-node bifurcation. A saddle-node bifurcation is also referred to as a blue-sky bifurcation or a fold catastrophe.

6. What is the bottleneck phenomenon and with which bifurcation it is associated?

The bottleneck phenomenon refers to a situation in which the closer the dynamics gets to the values where a fixed point once existed, the longer it takes to escape from this region of phase space. It is typically associated with a saddle-node bifurcation.

7. What does the term "symmetry" mean in the context of a pitchfork bifurcation?

Symmetry means that a change of variable $x \rightarrow -x$ does not change the underlying equation.

8. What is the slowing-down phenomenon and why is it important in the laboratory?

The slowing-down phenomenon is observed for every bifurcation in which the fixed point is not destroyed but changes its stability. It is observed for a variety of different bifurcations including transcritical, pitchfork, and Hopf bifurcations. As the parameter approaches the value at which the bifurcation occurs, the real part of the eigenvalue becomes smaller and smaller, and perturbations are damped more and more slowly. One consequence is that it is very difficult to pinpoint the value of the parameter where the bifurcation occurs experimentally (and in fact, it is also difficult using computer simulations).

9. What are critical phenomena and why are they important in biology?

The term "critical phenomena" refers to a collection of dynamical behaviors that are observed in systems undergoing a critical phase transition. These behaviors include the slowing-down phenomenon, increased variance of the fluctuations, increased autocorrelation, intermittency, and the presence of power-law behaviors [94, 351, 741]. Experimental observation of such behaviors suggests that the system is tuned close to an edge of stability. These observations may be useful for predicting an impending crisis.

10. What type of bifurcation corresponds to a first-order phase transition and what type corresponds to a critical phase transition?

A first-order phase transition corresponds to a subcritical pitchfork bifurcation. A critical phase transition corresponds to a supercritical pitchfork bifurcation.

11. What is the critical slowing-down phenomenon?

> The critical slowing-down phenomenon refers to a situation in which the slowing-down
> phenomenon occurs in a dynamical system undergoing a critical phase transition. Thus
> in addition to slowing down, the other critical phenomena listed above may also be
> observed.

5.7 Exercises for Practice and Insight

1. Write a XPPAUT program to integrate (5.1) from $x = -5$ to $x = 5$ and vary μ
   from 0.00001 to 0.1. Measure the time it takes to go between these two values of
   x. Does this time scale as $1/\sqrt{\mu}$?
2. The generic equation for a transcritical bifurcation is given by (5.3).

   a. Sketch a plot of dx/dt versus x when $\mu < 0$, $\mu = 0$, and $\mu > 0$.
   b. Use these observations to construct the bifurcation diagram shown in
      Figure 5.4.

3. The generic equation for a supercritical pitchfork bifurcation is given by (5.10).

   a. Sketch a plot of dx/dt versus x when $\mu < 0$, $\mu = 0$, and $\mu > 0$.
   b. Use these observations to construct the bifurcation diagram shown in
      Figure 5.6a.

4. The generic equation for a subcritical pitchfork bifurcation is given by (5.11).

   a. Sketch a plot of dx/dt versus x when $\mu < 0$, $\mu = 0$, and $\mu > 0$.
   b. Use these observations to construct the bifurcation diagram shown in
      Figure 5.6b.

5. Explain why we cannot write (5.12) as

$$\frac{dx}{dt} = \mu x + x^3 - x^4 .$$

Chapter 6
Transient Dynamics

Scientists, particularly those who investigate physiological and/or biomechanical systems, use inputs to explore the nature of biological dynamical systems (e.g., the black box shown in Figure 6.1). Since in the laboratory, we most often investigate dynamical systems that are stable (at least in some sense), we expect the observed responses to an input to contain two components: a transient component on short time scales that describes the immediate effects of the input and the subsequent recovery of the system, and a steady-state response in which the input to the system equals the output (as in the water-fountain example discussed in Section 3.3.1).

Whereas biomathematicians use sudden, very brief perturbations to examine the stability of fixed points, experimentalists also input time-dependent signals, such as ramp functions and periodic signals. Thus the starting point of experimentalists often differs from that of biomathematicians. To illustrate, consider our now-familiar first-order ODE

$$\frac{dx}{dt} + kx = 0 \,,$$

where k is a constant. Stability is determined by examining the time-dependent behavior for a particular initial condition $x(t_0)$ that is not equal to the fixed point $x^* = 0$. There is no mention of the mechanism that caused the change to $x(t_0)$. In contrast, at the benchtop, it is necessary to specify the mechanism by which the perturbation was introduced, and hence the more relevant model is

$$k^{-1}\frac{dx}{dt} + x = b(t) \,, \tag{6.1}$$

where $b(t)$ is a *forcing function*.[1] The forcing function describes how the system is perturbed away from its fixed point and whether the perturbation is brief or long-lasting.

The first part of this chapter examines the transient responses of a linear dynamical system in the time domain to three types of forcing functions: step functions,

---

[1] The reader should note that we have written this equation to be dimensionally correct.

© Springer Nature Switzerland AG 2021
J. Milton and T. Ohira, *Mathematics as a Laboratory Tool*,
https://doi.org/10.1007/978-3-030-69579-8_6

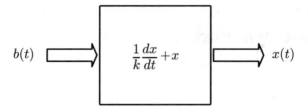

Fig. 6.1 Our black-box paradigm. The contents of the black box in this figure should be compared to the black box in Figure 2.4. In the figure shown here, the inside of the black box has been presented correctly. In particular, the units for each term in the box are the same, and they are the same as the units for $b(t)$.

ramp functions, and impulse responses. Surprisingly, confronting these problems leads to an array of powerful tools for the analysis of biological systems. Moreover, through using these tools, we can better understand what it means to make an experimental measurement. The following two chapters consider the analysis of biological systems in the frequency domain. We will see that in that case, the analysis focuses on the responses to sinusoidal inputs once all initial transients have died out.

The second part of this chapter introduces the transient responses of nonlinear dynamical systems. Compared to linear dynamical systems, the study of transients in nonlinear dynamical systems is very much a "work in progress." However, it is clear that as experimentalists make the transition from the study of isolated systems in the laboratory to dynamics in the real world, understanding the role of transients will become more and more important. Indeed, it can be argued that real-world dynamics are dominated by transients. Here we introduce two types of nonlinear transients, *excitability* and *mesoscopic states*. In later chapters, we introduce two other types of transient behaviors, namely *delay-induced transient oscillations (DITOs)*, in Section 13.6, and *intermittency*, in Section 15.3.

The questions we will answer in this chapter are:

1. What is a step function and why is it useful?
2. What is a ramp function and why is it useful?
3. What is a delta function and why is it useful?
4. What is an impulse response?
5. What is the relationship between step, ramp, and delta forcing functions?
6. How can an impulse response be measured?
7. What is a convolution integral?
8. Why are convolution integrals important?
9. What do we mean by the term "metastability"?
10. What do we mean by the term "mesoscopic state"?

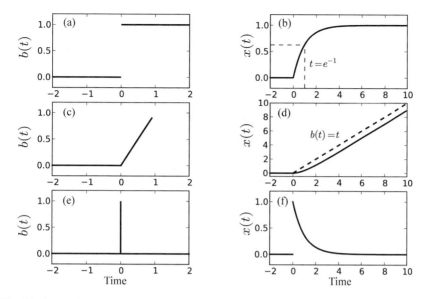

Fig. 6.2 Comparison of a forcing function with the response for a step function (a), (b); ramp function (c), (d); and delta function (e), (f). In all cases, the forcing function was applied at $t_0 = 0$. The dashed line in (d) is the applied ramp. In (e), it is impossible to represent a delta function in the time domain (it would be an infinitely high, infinitely brief perturbation). Therefore, we have shown a brief unit pulse that elicits a response that is indistinguishable from the response to the delta-function input shown in (f) (see Section 6.3.1).

6.1 Step Functions

Before we begin an experiment in which we hope to learn something about how a biological system changes over time, we need to have a mathematical model that describes the state of the dynamical system before the experiment begins, at the moment the experiment starts, and during the time the experiment is running.

An important question is whether we in fact need to know anything more about the past history of the system than the values of the variables at the time the experiment started. In other words, we can ask whether the system "remembers" its past history. For more on this question, see Chapters 9 and 10. In this chapter, we shall consider only cases in which nothing before the beginning of the experiment affects the evolution of the experiment.

Suppose that the dynamical system described by (6.1) is initially at its fixed point. A typical experimental scenario would be to start the experiment by turning on $b(t)$ with a switch. Before the switch is thrown, the input to the black box is zero. Then at time $t = t_0$, we turn the switch, and "suddenly," the dynamics of the black box are changed, or "forced," on account of the application of $b(t)$. Such functions are called *forcing functions*. The responses of this system are described by

$$k^{-1}\frac{dx}{dt} + x = \begin{cases} 0, & \text{if } t < t_0, \\ b(t) & \text{if } t \geq t_0, \end{cases} \tag{6.2}$$

with the initial condition $x(t_0) = 0$.

The Heaviside, or unit step, function shown in Figure 6.2a,

$$H(t - t_0) = \begin{cases} 0, & \text{if } t < t_0, \\ 1, & \text{if } t \geq t_0, \end{cases} \tag{6.3}$$

can be used to describe experiments conducted in this manner. Note that $H(t - t_0)$ is not continuous at $t = t_0$. From Figure 6.2a, we see that $H(t - t_0)$ equals 0 when t_0 is approached from the left, and it equals 1 when it is approached from the right. A function is said to exhibit a *jump discontinuity* at t_0 if it has finite but different left- and right-hand limits as t approaches t_0. In the context of a computer program, the $t - t_0$ term is referred to as the argument of the Heaviside function: Using $H(t - t_0)$, we can write (6.2) as

$$k^{-1}\frac{dx}{dt} + x = H(t - t_0)b(t) \tag{6.4}$$

with the initial condition $x(t_0) = 0$. As an example, take $b(t) = 1$. For $t \geq t_0$, (6.4) becomes

$$\frac{dx}{dt} + kx = k.$$

We can solve this equation using the integrating factor tool (Tool 6) (see Exercises 3–5). In this way, we obtain the solution

$$x(t) = 1 - e^{-kt}, \tag{6.5}$$

where we have made use of the initial condition $x(0) = 0$.

The response shown Figure 6.2b is referred to as a *simple lag* or *exponential lag*. The response contains two components: a transient component e^{-kt}, and a steady component 1. The transient nature of e^{-kt} follows from the fact that e^{-kt} approaches 0 as t approaches ∞. The constant $1/k$ is called the *time constant*. If we set $t = 1/k$, we see that the time constant is the time at which the value of x has increased to $1 - e^{-1}$ (dashed line in Figure 6.2b). Engineers use $1/e$ time to discuss the response of a dynamical system to a sudden perturbation. Equation (6.5) describes, for example, the charging of a capacitor in response to a sudden constant DC input or the rise in the mercury level of a thermometer that has been suddenly plunged into a beaker full of hot water. Since the response approaches the steady-state value asymptotically, it is impossible to say at what time, for example, the thermometer measures the steady-state value. Experimentalists use the estimate $t = 4k^{-1}$, since for this choice, $x(t)$ is within 2 % of its maximum value of 1. If we wanted to follow rapid changes in temperature more accurately, we would choose a thermometer whose time constant is suitably small, namely a thermometer having a large value of k.

6.2 Ramp Functions

The ramp forcing function $r(t - t_0)$ is used to investigate the response of a system to an input that changes continuously as a function of time (Figure 6.2c),

$$r(t - t_0) = \begin{cases} 0, & \text{if } t < t_0, \\ t - t_0, & \text{if } t \geq t_0. \end{cases} \tag{6.6}$$

A familiar example of ramp forcing is a thermometer placed in a heated beaker of water. In biology, ramp forcing functions are used to investigate a wide range of phenomena. One application occurs in the investigation of dynamical systems characterized by the presence of two antagonistic forces. A much-studied example concerns the agonist–antagonist arrangement of muscles that move a joint as in the stick insect [151] or the actions of the biceps and triceps in causing movements of the elbow joint. Ramp forcing is also used to study the viscoelastic properties of ligaments, tendons, and soft tissues in the body [184] and the dynamics of discontinuous switching networks which arise in the control of accommodation [426, 815, 831].

We immediately see that $r(t - t_0)$ is the integral of $H(t - t_0)$. Thus we anticipate that there will be connections between the responses of a system to these two types of forcing functions. The response of (6.1) to the ramp forcing function is the solution of

$$\frac{dx}{dt} + kx = kr(t), \quad \text{for } t \geq 0,$$

together with the initial condition $x(0) = 0$. Using the integrating factor tool (Tool 6), we obtain the solution

$$x(t) = t - \frac{1}{k}\left[1 - e^{-kt}\right], \tag{6.7}$$

shown in Figure 6.2d.

As before, we see that the solution contains a transient component and a steady-state component that increases in parallel with the applied input (dashed line in Figure 6.2d).[2] If we differentiate (6.7) with respect to t, we obtain $1 - e^{-kt}$, which is the same as (6.5), namely the response to a step forcing function. On the other hand, if we integrate (6.5) with respect to t, we obtain (6.7). It is very useful to remember that the response to a ramp forcing function can be determined by integrating the response to a step forcing function. This observation can be extended to higher-order linear ODEs using the Laplace transform methods introduced in the next chapter.

A step forcing function represents the most severe disturbance that a system can receive (sudden onset of the applied "force"). However, in some experimental contexts, a step forcing function is unreasonable (e.g., systems with high inertia), and

---

[2] This is an example of a situation in which a system attains a steady state but is not at equilibrium.

there may be concerns that it produces undesired side effects. The use of a ramp forcing function is gentler.[3]

6.3 Impulse Responses

An important possible input to a black box is a rectangular pulse with unit area, shown in Figure 6.3. This rectangular pulse $\Pi(t)$ has height Δ^{-1} and length Δ, so that its area is equal to 1. As we will see, the response of linear dynamical systems to pulses is of great importance. As we decrease the width of $\Pi(t)$ while keeping the area equal to 1, the height of the pulse must increase. What happens as Δ approaches zero? The answer is that $\Pi(t)$ approaches a *delta function* (Figure 6.2e), namely

$$\delta(t - t_0) = \lim_{\Delta \to 0} \int_{-\infty}^{\infty} \Delta^{-1} \Pi\left(\frac{s}{\Delta}\right) ds.$$

We see that the delta, or Dirac[4] delta, function is defined only by its integral

$$\int_{-\infty}^{\infty} \delta(s) \, ds = 1, \tag{6.8}$$

so that

$$\delta(t - t_0) = 0 \quad \text{for } t \neq t_0.$$

What is the response of a linear dynamical system to a delta-function input? In terms of our first-order ODE example, we need to solve

$$k^{-1} \frac{dx}{dt} + x = \delta(t - t_0). \tag{6.9}$$

Equation (6.9) can be written as

$$\frac{1}{k} \frac{dx}{dt} + x = \begin{cases} 0 & \text{if } t < t_0, \\ \int_{-\infty}^{\infty} \delta(s - t_0) ds & \text{if } t = t_0, \\ 0 & \text{if } t > t_0. \end{cases} \tag{6.10}$$

We can determine the response to a delta-function input by examining the three cases $t < t_0$, $t = t_0$, and $t > t_0$.

For $t < t_0$, the response is described by the equation

$$\frac{dx}{dt} = -kx. \tag{6.11}$$

---

[3] An even gentler change can be obtained with a function that is not only continuous at the point where it begins to increase but differentiable as well. A parabolic forcing function would accomplish this.
[4] Paul Adrien Maurice Dirac (1902–1984), English theoretical physicist.

However, if the system is to represent some real-world phenomenon, then we must require that it be physically realizable. The mathematical definition of physical realizability requires that the impulse response be zero for $t < t_0$. Consequently,

$$x(t) = 0, \quad \text{for } t < t_0. \tag{6.12}$$

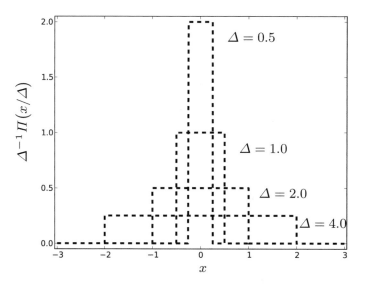

Fig. 6.3 Graphical illustration of the limiting process that gives rise to the delta function. The area of each rectangle is equal to 1.

What happens at $t = t_0$? The trick here is to keep in mind that the delta function is defined only by its integral, i.e., (6.8). Thus, we integrate (6.9) over a short time interval that includes t_0. For simplicity, assume that the time interval has width 2ε and that t_0 is the midpoint. Hence we have

$$\frac{1}{k} \int_{t_0-\varepsilon}^{t_0+\varepsilon} \frac{dx(s)}{ds} \, ds = -\int_{t_0-\varepsilon}^{t_0+\varepsilon} x(s) \, ds + \int_{t_0-\varepsilon}^{t_0+\varepsilon} \delta(s-t_0) \, ds. \tag{6.13}$$

Since integration is the opposite of differentiation, we can evaluate the left-hand side immediately. Assuming that ε tends to zero, the first integral on the right-hand side will vanish. The second integral yields 1 because of the property of the delta function. Thus we have

$$\frac{1}{k} [x(t+\varepsilon) - x(t-\varepsilon)] = 1. \tag{6.14}$$

Since $x(t_0 - \varepsilon) = 0$ to ensure realizability, we obtain

$$x(t_0) = k. \tag{6.15}$$

Now we consider the solution for $t > t_0$. Again we can use (6.11) and see that

$$x(t) = \tilde{X}e^{-kt},\tag{6.16}$$

where $\tilde{X}$ is the value of $x(t)$ when t is very, very close to t_0. In order to complete our solution, we need to determine $\tilde{X}$. From (6.16), we have

$$x(t_0) = \tilde{X}e^{-kt_0},$$

and thus using (6.15), we see that

$$\tilde{X} = ke^{kt_0}.$$

Consequently,

$$x(t) = ke^{kt_0}e^{-kt} = ke^{-k(t-t_0)} \quad \text{for } t \geq t_0.\tag{6.17}$$

The response of a linear dynamical system to a delta-function input is called the *impulse response*, where

$$I(t,t_0) = \begin{cases} ke^{-k(t-t_0)}, & t \geq t_0, \\ 0, & t < t_0. \end{cases}\tag{6.18}$$

The notation $I(t,t_0)$ means that the impulse is delivered at time t_0 and the response is measured at time $t \geq t_0$. The impulse response for (6.9) is shown in Figure 6.2f. It is easy to verify that if $b(t) = B\delta(t-t_0)$, where B is a positive constant, then (see Exercise 4)

$$I(t,t_0) = \begin{cases} Bke^{-k(t-t_0)}, & t \geq t_0, \\ 0, & t < t_0. \end{cases}\tag{6.19}$$

It is helpful to keep in mind that "functions" like the delta function are not really functions at all. Rather, they belong to the class of *generalized functions*, or *distributions*, which have many strange and often counterintuitive properties. Most of these properties are not used in this book. Here we make use of two properties of the delta function. First, the delta function is the derivative of the step function:

$$\delta(t-t_0) = \frac{d}{dt}H(t-t_0).$$

Thus we expect that the derivative of the response to the step function given by (6.5) will be the impulse response given by (6.18). This result can also be extended to higher-order linear dynamical systems using techniques based on the Laplace transform.

Second, the delta function has a sifting property, namely

$$\int_{-\infty}^{\infty} \delta(s-t_0)f(s)ds = f(t_0).\tag{6.20}$$

In other words, the operation of the delta function on $f(t)$ is to "sift out" a single value of $f(t)$. The importance of the sifting property of the delta function will soon become clear.

6.3.1 Measuring the Impulse Response

The impulse response is of considerable interest to biologists. What happens to a population in response to a sudden addition (e.g., invasion) or removal (e.g., predation) of a certain cohort? How does a tumor respond to a single dose of a chemotherapeutic agent? What is the change in blood flow to the brain in response to a single thought? All of these questions and more can be answered by measuring the impulse response.

A variety of methods have been developed to measure the impulse response in the laboratory. Although we have not yet developed the mathematical tools we need to understand how all of these methods work, and it is convenient to list them in one place for future reference.

Brief Pulses

Of course, it is impossible to construct an instrument that could generate $\delta(t - t_0)$, because measuring devices are limited by both their response time (e.g., Figure 6.2b) and their resolving power. However, suppose that we find that we cannot distinguish between the response of a dynamical system to a brief pulse and the response to a briefer pulse. Since it would be very difficult to distinguish the response of the black box to this pulse and to that described by $\delta(t - t_0)$, it is customary to use the response of the black box to a brief pulse as an estimate of the impulse response.

Response to a Step Input

It is often easier to measure the response of a system to a step input than to design a suitably brief pulse input. In principle, we can determine the impulse response by differentiating the step function response. However, the determination of the derivative from noisy experimental data is often problematic. Thus this approach for determining the impulse response requires the use of appropriate curve-fitting techniques; the impulse response is determined by taking the derivative of the best-fit curve.

Sinusoidal Inputs

The use of sinusoidal inputs allows the investigator to determine the transfer function of the system, which is the topic of the next chapter. The transfer function makes it possible to predict the output of a linear dynamical system to any input, including a delta-function input. The transfer function of a system is the Laplace transform of its impulse response (see Section 7.3).

Random-Noise Inputs

The approach using random-noise inputs is based on the statistical concept of cross-correlation and is discussed in Chapter 14.

6.4 The Convolution Integral

If the response of a linear dynamical system to a delta-function input, that is, the impulse response, is known, then the solution of the associated linear differential equation is known for all possible inputs. This amazing fact is based on the mathematical observation that the solution to any linear dynamical system that receives an input $b(t)$ is given by

$$x(t) = \int_{-\infty}^{t} b(t')I(t,t')dt'. \tag{6.21}$$

The integral on the right-hand side of this equation is called the *convolution integral*, and the process of calculating this integral is called *convolution*. Convolution integrals commonly arise in biology in situations in which inputs are summed and signals are measured by either sensory receptors or laboratory instruments. For example, a convolution integral can be used to determine the dosing schedule required to maintain an adequate blood level of an antibiotic medication to treat an infection (see Exercise 8 at end of chapter).

We can use the delta function to gain insight into (6.21). Let us assume that $b(t)$ is a discrete-time or digitized input. This means that we can represent $b(t)$ as a sequence of delta functions each of which corresponds to the time t_i of an input, and we scale the delta-function impulse response to account for the magnitude of the input at that time. Then we can write

$$k^{-1}\frac{dx}{dt} + x(t) = B(t_0)\int_{-\infty}^{\infty} \delta(s-t_0)\,ds + B(t_1)\int_{-\infty}^{\infty} \delta(s-t_1)\,ds + \cdots$$

$$+ B(t_i)\int_{-\infty}^{\infty} \delta(s-t_i)\,ds + \cdots, \tag{6.22}$$

where $B(t_i)$ is the magnitude of the ith input. This equation means that at each time t_i, the dynamical system responds to a delta-function impulse of magnitude $B(t_i)$. We know that the impulse response $I(t,t_i)$ describes the response of a linear dynamical system (our black box) to a delta-function input given at time t_i. Hence $B(t_i)I(t,t_i)$ describes the response to the product of B and the impulse response. Since (6.22) is a linear ODE, the solution is equal to the sum of the impulse solutions (the principle of linear superposition, Tool 1), and hence

$$x(t) = k[B(t_0)I(t,t_0) + B(t_1)I(t,t_1) + \cdots + B(t_i)I(t,t_i) + \cdots], \qquad (6.23)$$

where for all $t_i \leq t$, we must keep in mind that

$$I(t,t_i) = \begin{cases} I(t,t_i), & t \geq t_i, \\ 0, & t < t_i. \end{cases} \qquad (6.24)$$

If we make the time between successive delta-function inputs smaller and smaller, then we can replace the sum with an integral to obtain (6.21).

The convolution integral written in the form of (6.21) is also referred to as Green's method,[5] and $I(t,t')$ is called a Green's function. This interpretation of the convolution integral is most relevant to our goal of obtaining the solution to a linear inhomogeneous differential equation. In other applications, however, it is convenient to write the convolution integral in the form

$$x(t) = \int_{-\infty}^{t} b(u)I(t-u)\,du \qquad (6.25)$$

(see, for example, Section 16.9.1).

Every time one uses a device to make a measurement, one essentially performs a convolution. Thus the notion of convolution is one of the most important concepts in science. All physical observations are limited by the resolving power of instruments, and for this reason alone, convolution is ubiquitous. However, this principle does not apply only to laboratory instruments; inputs to our sensory systems, such as light intensity, vibration, smell, and sound, are also convolved by our sensory receptors, which translate these physical stimuli into neural spike trains. In order to determine the original input from the convolved result, we need to perform the reverse operation, i.e., a *deconvolution*. In Chapter 16, we will see that the convolution integral has many other applications.

Despite the importance of the convolution integral and the ease with which it can be calculated using a computer, it can be quite difficult to obtain an intuitive understanding of this process. We briefly mention a graphical interpretation of convolution, popularized by Brigham[6] [82], that is widely used. There are four steps:

1. *Folding:* Reflect $I(u)$ about the ordinate axis to obtain $I(-u)$.
2. *Displacement:* Shift $I(-u)$ by an amount t to obtain $I(t-u)$.

---

[5] George Green (1793–1841), British mathematical physicist.
[6] E. Oran Brigham (b. 1919), American applied mathematician.

3. *Multiplication:* Multiply $I(t-u)$ by $b(u)$. It is easiest to understand this step iteratively: for each time step, do the multiplication (and then step 4); then displace the
   reflected impulse response another time step to the right and repeat the process.
4. *Integration:* Determine the area under the curve produced by $I(t-u)$ and $b(u)$.
   This value is the value of the convolution at time t.

Figure 6.4 illustrates this procedure (see Exercise 7). It is very useful to keep this
graphical approach in mind in order to understand issues that arise in the evaluation
of the convolution integral using the fast Fourier transform.[7]

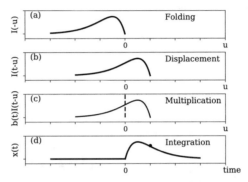

Fig. 6.4 Graphical approach for determining the convolution between a delta-function input and
the alpha function (the impulse response) given by (6.29): (a) folding; the impulse response is
folded in time by taking its mirror image about the vertical axis, while in (b), the folded impulse
response has been shifted by an amount t. (c) The folded and displaced alpha function is multiplied by the delta function (dashed line). (d) The area under the product of the displaced and folded
impulse response and the delta-function input (indicated by •) is equal to the value of the convolution integral, $x(t)$, at time t. By the shifting property of the delta function, the area is just the value
of the reflected alpha function at time t.

Those of us who do not enjoy computing integrals will be relieved to know
that it is quite uncommon to compute the convolution directly by solving (6.21)
using paper and pencil. Typically, a computer program is used. The convolution integral can be computed in the time domain in Python using the function
`np.convolve()` and in the frequency domain by making use of the fast Fourier
transform. By combining these techniques with digital computers, the evaluation of
convolution integrals becomes a readily usable tool in the laboratory. In this way,
we can easily deal with convolutions that are so complex that we cannot easily write
down the integral in closed mathematical form.

A potential point of confusion is that commands such as `np.convolve()` in
Python and `conv().m` in MATLAB do not evaluate the convolution integral. In

---

[7] Several applets demonstrating this graphical approach for computing the convolution integral can
be readily located on the Internet; see, for example, http://en.wikipedia.org/wiki/Convolution.

Python, the convolution of two discretely sampled functions s and r of, respectively, lengths N and M is

$$\text{conv}(s,r)(j) = \sum_{k=-N/2+1}^{N/2} s(j-k)r(k). \tag{6.26}$$

In terms of the Brigham's interpretation of convolution, this result corresponds to the first three steps. Missing is the final step, integration. If we interpret integration as computing the area under a curve, then we can approximate the convolution integral as a Riemann sum, namely the sum of the areas of rectangles of width Δt, the sample interval, and height (6.26). In other words, we multiply the results of np.convolve() by Δt in order the obtain the correctly scaled convolution integral.

A second point of confusion arises because commands, such as np.convolve() in Numpy, assume that the length of the impulse function, r, and the signal, s, is the same and that s is periodic [692]. The solution to these constraints involves padding the signal and the impulse response with zeros. This procedure is called *padding with zeros*. The first constraint is easily dealt with since it is typically the case that the length of the impulse response, M, is much shorter than the length of the signal, N. Thus we need to pad the r with zeros to make its length equal to that of s (in Numpy you can use np.zeros). The second constraint is subtler since not only must the signal be considered to be periodic, but also the impulse function. Thus it can happen that a portion of each end of s can be erroneously wrapped around with r during the calculation of the convolution. Thus it is necessary to set up a buffer zone of zero-padded values at the end of s in order to make the effects of this contamination equal to zero. How many zeros do we need to add? We need to pad s on one end with M zeros. Although these issues can sometimes be overcome by choosing the mode='full' option in Numpy overlap problems may still exist. Thus the above procedure can be useful.

Finally, we note that computer multiplication is faster than the evaluation of (6.26). Thus for large data sets, it is much more efficient to use the fast Fourier transform to evaluate the convolution integral (see Section 8.4.3).

6.4.1 Summing Neuronal Inputs

The input to a neuron takes the form of a neural spike train, namely a time series of action potentials generated by other neurons. In general, neurons receive many inputs. For example, it has been estimated that a cortical neuron in the human brain receives as many as 10^4 inputs [806]. We can use the convolution integral as a tool to obtain insight into how neural inputs are summed at the axon hillock.

To illustrate, let us reconsider the integrate-and-fire (IF) neuron that we introduced in Section 1.3.2. Assume that this neuron receives a periodic train of inputs each producing an excitatory postsynaptic potential (EPSP) and that the IF neuron

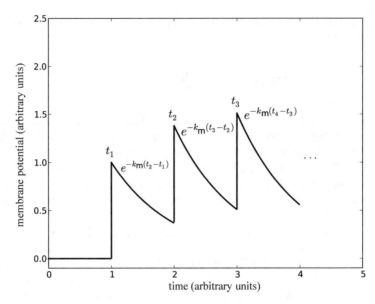

Fig. 6.5 Increase in the membrane potential of an integrate-and-fire neuron receiving periodic excitatory postsynaptic potentials of magnitude $\varepsilon = 1$, where $k_{\mathrm{m}} = 1$.

sums these inputs linearly. Without loss of generality, assume that at $t = 0$, the resting membrane potential is zero, i.e., $V_{\mathrm{m}}(t_0) = 0$. For this special case, we can obtain insight into how the membrane potentials sum (Figure 6.5). Suppose that the neuron receives its first EPSP at time t_1. The membrane potential at this time is

$$V_{\mathrm{m}}(t_1) = \varepsilon,$$

where ε is the amount by which the membrane potential is increased by a single EPSP. The second EPSP arrives at t_2, and the membrane potential at this time is

$$V_{\mathrm{m}}(t_2) = \varepsilon + V_{\mathrm{m}}(t_1)e^{-k_{\mathrm{m}}(t_2-t_1)} = \varepsilon\left[1 + e^{-k_{\mathrm{m}}(t_2-t_1)}\right].$$

The exponential decay term describes the fact that neurons are "leaky integrators." That is, they do not hold their charge. Each time the membrane is increased by an EPSP, it subsequently decays exponentially with membrane time constant $1/k_{\mathrm{m}}$. The exception to this rule occurs when the spiking threshold is exceeded, and the neuron generates an action potential. The third EPSP arrives at t_3, and we have

$$V_{\mathrm{m}}(t_3) = \varepsilon + V_{\mathrm{m}}(t_2)e^{-k_{\mathrm{m}}(t_3-t_2)} = \varepsilon\left[1 + e^{-k_{\mathrm{m}}(t_3-t_2)} + e^{-k_{\mathrm{m}}(t_3-t_1)}\right].$$

Continuing in this way, we have, after n inputs,

$$V_{\mathrm{m}}(t_n) = \varepsilon \sum_{i=1}^{n} e^{-k_{\mathrm{m}}(t_n-t_i)}.$$

Since the input is periodic, we have

$$F = \frac{1}{t_{i+1} - t_i} = \frac{1}{\Delta t}$$

and hence

$$V_m(t_n) = \varepsilon \sum_{i=0}^{n} e^{-ik_m/F} . \tag{6.27}$$

If we define $s = e^{-k_m/F}$, then we see that (6.27) is a geometric series.[8] Thus as the number of spike inputs becomes very large, we have

$$V_m(\infty) = \frac{\varepsilon}{1 - \exp(-k_m/F)} . \tag{6.28}$$

The same result can be obtained by convolving a periodic input of delta functions with an exponential impulse response, $I(t,t') = \varepsilon e^{-k_m(t-t')}$.

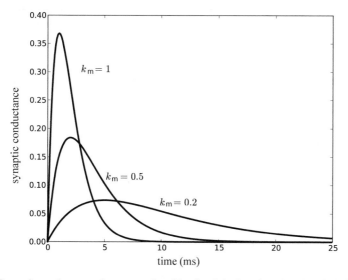

Fig. 6.6 Synaptic conductance changes produced by the alpha function given by (6.29) for different membrane rate constants.

Neuroscientists do not always use an exponential impulse response to model neuronal inputs. The reason is that it is sometimes necessary to account for the time it takes for a change in membrane potential that starts in the dendritic tree to propagate to the axon hillock. Wilfrid Rall[9] has suggested that a better choice of the impulse response is the *alpha function* [708]

---

[8] A geometric series is defined by $S_n = \sum_{i=0}^{n} s^i$, which converges for $|s| < 1$ to $1/(1-s)$.

[9] Wilfrid Rall (b. 1922), American neuroscientist.

$$I_\alpha(t,t') = \begin{cases} \pm k_{\mathrm{m}}^2(t-t')e^{-k_{\mathrm{m}}(t-t')}, & \text{if } t \geq t', \\ 0, & \text{if } t < t'. \end{cases} \tag{6.29}$$

Figure 6.6 shows how $I_\alpha(t,t')$ changes as the membrane time constant k_{m} changes. The maximum in $I_\alpha(t,t')$ occurs at $t = 1/k_{\mathrm{m}}$. Thus the conductance changes at the axon hillock peak after the end of the dendritic input, provided that k_{m} is sufficiently short (see Section 11.1 for a discussion of conductance). This result is observed experimentally. Figure 6.7 shows the result of convolving a periodic input of delta functions with an alpha function.

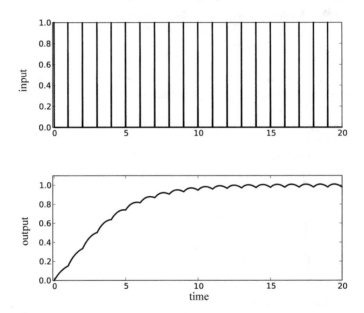

Fig. 6.7 The membrane potential of a neuron (bottom panel) in response to a pulsatile excitatory EPSP input whose timing corresponds to the timing of spikes in the spike train shown in the top panel. The changes in the neuron's membrane potential following the arrival of each EPSP are described by (6.29) with $k_{\mathrm{m}} = 0.5$.

Subsequently, Uwe an der Heiden[10] [16] extended this definition of the alpha function to account for the wide variety of dendritic tree morphologies that neurons possess (see [863] for examples):

$$I_{\alpha,n}(t,t') = \begin{cases} \pm \frac{k_m^2}{(n-1)!}(t^{n-1}-t')e^{-k_{\mathrm{m}}(t-t')}, & \text{if } t \geq t', \\ 0, & \text{if } t < t'. \end{cases} \tag{6.30}$$

Rall's definition of the alpha function corresponds to $n = 1$ and represents the situation in which the dendritic synapses are located not too far from the axon hillock.

---

[10] Uwe an der Heiden (b. 1942), German mathematician and philosopher.

For situations in which there is a large distance between dendritic synapses and axon hillock, take $n > 1$.

6.5 Transients in Nonlinear Dynamical Systems

Sufficiently near a fixed point, a nonlinear dynamical system behaves like a linear dynamical system. Thus it is possible to apply the techniques we have discussed in this chapter to study transients local to stable fixed points. Of course, one must choose the forcing functions carefully to ensure that their magnitude is small compared to the size of the associated basin of attraction. A limitation of such mathematical approaches is that the determination of stability is based on asymptotic behavior, namely the behavior observed after all transients have died out. Real dynamical systems are always responding to unexpected perturbations. Indeed, most experimentalists would agree that in the real world, it is very likely that biological dynamics are dominated by events that are transient in both time and space. Novel dynamical behaviors can exist that do not fit well into the classification schemes we have discussed up to this point [279, 614, 704, 915]. In the remainder of this section, we draw attention to two types of transients that occur in nonlinear dynamical systems that are of particular interest to biologists.

6.6 Neuron spiking thresholds

An important nonlinear behavior for life on this planet is *excitability*. The cardinal feature of an excitable system is the generation of a transient, "all or none" type response which depends on the magnitude of the input. An example is the generation of an action potential by a neuron when its membrane potential becomes sufficiently high. However, this excitable behavior is not restricted to neurons but is a property exhibited by cardiac, skeletal, and intestinal smooth muscle cells; insulin-producing pancreatic beta cells; and even certain plant cells. In Section 11.1 we show that the molecular engines for excitable behavior are the membrane ion channels. The purpose of this section is to examine the curious nature of the spiking threshold of a neuron.

Figure 6.8 shows the phase plane for the FHN equations we discussed in Section 4.5.2. We have chosen the parameters so that there is only one intersection between the v and w nullclines. What is the spiking threshold? A reasonable guess is that the spiking threshold corresponds to the point where the middle branch of the v-nullcline crosses $w = 0$. How should we measure the spiking threshold, T? Of course we can do this by applying an external current I_{ext}. The problem is that when we apply I_{ext} we vertically shift the v-nullcline up or down since

$$w = f(v) + I_{ext}$$

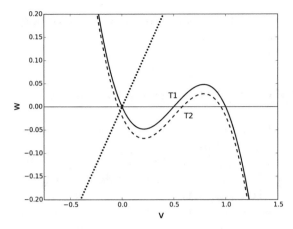

Fig. 6.8 Effect of changing the input current, I_{ext}, on the spiking threshold, T, of a Fitzhugh–Nagumo neuron. The dotted line is the w-nullcline. The solid line is the v-nullcline when $I_{\text{ext}} = 0$ and the dashed line is the v-nullcline when $I_{\text{ext}} = -0.1$. The spiking threshold, T_1, is determined when $I_{\text{ext}} = 0$ and is T_2 when $I_{\text{ext}} = -0.1$.

where $f(v) = v(a - v)(v - 1)$. Thus the threshold is a function of I_{ext}! The consequence is that we obtain different estimates of the threshold for different choices of $I_{\text{ext}}(t)$! A true threshold should be independent of how it is measured. Consequently, we see that the FHN neuron does not have a true spiking threshold. The same conclusion is obtained if we use other models for excitable cells. Indeed these models suggest that there is an intrinsic unpredictability to how a neuron responds to stimulation [293]. The major exceptions are the integrate-and-fire models discussed in Sections 1.3.2 and 6.4.1 and the reduced neuron to be discussed in Section 11.6.

In model neurons, such as the Fitzhugh-Nagumo model, the "spiking threshold" is never a number, but is generally a manifold [225]. A manifold is a curve in the phase plane of 2-D models and a plane or surface in higher-dimensional models. Benchtop neuroscientists have long recognized that the generation of an action potential by a living neuron depends on the state of the neuron as determined by, for example, the time since the neuron last generated an action potential. It also depends upon how $I_{\text{ext}}(t)$ is delivered to the neuron, for example, as a pulse, a step, a ramp, or periodic input. Of course in the organism $I_{\text{ext}}(t)$ represents the sum of the contributions of all of the active excitatory and inhibitory inputs to the neuron. Counter-intuitively action potentials can sometimes be generated by the sudden release of a prolonged hyperpolarizing current (so-called anodal break excitation) or by combining an inhibitory current with a subthreshold excitatory current. These phenomena cannot readily be accounted for by the integrate-and-fire models. The physiological explanation for the complexities associated with action potential generation is based on the observation that several ionic currents are involved which operate on widely different time scales. The triumph of modern mathematical neuroscience is the recognition that it is not the ionic currents per se that are important, but rather it is the type of bifurcation that occurs which is the key that unlocks the neuron's secrets. We return to this discussion in Section 11.3.

Resolving issues related to action potential generation is key for developing an understanding of how the nervous system processes information and computes. Indeed, when neurons are separated by more than a few millimeters the only manner of communication between them takes the form of discrete synaptic potentials driven by action potentials. In other words living neural networks are pulse-coupled [238, 398]. Moreover, the increasing interest in the development of implantable electronic devices ("electroceuticals") has driven a search to identify energetically efficient neuronal stimuli that conserve battery life [112–114].

6.6.1 Bounded Time-Dependent States

Dynamical systems in biology are very complex. On long time scales, say the lifetime of an organism, dynamics are bounded and time-dependent [239, 240, 569]. The term *time-dependent* distinguishes these dynamics from those associated with equilibria and steady states, both of which are time-independent states. The term *bounded* reflects the "checks and balances" of Mother Nature. In general, dynamical trajectories in living biological systems tend neither toward zero nor toward infinity, but fluctuate, sometimes precariously, between these bounds. For example, after birth, the size and weight of an organism typically increase, but the maximum size that the organism achieves is ultimately limited by its genetic code and the environment in which it lives. However, day by day, the weight of the organism continually changes as a function of its intake of food and expenditure of energy.

Much present-day biological research is directed toward identifying the nature of the control mechanisms that keep a biological system's dynamics in check. However, it may be even more relevant to focus on the nature of the events that lead to transient behaviors [704]. The term *mesoscopic state* has been used to describe these complex (aperiodic) bounded and time-dependent states [239, 240]. A variety of mechanisms help maintain a mesoscopic state. For example, in the human brain, neural activity does not approach zero, because excitatory inputs to each neuron greatly outnumber the inhibitory ones [36]. On the other hand, neural activity cannot become very large: neural refractoriness, accommodation, and limited energy resources ensure that the neural spiking frequency does not exceed about 700 Hz. Even in animal populations, there are mechanisms that maintain their mesoscopic nature. For example, litter sizes increase and gestation times decrease when population densities are small, and these trends reverse when population densities are large [41, 251]. In addition, the efficiency of disease mechanisms and the behavior of organisms change as a function of population density, even in the lowly slime mold [475].

The idea that a biological dynamical system can itself change dynamically has received insufficient attention. A possible relevant scenario arises in physiological control, since multiple feedback loops are typically involved [35, 277]. It has been suggested that the associated network topology may be organized in a nested, or "safety net," manner [290]. In other words, as deviations from the controlled set point increase in magnitude, more and more sensory feedback loops participate in restoring control. Such situations arise in the context of human balance control [209, 595, 596, 769] and the behavior of financial markets [783].

Transient behaviors can reflect, for example, an organism's response to sudden changes in its environment [108, 319], changes in neural coding [705, 706], transient instabilities in physiological control [162, 282, 569, 587, 663, 704, 854], and so on. For example, a typical partial complex seizure arising from the temporal lobe lasts 60–120 s. Thus, from a dynamical systems point of view, a seizure can be considered a transient event. In fact, all medical emergencies represent transient events, including infections, trauma, cardiac arrhythmias, and so on. The nature of transient behaviors cannot always be inferred from experiments conducted under "constant laboratory conditions," since they are often dominated by the nonlinearities in the dynamical system. The term *metastable state* is used to describe dynamic states that are long-lived but not permanent.

It is difficult to imagine how deterministic approaches can be used to explore such systems, especially given that transients and noisy perturbations likely dominate their behavior. Much more promising are statistical approaches that benefit not only from the large number of elements involved but also from the coexisting effects of noise [89]. Such approaches also provide insight into the nature of emergent behaviors, that is, those exhibited by a population but not by single elements of the population. We will return to this discussion once we have studied the dynamics of stochastic dynamical systems in Chapters 15 and 16.

6.7 What Have We Learned?

1. What is a step function and why is it useful?

   A step function has one constant value before a given time t_0 and another constant time thereafter. The response of a system to a step-function input can be used to estimate the impulse response.

2. What is a ramp function and why is it useful?

   A ramp function has a constant value before a certain time t_0 and then increases linearly as a function of time thereafter. It makes it possible to introduce a forcing function that is more gentle than the step function. The response to a ramp function can be calculated by integrating the response to a step function. Under the influence of a ramp forcing function, a dynamical system achieves a steady state, but not an equilibrium.

3. What is a delta function and why is it useful?

   A delta function describes a pulse of infinitely brief duration. Although mathematically, the delta function is not a true function and therefore belongs to the class of generalized functions, or distributions, it is a useful tool for experimentally oriented scientists. The delta function is defined by its integral.

4. What is an impulse response?

   An impulse response is the response of a linear dynamical system to a delta-function input. If the impulse response is known for a dynamical system, then the response of the dynamical system to all possible inputs is also known.

5. What is the relationship between step, ramp, and delta forcing functions?

   A ramp function is the integral of a step function. A delta function is the derivative of a step function.

6. How can an impulse response be measured?

   How an impulse response can be measured in the laboratory is discussed in Sections 6.3.1 and 14.8.1.

7. What is a convolution integral?

   A convolution integral is given by (6.21).

8. Why are convolution integrals important?

   Every time we make an experimental measurement, we perform a convolution. Convolution can also be said to be carried out by the sensory nervous system. Later in this book, we shall use convolution integrals to explore the dynamics of a random walk (see Chapter 16).

9. What do we mean by the term "metastability"?

   A dynamic state characterized by the presence of transient behaviors is said to be metastable. Such dynamical states can be long-lived, but they do not last forever. An example is critical slowing-down phenomena in neurons (see Figure 5.11). Another example is a delay-induced transient oscillation (Section 13.6).

10. What do we mean by the term "mesoscopic state"?

    A mesoscopic state is a complex aperiodic state in which time-dependent behaviors remain bounded; that is, none of the variables become too large or too small. An example of a mesoscopic state is the dynamics of the human brain.

6.8 Exercises for Practice and Insight

1. Suppose you require a thermometer that is capable of rapid detection of temperature change. Should you purchase one that has a large or a small time constant?
2. If the impulse response function is ke^{-kt}, then what is the response to a unit step input.
3. Using the integrating factor tool (Section 3.4.2), verify that the solution of (6.1) is (6.5) when $b(t) = H(t - t_0)$. What is the response if $b(t) = BH(t - t_0)$, where B is a constant? In this case, what is the impulse response?
4. Suppose that the step response is $x(t) = 1 - c_1 e^{-k_1 t} + c_2 e^{-k_2 t} + c_3 e^{-k_3 t}$. What is the impulse response function?

5. It is very useful to be able to generate a pulse or single square wave. For example, in Chapter 11, we use a carefully timed pulse in order to determine the phase-resetting curve of a neuron. In XPPAUT, such a pulse is produced by the command `pulse(t)=heav(t)*heav(sigma-t)`, where `heav()` is the Heaviside function: `heav(arg)=0` if `arg < 0` and 1 otherwise.

   a. Sketch `pulse(t)` as a function of time.
   b. Sketch `pulse(t-a)`.

6. Write a computer program to compute the subthreshold membrane potential of a neuron that receives a periodic spike train input. Using the following choices of the impulse response function, compare the subthreshold responses of a neuron that receives 2000 consecutive spike inputs with an interspike interval of 100 ms:

   a.
   $$I(t,t') = \begin{cases} \varepsilon e^{-k_m t}, & \text{if } t \geq t', \\ 0, & \text{if } t < t'. \end{cases}$$

   b.
   $$I_\alpha(t,t') = \begin{cases} \pm k_m^2 (t-t') e^{-k_m(t-t')}, & \text{if } t \geq t', \\ 0, & \text{if } t < t'. \end{cases}$$

   c.
   $$I_{\alpha,n}(t,t') = \begin{cases} \pm \frac{k_m^2}{(n-1)!} (t^{n-1} - t') e^{-k_m(t-t')}, & \text{if } t \geq t', \\ 0, & \text{if } t < t'. \end{cases}$$

   Then answer the following questions:

   a. What is the effect of changing the membrane time constant k_m on $V_m(\infty)$?
   b. For a given k_m, what is the effect of the different neuron impulse responses on $V_m(\infty)$?
   c. For a given k_m, which neuron impulse response results in $0.5 V_m(\infty)$ in the shortest time?
   d. For $I_{\alpha,n}(t,t')$, what is the effect of increasing n? For a given $n > 1$, can the effects of increasing n be ameliorated by changing k_m?

7. Use the graphical technique discussed in Section 6.4 to compute the convolution function for the following choices of $b(u)$ and $I(t-u)$:

   a. $b(u) = $ constant and $I(t) = e^{-t}$.
   b. $b(u)$ is a single square-wave pulse and $I(t) = e^{-t}$.
   c. Both $b(u)$ and $I(t)$ are square-wave pulses.

8. An impulse response can also be used to model the changes in blood concentration of a drug ingested orally. For example, if we take the impulse response in Exercise 3 as a simple impulse response for drug ingestion, then the sharp rising portion indicates the time course of the drug going from the stomach to the blood (typically minutes), and the falling phase describes the clearance of drug from the body by the actions of the liver and kidneys (typically hours to days). Consider

that a doctor prescribes a drug for which the $1/e$ time is $\approx 6\,$h and asks the patient to take the medication four times a day. Most patients do not take the medication every $6\,$h, but more often at around 8 a.m., noon, 5 p.m., and midnight. Write a computer program to look at the effects of different dosing schedules on blood levels to answer the following questions:

a. Would it be better for the patient to take the medication three times a day, every $8\,$h?
b. If the $1/e$ time is $\approx 4\,$h, what would be the difference between the two dosing schedules?
c. If the $1/e$ time is $\approx 12\,$h, what would be the difference between the two dosing schedules?

9. A popular model for a chemical oscillator is the Brusselator, which is described by the following system of first-order differential equations:

$$\frac{du}{dt} = 1 - (k_1 + 1)u + k_2 u^2 v, \qquad \frac{dv}{dt} = k_1 u - k_2 u^2 v.$$

a. What are the fixed points?
b. Take $k_1 = k_2 = 1$. What are the linearized equations about the fixed point?
c. What is the stability of this fixed point?
d. Using the nullcline approach described in Section 4.5.2, determine whether it is possible for the Brusselator to exhibit a (i) stable node, (ii) stable spiral, (iii) unstable node, (iv) unstable spiral, and (v) saddle point.
e. Write a program in Python to integrate these equations and check your answers. How does the solution change when the parameters are chosen to be $k_1 = 2.5$ and $k_2 = 1$ (take, for example, initial conditions $u(0) = 1$ and $v(0) = 0$)?

10. In her textbook *Mathematical Models in Biology*, Leah Edelstein-Keshet[11] considers the following model for population dynamics [186]:

$$\frac{dx}{dt} = -y, \qquad \frac{dy}{dt} = k_1 x(1 - x) - y.$$

a. What are the fixed points?
b. What are the linearized equations about each fixed point?
c. What is the stability of each fixed point?
d. Using the nullcline approach described in Section 4.5.2, determine the behavior in the (x, y) plane for different ranges of initial conditions.
e. Write a computer program to integrate these differential equations numerically and verify your conclusions.

11. Write a computer program to integrate the van der Pol equations given by (4.19)-(4.20).

---

[11] Leah Edelstein-Keshet (PhD 1982), Israeli–Canadian mathematical biologist.

a. Examine the plot of x versus t for different values of the damping coefficient ε, say $\varepsilon = 0.1$, $\varepsilon = 0.5$, and $\varepsilon = 1.0$. What seems to be happening to the waveforms as ε gets smaller?

b. Draw the phase plane of the van der Pol equation for the above three values of ε (do one value of ε at a time). Once you have the phase-plane representation, you can superimpose the nullclines. The nullclines are the curves along which $dx/dt = 0$ and $dy/dt = 0$. Does the relationship between the phase-plane trajectory and the nullclines change as ε changes?

c. In what parts of the phase plane does the trajectory seem to be moving the fastest, along the nullclines or as it jumps between them? (Hint: You can get an idea of the speed at which the trajectory is moving by looking at the data points, keeping in mind that the step size is constant. Change the step size and see how the spacing between the data points changes.)

d. For $\varepsilon = 0.1$, plot x versus t. The result shows the classic waveform for a relaxation oscillator. What part of the solution trajectory corresponds to the fast steps (the slow steps) that we identified in the phase plane? Modify your program so that the nullclines and the limit-cycle solution are superimposed in the same phase plane, and then label the part of the limit cycle that is traversed fast "quick" and the part of the limit cycle that is traversed slowly "slow."

12. Repeat Exercise 11 using the Fitzhugh–Nagumo equation.

Chapter 7
Frequency Domain I: Bode Plots and Transfer Functions

Up to now, we have focused on descriptions of the dynamics of biological systems in the *time domain*, that is, on how variables change as a function of time. However, for certain types of problems, it is often more convenient to analyze dynamics in the *frequency domain*. For example, we typically describe the heartbeat in terms of frequency, namely, the number of beats per minute, rather than in terms of its period, i.e., the time between successive beats. In this chapter, we focus on the response of linear dynamical systems to sinusoidal inputs in the frequency domain. Sinusoidal inputs are of interest since they are easy to implement in the laboratory. The responses provide an estimate of the linearity of the dynamical system, and they can be used to approximate the differential equation that is consistent with the observed input–output relationships.

A remarkable property of linear dynamical systems is that a sinusoidal input produces a sinusoidal output. Indeed, such an observation is sufficient to establish that the system is behaving like a linear dynamical system. Techniques based on this observation represent the frequency-domain complement to time-domain techniques based on Tool 1. There is a long history concerning the study of the response of biological systems to sinusoidal stimulation [696, 793]. A surprising observation is that a biological system composed of nonlinear elements, such as a population of neurons or a complex metabolic network, can act collectively as a linear dynamical system [615] (see also Section 15.2.1).

There are two additional reasons why biologists should familiarize themselves with the analysis of dynamical systems in the frequency domain. First, the tools are available in computer software packages such as MATLAB, Octave, and Python. These computer packages simplify the processing of experimental data. Second, frequency-domain approaches form the language that engineers use to understand how dynamical systems are controlled. Issues concerning the regulation of biological systems are of major concern. Examples include the regulation of cellular and genetic networks, neural populations, movements of multicellular organisms, animal populations, and climate. The experience gained by engineers in understanding how man-made machines are controlled provides a strong foundation for the investigation of how biological systems are controlled. This engineering perspective

© Springer Nature Switzerland AG 2021
J. Milton and T. Ohira, *Mathematics as a Laboratory Tool*,
https://doi.org/10.1007/978-3-030-69579-8_7

becomes particularly useful when the experimentalist's goal is to construct devices to control biological systems such as the artificial pancreas and implantable devices to prevent life-threatening cardiac arrhythmia and abort epileptic seizures (see Chapter 9).

This chapter is organized as follows. First we introduce the concept of a *low-pass filter*. Biological control mechanisms often act as low-pass filters. The identity of these filters can be determined using experimental measurements together with a graphical technique referred to as *Bode plot analysis*. Bode plot analysis is being used increasingly by biologists to investigate the properties of complex cellular [265], metabolic [21, 773], neural [793], physiological [56, 512, 559, 723], and genetic [471, 767] networks. We describe two mathematical tools, Euler's formula for complex numbers (Tool 8) and the Laplace integral transform (Tool 9), which are very useful for describing the responses of linear dynamical systems to sinusoidal inputs. Next, we introduce the *transfer function* and show how it can be evaluated by measuring the amplitude and phase of the response of a linear dynamical system to sinusoidal inputs using Bode plot analysis. The implication is that the differential equations that describe the input–output relationships for linear dynamical systems can be deduced experimentally.

The questions we will answer in this chapter are

1. What is the criterion for the linearity of a dynamical system in the frequency domain?
2. What is the Laplace transform, and why is it useful in the laboratory?
3. What are the limitations on the use of the Laplace transform in the laboratory?
4. What is the transfer function?
5. What is a Bode plot, and why it is important?
6. What is the effect of a time delay on the phase of the frequency response of a linear dynamical system that receives a sinusoidal input?

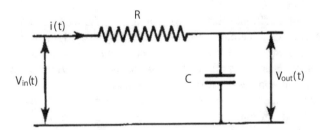

Fig. 7.1 An RC circuit with resistance R, capacitance C, current i, and voltage V. This configuration of the RC circuit corresponds to a low-pass filter.

7.1 Low-Pass Filters

The concept of a low-pass filter is illustrated by considering the behavior of an RC (resistor and capacitor) circuit (Figure 7.1). RC circuits arise in the description of the electrical properties of membrane ion channels, which form the basis of the Hodgkin–Huxley equation for the neuron (Section 11.1). Moreover, the RC circuit is a convenient and inexpensive way to construct a low-pass filter, a circuit that removes frequencies from a signal above a certain cutoff (see below). The low-pass filtering of a continuous, or analog, signal prior to digitization using an A/D (analog-to-digital) board is an essential step in time series analysis (Section 8.4). The differential equation that describes the dynamics of this circuit is

$$i(t) = \frac{V_i(t) - V_o(t)}{R} = C\frac{dV_o(t)}{dt},$$

or

$$RC\frac{dV_o(t)}{dt} + V_o(t) = V_i(t), \tag{7.1}$$

where R is the resistance in ohms, C is the capacitance in farads, i is the current in amperes, and V_i, V_o are, respectively, the input and output voltages in volts.

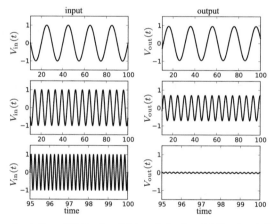

Fig. 7.2 Computer simulation of (7.1) when $V_i(t) = \sin 2\pi ft$ and $RC = 1 \text{ s}^{-1}$. From top to bottom, the input frequencies are 0.05 s^{-1}, 0.159 s^{-1}, and 5.0 s^{-1}. The left-hand column shows the input sine wave, and the right-hand column shows the RC circuit output.

Figure 7.2 illustrates the low-pass filtering properties of the RC circuit. Low-frequency sinusoidal inputs (top row) pass through the RC circuit with little or no attenuation of the amplitude. As the frequency of the input sine wave increases (middle and bottom rows), the amplitude of the output sine wave is attenuated: the larger the frequency, the greater the attenuation. However, on closer inspection, we can see that there is another difference between the input and output. The maximum and minimum of the sinusoids don't occur at the same time. In other words, the output sinusoid has been *phase-shifted* with respect to the input.

The concept of phase shift ϕ is illustrated in Figure 7.3. This figure compares the two functions $\sin(2\pi t - \pi/2)$ (dashed curve) and $\sin(2\pi t)$ (solid curve). Both of these functions have the same amplitude, shape, and frequency (period). The difference between them is that the dashed curve is shifted to the right in comparison to the solid curve. We say that the *phase shift* is $\pi/2$. Note that $\pi/2$ radians is one-fourth of a full cycle of 2π radians, and so the function is shifted by one-fourth of the period (which is equal to 1), or 0.25.

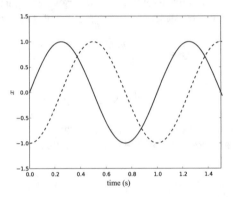

Fig. 7.3 The solid curve is a plot of $\sin 2\pi t$ as a function of time t, while the dashed curve is a plot of $\sin(2\pi t - \phi)$, where $\phi = \pi/2$.

We can predict the behaviors shown in Figure 7.2 by solving (7.1) using the convolution integral we developed in Section 6.4. Since the RC circuit is described by a linear first-order differential equation, the impulse function is

$$I_{RC}(t) = ke^{-kt}, \tag{7.2}$$

where $k = (RC)^{-1}$, and we have chosen $t_0 := t(0) = 0$. Using the convolution theorem, we have

$$V_o(t) = \int_0^t I_{RC}(t-u)V_i(u)\,du. \tag{7.3}$$

When

$$V_i(t) = \sin 2\pi ft,$$

(7.3) becomes

$$V_o(t) = \int_0^t ke^{-k(t-u)} \sin(2\pi fu)\,du,$$

whose solution is[1]

---

[1] In order to obtain this solution, it is necessary to evaluate an integral of the form

$$V_o(t) = k\frac{2\pi f}{k^2 + (2\pi f)^2}e^{-kt} + \frac{k^2\sin(2\pi ft) - 2k\pi f\cos(2\pi ft)}{k^2 + (2\pi f)^2}. \tag{7.4}$$

At steady state, namely as t approaches $+\infty$, (7.4) becomes

$$V_o(t) = \frac{k[k\sin(2\pi ft) - 2\pi f\cos(2\pi ft)]}{k^2 + (2\pi f)^2}.$$

This equation can be further simplified by noting that

$$k\sin(2\pi ft) - 2\pi f\cos(2\pi ft) = r\sin(2\pi ft - \phi),$$

where

$$r = \sqrt{k^2 + (2\pi f)^2}, \quad \phi = \tan^{-1}\frac{2\pi f}{k},$$

to obtain

$$V_o(t) = \frac{k}{r}\sin(2\pi ft + \phi), \tag{7.5}$$

where ϕ is a phase shift. This result tells us that the output of the RC circuit that receives a sinusoidal input is also a sinusoidal signal with the same frequency. The key point is that the amplitude kr^{-1} of the output is different from that of the input and the output sinusoid is phase-shifted from the input by an amount equal to ϕ.

Figure 7.4a shows the amplitude of $V_0(t)$ as a function of frequency. The shape of this amplitude plot can be readily appreciated by noting that kr^{-1} resembles the Hill-type function that we introduced in Section 3.2.1. Thus

$$kr^{-1} = \begin{cases} 1 & \text{when } k^2 \gg 4\pi^2 f^2, \\ \frac{1}{\sqrt{2}} & \text{when } k^2 = 4\pi^2 f^2, \\ 0 & \text{when } 4\pi^2 f^2 \gg k^2. \end{cases}$$

Figure 7.4b shows the frequency dependence of ϕ. The negative phase shift means that the output sine wave lags behind the input sine wave: the higher the input frequency, the greater the lag.

An RC circuit can also be configured to act as a *high-pass filter*, namely a filter that passes through only frequencies that are higher than a preset value (see Exercise 8). In addition, it is possible to combine a low-pass with a high-pass filter to construct a *band-pass filter*, a filter that allows frequencies to pass within a specified frequency band.

A neuron acts as a band-pass filter. The high-pass component arises because the frequency of an input spike train must be high enough that the membrane

$$\int e^{ax}\sin(bx)\,dx.$$

For those who are not adept at integration, the answer can either be looked up on the Internet or obtained from a symbolic manipulation software package, such as Mathematica, Sage, and WolframAlpha.

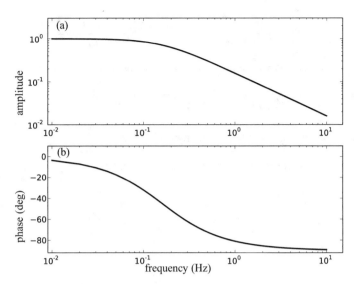

Fig. 7.4 Plot of the (a) amplitude and (b) phase of an RC circuit as a function of the frequency of the input sine wave. Note that (a) is a loglog plot, whereas (b) is a semilog plot.

potential exceeds the spiking threshold. The low-pass component arises because neuron refractoriness limits the neuron output spiking frequency to ≈ 700–800 Hz.

From a mathematical point of view, all differential equations in some way filter a sinusoidal input. Since it is impossible to construct an ideal filter [73], compromises must be made based on the nature of the data at hand and the questions to be answered. For example, the RC circuit causes distortion of the signal through its effects on phase near $f = k/2\pi$. Consequently, a variety of filters have been developed that differ with respect to their effects on the phase response and the steepness of the cutoff in the amplitude response. Examples include Bessel filters, Butterworth filters, Chebyshev filters, and elliptic filters.[2] It should be noted that an RC circuit corresponds to a Butterworth filter of order 1. For a practical discussion of the pros and cons of various filters, see [846].

Figure 7.4a and b is collectively referred to as a *Bode plot*.[3] Is it possible to work backward from the observations in Figure 7.4 to obtain (7.1)? The answer is yes. The importance of this observation is that it implies that we may be able to deduce a mathematical model describing a linear biological system directly from experimental observations. However, in order to be able to do this easily and efficiently, we need mathematical tools appropriate for studying dynamical systems in the frequency domain.

---

[2] Stephen Butterworth (1885–1958), British physicist; Pafnuty Lvovich Chebyshev (1821–1894), Russian mathematician; Friedrich Wilhelm Bessel (1784–1846), German mathematician and astronomer who systematized the functions discovered by Daniel Bernoulli (1700–1782), Swiss mathematician and physicist.

[3] Hendrik Wade Bode (1905–1982), American engineer.

7.2 Laplace Transform Toolbox

Four tools are useful for studying dynamical systems in the frequency domain: Euler's formula for complex numbers (Tool 8), the Laplace integral transform (Tool 9), the Fourier series (Tool 10), and the Fourier integral transform (Tool 11). Here we describe Tools 8 and 9; Tools 10 and 11 are discussed in the following chapter.

7.2.1 Tool 8: Euler's Formula

Recall from Section 4.3.2 that a complex number z can be thought of as the sum of a real part x, denoted by $x = \Re(z)$, and an imaginary part y, denoted by $y = \Im(z)$, multiplied by j, the imaginary unit. That is, we can write

$$z = x + jy,$$

where

$$j = \sqrt{-1}.$$

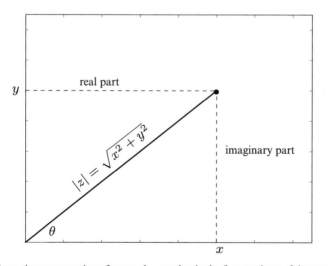

Fig. 7.5 Schematic representation of a complex number in the first quadrant of the complex plane.

It is helpful to think of complex numbers geometrically. This is accomplished by plotting them in the complex plane (Figure 7.5). From this perspective, a complex number can be represented as a point in the plane with a horizontal real axis and vertical complex axis. For our purposes, it will generally be convenient to graph a complex number in polar coordinates, whereby we represent the complex number in terms of its *argument* θ and modulus $|z|$. The argument is the angle taken counterclockwise from the positive real axis to the ray from the origin to z, and the modulus is the absolute value of z, denoted by $|z|$, which is the distance from z to the origin. From Figure 7.5, it can be seen that

$$|z| = \sqrt{x^2 + y^2}.$$

The most important property of the modulus $|z|$ is that it is a nonnegative real number. The *complex conjugate* of $z = x + jy$, denoted by z^*, is by definition equal to $x - jy$. We immediately see that

$$zz^* = (x + jy)(x - jy) = x^2 + y^2 = |z|^2. \tag{7.6}$$

The argument θ can be determined as

$$\tan \theta = \frac{y}{x},$$

or equivalently,

$$\theta = \tan^{-1} \frac{y}{x}.$$

If we set $r = |z|$, then we see from Figure 7.5 that

$$x = r \cos \theta, \quad y = r \sin \theta,$$

and therefore,

$$z = r(\cos \theta + j \sin \theta). \tag{7.7}$$

This relationship is often referred to as the *polar form* of a complex number.

A very useful way to express a complex number is through *Euler's formula*,

$$e^{j\theta} = \cos \theta + j \sin \theta. \tag{7.8}$$

It makes the association between complex numbers and sinusoidal oscillations transparent. It is not difficult to derive this relationship: The series representation of e^t is

$$e^t = 1 + t + \frac{t^2}{2!} + \frac{t^3}{3!} + \cdots + \frac{t^n}{n!}.$$

Let $t = j\theta$ and recall that $j^2 = -1$. We then obtain

$$e^{j\theta} = 1 + j\theta + \frac{(j\theta)^2}{2!} + \frac{(j\theta)^3}{3!} + \cdots + \frac{(j\theta)^n}{n!}$$

$$= 1 + j\theta - \frac{\theta^2}{2!} - \frac{j\theta^3}{3!} + \frac{\theta^4}{4!} + \frac{j\theta^5}{5!} + \cdots. \tag{7.9}$$

Since the series expansions of $\cos \theta$ and $\sin \theta$ are

$$\cos\theta = 1 - \frac{\theta^2}{2!} + \frac{\theta^4}{4!} - \frac{\theta^6}{6!} + \cdots \quad \text{and} \quad \sin\theta = \theta - \frac{\theta^3}{3!} + \frac{\theta^5}{5!} - \cdots,$$

we see that (7.9) can be rearranged to yield (7.8).

There are three useful results that we can obtain using Euler's formula. First, the polar form of a complex number can be rewritten as

$$z = x + jy = |z|\left[\cos\theta + j\sin\theta\right] = |z|e^{j\theta}. \tag{7.10}$$

Second, it is easy to see that for all θ, the point $e^{j\theta}$ in the complex plane lies on the unit circle (that is, the circle of radius 1), since

$$\left|e^{j\theta}\right| = \sqrt{\cos^2\theta + \sin^2\theta} = 1. \tag{7.11}$$

Finally, we can write $\cos\theta$ and $\sin\theta$ as

$$\cos\theta = \frac{1}{2}\left(e^{j\theta} + e^{-j\theta}\right), \quad \sin\theta = \frac{1}{2j}\left(e^{j\theta} - e^{-j\theta}\right). \tag{7.12}$$

Example 7.1 We can use Euler's formula to write the solution of

$$\frac{d^2x}{dt^2} + k_1\frac{dx}{dt} + k_2x = 0$$

when the λ are a pair of complex numbers $\lambda_{1,2} = \Re(\lambda) \pm j\Im(\lambda)$ in terms of real quantities, namely

$$x(t) = Xe^{\Re(\lambda)t}\sin(\Im(\lambda)t + \phi), \tag{7.13}$$

where ϕ is the phase shift (Section 4.3.2). To see how this is done, we first note that the general solution

$$x(t) = c_1e^{\lambda_1 t} + c_2e^{\lambda_2 t},$$

where c_1, c_2 are two constants to be determined using the two initial conditions, can be rewritten using Euler's formula as

$$x(t) = e^{\Re(\lambda)t}\left[(c_1 + c_2)\cos\omega t + j(c_1 - c_2)\sin\omega t\right]. \tag{7.14}$$

It is possible that c_1 and c_2 may be complex numbers. However, if they were complex numbers, that would mean that there would be four constants to determine from only two equations. Since this is impossible, it must be true that not all of these constants are independent. A convenient way to reduce the number of constants to the required two is to make c_1 and c_2 complex conjugates, say $c_{1,2} = \alpha \pm j\beta$. Hence, we can write

$$A = c_1 + c_2 = \alpha + j\beta + \alpha - j\beta = 2\alpha,$$
$$B = j(c_1 - c_2) = j(-2j\beta) = 2\beta,$$

and (7.14) becomes

$$x(t) = e^{\Re(\lambda)t}(A\cos\Im(\lambda)t + B\sin\Im(\lambda)t) \,. \tag{7.15}$$

Equation (7.13) can be obtained in three steps:

1. Define $X = \sqrt{A^2 + B^2}$.
2. Multiply both sides of (7.15) by X/X to obtain

$$x(t) = Xe^{\Re(\lambda)t}\left[\frac{A}{X}\cos\Im(\lambda)t + \frac{B}{X}\sin\Im(\lambda)t\right],$$

and define ϕ by the relations

$$\sin\phi := \frac{A}{X}, \quad \cos\phi := \frac{B}{X}, \quad \tan\phi := \frac{A}{B} \,.$$

3. Make use of the trigonometric relationship $\sin(x+y) = \sin x\cos y + \cos x\sin y$.

$\diamondsuit$

7.2.2 Tool 9: The Laplace Transform

The Laplace[4] integral transform $F(s)$, or Laplace transform for short, of a function of time $f(t)$ is defined as

$$F(s) = \mathscr{L}[f(t)] = \int_0^\infty f(t)e^{-st}dt \,, \tag{7.16}$$

where $s = \gamma + j\omega$ is a complex variable with real part γ and angular frequency $\omega = 2\pi f$, in units of radians per second. Our goal in this section is not to give a rigorous discussion of this transform, but merely to provide enough background to enable the reader to understand the Bode plot and its interpretation.

The Laplace transform is thus an integration process applied to $f(t)$. The transform $F(s)$ is finite even when $f(t)$ does not tend to zero. The only requirements are that $f(t)$ be defined for all $t > 0$ and that s be sufficiently large to ensure that the integral converges. In our studies, we will restrict our attention to real dynamical systems, that is, systems in which the variables exist and can be measured using equipment available in the laboratory. The Laplace transform exists for all such functions, even if the function is only piecewise continuous. However, this definition of the Laplace integral is insufficient to include the delta function and the step function. One solution that makes it possible to compute the Laplace transform of these functions it to require that $f(t) = 0$ for $t < 0$ (see also Section 6.3). Alternatively, we can rewrite (7.16) as

$$F(s) = \mathscr{L}[f(t)] = \lim_{\substack{T\to\infty\\\varepsilon\to 0}}\int_\varepsilon^T f(t)e^{-st}\,dt \,.$$

---

[4] Pierre-Simon, Marquis de Laplace (1749–1827), French mathematician and astronomer.

The following two examples illustrate how the Laplace integral transform $\mathscr{L}(s)$ is calculated for different choices of $f(t)$.

Example 7.2 What is the Laplace transform of $f(t) = 1$?

Solution. From (7.16), we have

$$\mathscr{L}(1) = \int_0^\infty e^{-st} dt = \frac{1}{s}.$$

◇

Example 7.3 What is the Laplace transform of $f(t) = e^{at}$?

Solution. From (7.16), we have

$$\mathscr{L}(e^{at}) = \int_0^\infty e^{-(s-a)t} dt = \frac{1}{s-a}.$$

◇

Most laboratory researchers do not compute the Laplace transform but simply look it up in a table on a need-to-know basis. Some examples are given in Table 7.1. The last row of the table shows why those of us who do not enjoy performing integration will love the Laplace transform for doing convolution. We see that

$$\mathscr{L}\left(\int_0^t f(t-u)g(u)du \right) = F(s)G(s). \tag{7.17}$$

In other words, we can evaluate a convolution integral as the product of two Laplace transforms! This relationship is also true for the Fourier integral transform (see the following chapter).

The Laplace transform of a derivative is equal to

$$\mathscr{L}\left[\frac{d^n f(t)}{dt^n} \right] = s^n \mathscr{L}[f(t)] - s^{n-1}\left(\frac{df}{dt} \right)_0 - \cdots - s\left(\frac{d^{n-2}f}{dt^{n-2}} \right)_0 - \left(\frac{d^{n-1}f}{dt^{n-1}} \right)_0,$$

where $d^0 f/dt^0 = f(t)$. An alternative way of writing this relationship is given in Table 7.1. It should be noted that the Laplace transform of the nth-order derivative of a function includes information related to the initial value of the function as well as the initial values of all derivatives up to the $(n-1)$st.

In the discussions that follow, we are most interested in applications to first- and second-order ordinary differential equations. Thus, it is helpful to know the Laplace transforms for the first and second derivatives of a function. When $n = 1$, we have

$$\mathscr{L}\left[\frac{df(t)}{dt} \right] = s\mathscr{L}[f(t)] - f(0), \tag{7.18}$$

where $f(0)$ is the function evaluated at $t = 0$. Similarly, when $n = 2$, we have

$$\mathscr{L}\left[\frac{d^2f(t)}{dt^2}\right] = s^2\mathscr{L}[f(t)] - sf(0) - \left(\frac{df}{dt}\right)_0, \qquad (7.19)$$

where the notation $\left(\frac{df}{dt}\right)_0$ means that the derivative is evaluated at $t = 0$

Table 7.1 Some useful Laplace transforms. More extensive tables are available at various Internet sites.

$f(t) = \mathscr{L}^{-1}\{F(s)\}$	$F(s) = \mathscr{L}\{f(t)\}$
1	$\frac{1}{s}, \quad \Re(s) > 0$
e^{kt}	$\frac{1}{s-k}, \quad \Re(s) > k$
t^n, n a positive integer	$\frac{n!}{s^{n+1}}, \quad \Re(s) > 0$
t^p, $p > -1$	$\frac{\Gamma(p+1)}{s^{p+1}}, \quad \Re(s) > 0$
$t^n e^{kt}$, n a positive integer	$\frac{n!}{(s-k)^{n+1}}, \quad \Re(s) > k$
$\sin 2\pi f t$	$\frac{2\pi f}{s^2 + (2\pi f)^2}, \quad \Re(s) > 0$
$\cos 2\pi f t$	$\frac{s}{s^2 + (2\pi f)^2}, \quad \Re(s) > 0$
$e^{kt}\sin 2\pi f t$	$\frac{2\pi f}{(s-k)^2 + (2\pi f)^2}, \quad \Re(s) > k$
$e^{kt}\cos 2\pi f t$	$\frac{s-k}{(s-k)^2 + (2\pi f)^2}, \quad \Re(s) > k$
$H(t_0)$	$\frac{e^{-t_0 s}}{s}, \quad \Re(s) > 0$
$e^{t_0 t} f(t)$	$F(s - t_0)$
$\delta(t - t_0)$	$e^{-t_0 s}$
$\frac{d^n f(t)}{dt^n}$	$s^n \mathscr{L}[f(t)] - \sum_{k=1}^{n} s^{k-1}\left(\frac{d^{n-k}f(t)}{dt^{n-k}}\right)_0$
$\int_0^t f(u)du$	$\frac{F(s)}{s}$
$\int_0^t f(t-u)g(u)gu$	$F(s)G(s)$

Laplace transforms can be used to solve differential equations. To do so, we perform the following steps:

1. Transform the equation into the Laplace domain by changing the variable from time to a new complex variable s.

2. Solve the equation in the Laplace domain by making simple algebraic manipulations, yielding a solution in the s domain.
3. Invert the transform of the solution from the s domain to the time domain using tables of Laplace transforms and their inverses, such as Table 7.1.

Example 7.4 Solve

$$\frac{dx}{dt} + kx = 0$$

with $x(0) = 1$ using the Laplace transform.

Solution. Using (7.18), we have

$$s\mathscr{L}(s) - x(0) + k\mathscr{L}(s) = 0 \Rightarrow (s+k)\mathscr{L}(s) = x(0) \Rightarrow \mathscr{L}(s) = \frac{x(0)}{s+k}.$$

From Table 7.1, we see that

$$x(t) = x(0)e^{-kt}.$$

◇

There are two major limitations on the use of the Laplace transform in the laboratory setting [201]. First, it is typically difficult to obtain the inverse transform, and second, there are presently no efficient numerical algorithms to evaluate the Laplace transform directly from experimental data. As we will see, the importance of this transform for laboratory researchers is related not so much to solving differential equations as to interpreting the response of linear dynamical systems to sinusoidal inputs using the Bode plot.

7.3 Transfer Functions

Consider the first-order differential equation

$$k^{-1}\frac{dx}{dt} + x = b(t), \tag{7.20}$$

where $b(t)$ is an input and $x(0) = x_0$ is the initial condition. Using (7.18), we can rewrite this equation in terms of the Laplace transform as

$$k^{-1}s\mathscr{L}(x(s)) + \mathscr{L}(x(s)) - x(0) = \mathscr{L}(b(s)),$$

or

$$\mathscr{L}(x(s))[1 + k^{-1}s] - x(0) = \mathscr{L}(b(s)). \tag{7.21}$$

Let us assume that $x(0) = 0$. This assumption will be valid if the dynamical system is stable and has been at rest for a sufficiently long time before the sinusoidal input is applied. These conditions can be readily satisfied at the benchtop.

Fig. 7.6 Schematic representation of the black-box analogy to experimental science in the s domain. The Laplace transforms of the input, output, and the transfer function are, respectively, B(s), X(s), and C(s).

It is useful to connect (7.21) with the black-box analogy shown schematically in Figure 7.6: $\mathcal{L}(x(s))$ is the output, $\mathcal{L}(b(s))$ is the input, and $1 + k^{-1}s$ describes the black box. With this in mind, we can rewrite (7.21) as

$$X(s) = C(s)B(s), \tag{7.22}$$

where

$$X(s) := \mathcal{L}(x(s)), \quad C(s) := \mathcal{L}(c(s)), \quad B(s) := \mathcal{L}(b(s)).$$

Note that the right-hand side of (7.22) is the Laplace transform of a convolution integral. The quantity $C(s)$ is the *transfer function*,

$$C(s) = \mathcal{L}(c(s)) := \frac{\mathcal{L}(x(s))}{\mathcal{L}(b(s))} = \frac{\mathcal{L}(\text{output})}{\mathcal{L}(\text{input})} = \frac{1}{1 + s/k}.$$

The transfer function is always equal to the Laplace (Fourier) transform of the output divided by the Laplace (Fourier) transform of the input. Thus, the transfer function relates a change in the input to a change in the output when all of the initial conditions are equal to zero. When the differential equation that describes the dynamical system is known, it is straightforward to determine its transfer function.

Example 7.5 Suppose the input–output relations of a black box are described by the differential equation

$$\frac{d^2x}{dt^2} + k_2\frac{dx}{dt} + k_1x = k_2(t) + \frac{db}{dt}.$$

What is the transfer function?

Solution. The Laplace transform of this equation is

$$s^2X(s) + k_2sX(s) + k_1X(s) = B(s) + sB(s),$$

where we have assumed that all terms related to the initial conditions are equal to zero. We can rewrite this equation as

$$X(s) = \frac{1+s}{s^2 + k_2s + k_1}B(s),$$

and hence the transfer function is

$$C(s) = \frac{X(s)}{B(s)} = \frac{1+s}{s^2 + k_2 s + k_1} .$$

◊

The transfer function has a number[5] of useful properties.

1. The denominator of the transfer function set to zero is the characteristic equation of the corresponding homogeneous differential equation (the differential equation when $b(t) = 0$). The reason we denoted the transfer function by $C(s)$ was to remind the reader of this fact.

Example 7.6 An investigator has suggested that the transfer function for a black box is

$$C(s) = \frac{s+16}{s^2 - 3s + 1} .$$

Does this transfer function represent a stable or an unstable dynamical system?

Solution. The characteristic equation is

$$\lambda^2 - 3\lambda + 1 = 0 .$$

Thus the dynamical system is unstable. In fact, the fixed point is an unstable node (see Table 4.1). ◊

2. The transfer function of a system is the Laplace transform of its impulse function. This result can readily be understood by using the convolution theorem to solve (7.20) and then taking the Laplace transform to obtain

$$X(s) = B(s)I(s) ,$$

where $I(s)$ is the Laplace transform of the impulse function. Since the solution of this equation must be the same as that of (7.22), it follows that $C(s) = I(s)$.

Example 7.7 It is observed that the impulse function for a black box is ke^{-kt}. What is the transfer function?

Solution. When $B(s)$ is a delta function, we have from (7.22) that

$$C(s) = X(s),$$

and hence

$$C(s) = \mathscr{L}(ke^{-kt}) = \frac{k}{s+k} .$$

◊

---

[5] Strictly speaking, we should take the inverse Laplace transform to recover the homogeneous differential equation and then use Tool 4 to obtain the characteristic equation. However, it is much easier to make the substitution $s \mapsto \lambda$.

3. If $C(s)$ is known, then we can determine the response to all possible inputs, since

$$C(s)B(s) = \frac{X(s)}{B(s)} \cdot B(s) = X(s).$$

For example, the *impulse response function* is the response to a delta-function input. Since we know that $\mathscr{L}(\delta(s)) = 1$ (Table 7.1), we see that the impulse response function will be the inverse Laplace transform of $C(s)$. Similarly, the frequency response corresponds to the solution obtained when $b(t) = \sin \omega t$.

4. In typical laboratory applications, the transfer function can be written as the ratio of two polynomials [748]:

$$C(s) = \frac{P(s)}{Q(s)} = \frac{k(s-z_1)(s-z_2)\cdots(s-z_m)}{(s-p_1)(s-p_2)\cdots(s-p_n)}. \tag{7.23}$$

The condition $n > m$ ensures that the system is physically realizable (see also Section 6.3). The roots z_i are referred to as the *zeros*, and the roots p_i are the poles. This relation follows from the observations that the dynamical system is linear and the Laplace transforms of the inputs that most commonly arise in laboratory settings can be expressed in terms of polynomials (see Table 7.1).

We can use (7.23) to generalize the results we obtained for the RC circuit to arbitrary dynamical systems that are linear and stable. Under these conditions,

$$X(s) = B(s)C(s) = \frac{k\omega(s-z_1)(s-z_2)\cdots(s-z_m)}{(s^2+\omega^2)(s-p_1)(s-p_2)\cdots(s-p_n)}. \tag{7.24}$$

Since the numerator and denominator of (7.24) are both polynomials, we can rewrite this equation, at least in principle, using a partial fraction expansion to get

$$X(s) = \frac{A_1}{s-j\omega} + \frac{A_2}{s+j\omega} + \frac{D_1}{s-p_1} + \frac{D_2}{s-p_2} + \cdots + \frac{D_n}{s-p_n}, \tag{7.25}$$

where A_1, A_2 and $D_1, D_2, \ldots, D_n$ are constants. We do not need to determine the values of the D_i. In order to understand why this is possible, take the inverse transform to obtain

$$x(t) = A_1 e^{j\omega t} + A_2 e^{-j\omega t} + D_1 e^{p_1 t} + D_2 e^{p_2 t} + \cdots + D_n e^{p_n t}. \tag{7.26}$$

However, we have assumed that the system is stable. This means, by definition, that the real parts of all of the p's must be negative. Since

$$\lim_{t\to\infty} e^{\Re(p)t} = 0$$

when $\Re(p) < 0$, we see that the steady-state solution is simply

$$\lim_{t\to\infty} x(t) = A_1 e^{j\omega t} + A_2 e^{-j\omega t}. \tag{7.27}$$

Thus,

$$\frac{C(s)\omega}{s^2 + \omega^2} = \frac{A_1}{s - j\omega} + \frac{A_2}{s + j\omega}, \tag{7.28}$$

where we have taken the Laplace transform of the right-hand side of (7.27).

To determine A_1, we multiply both sides of (7.28) by $(s - j\omega)$ to obtain

$$\frac{C(s)(s - j\omega)\omega}{(s - j\omega)(s + j\omega)} = A_1 + \frac{A_2}{s + j\omega}(s - j\omega),$$

or

$$\frac{C(s)\omega}{s + j\omega} = A_1 + \frac{A_2}{s + j\omega}(s - j\omega).$$

We can evaluate A_1 by setting $(s - j\omega) = 0$, or equivalently, $s = j\omega$, so that

$$A_1 = \frac{1}{2j}C(j\omega). \tag{7.29}$$

Using Tool 8, we can rewrite (7.29) as

$$A_1 = \frac{1}{2j}|C(j\omega)|e^{j\theta}. \tag{7.30}$$

In a similar way, we can obtain A_2 by multiplying both sides of (7.28) by $(s + j\omega)$ and setting $s = -j\omega$:

$$A_2 = -\frac{1}{2j}|C(j\omega)|e^{-j\theta}. \tag{7.31}$$

In this way, we obtain

$$\lim_{t \to \infty} x(t) = |C(j\omega)|\frac{1}{2j}\left[e^{j\omega t + j\theta} - e^{-j\omega t - j\theta}\right]. \tag{7.32}$$

Using Tool 8, we obtain using (7.12) and a few trigonometric manipulations,

$$\lim_{t \to \infty} x(t) = |C(j\omega)|\sin(\omega t + \angle C(j\omega)), \tag{7.33}$$

where the symbol $\angle$ means that we have written the phase in degrees. The reader should note that the engineering convention for the analysis of Bode plots is to measure phase in degrees rather than radians.

Equation (7.33) is of the same form as the result we obtained in Section 7.1. Thus, we have shown that for all dynamical systems that are stable and linear, the output from a sinusoidal input is also a sinusoid with the same frequency, but possibly having a different amplitude and a shifted phase.

The practical importance of these observations is that we can use responses to sinusoidal inputs to assess whether a dynamical system is behaving linearly. Often, the frequency of a dynamical system depends on its linear properties, whereas the shape of the waveform depends on its nonlinear properties [792]. Thus, if a sinusoidal input produces a sinusoidal output, then we can infer, at least to a first approximation, that the feedback system is operating in a linear range. Of course, this conclusion is valid only over the range of amplitudes and frequencies used to test for linearity. However, it is usually possible using a signal generator to generate sine wave inputs whose amplitudes and frequencies span the normal physiological range.

7.4 Biological Filters

Many biological dynamical systems act as low-pass filters. The reader can readily demonstrate the low-pass filtering properties of their own musculoskeletal system using a whole-body vibrating platform, an apparatus available in many college and university fitness centers.[6] Whole-body vibration is used for warming up prior to strenuous exercise and to enhance recovery from overexercise and injury [540]. The peak-to-peak amplitude of the vibration is in the range of $\approx$ 3–5 mm measured at the soles of the feet, and the vibration frequencies are typically in the range of 15–50 Hz. In the terminology of this chapter, the vibrating platform produces a sinusoidal vibratory input to the body, and the output is the vibrations recorded from the body. It is very uncomfortable to stand on a vibrating platform with legs straight; however, this discomfort is greatly minimized by bending the legs slightly. You can assess the effect of vibration on your body by fixing your gaze on some object in your range of view: the apparent movement of this object is greatly reduced when your knees are slightly flexed. This observation means that the amplitude at which your body is vibrating up and down is less when your knees are flexed than when your legs are straight, even though the vibrating platform's vibrational amplitude remains constant. For example, measurements of the movements of the index finger using high-speed motion-capture cameras indicate a peak-to-peak amplitude of 0.1 mm when the platform vibrates with a peak-to-peak amplitude of 3 mm at a frequency of 50 Hz [595], which amounts to a 10-fold reduction in vibrational amplitude.

The observed reduction in vibrational amplitude of the body through flexing the knees arises because in this situation, the musculoskeletal system functions as a powerful low-pass filter with respect to frequencies greater than about 10 Hz. The observation that frequencies lower than about 2–3 Hz are not filtered out by the musculoskeletal system has an impact on the design of seats in moving vehicles such as cars, jeeps, trucks, and tanks [624]. This is because low-frequency vibration is harmful to people and can cause, for example, motion sickness.

In order to characterize the low-pass filtering properties of biological systems quantitatively, it is necessary to obtain the Bode plot. Here, we give two examples (other examples are discussed throughout this chapter, including the exercises, and in the following chapter).

---

[6] Some whole-body vibrating platforms do not produce vertical vibrations and hence would not be suitable for this demonstration [595].

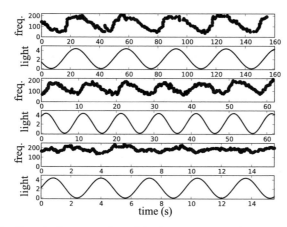

Fig. 7.7 Neural spike train frequency of crayfish photoreceptor as a function of the frequency of sinusoidally modulated light. The astute reader will observe that there is some harmonic distortion in the waveform for the neural spiking rate; that is, the waveform is not perfectly sinusoidal. The reason for this distortion is not known. Stark measured the power spectrum (see the following chapter) for the fluctuations in neural spike frequency. Since the power is overwhelmingly in the frequency band of the sinusoidally varying light intensity, he felt that a linear approximation was justified. However, it is perhaps more correct to think of the crayfish photoreceptor as an *almost linear dynamical system*. Figure redrawn using data from.[793] with permission

7.4.1 Crayfish Photoreceptor

Lawrence Stark[7] was an early pioneer in the measurement of transfer functions for biological preparations [793]. For this purpose, he chose to study the terminal, or sixth, abdominal ganglion of the crayfish. The ganglion contains 600–700 neurons [437]. It contains photoreceptors and plays a role in the light-avoidance reflex of the crayfish. Since the output of the photoreceptor does not significantly change with repeated stimulation, it is an ideal system in which to measure the transfer function.

Stark's experimental approach exemplifies the black-box model. Surgical dissection was used to isolate the photoreceptor so that the only input it received was light. Then Stark varied the light input intensity sinusoidally at frequency f. The measured output is the neural spike rate measured using recording electrodes placed on the spinal cord just anterior to the photoreceptor. Figure 7.7 shows the frequency of the nerve impulses recorded from the photoreceptor ganglion for three different frequencies of light modulation. A sinusoidal light input produces a sinusoidal output in neural spike frequency. This observation suggests that despite the fact that the ganglion is composed of many neurons, the collective behavior with respect to its response to light can be approximated by a linear dynamical system.

As the frequency of the light modulation increases, the amplitude of the frequency of the spike-rate fluctuations decreases. In addition, there is a phase lag between the light modulation and spike-rate response that increases as light modulation

---

[7] Lawrence W. Stark (1926–2004), American neurologist.

frequency increases. Figure 7.8 shows the Bode plot for the crayfish receptor obtained by Stark [793]. For these experiments, he defined the *gain* as the amplitude of the output (spike-rate sinusoid measured as spikes per second) divided by the amplitude of the input (light intensity measured as millilumens per square millimeter (mlm · mm^{-2})). Overall, this Bode plot resembles the one that we obtained for an RC circuit (Figure 7.4). Thus, we can say that the isolated crayfish photoreceptor functions like a low-pass filter with respect to light inputs of different frequencies. However, there are some subtle but important differences (see Section 7.6).

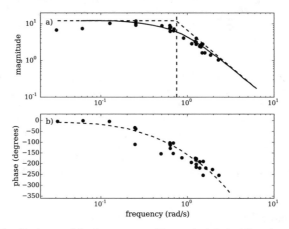

Fig. 7.8 Bode plot for the crayfish photoreceptor. The vertical dashed line shows the corner frequency. Redrawn using data from [793] with permission.

7.4.2 Osmoregulation in Yeast

In response to osmotic stress caused, for example, by drought, heavy rain, or flood, plant cells trigger a signal transduction cascade that involves multiple feedback loops, each of which operates on a different time scale. The purpose of this response is to enable the plant to make adjustments to withstand the osmotic shock. An example is the response of the budding yeast *Saccharomyces cerevisiae* to a hyperosmolar shock [339, 559]. This osmotic stress triggers a signal transduction cascade that terminates in the activation of the kinase Hog1. This activation causes Hog1, initially localized in the cytoplasm, to accumulate in the nucleus. This Hog1 response occurs even if protein synthesis is inhibited by cycloheximide, implying that it does not require transcription [559].

Figure 7.9 shows the Bode plot for the changes in nuclear accumulation of Hog1 in response to periodically varying changes in osmotic stress (for more details concerning this input stimulus, see Section 8.1.1). Again, we see that the overall pattern is consistent with a low-pass filter; however, there are two differences from what we observed for the RC circuit and the crayfish ganglion: the amplitude of the output increases before decreasing, and there are both phase advances and phase delays.

What do the properties of the Bode plots shown in Figures 7.8 and 7.9 indicate? In order to answer this question, we need to understand first how to interpret the Bode plot to obtain the transfer function and then how to relate the transfer function to the underlying control mechanism. We tackle the first task in the remainder of this chapter, and the second in Chapter 9.

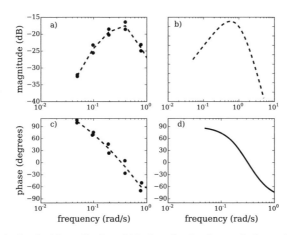

Fig. 7.9 Bode plot for the (a) amplitude and (c) phase for the changes in the nuclear accumulation of Hog1 to periodic step changes in osmolarity (see Section 8.1.1 for more details). Figure drawn using data from [559] with permission. Graphs (b) and (d) show the expected Bode plot for the amplitude and phase, respectively, when the transfer function is given by (7.51).

7.5 Bode Plot Cookbook

The Bode plot can be used to determine the transfer function $C(s)$. Here we list the properties of the Bode plot.

1. The Bode plot utilizes measurements of the response of a linear system to a sinusoidal input in order to construct the transfer function. All measurements are made under steady-state conditions, that is, after transients have died out. Since the input and output of the dynamical system at steady state is a pure sinusoid, we take $s = j\omega$.
2. The transfer function for a linear dynamical system can typically be written as (7.23). Taking the logarithm, we obtain

$$\log(C(s)) = \log K + \log(s - z_1) + \log(s - z_2) + \cdots + \log(s - z_n)$$
$$- \log(s - p_1) - \log(s - p_2) - \cdots - \log(s - p_n). \qquad (7.34)$$

The advantage of taking the logarithm is that $C(s)$ can be built up by adding the contributions related to the individual terms. Making use of (7.10), we see that

$$C(j\omega) = |C(j\omega)|e^{j\angle C(j\omega)},$$

or

$$\log(C(j\omega)) = \log(|C(j\omega)|) + j\angle C(j\omega).$$

Thus, we can treat the amplitude and phase of the output as independent variables.

3. The interpretation of Bode plots for biological systems is easier if we use the same units like those used in engineering applications. We have already pointed out that phase is measured in degrees. The magnitude x of the ratio of power in the output sinusoid and the power in the input sinusoid of a sinusoidal oscillation is measured on a logarithmic scale in *decibels* (dB). Since the power of a time-varying signal is proportional to the square of its amplitude x (see Section 8.3.3), the power ratio P_{dB} expressed in decibels is given by

$$P_{\mathrm{dB}} = 20\log_{10}x.$$

Thus, for example, $x = 0.1$ corresponds to $P_{\mathrm{dB}} = -20$ dB. Finally, frequency is expressed in terms of the angular frequency $\omega = 2\pi f$ and has units of radians per second [$\mathrm{rad \cdot s^{-1}}$]. There is a tendency of experimental biologists to express frequency in Hertz, i.e., cycles per second. The reader should be cautious since this can affect the determination of important parameters from the Bode plot (see Section 7.5.3).

4. The steady-state solutions are sinusoidal with the same frequency as the input. Thus, we plot the amplitude $|C(j\omega)|$ in dB and the phase $\angle C(j\omega)$ in degrees of the output as a function of the angular frequency of the sinusoidal input. These two plots together are referred to as the Bode plot.

5. There are only five possible types of terms: K, s, $(s - r_i)$, and $s^2 + r_i^2$, where r_i is either a pole or a zero, and finally $e^{\tau s}$, which occurs when a time delay τ is present. In order to interpret a Bode plot, we need to understand the contribution of each term.[8]

7.5.1 Constant Term (Gain)

The Bode plot for a constant K is shown in Figure 7.10. From (7.34), when $C(s) = K$ and $s = j\omega$, we have

---

[8] In our experience, it is much easier to learn how to calculate the contributions of each term than to memorize them.

$$C(j\omega) = K.$$

It is convenient to think of K, a real number, as a complex number whose imaginary part is equal to zero (Figure 7.10a). The amplitude, or modulus, of $C(j\omega)$ is K, and the phase, or argument, is 0. Neither the amplitude nor the phase depends on frequency. Thus, the effect of the gain term is merely to shift the overall magnitude of the amplitude plot up or down by a certain number of decibels: $20\log_{10}K$ dB if $K > 0$, and $-20\log_{10}K$ dB if $K < 0$. Although the magnitude of K has no effect on phase, its sign determines whether the phase is $0°$ or $-180°$. The Bode plot for the case $K > 0$ is shown in Figure 7.10b.

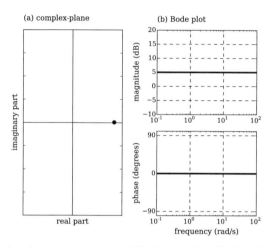

Fig. 7.10 (a) Complex plane representation and (b) Bode plot for $C(j\omega) = K > 0$.

7.5.2 Integral or Derivative Term

Pole at the Origin (Integral Term)

The Bode plot for a simple pole is shown in Figure 7.11. In this case, (7.34) becomes

$$C(j\omega) = \frac{1}{j\omega} = -\frac{j}{\omega}.$$

As before, we plot $C(j\omega)$ in the complex plane (Figure 7.11a). The amplitude is

$$|C(j\omega)| = \sqrt{\Re(C(j\omega))^2 + \Im(C(j\omega))^2} = \sqrt{0^2 + \left(\frac{-1}{\omega}\right)^2} = \frac{1}{\omega},$$

and the phase is

$$\angle C(j\omega) = -90°.$$

This means that the contribution of a simple pole to the amplitude plot is a straight line with slope equal to $-20\log_{10}(\omega)$ dB. In particular, when $\omega = 1$ rad $\cdot$ s^{-1}, the amplitude is zero dB, since $\log_{10} 1 = 0$. The amplitude decreases by 20 dB for a 10-fold increase in ω. The phase $\angle C(j\omega) = -90°$ does not change with frequency (Figure 7.11b). If we have a simple pole of the form $1/s^n$, then the amplitude of the transfer function is $-20n\log_{10}(\omega)$ dB, and $\angle C(j\omega) = -n \cdot 90°$, where n is the order of the integral term.

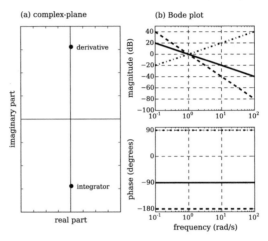

Fig. 7.11 (a) Complex plane representation and (b) Bode plot when $C(j\omega)$ is either a derivative or integral term. The solid line gives the amplitude (top) and phase (bottom) when $C(j\omega) = -j/\omega$, the dashed line when $C(j\omega) = -1/\omega^2$, the dash-dotted line for $C(j\omega) = j\omega$.

A pole at the origin is also referred to as an *integral* term. We can understand this terminology by using the derivative df/dt of a function $f(t)$ as input into a device that performs integration to obtain

$$C(s) = \frac{\text{output}}{\text{input}} = \frac{F(s)}{sF(s)} = \frac{1}{s}.$$

Zero at the Origin (Derivative Term)

The Bode plot for a first-order zero at the origin is shown in Figure 7.11 (dash-dotted line). In this case,

$$C(j\omega) = j\omega.$$

Repeating the above arguments, we find that the amplitude is $|C(j\omega)| = \omega$, and the phase is either a $-270°$ lag or a $+90°$ lead. In other words, a zero at the origin contributes a straight line of slope $+20$ dB per decade, i.e., for every 10-fold change in frequency. This line passes through 0 dB at 1 rad·s^{-1} in the amplitude plot. The phase is independent of the frequency. If we have a simple zero of the form s^n, then the amplitude of the transfer function is $+20n\log_{10}(\omega)$ dB, and $\angle C(j\omega) = +n \cdot 90°$, where n is the order of the integral term. Since the phase is positive, the output sinusoid leads the input.

A zero at the origin is also referred to as a *derivative* term. We can understand this terminology by inputting a function $f(t)$ into a device that performs differentiation to obtain

$$C(s) = \frac{sF(s)}{F(s)} = s.$$

7.5.3 Lags and Leads

Simple Lag

The Bode plot for a simple lag is shown in Figure 7.12. Simple lag contributions to the Bode plot have the form

$$C(j\omega) = \frac{1}{1 + j\omega/k}. \tag{7.35}$$

An example of a dynamical system whose transfer function corresponds to a simple lag is the RC circuit discussed in Section 7.1.

Amplitude

The amplitude of the Bode plot is

$$|C(j\omega)| = \frac{1}{\sqrt{1^2 + (\omega/k)^2}}.$$

We note that $C(j\omega)$ has the form of a Hill function (see Section 3.2.1). This observation motivates consideration of three special cases, namely

$$\log_{10}|C(j\omega)| \begin{cases} \approx 0 \text{ dB when } \left(\frac{\omega}{k}\right)^2 \ll 1, \\ = -10\log_{10}2 \text{ dB when } \left(\frac{\omega}{k}\right)^2 = 1, \\ = -20\log_{10}\frac{\omega}{k} \text{ dB when } \left(\frac{\omega}{k}\right)^2 \gg 1. \end{cases}$$

Engineers use a rule-of-thumb approximation to interpret this Bode plot. They fit a straight line to the low-frequency range (i.e., amplitude zero) and a straight line to the high-frequency portion (i.e., slope n times -20 dB per decade), and then extrapolate the lines so that they intersect (dashed lines in Figure 7.12a). The point

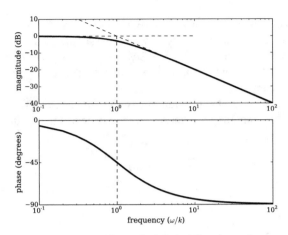

Fig. 7.12 Bode plot for a simple lag. The vertical dashed line shows the corner frequency. The normalized frequency is ω/k.

at which these lines intersect is called the *break point*. The frequency at the break point is called the corner frequency, ω_c, where

$$\omega_c = k.$$

The reader should note that if the frequency is expressed in cycles per second (Hertz), the corner frequency, f_c, is equal to $k/2\pi$.

A practical problem that arises when one is working with biological systems is that the range of frequencies that can be studied is often small, and hence it can be difficult to estimate the slope of frequencies greater than ω_c. A convenient way to estimate the order of the lag is to measure the difference between the amplitude of the Bode plot and the straight-line approximation used to estimate ω_c. For a first-order lag, this difference is ≈ 3 dB, and for an nth-order lag, it is $\approx 3n$ dB.

Phase

The phase is given by the expression

$$\angle C(j\omega) = -\tan^{-1}\frac{\omega}{k}.$$

When $k = \omega$, we have $\angle C(j\omega) = -45°$; when $\omega \ll 0.1k$, we have $\angle C(j\omega) \approx 0°$; and when $\omega \gg 10k$, we have $\angle C(j\omega) \approx -90°$. It is typically assumed that $\angle C(j\omega)$ varies linearly between these endpoints.

Simple Lead

The Bode plot for a simple lead is not shown. In this case, we have $C(s) = 1 + s/k$, or

$$G(j\omega) = 1 + j\frac{\omega}{k}. \tag{7.36}$$

The expressions for magnitude and phase are identical to those for a real pole except that they have the opposite sign. The curves on the Bode plot are thus mirror images about the 0 dB and $0°$ lines. The magnitude and phase, therefore, both increase with frequency, the latter tending toward $90°$ (a phase lead) for frequencies above $10k$.

7.5.4 Time Delays

In Chapter 9, we will discuss the role of time delays in feedback control. A time delay introduces terms of the form $e^{-\tau s}$ to the transfer function. The effect of a time delay on the amplitude is

$$|e^{-j\tau\omega}| = |\cos\omega\tau - j\sin\omega\tau| = 1. \tag{7.37}$$

Hence a time delay has no frequency-dependent effect on the amplitude. For the phase, we have

$$\angle C(j\omega) = \tan^{-1}\left[\frac{\Im(e^{-j\omega\tau})}{\Re(e^{-j\omega\tau})}\right] = \tan^{-1}\left(-\frac{\sin\omega\tau}{\cos\omega\tau}\right) = -\omega\tau.$$

Thus, a time delay introduces a phase lag whose magnitude is proportional to the frequency of the sinusoidal input.

7.5.5 Complex Pair: Amplification

Figure 7.13 shows the contribution of a complex pair of poles

$$s^2 + \omega^2 = (s + j\omega)(s - j\omega)$$

to the Bode plot. We note that over a small range of frequencies, the amplitude of the output is larger than that of the input. In other words, the dynamical system can amplify the input. Electronic amplifiers are used to amplify weak signals. In biology, Bode plots of this form are often obtained for the response of joints, such as the human ankle and knee, to mechanical perturbations [874].

In order to understand the Bode plots in Figure 7.13, it is necessary to determine the transfer function for the second-order differential equation

$$\frac{1}{\omega_n^2}\frac{d^2x}{dt^2}+\frac{2\varepsilon}{\omega_n}\frac{dx}{dt}+x=b(t)\,, \tag{7.38}$$

where ω_n is the natural frequency of the oscillator (see below). The damping ratio $\varepsilon \geq 0$ determines the nature of the responses of the oscillator: overdamped ($\varepsilon > 1$), critically damped ($\varepsilon = 1$), underdamped ($1 > \varepsilon > 0$), and undamped ($\varepsilon = 0$). A requirement for amplification to occur is that the system be underdamped. This means that the roots of the characteristic equation are a pair of complex numbers with negative real part (Section 4.3.2). Hence in the following discussion, we take $\varepsilon < 1$.

A practical question is how to compare different amplifiers. Typically, engineers compare an amplifier to a harmonic oscillator, i.e.,

$$\frac{d^2x}{dt^2}+\omega_n^2x(t)=0 \tag{7.39}$$

where ω_n is the natural frequency of the harmonic oscillator. The characteristic equation is

$$\lambda^2+\omega_n^2=0\,, \tag{7.40}$$

and the eigenvalues are purely imaginary. For a harmonic oscillator, we can write the solution as

$$x(t)=K\sin(\omega_n t+\phi)\,, \tag{7.41}$$

where K is a constant.

Now consider the damped harmonic oscillator

$$\frac{d^2x}{dt^2}+2\varepsilon\omega_n\frac{dx}{dt}+\omega_n^2x=0\,, \tag{7.42}$$

where ε is the damping ratio (see below). To appreciate why it is useful to write 2ε instead of ε, determine the roots of the characteristic equation for (7.42):

$$\lambda_{1,2}=\frac{-2\varepsilon\omega_n\pm\sqrt{(2\varepsilon\omega_n)^2-4\omega_n^2}}{2}=-\varepsilon\omega_n\pm\omega_n\sqrt{\varepsilon^2-1}\,. \tag{7.43}$$

We see that by rewriting the damping ratio as 2ε, we simplify (7.43) by getting rid of the 2's. Second, since $\varepsilon < 1$ at steady state, the solution will be of the form

$$x(t)\approx\sin\left[\omega_n\sqrt{1-\varepsilon^2}t+\phi\right]. \tag{7.44}$$

By comparing (7.44) to (7.41), we see that the frequency has been reduced by a factor of $\sqrt{1-\varepsilon^2}$. For this reason, the term ε is called the *damping ratio*. As ε approaches zero, the frequency of the underdamped oscillator approaches that of the harmonic oscillator.

Equation (7.38) is the dimensionally correct form of (7.42) for determining the transfer function

$$C(s) = \frac{\omega_n^2}{s^2 + 2\varepsilon\omega_n s + \omega_n^2}, \tag{7.45}$$

and hence setting $s = j\omega$ yields

$$C(j\omega) = \frac{\omega_n^2}{-\omega^2 + 2j\varepsilon\omega\omega_n + \omega_n^2} = \frac{1}{(1 - \omega^2/\omega_n^2) + 2\varepsilon j\omega/\omega_n}. \tag{7.46}$$

Note that ω_n is a constant, while ω is a variable.

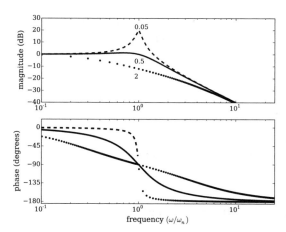

Fig. 7.13 Bode plot for a quadratic lag for three values of the damping ratio ε. The normalized frequency is ω/ω_n.

Amplitude

The amplitude contribution to the Bode plot is

$$|C(j\omega)| = \frac{1}{\sqrt{(1 - \omega^2/\omega_n^2)^2 + (2\varepsilon\omega/\omega_n)^2}}. \tag{7.47}$$

Again, we recognize a Hill-type function and hence consider three cases:

$$|C(j\omega)| = \begin{cases} -20\log_{10} 1 = 0\ dB & \text{when } \frac{\omega}{\omega_n} \ll 1; \\ -20\log_{10} 2\varepsilon\ dB & \text{when } \frac{\omega}{\omega_n} = 1; \text{thus we see that the condition for amplification is } \varepsilon < 0.5; \\ -40\log_{10}\frac{\omega}{\omega_n}\ dB & \text{when } \frac{\omega}{\omega_n} \gg 1; \text{this is twice the limiting slope that we obtained for a simple lag.} \end{cases}$$

Phase

The contribution to the phase is

$$\angle C(j\omega) = -\tan^{-1}\frac{2\varepsilon\omega/\omega_n}{1-\omega^2/\omega_n^2}. \qquad (7.48)$$

The argument of the inverse tangent is a Hill-type function, and hence we can sketch $\angle C(j\omega)$ by noting that

$$\angle C(j\omega) = \begin{cases} 0 & \text{when } \omega/\omega_n \ll 1, \\ -90° & \text{when } \omega/\omega_n = 1, \\ -180° & \text{when } \omega/\omega_n \gg 1. \end{cases}$$

7.6 Interpreting Biological Bode Plots

We now have the tools to interpret the Bode plots shown in Figures 7.8 and 7.9.

7.6.1 Crayfish Photoreceptor Transfer Function

Since the overall shape of the Bode plot resembles that of an RC circuit, we anticipate the presence of a lag. The corner frequency ω_c is ≈ 0.75 rad $\cdot$ s^{-1}. The order of the lag is determined from the limiting negative slope of the response for frequencies higher than ω_c. However, given the scatter of the data points and the small frequency range, it is difficult to determine the slope with certainty. Stark noted that at ω_c, there is a decrease in the gain from the straight-line approximation of the lag (dashed line in Figure 7.8a) of ≈ 6 dB. Since 6 dB is twice 3 dB, he concluded that there is a second-order lag. Finally, Stark observed for many trials that the initial decrease of phase with frequency (Figure 7.8b) was linear with a slope of ≈ 1, suggesting the presence of a time delay of ≈ 1 s.

Putting all of this together, we obtain the following transfer function:

$$C(s) \approx \frac{12e^{-1.0s}}{(1+1.3s)^2} \text{ spikes} \cdot \text{s}^{-1} \cdot \text{mlm}^{-1} \cdot \text{mm}^2. \qquad (7.49)$$

The units for $C(s)$ can be understood as follows. The input to the crayfish photoreceptor is light. Illuminance has units of millilumens per square millimeter (mlm/mm^2). The output is neural spike frequency, namely spikes/s. As ω approaches 0, the gain approaches ≈ 12 spikes $\cdot$ s$^{-1} \cdot$ mlm$^{-1} \cdot$ mm^2 (Figure 7.8a).

The Bode plots obtained for an isolated photoreceptor neuron and for a small population of photoreceptor neurons (three) were observed to be very similar. The only difference was that the gain of the population response was higher. To test the validity of this transfer function, Stark used it to predict the response of the crayfish abdominal photoreceptor to a step change in light intensity [793]. The denominator of $C(s)$, given by (7.49), yields the differential equation

$$1.69\frac{d^2P}{dt^2} + 2.6\frac{dP}{dt} + P = 12H(t-1),\tag{7.50}$$

where $H(t)$ describes the step input in light intensity. The step function $H(t-1)$ introduces a delay of 1 s between the step change in light intensity and the response of the photoreceptor. Figure 7.14 compares the prediction of (7.50) to that observed experimentally. The agreement is quite good. The differences between prediction and observation were thought by Stark to arise because the step change was not sufficiently small, and hence nonlinearities in the response of the photoreceptor influenced the input–output relationship.

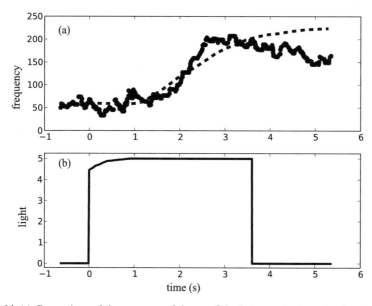

Fig. 7.14 (a) Comparison of the response of the crayfish photoreceptor to a step input of light (solid line) to that predicted by (7.50) (dashed line). (b) The experimentally generated step input of light. Figure redrawn using data from [793] with permission.

7.6.2 Yeast Osmotic Stress Transfer Function

The unusual feature of the Bode amplitude plot shown in Figure 7.9a is that the amplitude grows at lower frequencies and declines at higher frequencies. The simplest choice of $C(s)$ consistent with this observation is

$$C(s) \approx \frac{s}{1+k^{-1}s}.$$

However, this choice of $C(s)$ does not explain the phase plot. A derivative term s in the numerator of $C(s)$ contributes $+90°$ to the phase plot at low frequencies. The simple lag $1 + s/k$ contributes $0°$ at low frequencies to the phase plot and $-90°$ at high frequencies. However, this implies that the high-frequency phase response should be $0°$ (namely, $+90° - 90° = 0°$), which is inconsistent with the observed phase of $-90°$. This observation suggests that there must be present at least a second-order lag. A choice of $C(s)$ consistent with the measured Bode plot is (Figure 7.9b, d)

$$C(s) \approx \frac{s\omega_n^2}{s^2 + 2\varepsilon 2\omega_n s + \omega_n^2}. \tag{7.51}$$

The distinctive shape of the amplitude component of the Bode plot shown in Figure 7.9 has also been observed for chemotaxis in bacteria [11, 52, 54, 76, 314, 554, 555] and eukaryotic cells [473] and also in certain photoreceptor responses [333, 712]. This Bode amplitude–frequency response is interpreted in terms of *adaptation*, and in particular, is suggestive of *perfect adaptation* [179, 554, 559, 903].

Adaptation refers to the phenomenon that in the presence of a constant stimulus, the system first shows an initial transient response and then gradually moves toward its behavior before the stimulus was present. If in the presence of prolonged stimulus, the system's behavior returns exactly to that observed prestimulus, the system is said to exhibit perfect adaptation. A feature of this adaptive control mechanism is a transfer function that has a derivative term (zero) in the numerator [179] (see Exercise 16).

However, it is important to understand that although the Bode plot identifies the transfer function, this is insufficient for identifying the nature of the control mechanism. Thus, for example, the presence of a derivative term in the transfer function does not mean that the osmotic stress response of a yeast cell is controlled by a derivative-type feedback controller. In Section 9.2.1, we will show that perfect adaptation can be achieved by an integral-type feedback controller. Recently a biologically plausible mechanism for perfect adaptation, referred to as antithetic integral feedback, has been identified [12, 80, 241].

It is of interest to reexamine the Bode plot for the crayfish photoreceptor (Figure 7.8) in view of our discussion of adaptation. As Stark himself pointed out [793], there is a clear divergence between the low-frequency gain and that predicted by the transfer function given by (7.49). Although he suggested the addition of a derivative term to the transfer function, he did not pursue this line of inquiry further. In contrast to the yeast osmoregulatory response, the crayfish photoreceptor shows little or no adaptation to a constant light stimulus. These observations point to the possibility that there exist other mechanisms to explain the Bode plot shown in Figure 7.9.

7.7 What Have We Learned?

1. What is a criterion for the linearity of a dynamical system in the frequency domain?

   A criterion for linearity is that the output be sinusoidal with the same frequency as the input sinusoid. However, the amplitude of the output sinusoid is not necessarily the same as that of the input. Moreover, the output sinusoid will generally be phase-shifted with respect to the input sinusoid.

2. What is the Laplace transform and why is it useful in the laboratory?

   The Laplace transform is given by (7.16). It provides the theoretical basis for the Bode plot method for identifying the transfer function of a linear dynamical system.

3. What are the limitations on the use of the Laplace transform in the laboratory?

   The usefulness of the Laplace transform is limited by the fact that an efficient computer algorithm for determining the Laplace transform of an experimentally measured time series is not yet available. Moreover, obtaining the inverse Laplace transform is often problematic.

4. What is the transfer function?

   The transfer function of a linear dynamical system is the ratio of the integral transform (Laplace or Fourier) of the output to that of the input. It plays the same role in the frequency domain that the impulse function plays in the time domain.

5. What is a Bode plot and why it is important?

   The Bode plot is a graphical method for determining the transfer function from the frequency dependence of the amplitude and phase shift of the output of a linear dynamical system to a sinusoidal input. The Bode plot makes it possible to determine the transfer function for a stable linear dynamical system from experimental measurements.

6. What is the effect of a time delay on the phase of the frequency response of a linear dynamical system that receives a sinusoidal input?

   The effect of a time delay is to produce a phase shift between the sinusoidal input and output of a dynamical system that is directly proportional to the frequency. The presence of a time delay does not affect the amplitude of the sinusoidal output.

7.8 Exercises for Practice and Insight

1. A surprising observation is that in the awake monkey, cortical neurons spend $\approx 98.6\%$ of their time idly [1]. This statement is a colorful way of stating that the average neural firing frequency is low. For example, if we assume that the duration of an action potential is 2 ms, then a neuron spiking at a frequency of 1 Hz spends 99.8 % of its time idly. What is the mean spiking frequency of a neuron that spends 98.6 % of its time idly?

2. If resistance R is measured in ohms and capacitance C is measured in farads, show that the unit of RC is seconds.
3. Show that

$$\cos\theta = \frac{1}{2}\left(e^{j\theta} + e^{-j\theta}\right), \quad \sin\theta = \frac{1}{2j}\left(e^{j\theta} - e^{-j\theta}\right).$$

4. The transfer function $C(s)$ for a thermometer has the form

$$C(s) = \frac{1}{1+\alpha s},$$

where α is a constant.

a. Sketch the Bode plot for the thermometer.
b. If the fluctuations in temperature (the input) occur in the frequency range of $0.1 \text{ cycles} \cdot \text{s}^{-1}$, how should α be chosen so that the thermometer will be able to monitor the temperature changes? (Hint: what should the corner frequency be?)

5. What is the Laplace transform of the output (solution) of

$$k^{-1}\frac{dx}{dt} + x = b(t)$$

when $b(t)$ is

a. a delta function.
b. a Heaviside (step) function.
c. a ramp function.
d. a parabolic forcing function (e.g., t^2).

6. What is the transfer function for

$$k^{-1}\frac{dx}{dt} + x = b(t)$$

when $b(t)$ is

a. a delta function.
b. a Heaviside (step) function.
c. a ramp function.
d. a parabolic forcing function (e.g., t^2).

7. What is the Laplace transform of the output (solution) of

$$\frac{1}{\omega_n^2}\frac{d^2x}{dt^2} + \frac{2\varepsilon}{\omega_n}\frac{dx}{dt} + x = b(t)$$

when $b(t)$ is

a. a delta function.

b. a Heaviside (step) function.

c. a ramp function.

d. a parabolic forcing function (e.g., t^2).

8. What is the transfer function of

$$\frac{1}{\omega_n^2}\frac{d^2x}{dt^2} + \frac{2\varepsilon}{\omega_n}\frac{dx}{dt} + x = b(t)$$

when $b(t)$ is

a. a delta function.

b. a Heaviside (step) function.

c. a ramp function.

d. a parabolic forcing function (e.g., t^2).

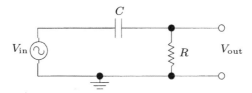

Fig. 7.15 An RC circuit configured to function as a high-pass filter where V_{in}, V_{out} are, respectively the input and output voltages.

9. Many plants release their pollen into the wind. A differential equation that approximates the rotational dynamics of a rigid pollen particle is

$$J\frac{d^2\theta}{dt^2} - NL\theta = T,$$

where θ is the pollen attitude angle, T is an applied torque, J is its inertia, N is the normal force coefficient, and L is the distance from the center of gravity.

a. What is the transfer function?

b. For what values of N is the pollen trajectory unstable?

c. What kind of stability does this system exhibit? How does this translate into the movement of the pollen particle through the air?

10. When an RC circuit is configured as in Figure 7.15, it acts as a high-pass filter (compare with Figure 7.1).

a. Show that the transfer function has the form

$$C(s) = \frac{s}{s + 1/RC}.$$

b. Show that the Bode amplitude plot is that expected for a high-pass filter.

11. The transfer function for the human ankle joint is given by [420]

$$C(s) = \frac{G_0 \omega_n^2}{s^2 + 2\varepsilon \omega_n s + \omega_n^2},$$

where $G_0 = -0.0028$, $\omega_n = 163.4 \text{ rad} \cdot \text{s}^{-1} = 26$ Hz, and $\varepsilon = 0.35$.

a. Draw the Bode plot and compare it with Figure 7.13.
b. Write a computer program to compute the impulse response function.

12. The noted dingbatologist Xuan Kirwan is trying to capture a rare man-eating Australian dingbat[9] for her research. While Prof. Kirwan is an expert at the benchtop, she has not had much experience with a tranquilizer gun, and she is worried that if she misses the first shot, the dingbat will eat her before she gets off the next one. Therefore, Xuan has designed two special blowguns that she thinks the dingbat will be unable to hear. One of them emits a sound at 0.23 Hz, the other at 5 Hz. She has determined that the transfer function for the tympanic membrane (part of the ear) is equal to

$$C(s) = \frac{13}{(1 + 2.45s)^2}.$$

Which blowgun should she use and why?

13. A scientist wants to design a second-order low-pass filter using the differential equation

$$\frac{1}{\omega_n^2} \frac{d^2 x}{dt^2} + \frac{k_2}{\omega_n} \frac{dx}{dt} + k_1 x = I(t),$$

where k_1, k_2 are constants and $I(t)$ is the signal to be filtered.

a. What values of a, b would you choose so that the cutoff frequency is 2 Hz?
b. Sketch the Bode plot for this filter.
c. My computer makes a high-pitched 19000-Hz screeching noise. Would this filter be useful to give me some peace and quiet?

14. Write a computer program to construct the Bode phase for a complex pair of roots described by (7.48). This program is more difficult than it looks at first. Be careful how you compute the arctangents.

15. When a biological system is exposed to a constant stimulus, it often adapts, that is, it shows an initial transient response and then gradually moves back toward its behavior before the stimulus was presented. An example arises in the dynamical behavior of the mechanoreceptors of the large tactile spine on the femur of the cockroach *Periplaneta americana*. The step-function transient responses of the mechanoreceptor are described by a power-law relation of the form [116]

$$y = Kt^{-k},$$

---

[9] To our knowledge, this animal does not exist except in the imagination of the authors. Rarer than rare, this animal is in fact nonexistent.

where y is the response frequency, and K and k are constants.

- Assuming that the observations are obtained under open-loop conditions, show that the transfer function $C(s)$ is given by

$$C(s) = K\Gamma(1-k)s^k,$$

where Γ is the gamma function (see Table 7.1).
- What is the frequency response function?
- Would the output be expected to lead or lag a sinusoidal input, and if so, by how much?

Power-law adaptation in sensory receptors of this type is frequently observed; however, its explanation remains elusive. For a review of possible explanations, see [243].

16. It has been argued that for bacterial chemotaxis [179], perfect adaptation requires a transfer function that combines a third-order lag with a derivative term. Evaluate this hypothesis by determining the amplitude component for the Bode plots described by the following transfer functions:

a.

$$\frac{1.5}{(1+0.2s)(1+0.3s)(1+0.45s)}.$$

b.

$$\frac{s+1}{(1+0.2s)(1+0.3s)(1+0.45s)}.$$

c.

$$\frac{s}{(1+0.2s)(1+0.3s)(1+0.45s)}.$$

d.

$$\frac{s-1}{(1+0.2s)(1+0.3s)(1+0.45s)}.$$

e. Check your results by examining [179]. Which of these four transfer functions generates a Bode frequency plot that most closely resembles that observed for the yeast response to osmotic stress? For the best-fit transfer function, what is the Bode phase response? Does it resemble that shown in Figure 7.9c?

Chapter 8
Frequency Domain II: Fourier Analysis and Power Spectra

There are several practical problems associated with the use of the Laplace transform to study input–output relationships in the laboratory. In particular, it is extremely difficult to obtain the Laplace integral transform for measured signals, and even if the transform is known, obtaining the inverse transform can be problematic. At the root of these problems is the lack of efficient numerical methods to calculate the Laplace transform and its inverse [201]. The solution is to use another integral transform, the[1] *Fourier transform* [73, 846]. Most of what we have discussed for the Laplace transform is also true for the Fourier transform. So where is the advantage? The advantage comes from the fact that two mathematicians, James Cooley and John Tukey,[2] developed a fast and efficient computer algorithm to calculate the Fourier transform numerically and obtain its inverse. This algorithm is called the *fast Fourier transform* and is abbreviated FFT [144].

Fourier analysis is the area of mathematics that studies how best to represent an arbitrary function as a sum of simple trigonometric functions such as sines and cosines. The two tools of Fourier analysis that we discuss in this chapter are the Fourier series (Tool 10) and the Fourier transform (Tool 11). The impact of Fourier analysis on human activities, both scientific and social, has been enormous, and in particular, has been even more greatly enhanced by the widespread availability of inexpensive high-speed personal computers. Indeed, it is difficult to imagine what daily life would be like without Fourier analysis: no television or radio, no smart phones, no global positioning systems (GPS), no ATMs, and no credit card transactions. Biological research applications range from the development of modern imaging devices, such as CAT and MRI scanners, to the techniques used by biologists to characterize the vocalizations of birds and humans. These accomplishments become possible because the FFT enables the mathematical ideas of Fourier analysis to be translated into practical applications.

---

[1] Jean Baptiste Joseph Fourier (1768–1830), French mathematician and physicist.

[2] James William Cooley (b. 1926), American mathematician; John Wilder Tukey (1915–2000), American mathematician.

© Springer Nature Switzerland AG 2021
J. Milton and T. Ohira, *Mathematics as a Laboratory Tool*,
https://doi.org/10.1007/978-3-030-69579-8_8

The main goal of this chapter is the calculation of the power spectrum, a plot of the power in a signal as a function of frequency. Historically, measurement devices output their information in the form of *analog signals*, such as a continuous voltage as a function of time. Consequently, the focus of research was on the determination of the power spectra for analog signals. However, with the increasing impact of digital computers on laboratory research, most present-day measurement devices output signals in the form of digital, or discrete time, signals. Internally, these devices use an A/D (analog-to-digital) board to digitize the analog signal (Section 8.4.5). Consequently, an important, but admittedly technical, issue concerns how best to ensure that the power spectrum of a digitized signal faithfully represents that of the original analog signal.

This chapter is organized as follows. First, we motivate the use of Fourier analysis by discussing three examples that frequently arise in the research laboratory. Next, we develop a Fourier analysis toolbox to introduce Tools 10 and 11. Although at first glance, the mathematical notation looks forbidding, we suspect that with a little patience, the reader will realize that the presentation actually builds on topics covered in earlier chapters. Then we discuss how the power spectrum is calculated. A number of practical issues that affect the determination of the power spectrum from a digital signal are discussed, including the effects of sampling frequency and the length of the time series. Many researchers will be tempted to skip the descriptions of these effects on the power spectrum and go directly to the last section of this chapter, which demonstrates how easy it is to calculate a power spectrum using readily available computer algorithms, such as `matplotlib.mlab.psd()`. Nonetheless, we encourage readers to refer to the skipped sections on a need-to-know basis as problems arise in the analysis of their data.

The questions we will answer in this chapter are

1. What is the fundamental conclusion of Fourier's theorem?
2. What are the important examples of a Fourier transform pair?
3. How can a sine wave be obtained from a periodic square-wave time series?
4. What is the power spectrum?
5. What are the effects of the length of a time series on the power spectrum?
6. Why is it important to use a windowing function for the calculation of a power spectrum and which windowing functions are recommended for routine laboratory use?
7. What is the effect of an abrupt change in a time series on its power spectrum?
8. What is the fundamental uncertainty that underlies the representation of trajectories in the time and frequency domains?
9. What factors determine the resolution of the power spectrum?
10. How can the power spectrum be determined from the Fourier transform of the output of a dynamical system?

11. Under what conditions can a continuous time series be faithfully represented by a discrete one?
12. The Shannon–Nyquist theorem stresses the importance of low-pass filtering. What are the important constraints on this low-pass filtering process?
13. Why is calculating a convolution using the FFT typically faster than using, for example, np.convolve.

8.1 Laboratory Applications

Techniques based on Fourier analysis are part of the everyday life of scientists who study dynamical systems in the laboratory. Such techniques are used because it is appropriate for the problem at hand, such as the investigation of sensory systems or spatial structures. In addition, tools, such as the measurement of a power spectrum, are used to address an often vexing experimental challenge, namely to ensure that measurements are not contaminated by artifacts. Here, we present three examples to illustrate the uses of Fourier analysis in the laboratory.

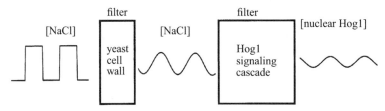

Fig. 8.1 Schematic representation of the input–output relationships for the response of the yeast Hog1 system to a square-wave periodic change in the osmolarity of the suspension medium. The square-wave input is filtered twice: first by the cell wall of the yeast cell and then by the Hog1 signaling cascade.

8.1.1 Creating Sinusoidal Inputs

In Section 7.4.2, we discussed the responses of yeast cells to hyperosmolar shock. A key point was the observation that a sinusoidal variation in the osmolarity of the medium produced a sinusoidal variation in the intracellular location of Hog1. How were the investigators able to produce a sinusoidal variation in the osmolarity of the medium? The answer is that they created a periodic change in osmolarity by rapidly changing the medium in a stepwise manner (Figure 8.1). The mathematical point is that a sine wave (middle of Figure 8.1) can be generated by appropriately filtering a periodic square wave (left-hand side of Figure 8.1). The fact that this method produces the desired result will become clear once we discuss, in Section 8.3.2, the Fourier series representation of a periodic square wave in the frequency domain.

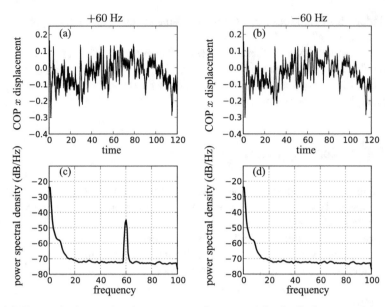

Fig. 8.2 Time series (top row) and power spectra (bottom row) for the displacement of the center of pressure (COP) measured using a force platform in the x (or anterior–posterior) direction for a subject standing quietly with eyes closed (Section 16.7). A low-amplitude 60-Hz component was added to the signal (left column), and the effects are compared to the same signal without the 60-Hz component (right column). The digitization frequency was 200 Hz.

Thus, the investigators relied on the properties of the cell wall and plasma membrane of the yeast cell [431] to perform the required low-pass filtering.

This method of producing a sine wave is quite useful in the laboratory setting in which a sinusoidal input of fixed frequency is required. In general, a two-pole low-pass filter is the least that is required. For example, one can obtain a reasonably good sine wave by passing the signal through the same RC (resistor and capacitor) circuit twice. An important question is how to choose the cutoff frequency of the low-pass filter (see Exercise 4 at the end of the chapter).

Example 8.1 A pure tone is composed of a sinusoidal waveform (sine or cosine) with a single frequency. For example, a musician hears a sine wave with a frequency of 440 Hz as the note A above middle C on the piano. The complex sounds that we hear about us are combinations of many pure tones of various frequencies. Pure tones are used clinically to test a person's hearing. You can hear a pure sine tone by downloading the appropriate `.wav` files from the Internet and playing them on your laptop. You can also generate them with commercially available software, such as Mathematica and MATLAB. It is also possible to convert a binary `.wav` file into a text file using, for example, Python's `scipy.io.wavfile.read()`.   ◇

8.1.2 Electrical Interference

Typically, experiments are performed in a laboratory setting that is engulfed by electric fields having the frequency of the household alternating current (either 50 or 60 Hz, depending on which country you live in). For the most part, these electric fields are generated by the electric currents used to light the laboratory and power the equipment used to make measurements. When the electrical equipment used in an experiment is not properly grounded, it can act as an antenna and record these alternating current signals.

Common causes of such problems are loose connections, improper grounding, multiple ground loops, and poor electrode contact [639]. Thus, it is not surprising that electrophysiologists spend a great amount of effort trying to minimize the effects of "60-cycle" and other sources of electrical artifact. For example, an experimental preparation may be electrically shielded from the laboratory environment through the use of a Faraday[3] cage. When a light source is needed inside a Faraday cage to enable the use of a microscope, the light source is powered by a direct current (DC) power supply. In addition, it may be necessary to filter out the 60-Hz component from the time series using a special kind of band-pass filter called a *notch filter*. It is important to ensure that the filtering is done prior to digitization to avoid aliasing artifacts (Section 8.4.5).

Sometimes, it is helpful for the experimentalist to estimate the magnitude of the 60-cycle artifact. This can be done by obtaining the power spectrum. To illustrate, consider the estimate of the presence of a 60-Hz electrical artifact in the measurement of the changes in the center of pressure (COP) while a subject stands on a force platform quietly with eyes closed. The significance of the measurement of COP is discussed in Section 16.7. Here, our focus is on the identification of a 60-cycle electrical artifact. It is clearly difficult to identify by eye the presence of an artifact with a period of 16.67 ms (compare Figure 8.2a and b). However, by calculating the power spectrum, we can see much more clearly the presence of the 60-cycle artifact (compare Figure 8.2c and d). Given that the magnitude of the 60-Hz artifact is quite small (about −45 dB), some investigators might choose to ignore it. However, most investigators will check their apparatus to see whether they can identify the cause (such as a loose connection or a faulty connector or ground).

Electrical interference can also come from the organism itself. For example, the electric field generated by the beating heart can be very troublesome when an electrophysiologist is interested in measuring the signals generated by the brain. The electric field of the heart is measured by physicians as the difference in potential between two electrodes placed at different locations on the body. A plot of this difference in potential as a function of time is called an *electrocardiogram* (EKG). When an electrical recording device used to measure brain signals is not properly grounded, it acts as an EKG machine and records an EKG artifact. Since the differences in electrical potential for brain signals are in the microvolt range, whereas those for the heart rate are a thousand times larger (millivolts), brain signals easily

---

[3] Michael Faraday (1791–1867), English physicist and chemist.

get overwhelmed by the EKG artifact. Thus, it is not surprising that electrophysiologists spend a great deal of time trying to minimize "EKG artifact" and other sources of an electrical artifact.

8.1.3 The Electrical World

Up to this point, we have discussed applications of Fourier analysis for conducting experiments. However, the determination of power spectra quite naturally arises in investigations of biological sensory systems. A case in point concerns the electric fish. Electric fish are familiar to tropical-fish fanciers and are a source of fascination for biologists and neuroscientists. These fish, the gymnotiforms from South America and the mormyniforms from Africa, possess an electrogenerative organ in the caudal part of their body [601]. They also possess an array of electroreceptors on their skin. This arrangement allows these fish to sense perturbations in their self-generated electric field caused by objects and other electric fish in their environment. This electrosense is used both to locate objects in space and to interact with other electric fish.

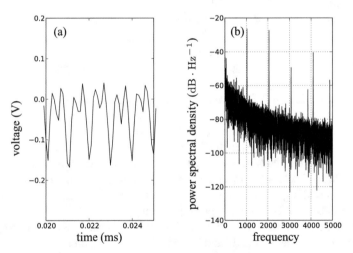

Fig. 8.3 (a) Time series and (b) power spectral density for the signal emitted by the electric fish *Apteronotus albifrons*. The digitization rate was 40 kHz, and the low- and high-pass filter settings were, respectively, 100 Hz and 100 kHz. The power spectral density was calculated using `matplotlib.mlab.psd()`, as discussed in Section 8.4.6.

Figure 8.3 shows the time series and power spectrum for the electric signal generated by the black ghost fish *Apteronotus albifrons*. The waveforms of the electric signals produced by different species of electric fishes differ from one another [247]. Thus, we can use the power spectrum to confirm the identity of a particular species of electric fish. Within a given species, individuals tend to generate signals with slightly different frequencies. Moreover, it is possible to communicate with certain species of captive electric fish by injecting an external periodic signal into its aquarium. For example, an electric fish often responds with "small chirps" to external signals with frequencies approximately 5 Hz lower than its own frequency and with "big chirps" if the signal is about 200 Hz lower (note that male frequencies are typically ≈ 200 Hz higher than those of females) [910]. It is thought that small chirps are important for defending personal space, while big chirps are important in courtship.

8.2 Fourier Analysis Toolbox

The idea behind Fourier's theorem is that a "reasonably behaved" function $g(t)$ (and keep in mind that in the laboratory, we are interested primarily in reasonably behaved functions) that is periodic with some period T, that is, a function satisfying the periodicity condition

$$g(t) = g(t+T),$$

can be decomposed into contributions of $\sin nt$ and $\cos nt$ for $n = 1, 2, 3, \ldots$, where the variable t stands for time. (In this book, we are interested primarily in functions whose independent variable is time. However, the independent variable could equally well be a spatial dimension.)

We shall begin by assuming that the period T is equal to 2π, which makes the calculations easier. We can easily modify our results for an arbitrary period T.

Note, first of all, that $\sin(nt)$ and $\cos(nt)$ are periodic on the interval 2π for every integer n, and moreover, $\sin(nt)$ and $\cos(nt)$ make n complete cycles on every interval of length 2π. And recall that the sine is an odd function, since $\sin(-t) = -\sin(t)$ for all t, and that the cosine is an even function, since $\cos(t) = \cos(-t)$. The idea behind such a decomposition is that we can think of a periodic function as consisting of a number of components of various frequencies and amplitudes. Those frequencies will be integer multiples n of the fundamental frequency (equal to $1/T$), and the amplitude of a component will indicate the size of the contribution that a particular frequency makes to the overall function. Therefore, the contribution from frequency n can be expressed as an even component $a_n \cos nt$ plus an odd component $b_n \sin nt$, where a_n and b_n are the amplitudes.

Fourier analysis is actually not terribly difficult as long as you keep Tool 8 (Euler's formula) and the properties of the sine and cosine functions at your fingertips. We have at our disposal two important tools of Fourier analysis: the Fourier series (Tool 10) and the Fourier transform (Tool 11). Although Tool 10 is frequently emphasized in courses in mathematics and physics, we will see that Tool 11 is actually the more useful tool in the laboratory setting. Thus, our presentation of Tool 10 is limited to providing a basis for understanding Tool 11.

8.2.1 Tool 10: Fourier Series

Fourier's brilliant idea was to represent a 2π-periodic function $g(t)$ as an infinite sum of sine and cosine functions:

$$g(t) = a_0 + \sum_{n=1}^{\infty} a_n \cos(nt) + \sum_{n=1}^{\infty} b_n \sin(nt), \quad -\pi \le t \le \pi, \tag{8.1}$$

known today as a *Fourier series*.

Given a function $g(t)$, how can we find the coefficients a_n and b_n that determine the Fourier series? We begin with the integrals of the sine and cosine functions:

$$\int \sin(nt)\,dt = -\frac{\cos nt}{n}, \quad \int \cos(nt)\,dt = \frac{\sin nt}{n}.$$

Our life is made easier by the fact that the functions $\cos(nt)$ and $\sin(nt)$ satisfy the following orthogonality conditions:

$$\int_{-\pi}^{\pi} \cos(mt)\sin(nt)\,dt = 0 \quad \text{for all } m, n, \tag{8.2a}$$

$$\int_{-\pi}^{\pi} \cos(mt)\cos(nt)\,dt = \pi\delta(m,n), \tag{8.2b}$$

$$\int_{-\pi}^{\pi} \sin(mt)\sin(nt)\,dt = \pi\delta(m,n), \tag{8.2c}$$

where

$$\delta(m,n) = \begin{cases} 1 & \text{if } m = n, \\ 0 & \text{otherwise} \end{cases} \tag{8.3}$$

is the Kronecker[4] delta function. It is these orthogonality conditions that allow us to determine the coefficients a_n and b_n.

The value of a_0 can be obtained by integrating both sides of (8.1). Assuming that it is permissible to integrate the series term by term, we obtain

$$\int_{-\pi}^{\pi} g(t)\,dt = \int_{-\pi}^{\pi} a_0\,dt + \int_{-\pi}^{\pi} \sum_{n=1}^{\infty} (a_n \cos nt + b_n \sin nt)\,dt \tag{8.4}$$

$$= 2\pi a_0 + \sum_{n=1}^{\infty} a_n \int_{-\pi}^{\pi} \cos nt\,dt + \sum_{n=1}^{\infty} b_n \int_{-\pi}^{\pi} \sin nt\,dt.$$

But

$$\int_{-\pi}^{\pi} \cos nt\,dt = 0, \quad \int_{-\pi}^{\pi} \sin nt\,dt = 0, \tag{8.5}$$

and thus

---

[4] Leopold Kronecker (1823–1891), German mathematician.

$$a_0 = \frac{1}{2\pi} \int_{-\pi}^{\pi} g(t)\, dt \,. \tag{8.6}$$

So much for a_0. Now we need to evaluate a_n and b_n for integer values of $n \geq 1$. In order to obtain a_n, we multiply both sides of (8.1) by $\cos(n't)$, where n' is a particular choice of n, and integrate to get

$$\int_{-\pi}^{\pi} g(t) \cos n't\, dt = \sum_{n=1}^{\infty} \int_{-\pi}^{\pi} a_n \cos(nt)\cos(n't)\, dt \,, \tag{8.7}$$

where the terms of the form $\int_{-\pi}^{\pi} b_n \sin nt \cos n't$ are equal to zero as a result of (8.2a), and the term $\int_{-\pi}^{\pi} a_0 \cos n't$ is equal to zero by (8.5). From (8.2b), we see that the solution of (8.7) is

$$\int_{-\pi}^{\pi} g(t) \cos(n't)\, dt = \pi \sum_{n=0}^{\infty} a_n \delta(n,n') = a_{n'} \pi \,.$$

We can repeat this argument by choosing different values of n' until we have covered the whole range of n. Hence

$$a_n = \frac{1}{\pi} \int_{-\pi}^{\pi} g(t) \cos nt\, dt \,. \tag{8.8}$$

We can use an identical procedure to evaluate the b_n. Multiply each side of (8.1) by $\sin(n't)$, where n' is a particular choice of n, and integrate to obtain

$$\int_{-\pi}^{\pi} g(t) \sin(n't)dt = \sum_{n=1}^{\infty} \int_{-\pi}^{\pi} b_n \sin(nt)\sin(n't)\, dt \,. \tag{8.9}$$

Hence the solution of (8.9) is

$$\int_{-\pi}^{\pi} g(t) \sin(n't)\, dt = \sum_{n=1}^{\infty} b_n \pi \delta(n,n') = b_{n'} \pi \,,$$

where we have used (8.2c). Finally, we obtain

$$b_n = \frac{1}{\pi} \int_{-\pi}^{\pi} g(t) \sin nt\, dt \,. \tag{8.10}$$

In summary, the Fourier series representation of a function $g(t)$ is

$$g(t) = a_0 + \sum_{n=1}^{\infty} (a_n \cos nt + b_n \sin nt) \,, \tag{8.11}$$

where

$$a_0 = \frac{1}{2\pi} \int_{-\pi}^{\pi} g(t)\, dt,$$

$$a_n = \frac{1}{\pi} \int_{-\pi}^{\pi} g(t) \cos nt\, dt, \qquad (8.12)$$

$$b_n = \frac{1}{\pi} \int_{-\pi}^{\pi} g(t) \sin nt\, dt,$$

for $1 \leq n \leq \infty$.

Note that (8.6) computes the mean value of $g(t)$. For this reason, a_0 is commonly referred to as the direct current (DC) component of the signal, and the remainder of the series as the alternating current (AC) component. In time series analysis, the DC component is typically subtracted from the signal before various statistical analyses are performed. Thus, some authors write the Fourier series with the mean already subtracted.

Finally, as we mentioned above, we can allow an arbitrary period for our function, so that we may consider functions that are periodic on an interval $[-L, L]$, where L is an arbitrary positive real number. This requires that we appropriately change the limits of integration for (8.12). Let

$$t = \frac{\pi t'}{L}, \quad dt = \frac{\pi dt'}{L},$$

so that

$$t' = \frac{Lt}{\pi},$$

and (8.11) becomes

$$g(t) = \frac{a_0}{2} + \sum_{n=1}^{\infty} a_n \cos\left(\frac{n\pi t}{L}\right) + b_n \sin\left(\frac{n\pi t}{L}\right), \qquad (8.13)$$

where, on dropping the "prime" notation,

$$a_0 = \frac{1}{L} \int_{-L}^{L} g(t)\, dt, \qquad (8.14a)$$

$$a_n = \frac{1}{L} \int_{-L}^{L} g(t) \cos nt\, dt, \qquad (8.14b)$$

$$b_n = \frac{1}{L} \int_{-L}^{L} g(t) \sin nt\, dt. \qquad (8.14c)$$

Example 8.2 In physics, the frequency content of sounds made by musical instruments and the vocalizations made by humans and other animals is studied in the context of a famous linear partial differential equation called the *wave equation*. The important point for our discussion is that the solution of the wave equation can be obtained in terms of a Fourier series [72]. The theory predicts that when, for example, an instrument sounds a note A at 440 Hz, the Fourier series will consist of one term with frequency 440 Hz and other terms related to its harmonics. The 440 Hz frequency term is called the *fundamental frequency*. The *harmonics* are terms with frequencies that are integer multiples of the fundamental frequencies, for example 800 Hz, 1320 Hz, and so on. As we will see in Section 8.3.2, this prediction holds up very well for the common string and wind instruments of the orchestra. Curiously, when sounds are amplified through a loudspeaker, *subharmonics* arise with frequencies $1/n$ times the fundamental frequency, e.g., 220 Hz, 110 Hz, and so on. Subharmonics are naturally present in the sounds made by bells and animal voices, and they can be created by certain instruments developed by mathematicians, including the *tritare* and the chaotic Chua[5] circuit [561, 725]. The subharmonics arise because of nonlinearites, and they are thought to be particularly important for emotive speech in humans [655] and for coordinating affect across animal groups [86]. Issues related to subharmonics are particularly relevant for the design of hearing aids and the development of voice recognition programs. ◇

8.2.2 Tool 11: The Fourier Transform

We saw in the previous section how to find the various frequency contributions that make up a periodic function. In this section, we shall see how those ideas can be extended to a much larger class of functions including aperiodic functions. We are going to derive a pair of functions that will allow us to convert a signal in the time domain to its representation in the frequency domain, and vice versa. We shall begin by replacing the sine and cosine functions in the Fourier series by their complex representation in terms of the exponential function (Euler's formula) and then letting the period interval grow from $[-L,L]$ to $(-\infty, +\infty)$.

The complex series representation of (8.11) can be obtained by first recalling that (Euler's formula, Tool 8)

$$\cos\theta = \frac{1}{2}\left(e^{j\theta} + e^{-j\theta}\right), \quad \sin\theta = \frac{1}{2j}\left(e^{j\theta} - e^{-j\theta}\right).$$

Using these relationships and replacing θ with $n\omega t$, we can rewrite (8.11) as

$$g(t) = \frac{a_0}{2} + \frac{1}{2}\sum_{n=1}^{\infty} a_n\left(e^{jn\omega t} + e^{-jn\omega t}\right) + \frac{1}{2j}\sum_{n=1}^{\infty} b_n\left(e^{jn\omega t} - e^{-jn\omega t}\right),$$

which can be rearranged to yield

$$g(t) = \sum_{n=-\infty}^{\infty} c_n e^{jn\omega t}. \tag{8.15}$$

In exponential notation, the orthogonality conditions are

---

[5] Leon Ong Chua (b. 1936), American engineer.

$$\int_{-\pi}^{+\pi} e^{-jn\omega t} e^{jn'\omega t} dx = 2\pi\delta(n,n'),$$

and hence

$$c_n = \frac{1}{2\pi} \int_{-\pi}^{+\pi} e^{-jn\omega t} g(t)\, dt. \tag{8.16}$$

Since the real and complex notations are equivalent, we have, using (8.11),

$$c_0 = \frac{1}{2}a_0, \quad c_n = \frac{1}{2}(a_n - jb_n), \quad c_{-n} = \frac{1}{2}(a_n + jb_n).$$

The Fourier transform is an extension of the sum of discrete frequency components in (8.15) to a continuous function of frequency. It is given by

$$G(\omega) = \int_{-\infty}^{\infty} g(t)e^{-j\omega t}\, dt, \tag{8.17}$$

where ω is the angular frequency. Although the derivation of the Fourier transform is beyond the scope of this book, it is useful to be aware of the general idea.[6] The first step is to let $g(t)$ be periodic on an interval of arbitrary length $2L$ instead of 2π. Thus, (8.15) becomes

$$g(t) = \sum_{n=-\infty}^{\infty} c_n e^{jn\pi t/L}, \tag{8.18}$$

where

$$c_n = \frac{1}{2L} \int_{-L}^{+L} e^{jn\pi t/L} g(t)\, dt.$$

We can now let L get larger and larger, and in the limit as L approaches infinity, the period becomes infinite (in other words, the function is aperiodic). And as L gets large, the frequency contributions for n/L get closer and closer together, and in the infinite limit, the frequency spectrum becomes essentially continuous. This observation justifies the replacement of the summation in (8.18) by an integral.

The Fourier transform (8.17) allows us to pass from the representation of a function in the time domain to its representation in the frequency domain. But we can also go the other way: the Fourier transform can be run backward, from the frequency domain to the time domain. The inverse Fourier transform is given by

$$g(t) = \frac{1}{2\pi} \int_{-\infty}^{\infty} G(\omega)e^{j\omega t}\, d\omega. \tag{8.19}$$

We call (8.17) and (8.19) a *Fourier transform pair*. Our notational convention is to use a lowercase letter when the signal is represented in the time domain and an uppercase letter when it is represented in the frequency domain. It is important to observe that the sign in the exponent changes depending on whether we are calculating $G(\omega)$ or $g(t)$. Certain Fourier transform pairs will be of particular interest to us in later chapters. For example, the power spectrum and the autocorrelation function

---

[6] For a very accessible account of this derivation, see [846].

are a Fourier transform pair. This result is known as the Wiener–Khinchin[7] theorem. It is discussed in Section 14.6.

Most biologists find it more appealing to think of frequency f rather than angular frequency ω. Since $\omega = 2\pi f$, we can rewrite (8.17)–(8.19) as

$$G(f) = \int_{-\infty}^{\infty} g(t) e^{-j2\pi ft} dt, \quad g(t) = \int_{-\infty}^{\infty} G(f) e^{j2\pi ft} df.$$

An advantage in writing the Fourier transform pair in this way is that the pair is symmetric in the sense that the term in front of the integral is the same, in contrast to (8.17)–(8.19). Thus, we will use the frequency representation of the Fourier transform pair throughout the remainder of this book.

We will make much use of the Fourier transform of a derivative, particularly in our discussion of stochastic processes, such as the random walk; see Tool 12 (generating functions). The Fourier transform of the derivative of $g(t)$ can be obtained by solving[8]

$$G\left(\frac{dg}{dt}\right) = \int_{-\infty}^{\infty} \frac{dg}{dt} e^{-2\pi jft} dt. \tag{8.21}$$

We can evaluate this integral using integration by parts:

$$\int v \, du = uv - \int u \, dv.$$

If we make the identifications

$$du = \frac{dg}{dt} dt, \quad v = e^{-2\pi jft},$$

so that

$$u = g(t), \quad dv = -2\pi jf e^{-2\pi jft} dt,$$

we obtain

$$G\left(\frac{dg}{dt}\right) = g(t) e^{2\pi jft} \Big|_{t=-\infty}^{\infty} + \int_{-\infty}^{\infty} g(t)(2\pi jf) e^{-2\pi jft} dt. \tag{8.22}$$

The term

$$g(t) e^{2\pi jft} \Big|_{-\infty}^{\infty}$$

is composed of two parts: a real signal $g(t)$ and an oscillatory component $e^{2\pi jft}$ (recall Tool 8, Euler's formula). Since $g(t)$ is a real signal, it is necessarily bounded, and hence

$$\lim_{t \to +\infty} g(t) = 0.$$

---

[7] Norbert Wiener (1894–1964), American mathematician; Aleksandr Yakovlevich Khinchin (1894–1959), Soviet mathematician.

[8] For an alternate way of obtaining this result, see [73].

In other words, this term vanishes. Consequently, we have

$$G\left(\frac{dg}{dt}\right) = 2\pi j f G(g(t)). \tag{8.23}$$

By following the above procedure, we can extend this result to the nth derivative and obtain

$$G\left(\frac{d^n g}{dt^n}\right) = (2\pi j f)^n G(g(t)). \tag{8.24}$$

Example 8.3 Just as for the Laplace transform, we can use the Fourier transform to determine the impulse function of a linear differential equation. To illustrate this point, consider the now familiar differential equation

$$\frac{dx}{dt} + kx = b(t), \tag{8.25}$$

where $k > 0$. Taking the Fourier transform of (8.25), we have

$$2\pi j f X(f) + k X(f) = B(f),$$

or equivalently,

$$X(f) = \frac{B(f)}{k + 2\pi j f}.$$

Recall that the solution $X(f)$ is the Fourier transform of a particular solution of (8.25). The complete solution of (8.25) is the sum of $X(f)$ and the solution of the corresponding homogeneous differential equation [72]. We know that the solution to the homogeneous equation is $X_0 \exp(-kt)$, where X_0 is determined by the initial condition. However, as we did for Bode analysis, we are assuming that the system is stable and initially at rest. Hence $X_0 = 0$. If we specify $B(f)$, we can determine $X(f)$. For example, suppose that $B(f)$ is the Fourier transform of a delta function:

$$G(\delta(t - t_0))(f) = \int_{-\infty}^{\infty} \delta(t - t_0) e^{-2\pi j f t}\, dt = e^{-2\pi j f t_0},$$

where $\delta(t - t_0)$ is the delta function. If we take $t_0 = 0$, then we conclude that $G(\delta(t - t_0))(f) = 1$. For this choice, we have

$$X_\delta(f) = \frac{1}{k + 2\pi j f}, \tag{8.26}$$

where $X_\delta(f)$ is the *frequency response function*. It is the Fourier transform of the impulse function.   ◇

8.3 Applications of Fourier Analysis

Here, we discuss four applications of the Fourier analysis of continuous time signals.

8.3.1 Uncertainties in Fourier Analysis

In Chapter 1, we emphasized the importance of choosing the appropriate scale for the independent variable to be able to view the phenomena of interest optimally. A familiar example is the use of the "zoom in" and "zoom out" options in pinpointing a location of interest on an Internet map. Scale is introduced mathematically by multiplying the independent variable, here the space coordinate, by a constant, converting $g(x)$, say, to $g(kx)$. If $k > 1$, then the graph of $g(kx)$ as a function of x is squeezed horizontally compared to the graph of $g(x)$. When $k < 1$, the opposite occurs, and the graph of $g(kx)$ is stretched out compared to $g(x)$.

What happens to the scaling in the frequency domain when we change the scaling in the time domain? The answer is quite surprising. It is encompassed by the famous *Fourier scaling theorem*, also known as the *stretch theorem*,

$$g(kt) \Longleftrightarrow \frac{1}{|k|} G\left(\frac{f}{k}\right). \tag{8.27}$$

When $k > 1$, $G(f/k)$ is both stretched out horizontally and vertically squeezed compared to $G(f)$, and when $k < 1$, $G(f/k)$ is both compressed horizontally and stretched vertically compared to $G(f)$. Thus, the changes in the frequency domain are in the opposite direction from those observed in the time domain.

The scaling theorem can be interpreted as showing that there is no value of k that simultaneously localizes a signal in both the time and frequency domains optimally. This conclusion is typically presented in the form of an *uncertainty principle*, such as

$$\Delta t \, \Delta f \geq \frac{1}{4\pi}, \tag{8.28}$$

where Δt is the time duration and Δf is the frequency bandwidth.

Example 8.4 Science students first learn about the uncertainty principle in the study of quantum mechanics, where it turns out that the momentum and position wave functions are a Fourier transform pair to within a factor of Planck's[9] constant. Thus, one can think of the Heisenberg[10] uncertainty principle in quantum mechanics as a consequence of the wavelike properties of matter.   ◇

Example 8.5 Let us use (8.28) to learn something about echolocation in bats. Bats emit sound waves to locate objects in their vicinity [772]. Much like sonar is used

---

[9] Max Planck (1858–1947), German theoretical physicist.
[10] Werner Karl Heisenberg (1901–1976), German theoretical physicist.

by a submarine, the bat's nervous system measures the time interval Δt between the time a sound is emitted and the time it returns after being reflected off an object. Depending on the species of bat, Δf is 5–200 kHz. Using (8.28), we see that the uncertainty in Δt is 0.1–10 μs.   ◇

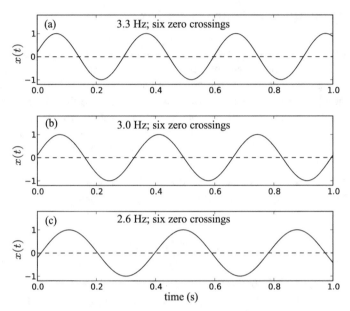

Fig. 8.4 Plot of $x(t) = \sin(2\pi f t + \phi)$, showing that different choices of frequency f can have the same number of zero crossings in a one-second interval. The phase shift ϕ is (a) 0.2 radians, (b) 0.1 radians, and (c) −0.2 radians.

A second factor that introduces uncertainty into Fourier analysis is the length L of the time series available for analysis. The frequency can be determined by counting the number z_c of *zero crossings* (Figure 8.4). A zero crossing corresponds to a half-cycle. Thus for a 3-Hz sine wave with $L = 1$ s, we have $z_c = 6$. However, since we can change the phase, it is possible that sine waves with frequencies other than 3 Hz also have $z_c = 6$ for time intervals of length 1 s. Indeed, the range of frequencies f_z having the same number of zero crossings is the open interval

$$\left(\frac{z_c - 1}{2}, \frac{z_c + 1}{2}\right) = (2.5\text{ Hz}, 3.5\text{ Hz}),$$

that is, all frequencies greater than 2.5 Hz and less than 3.5 Hz. If we increase L, then the range of

$$f_z \in \left(\frac{z_c - 1}{2L}, \frac{z_c + 1}{2L}\right)$$

having the same number of zero crossings becomes smaller: $(2.75 \text{ Hz}, 3.25 \text{ Hz})$ when $L = 2$ s, $(2.83 \text{ Hz}, 3.17 \text{ Hz})$ when $L = 3$ s, and so on.

The above observations demonstrate that there is a fundamental uncertainty in the resolution of frequencies related to the reciprocal of the length of the time series. For biological data, this limitation cannot be overcome simply by recording longer time series, because of issues related to nonstationarity (see Section 8.4.5). A practical solution is to append a string of zeros to the end of the time series to increase its length, a procedure referred to as *zero padding* [692]. In this way, resolution can be arbitrarily increased at the expense of increased data-processing time.

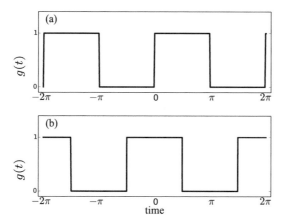

Fig. 8.5 (a) An odd-function representation of a square wave and (b) an even-function representation.

8.3.2 Approximating a Square Wave

Let us apply the theory of the Fourier series to approximate a square wave (Figure 8.5) by the first few terms of its Fourier series. We can make things easier on ourselves by defining the square wave in such a way that it is either an odd function or an even function, since then we will have only sine terms or only cosine terms. With that in mind, we define the odd function

$$g(t) = \begin{cases} 0 & \text{if } -\pi \le t < 0, \\ 1 & \text{if } 0 \le t < \pi, \end{cases}$$

and

$$g(t + 2\pi) = g(t).$$

This function is shown in Figure 8.5a. The even-function representation of the square wave (Figure 8.5b) appears as Exercise 1c.

First of all, what is a_0? We have from (8.6) that

$$a_0 = \frac{1}{2\pi} \int_{-\pi}^{\pi} dt = \frac{1}{2\pi} \left[\int_{-\pi}^{0} 0\, dt + \int_{0}^{\pi} 1\, dt \right] = \frac{1}{2}.$$

Since we have taken the odd-function form of $g(t)$, we have that $a_n = 0$ for all $n \geq 1$. The b_n can be readily determined using (8.10):

$$b_n = \frac{1}{\pi} \int_{-\pi}^{\pi} g(t) \sin nt\, dt = \frac{1}{\pi} \int_{-\pi}^{0} 0\, dt + \frac{1}{\pi} \int_{0}^{\pi} \sin nt\, dt = \frac{1}{n\pi}(1 - \cos n\pi).$$

Thus, we have

$$b_n = \begin{cases} 0 & \text{if } n \text{ is even}, \\ \dfrac{2}{n\pi} & \text{if } n \text{ is odd}. \end{cases} \tag{8.29}$$

Putting all this together, we see that the Fourier series representation of the square wave in Figure 8.5a is

$$g(t) = \frac{1}{2} + \frac{2}{\pi} \sin t + \frac{2}{3\pi} \sin 3t + \frac{2}{5\pi} \sin 5t + \cdots. \tag{8.30}$$

Since odd numbers can be written as $n = 2k - 1$, where k is a positive integer, we can rewrite (8.30) as

$$g(t) = \frac{1}{2} + \sum_{k=1}^{\infty} \frac{2}{(2k-1)\pi} \sin(2k-1)t \tag{8.31}$$

(see also Exercise 2 at the end of the chapter).

Figure 8.6a shows the approximations to the square wave produced by one, five, and ten terms of the Fourier series, while Figure 8.6b shows the approximation produced by the first 50 terms. Observe the little spikes at the edge of the square-wave approximation (indicated by the arrows in the figure). Such spikes are present even after many hundreds of terms have been added (indeed, they never go away, though they do become finer and finer). These spikes are a consequence of the attempt to describe a discontinuous function with smooth sine waves. They are referred to as the *Gibbs phenomenon*.[11] The importance of the Gibbs phenomenon is that it demonstrates that an abrupt change in a time series generates high-frequency components in the Fourier series (and, of course, also in the Fourier transform). In practical applications, this phenomenon arises when we try to compute the power spectrum of a signal that contains abrupt changes such as neural spikes [242, 681], or more commonly, a time series in which the two endpoints are different. Another

---

[11] This phenomenon was in fact discovered by Henry Wilbraham (1825–1883), English mathematician, and rediscovered by Josiah Willard Gibbs (1839–1903), American physicist, chemist, and mathematician.

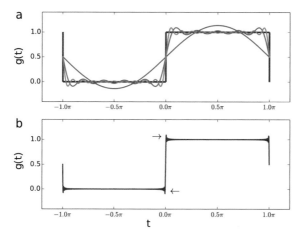

Fig. 8.6 (a) Approximations to a square wave by summing the first one (blue), five (red), and ten (green) terms of the Fourier series. (b) Approximation to the square wave generated by the first 50 terms of its Fourier series. The little spikes at the corners of the square waves (indicated by arrows) are known as the Gibbs phenomenon.

situation occurs when a figure is rendered digitally in the JPEG format [538]. The JPEG compression standard applies the discrete cosine transform to 8×8 blocks of pixels. The little dots that can be seen on the edges of a magnified JPEG figure are due to the Gibbs phenomenon.

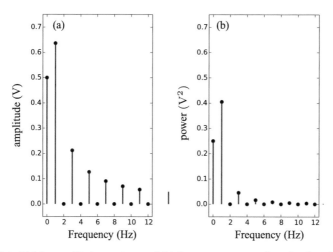

Fig. 8.7 Plot of (a) the amplitude spectrum and (b) the power spectrum for a 1-volt periodic square wave. The values of the amplitude and power calculated using (8.29) are represented by a bullet (•), and the solid vertical bars are those determined for a one-sided periodogram.

An alternative way to represent (8.31) graphically is to construct an *amplitude spectrum*. The amplitude spectrum for a periodic square wave is a plot of the amplitude given by (8.29) as a function of n (Figure 8.7a). Note that the amplitude when $n = 0$ is the mean value. Scientists prefer to plot these observations in the form of a *power spectrum*, namely a plot of the square of the amplitude of a frequency component (power) versus frequency (Figure 8.7b). We can extend this concept to other periodic waves $g(t)$; the only limitation is that we must be able to solve (8.12), at least numerically. The importance of the power spectrum is that it makes it possible at a glance to identify the frequency components of $g(t)$.

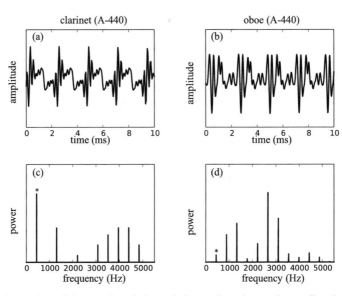

Fig. 8.8 Comparison of the sound made by a clarinet and an oboe each sounding the note A at 440 Hz. Figure (a) and (b) shows plots of the amplitude of the sound as a function of time. The corresponding power spectra are shown, respectively, in (c) and (d). The power spectra have been calculated as one-sided periodograms (see Section 8.4.4). The asterisk (∗) denotes the fundamental frequency ($f_0 = 440$ Hz). The time series data were extracted from ∗.wav files obtained from the Internet and downloaded using `scipy.io.wavfile.read()`.

Example 8.6 When a clarinet or an oboe or any other musical instrument sounds a note, it has a particular tone quality, or *timbre*. The timbre makes it possible for a listener to distinguish between two different instruments even though each is playing the same note. What is going on? Recall that the frequencies of sound waves generated by string and wind instruments are modeled using the *wave equation*, the solutions of which take the form of a Fourier series. Thus, just as we did for the periodic square wave, we can construct a power spectrum for the sound produced by a musical instrument. Figure 8.8 shows the power spectrum produced when a clarinet and an oboe play the note A at 440 Hz. The power spectrum for each instrument has the same general form. The leftmost peak corresponds to the fundamental frequency $f_0 = 440$ Hz. The harmonics occur at positive integer multiples of

f_0. The difference between the sounds produced by the two instruments is that the relative power of f_0 and the harmonics is different for the two instruments. In other words, the timbre of the sound depends on its spectrum. ◇

8.3.3 Power Spectrum

More insight into the concept of the power spectrum introduced above can be obtained using the Fourier integral. It is important to note that the only quantities that can be measured in the laboratory are those related to $|G(f)|$. This is because $|G(f)|$ is a real number. The reason that the power spectrum involves $|G(f)|^2$ and not $|G(f)|$ stems from physical considerations. Intuitively, the energy of a time-varying signal should be related to its amplitude $g(t)$. The power of a signal measured at an instant in time is $|g(t)|^2$ when the mean value of $g(t)$ has been subtracted. Thus, the total energy of the signal is

$$\int_{-\infty}^{\infty} |g(t)|^2 dt .$$

As a consequence of these observations, we see that the energy of a signal is proportional to its variance.

The energy, or variance, of a time-varying signal should be the same in the time and frequency domains, i.e.,

$$\int_{-\infty}^{\infty} |g(t)|^2 dt = \int_{-\infty}^{\infty} |G(f)|^2 df . \tag{8.32}$$

This observation is most commonly referred to as Lord Rayleigh's energy theorem, or *Parseval's theorem*.[12]

Since $G(f)$ is known, we can readily calculate $W(f)$ as

$$W(f) = \left| \int_{-\infty}^{\infty} g(t)e^{-j2\pi ft} dt \right|^2 = |G(f)|^2 = G(f)G^*(f) , \tag{8.33}$$

where $G^*(f)$ is the complex conjugate of $G(f)$ (Tool 8). Although analogies to the physical concepts of energy and power (energy per unit time) can be confusing in many applications, including, for example, the analysis of fluctuations in population densities, the term "power spectrum" remains widely used. As we will see, the properties of the power spectrum differ for discrete time signals (Section 8.4.4) and for deterministic versus stochastic signals (Section 14.6).

For real signals, $g(t)$ approaches zero as t approaches infinity, and consequently, the total energy is finite. There are two types of signals for which the total energy calculated as above would be infinite: (1) periodic signals and (2) stochastic signals (Chapter 14). In those situations, it is convenient to calculate the energy per unit time as

---

[12] Marc-Antoine Parseval des Chênes (1755–1836), French mathematician.

$$\lim_{\Delta t \to \infty} \frac{1}{2\Delta t} \int_{-\Delta t}^{\Delta t} |g(t)|^2 dt \,.$$

Parseval's theorem can also be written in terms of energy per unit time.

Example 8.7 What is the power spectrum of $\text{sinc}(f) = \sin \pi f / \pi f$? Using (8.33), we see that

$$W(f) = \frac{\sin^2(\pi f)}{(\pi f)^2} \,. \tag{8.34}$$

The significance of $\text{sinc}(f)$ is explained in Sections 8.4.1 and 8.4.4. $\diamondsuit$

8.3.4 Fractal Heartbeats

In nature, power spectra very frequently have the form (see also Section 14.7)

$$W(f) \approx \frac{1}{f^\beta} \,,$$

where β is positive and typically not integer-valued. The explanation for $1/f$ noise and its widespread occurrence is still not known. However, the Harvard cardiologist Ary Goldberger[13] and colleagues suggested that because of the complex branching structure of the electrical conduction pathways in the heart, called the Purkinje[14] fibers, the branching structure would be "self-similar" and hence would produce a $1/f^\beta$ power spectrum [280]. In the electrocardiogram (EKG), the QRS complex of a single heartbeat (○ in Figure 8.9a) corresponds to the depolarization of the ventricles as impulses generated by the sinoatrial node are distributed through the Purkinje fibers. The power spectrum shown in Figure 8.9b yields $\beta = 3.8$ (dashed line associated with ○). This observation was taken as supporting the fractal hypothesis for the variability in the QRS complex.

A few years later, this interpretation was challenged by two mathematical physiologists at McGill University, Tim Lewis and Michael Guevara[15] [479]. They argued as follows: The QRS complex of the EKG looks like a triangular function (solid line in Figure 8.9a), i.e.,

$$\text{tri}(t) = \begin{cases} 1 - |t|, & |t| < 1, \\ 0, & \text{otherwise}. \end{cases}$$

The Fourier transform of the triangular function is

---

[13] Ary L. Goldberger (b. 1949), American physician–scientist and fractal physiologist.

[14] Jan Evangelista Purkyně, or the Latinized Johannes Evangelist Purkinje (1787–1869), Czech anatomist and physiologist.

[15] Michael R. Guevara (PhD 1984), Canadian physiologist; Tim J. Lewis (PhD 1998), Canadian biomathematician.

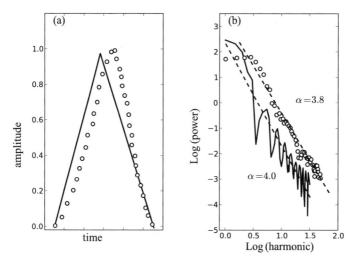

Fig. 8.9 (a) Comparison of the QRS complex in the electrocardiogram of a single heartbeat (○) with a triangular wave (solid line). (b) Comparison of the power spectrum W for the QRS complex (○) with that obtained for $sinc^2$ (solid line). One harmonic equals (window width)$^{-1}$, or ≈ 7.8 Hz. Data are from [479].

$$\text{TRI}(f) = sinc^2(f) = \frac{\sin^2(\pi f)}{(\pi f)^2},$$

and hence the power spectrum should resemble that of $sinc^2(f)$ (solid line in Figure 8.9b), i.e.,

$$W(f) \approx \frac{1}{f^4}.$$

Indeed, within experimental error, the slope measured by Goldberger and his coworkers is $4.0 \approx 3.8$. We see that the power spectrum of the QRS complex is not sufficient alone to argue in favor of a $1/f^\beta$ process (though we cannot, of course, conclude that it is impossible).

8.4 Digital Signals

In today's laboratories, it is usual to record a signal digitally. This is generally accomplished using an A/D board attached to a computer. This means that the time series is a discrete sequence of equally spaced points (the spacing is determined by the sampling frequency). The reason that we can represent a continuous time series by a discrete one is a consequence of[16] the famous *Nyquist–Shannon sampling theorem*.

Theorem 8.1 (Nyquist–Shannon). *The exact construction of a continuous-time baseband signal from its samples is possible if the signal is band-limited and the sampling frequency is greater than twice the signal bandwidth.*

Although the proof of this theorem is beyond the scope of this book, it is important to understand how this theorem is implemented. The important words in its statement are *baseband* and *band-limited*. The term *baseband* means that the signal must be low-pass filtered before it is digitized. The Nyquist frequency f_{Nyq} is half the sampling frequency. For example, if we want to include signals up to 500 Hz, we must sample at 1000 Hz. The condition that the time series must be low-pass filtered requires that the cutoff frequency be at least f_{Nyq} (see below). *Band-limited* means that the signal has significant power only over a finite range of frequencies. This requirement will generally be met for real signals, especially if they have been low-pass filtered. Thus, we see why the fact that differential equations can function as low-pass filters is so important.

Example 8.8 Music is recorded on a standard CD using a sampling rate of 44 100 Hz. Thus $f_{Nyq} = 22\,050$ Hz. A 16-bit A/D board is employed (Section 8.4.5) corresponding to $2^{16} = 65\,536$ distinct levels of amplitude, which, in turn, corresponds to a dynamic range of 96 dB. $\diamond$

8.4.1 Low-Pass Filtering

The importance of low-pass filtering for investigating the frequency content of digital signals is established by the Nyquist–Shannon theorem. In the frequency domain, the effects of filtering can be seen by solving the equation

$$X(f) = B(f)\text{Filter}(f),   (8.35)$$

where $\text{Filter}(f)$ is the frequency response function of the filter, $B(f)$ is the input to the filter, and $X(f)$ is the output of the filter. This equation is simply the Fourier transform of the convolution integral we introduced in Chapter 6. Often, as was the case in the previous example, $\text{Filter}(f)$ is determined by the relevant ordinary differential equation. Indeed, most of the common analog filters used in the laboratory setting, namely the Butterworth, Bessel, and Chebyshev filters, correspond to electrical circuit implementations of differential equations that bear the same name. For example, a linear first-order ordinary differential equation corresponds to a Butterworth filter of order 1. These observations are illustrated in the following example.

---

[16] Claude Elwood Shannon (1916–2001), American mathematician, engineer, and cryptographer; Harry Nyquist (1889–1976), American electronic engineer.

Example 8.9 The frequency response function for a Butterworth filter of order 1, a very commonly used low-pass filter in the laboratory, is given by (8.26). What are the effects of low-pass filtering a white noise input? White noise has the property that each frequency makes the same contribution to the total power, which implies that white noise has infinite power (for a more complete description of white noise, see Chapter 14). However, it is often useful to approximate noise as "white" over a finite frequency range. The Fourier transform of white noise is a constant B. Thus, the Fourier transform of the output $X_{wn}(f)$ is (see also Example 8.3)

$$X_{wn}(f) = \frac{B}{k + 2\pi j f}. \tag{8.36}$$

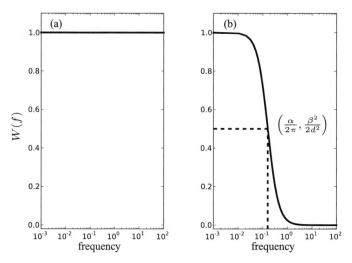

Fig. 8.10 (a). The power spectrum $W(f)$ of white noise. (b) Power spectrum of low-pass filtered white noise given in (8.37). It should be noted that the power spectrum for low-pass-filtered white noise is constant for all frequencies (Section 14.7). However, the power spectrum must be determined using the Wiener–Khinchin theorem as discussed in Section 14.6. If this is not done as shown here, the estimated power spectrum will be higher ($d > 1$) or lower ($d < 1$) than its true value. When $d = 1$, $\alpha = k$ and $\beta = B^2/k^2$.

We can use (8.33) to obtain $W(f)$ as

$$W_{wn}(f) = X_{wn}(f)X_{wn}^*(f) = \frac{B^2}{k^2 + (2\pi f)^2}. \tag{8.37}$$

Figure 8.10 compares the power spectrum of unfiltered white noise to low-pass-filtered noise described by (8.37). The shape of $W_{wn}(f)$ can be seen by rewriting (8.37) as

$$W_{wn}(f) = \frac{B^2}{k^2(1 + (\frac{2\pi f}{k})^2)} \tag{8.38}$$

to demonstrate more clearly its resemblance to a Hill function.[17] Thus we see that in the frequency domain, the effect of a first-order ODE is to truncate, or low-pass filter, an input. This is the same result that we obtained using the Laplace transform in Chapter 7.   ◊

Example 8.10 We can now understand why a sine wave can be generated from a periodic square wave (Section 8.1.1). In particular, the first term in (8.31) is a sine wave with the same frequency as the periodic square wave. The role of the low-pass filter is to remove all of the sine terms with higher frequencies.   ◊

There is a second type of filtering that impacts all experimentally measured time series. In theory, the Fourier transform is defined for an infinitely long time series. However, by necessity, in the laboratory, all time series have finite length. In other words, without knowing it, we have actually applied a filter, referred to as $rect(t)$, where (Figure 8.11a)

$$rect(t) = \begin{cases} 1 & \text{if } |t| < 1/2, \\ 0 & \text{otherwise}. \end{cases}$$

In order to determine the effects of this filter in the frequency domain, we use the Fourier transform to obtain

$$Rect(f) = \int_{-\infty}^{\infty} rect(t)e^{-j2\pi ft}\,dt = \frac{\sin(\pi f)}{\pi f} := Sinc(f). \qquad (8.39)$$

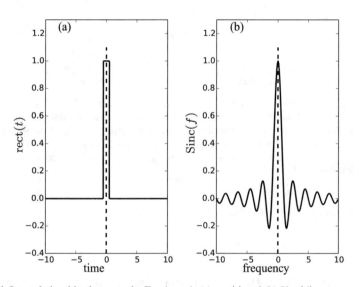

Fig. 8.11 Interrelationships between the Fourier pair (a) $rect(t)$ and (b) $Sinc(f)$.

---

[17] Note that we are just using one of our favorite tricks.

The function $\sin(\pi f)/\pi f$ arises so frequently in mathematical discussions [73, 264] that it has been given the name $\mathrm{Sinc}(f)$ (Figure 8.11b).[18]

Putting this all together, we see that the frequency-domain representation of the time series used to calculate the power spectrum has the form

$$X(f) = B(f)\,\mathrm{Sinc}(f)\text{low-pass filter }(f). \qquad (8.40)$$

8.4.2 Introducing the FFT

It is impractical to calculate by hand the Fourier transforms of time series measured in the laboratory. Thus, it becomes necessary to employ computerized numerical methods. Discrete approximations to integrals can be used to estimate a_0, a_n, and b_n. Here, we illustrate this procedure for the special case $g(t) = \sin 2\pi f_0 t$. Our reason for choosing $\sin 2\pi f_0 t$ is that the Fourier transform can be easily determined analytically, and thus it is possible to directly compare the theoretical and numerical results.

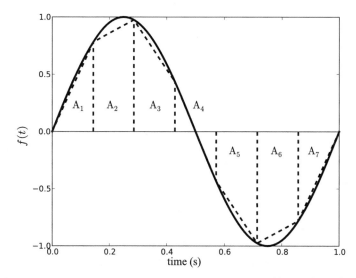

Fig. 8.12 Diagram of the trapezoidal rule applied to the function $g(t) = \sin(2\pi f_0 t)$ sampled at eight points. The trapezoidal rule computes the area of each block and then sums those areas.

The trapezoidal rule computes the integral by breaking down a function $g(t)$ into several trapezoids. Figure 8.12 shows $\sin(2\pi f t)$ sampled at eight time points t_i,

---

[18] This definition of $\mathrm{Sinc}(f)$ is the one most used in the signal-processing literature. In the mathematical literature, this function is written as $\mathrm{sinc}(t) = \sin(t)/t$. However, (8.39) is the definition that will be encountered most often by laboratory investigators.

$i = 0, \ldots, 7$, over one period. There are seven trapezoids A_i, and the area of the ith trapezoid is equal to

$$A_i = \frac{f(t_i) + f(t_{i+1})}{2} (t_{i+1} - t_i).$$

If we assume that the points t_i are equally spaced (in most applications, they are) such that $t_{i+1} - t_i = \Delta t$, then the total area is

$$A = \Delta t \left(\frac{g(t_0) + g(t_1)}{2} + \frac{g(t_1) + g(t_2)}{2} + \cdots + \frac{g(t_6) + g(t_7)}{2} \right),$$

which can be simplified to

$$\int_{-\infty}^{\infty} g(t)\,dt = 2 \int_{0}^{\infty} g(t)\,dt = \Delta t \left(g(t_1) + g(t_n) + 2 \sum_{i=2}^{N-1} g(t_i) \right). \tag{8.41}$$

Thus, the numerical approximation to (8.14b) that computes a_n for the Fourier transform of $g(t) = \sin 2\pi f_0 t$ becomes

$$a_n = \Delta t \left[\cos(2\pi n t_1) f(t_1) + \cos(2\pi n t_N) f(t_N) + 2 \sum_{i=2}^{N-1} \cos(2\pi n t_i) f(t_i) \right].$$

Similarly, for b_n, we have from (8.14c),

$$b_n = \Delta t \left[\sin(2\pi n t_1) f(t_1) + \sin(2\pi n t_N) f(t_N) + 2 \sum_{i=2}^{N-1} \sin(2\pi n t_i) f(t_i) \right]. \tag{8.42}$$

In this way, we compute $c_n = a_n + j b_n$.

How many operations must be done to calculate the Fourier transform using the trapezoidal rule? Equation (8.42) indicates that for each value of n, we would need to do N multiplications and N additions to calculate b_n. We would need to do this for all N Fourier modes and for both coefficients a_n and b_n. Thus, the total computation cost scales as $(2N)^2$. This means that if you double the number of data points, then the number of operations is quadrupled: a data set of $1,000$ points translates into $4\,000\,000$ operations. For signals of interest, the number of computations can rapidly become very large. For example, to sample music at a rate that sounds pleasing, one usually uses a sampling frequency of $44\,100$ Hz. In order to obtain $W(f)$, we would need $\approx 8 \times 10^9$ calculations for each second of data.

The fast Fourier transform (FFT) uses only $N \log_2(N)$ operations. For one second of data sampled at $44\,100$ Hz, this reduces the number of operations by a a factor of 10^4. It is important to keep in mind that the cost savings of the FFT can be fully realized only if the number of data points is a power of two, i.e., 2^n. However, this can always be accomplished by, for example, padding the signal with zeros.

Let us compare the Fourier transform of $\sin(2\pi f_0 t)$ computed analytically to that determined using the FFT. By definition, the Fourier transform of $\sin(2\pi f_0 t)$ is

$$G(\sin 2\pi f_0 t)(f) = \int_{-\infty}^{\infty} e^{-2\pi j f t} \sin 2\pi f_0 t \, dt, \tag{8.43}$$

where we have distinguished a particular value of the frequency as f_0. Since

$$\sin 2\pi f_0 t = \frac{e^{2\pi j f_0 t} - e^{-2\pi j f_0 t}}{2j},$$

we obtain

$$
\begin{aligned}
G(\sin 2\pi f_0 t)(f) &= \int_{-\infty}^{\infty} e^{-2\pi j f t} \left[\frac{e^{2\pi j f_0 t} - e^{-2\pi j f_0 t}}{2j} \right] dt \\
&= \frac{1}{2j} \int_{-\infty}^{\infty} \left[e^{-2\pi j (f - f_0) t} - e^{-2\pi j (f + f_0) t} \right] dt \\
&= \frac{1}{2j} [\delta(f - f_0) - \delta(f + f_0)] \\
&= \frac{j}{2} [\delta(f + f_0) - \delta(f - f_0)].
\end{aligned}
\tag{8.44}
$$

Thus the Fourier transform of $\sin(2\pi f_0 t)$ is composed of two delta functions, one centered at $+f_0$ and the other at $-f_0$.

Many computer programming languages have a command for computing the FFT of a discrete time series. It is generally of the form something like

```
fft(g(t)).
```

Let $g(t) = \sin(2\pi f_0 t)$ and write a computer program to generate the FFT when $f_0 = 1$ Hz and the sampling rate is 8 Hz. The output that appears on the computer screen is

```
   0.0000
  -0.0000          -4.0000j
   0.0000           0.0000j
   0.0000           0.0000j
   0.0000
   0.0000           0.0000j
   0.0000           0.0000j
  -0.0000          +4.0000j
```

where we have set all numbers less than 10^{-3} equal to 0. This output is in the standard form of the output generated by $\texttt{fft()}$, referred to as *reverse-wraparound ordering*. The ordering of the frequencies is

$$[0, 1, 2, 3, 4, -3, -2, -1].$$

Note that the eight discrete data points yield eight Fourier coefficients and that the highest frequency that will be resolved is the Nyquist frequency $f_{\text{Nyq}} = N/2$. The first half of the list of numbers corresponds to the positive frequencies, and the second half to the negative frequencies. Thus, we see that the Fourier transforms of $\sin(2\pi f_0 t)$ computed analytically and numerically are in excellent agreement.

8.4.3 Convolution using the FFT

In Section 6.4, we mentioned that it is much faster to calculate the convolution integral of a signal, s, of length N and an impulse function, r, of length M using the convolution integral. As we discussed previously, it is necessary to pad the FFT with zeros up to length $N + M$. Thus, the Python program for a convolution takes the form

```
R = np.fft.fft(r,len(s)+len(r))
S = np.fft.fft(s,len(s)+len(r))
Y=R*S
yi = np.fft.ifft(Y)[:len(s)].real
plt.plot(yi,'k-')
```

8.4.4 Power Spectrum: Discrete Time Signals

The W in our notation for $W(f)$ introduced in Section 8.3.3 indicates the power spectrum obtained when a mathematical expression for $G(f)$ is known. In the language of the benchtop, it is the power spectrum obtained for an infinitely long and continuous time series of a variable measured with infinite precision. However, time series measured in the laboratory are of finite length, contaminated with stochastic contributions, and discretely and imperfectly sampled. Thus, the experimentally determined power spectrum $w(f)$ provides only an estimate of $W(f)$ (hence the lowercase w). An obvious problem is how best to ensure that $w(f)$ provides a good estimate of $W(f)$.

Two terms are used to describe power spectra estimated from experimental observations: the *power spectrum* and the *power spectral density*. Spend a day searching the Internet for these terms, and you will quickly become thoroughly perplexed: some authors use these terms interchangeably [692], whereas others maintain a distinction. We believe that a brief historical perspective is useful.

Prior to the availability of A/D boards, the power spectrum of a continuous time signal was measured using a device called a spectrum analyzer. The basic component of this device is a band-pass filter. A band-pass filter allows the passage of only a narrow bandwidth of frequencies $f_c \pm \Delta f/2$, where f_c is the center frequency and Δf is the bandwidth. The output of the spectrum analyzer for a given f_c and Δf is squared and averaged over the whole signal to obtain a measure of power. This average value was divided by Δf to obtain an estimate of the change in the mean squared amplitude with frequency. Since this quantity is related to the statistical concept of density (see Section 14.2), the *power spectral density* has units of variance/frequency. For an electrical signal, the units are $V^2 \cdot Hz^{-1}$.

With the development of A/D boards, fast computers, and the FFT algorithm, it became possible to estimate power spectra numerically. The complex coefficients c_n determined using the FFT are multiplied by their complex conjugates to obtain a plot of $|c_n|^2$ as a function of frequency. The quantity $|c_n|^2$ is related to the variance. There are a variety of names that are used to refer to this plot, including the Fourier line spectrum, the variance spectrum, and, unfortunately, the power spectrum. In an effort to minimize confusion, we will refer to a plot of $|c_n|^2$ versus frequency as the *periodogram*. The important point for our discussion is that the units of power in the periodogram are those of variance, for example, V^2 for an electrical signal. We will reserve the term *power spectra* to refer collectively to the periodogram and the power spectral density. In order to appreciate how power spectra are calculated, it is important to consider the following questions:

1. How does the discrete Fourier transform differ from the continuous Fourier transform?

The discrete Fourier transform $G(n)$ of a time series of length N sampled at regular intervals Δ is

$$G(n) = \sum_{k=0}^{N-1} g(k)e^{-2\pi jnk/N}, \tag{8.45}$$

where we evaluate the transform only at the discrete frequencies

$$f_n = \frac{n}{N\Delta}, \quad n = -\frac{N}{2}, \ldots, \frac{N}{2},$$

so that $G(f_n) \approx \Delta G(n)$. It should be noted that k varies from 0 to $N-1$. The zero frequency corresponds to $n = 0$, the positive frequencies $0 < f < f_{\text{Nyq}}$ to $1 \leq n \leq N/2 - 1$, and the negative frequencies $-f_{\text{Nyq}} < f < 0$ to $N/2 + 1 \leq n \leq N - 1$. The value $n = N/2$ corresponds to both $f = f_{\text{Nyq}}$ and $f = -f_{\text{Nyq}}$.

The discrete Fourier transform shares all of the properties of the continuous Fourier transform that we have discussed up to this point. However, there is a subtle difference between these two forms of the Fourier transform in relation to the calculation of the power spectrum. The inverse discrete Fourier transform is

$$g(k) = \frac{1}{N} \sum_{n=0}^{N-1} G(n)e^{2\pi jkn/N}, \tag{8.46}$$

and we see that the right-hand side contains a factor $1/N$. This factor is sometimes overlooked in discussions of the discrete power spectrum (for a notable exception, see [692]), but it is important for ensuring that the discrete power spectrum is properly scaled.

2. What is the relationship between $|c_n|^2$ and the variance of an experimentally measured signal?

In other words, what does the mathematical quantity $|c_n|^2$ correspond to in the measured time signals? To answer this, we choose the constant of proportionality to be consistent with Parseval's theorem. From (8.46), we see that the discrete form of Parseval's theorem is

$$\sum_{k=0}^{N-1} |g(k)|^2 = \frac{1}{N} \sum_{n=0}^{N-1} |G(n)|^2, \tag{8.47}$$

where N is the number of points in the discrete time series. Thus, we have

$$\sum_{n=0}^{N-1} |\text{data amplitude}(n)|^2 = \sum_{k=0}^{N-1} |\text{FFT amplitude}(k)|^2 = \frac{1}{N} \sum_{n=0}^{N-1} |G(n)|^2$$

$$= \frac{1}{N} \sum_{n=0}^{N-1} |c_n|^2. \tag{8.48}$$

There are two subtleties that can cause confusion. The first concerns how the contribution of the negative frequencies in the Fourier transform is to be handled. The power spectra for an arbitrary signal $w(f)$ is

$$w(f) = |G(f)|^2 + |G(-f)|^2.$$

However, for real signals, the two terms on the right-hand side are equal, and hence

$$w(f) = 2|G(f)|^2.$$

Power spectra calculated when the 2 is included are called *two-sided power spectra*, and their use is favored by mathematicians and investigators working in communication theory. If the 2 is omitted, then we have *one-sided power spectra*, whose use is favored in the stochastic signals literature (see Section 14.6). Thus, (8.47) is the definition of the one-sided power spectrum: multiplying the right-hand side by 2 yields the two-sided power spectrum.

The second subtlety arises from the fact that there are several ways in which the power of a signal can be defined [692]. Equations (8.47) and (8.48) define power as the sum of the squared amplitudes. However, physicists often define power spectra by their relationship to the autocorrelation function as specified by the Wiener–Khinchin theorem. In this case, power is defined as the mean squared amplitude, and (8.48) becomes

$$\frac{1}{T} \int_0^T |g(t)|^2 dt \approx \frac{1}{N^2} \sum_{n=0}^{N-1} |c_n|^2,$$

and thus the right-hand side of (8.48) becomes $\frac{1}{N^2} \sum_{k=0}^{N-1} |c(n)|^2$. When this convention is used to measure power, a power spectrum is referred to as the autospectral density function or autospectrum [49].

3. What are the effects of discretization on the properties of power spectra?

A curious property of the periodogram is that the variance at each frequency is independent of the length N of the time series [692]. This observation is counterintuitive, since we might anticipate that as N approaches infinity, $w(f)$ should approach $W(f)$. The explanation is based on the fact that each frequency, defined as f_n, is not representative of the frequency bin. the frequency bin extends from halfway from the preceding discrete frequency bin to halfway in the next frequency bin. In contrast, f_n is only the value at the midpoint of the frequency bin. If we increase L, we simply increase the number of distinct frequencies at which the variance is determined; however, for a given discrete frequency, the variance remains the same. The simplest strategy for improving the estimate of $|c(n)|^2$ is to divide the original discrete time signal into n_d smaller segments each having the same length and sampled at the same frequency. The $|c(n)|^2$ determined for the n_d segments are average, and hence the variability of the estimate of $|c(n)|^2$ is reduced by n_d^{-1}.

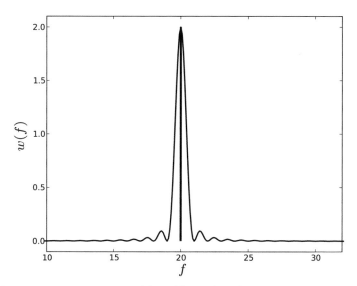

Fig. 8.13 Two-sided power spectrum $w(f)$ of $x(t) = \sin(2\pi f t)$ sampled at 128 Hz for one second when $f = 20$ Hz. The vertical dark line centered at 20 Hz is the two-sided delta-function power spectrum $W(f)$ calculated from (8.44).

4. What are the effects of the finite length of the data set on the determination of a power spectrum?

In the laboratory, power spectra are determined using a segment of finite length that is extracted from a signal that theoretically has infinite length. Mathematically, the procedure of extracting a finite segment from an infinitely long signal is equivalent to convolving the infinitely long time signal with the rectangular sample function, or window function, rect(t) that we introduced in Section 8.4.1. We know from (8.35) that for a continuous signal $g(t)$, the effects of rect(t) can be described in the frequency domain by the operation

$$X(f) = G(f) \operatorname{Sinc}(f) = G(f) \frac{\sin(\pi f)}{\pi f}, \tag{8.49}$$

where $G(f)$ is the Fourier transform of the low-pass-filtered signal. However, for discrete signals, we need to calculate the discrete Fourier transform of rect(t). Thus, (8.49) becomes

$$X(k) = \frac{G(k)}{N^2} \frac{\sin[\pi k]}{\sin[\pi k/N]}. \tag{8.50}$$

Suppose we take $g(t) = \sin(2\pi f_0 t)$. Then the two-sided periodogram becomes

$$W(f) = \frac{2}{N^2} \frac{\sin^2[\pi(f - f_0)]}{\sin^2[\pi(f - f_0)/N]}.$$

Figure 8.13 shows the periodogram $W(f)$ calculated for a 20-Hz sine wave sampled at 128 Hz for 1 second. Although most of the power is concentrated near $f_0 = 20$ Hz, there are ripples of power that occur at frequencies above and below f_0. These ripples correspond to leakage of power away from f_0. In contrast, for a continuous infinitely long sine wave time series, the two-sided $W(f)$ would consist of a single delta function at 20 Hz.

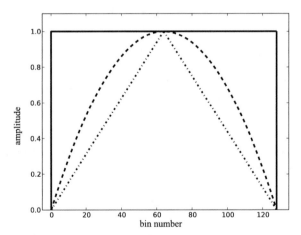

Fig. 8.14 Three commonly used windowing functions: rectangular (solid line), Welch (dashed line), and Bartlett (dash-dotted line).

A number of data windowing functions are used to minimize power leakage and the presence of ripples on the graphs of power spectra. Figure 8.14 compares rect(t) to two commonly used data windowing functions. From a practical point of view, both of these windowing functions are better than rect(t). However, the differences between the various nonrectangular windowing functions are not large [692]. These window functions are multiplied bin by bin with the data segment before the FFT is calculated. A problem with the use of windowing functions is that they go to zero at their endpoints. Thus valuable data can be lost, resulting in an increase in variability of the variance estimate. Fortunately, this problem is easily solved: multiple data segments are collected, and then the segments are overlapped by one-half their length.

8.4.5 Using FFTs in the Laboratory

A number of caveats must be kept in mind in performing an FFT on experimentally obtained time series (see also [692]).

Low-Pass Filter

A time series of finite length cannot be perfectly band-limited. Thus, it must be low-pass filtered to ensure that frequencies higher than the Nyquist frequency are not present.[19] The filtering procedure must be performed before the time series is digitized. If you fail to do this, you run the risk that the power spectrum will be contaminated by frequencies higher than the Nyquist frequency, a phenomenon referred to as *aliasing*. In other words, frequencies higher than the Nyquist frequency will be falsely represented as frequencies below the Nyquist frequency. The aliased position of the frequencies higher than the Nyquist frequency can be predicted from the relationship

$$\sin(2\pi[2f_{\mathrm{Nyq}} - f]t_n + \phi) = -\sin(2\pi f' t_n - \phi),$$

where ϕ is the phase and $f' = 2f_{\mathrm{Nyq}}f$. For example, $f_{\mathrm{Nyq}} = 22050$ for a compact disk. Hence a signal at a frequency of 34 100 Hz will look exactly like a signal at a frequency of $44\,100 - 34\,100 = 10\,000$ Hz. Developments in technology have made it possible to use an alternative strategy to avoid aliasing: the signal is sampled so frequently (e.g., 800 MHz) that it is unlikely that frequencies higher than the Nyquist frequency are present. In this case, it is not necessary to filter the signal before digitization.

---

[19] We do not consider the important topic of digital filters. We have kept our focus on analog filters because we wish to draw analogies to the filtering properties of biological dynamical systems.

Quantization Error

The analog-to-digital (A/D) board has become an essential tool for the study of time-dependent biological phenomena. It converts a continuous, or analog, signal, such as a voltage or current, into a series of numbers that are proportional to the magnitude of the signal. Since modern computers are digital devices, A/D conversion is essential in enabling computers to process the data for analysis and, in some cases, to control the instruments that collect the data. (For interesting applications, see [361, 809].) This latter process requires digital-to-analog (D/A) conversion. A/D conversion involves two steps: time discretization and amplitude quantization. Time discretization is the relatively error-free step, while the amplitude quantization step involves several issues that must be considered (see below). The *resolution* of an A/D board is the number of discrete values that it can produce over the input range of analog values. Computers are binary devices (0's and 1's), and hence the resolution of the A/D board is expressed in *bits*.

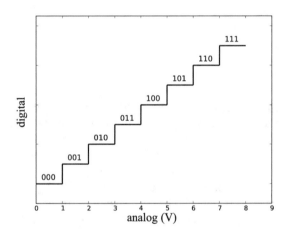

Fig. 8.15 Schematic illustration of the relationship between analog input to an A/D board and its digital output. This 3-bit A/D board has $2^3 = 8$ quantization levels.

Figure 15.4 shows the encoding scheme for a three-bit A/D board, namely an A/D board that has $2^3 = 8$ discrete levels. If we examine this figure closely, we see that an input between 1 V and 2 V would be encoded as 001 (the number 1 represented as a three-digit binary number). An input between 2 V and 3 V would appear as 010 (the three-digit binary representation of 2), and so on. The quantization error can be minimized by increasing the resolution of the A/D board. Standard A/D boards have 16-bit resolution ($2^{16} = 65\,536$ levels); however, higher-resolution A/D boards are available. Curiously, noise plays an important role in the functioning of an A/D board (see Section 15.2.1).

Digitization Frequency

At first glance, it is tempting to choose a low-pass filter whose corner frequency is equal to f_{Nyq}. However, it is important to keep in mind that analog filters, including the RC circuit, typically exhibit characteristics near the corner frequency that are less than ideal. For this reason, it is prudent to filter the signal using a filter whose corner frequency is greater than f_{Nyq}. The difference between the Nyquist frequency and the higher corner frequency of the filter is called the *guard band*.

Abrupt Changes

The presence of abrupt changes in the time series increases the presence of fast frequencies in the power spectrum (Gibbs phenomenon) that can cause aliasing. Another scenario that produces additional high-frequency components in a time series arises when the first and last data points in the signals are unequal. Thus, it is best to analyze time signals in which such abrupt changes do not occur. Digital filtering techniques can be used to overcome this problem. An example in which this issue arises is the calculation of the power spectrum for a neural spike train [242, 681].

Stationarity

When the statistical properties of a signal such as the mean and variance do not change significantly over the length of the chosen segment, the signal is said to exhibit *stationarity*. An assumption made in using the FFT is that the time series is stationary. Biological time series, however, are notoriously nonstationary. Indeed, it has been estimated that for a relaxed subject with eyes closed, the human electroencephalogram (EEG) is stationary for only about 10 seconds [138]. Well-known causes of nonstationarity in electrophysiology include slow DC-baseline drifts in the performance of measurement devices and recording artifacts related to movement and sweating. Of course, we cannot ignore the possibility that the biological process may itself be intrinsically nonstationary (see Section 8.4.6).

Issues of nonstationarity are particularly troublesome in situations in which it is necessary to characterize low-frequency oscillation, namely oscillations whose frequency is less than about 1 Hz. As the frequency to be resolved becomes smaller, the length of the time series needed to resolve it becomes greater, and as the length of the required signal becomes greater, the more important the problems related to nonstationarity become. This observation often frustrates biologists, since it is often the case that the low-frequency signals attract the most interest. Examples include the resting adult respiratory rhythm (0.13–0.33 Hz), circadian rhythms ($\approx 10^{-5}$ Hz), and a 10-year population cycle ($\approx 10^{-9}$ Hz). Sometimes, the nonstationarity takes the form of a clear trend in the time series, such as a linear increase in the mean. In such cases, it may be possible to subtract away the trend and then analyze the properties of the detrended time series. However, most of the time, the investigator is faced with much more complex nonstationarities that are very difficult to deal with (see Section 8.4.6).

Periodogram Versus Power Spectral Density

Few studies have addressed whether it is better in routine laboratory work to calculate the periodogram or the power spectral density. The key issue concerns the effects of random perturbations ("noise") on the produced power spectra.

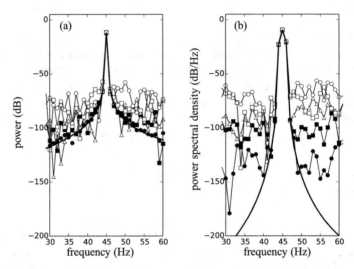

Fig. 8.16 Comparison of (a) the periodogram and (b) the power spectral density of a noisy sine wave $x(t) = \sin(2\pi f t) + \sigma^2 \xi(t)$, where $f = 45$ Hz and $\sigma^2 \xi(t)$ is white noise uniformly distributed on the interval $[a, b]$. The variance σ^2 of the noisy perturbations (see Section 14.2.3) is 0.0 (solid line), 0.01 (filled circle), 0.02 (filled box), 0.05 (open triangle), 0.1 (open square), and 0.2 (open circle). The digitization frequency was 1024 Hz.

Figure 8.16 compares the periodogram and the power spectral density for a 45-Hz sine wave of amplitude 1 volt as the intensity of added white noise is varied. The solid lines without symbols are the periodogram (Figure 8.16a) and the power spectral density (Figure 8.16b) calculated in the absence of noise. We make the following observations. First, both methods of calculating the power spectra work surprisingly well for identifying the presence of a 45-Hz component in the noisy signal up to a noise variance of $\approx 20\%$, the highest variance shown. However, there are some trade-offs. On the one hand, if we estimate the resolution as the width of a spectral peak at half height, then the periodogram has greater resolving power than the power spectral density. On the other hand, if we consider the height of the spectral peak above baseline as an estimate of the reliability that a spectral peak can be identified, then the power spectral density could be considered more reliable at low noise intensities. Overall, we suspect that there is little practical difference between the two types of power spectra. Presumably, an investigator should try both methods, and then choose the method that is better suited to the situation at hand.

8.4.6 Power Spectral Density Recipe

In the not too distant past, obtaining proper power spectra was a task that was beyond the capacity of most biologists. However, thanks to the development of computer software packages such as MATLAB, Mathematica, Octave, and Python, this task has become much easier. Although it is relatively easy to write a computer program for computing periodograms, it is not so easy to write such a program in a manner that is convenient for many of the scenarios encountered by an experimentalist. For this reason, we strongly recommend the use of Python's `matplotlob.mlab.psd()`, developed by the late John D. Hunter[20] using the method described by Bendat and Piersol [49].[21] This function has the form

```
power, freqs = matplotlib.mlab.psd(x, NFFT, Fs,
detrend, window, noverlap=0, pad_to, sides=,
scale_by_freq)
```

The output is two one-dimensional arrays, one of which gives the values of the power in $dB \cdot Hz^{-1}$, the other the corresponding values of the frequency.

The advantage of `matplotlib.mlab.psd()` is its versatility. Assuming that an investigator has available a properly low-pass-filtered time signal, the steps for obtaining the power density are as follows:

1. Enter the filtered and discretely sampled time signal $x(t)$.
2. The number of points per data block is NFFT, which must be an even number. The computation is most efficient if $NFFT = 2^n$, where n is a positive integer; however, with modern laptops, the calculations when NFFT is not a power of 2 are performed quickly.
3. Enter Fs. This parameter scales the frequency axis on the interval $[0, Fs/2]$, where $Fs/2$ is the Nyquist frequency.
4. Enter the detrend option. Recall that it is a standard procedure to remove the mean from the time series before computing the power spectrum.
5. Enter the windowing option. Keep in mind that using a windowing function is better than not using one, and from a practical point of view, there is little difference among the various windowing functions. The default is the Hanning[22] windowing function.
6. Typically accept the `noverlap` default choice of 0 (see below for more details).
7. Choose `pad_to` (see below). The default is 0.

---

[20] John D. Hunter (1968–2012), American neuroscientist and developer of `matplotlib`.

[21] Do not use PyLab's version of `psd()`. The function `matplolib.mlab.psd()` produces the same power spectrum as that obtained using the corresponding programs in MATLAB.

[22] Julius Ferdinand von Hann (1839–1921), Austrian meteorologist.

8. Choose the number of sides. The default for real data is a one-sided power spectrum.
9. Choose `scale_by_freq`. Typically, you will choose the default, since that makes the power spectrum compatible with that obtained using MATLAB.

The versatility of `mlab.psd()` resides in the interplay between the options NFFT, `noverlap`, and `pad_to`. The following scenarios illustrate the salient points for signals whose means have been removed.

1. Frequencies greater than 1 Hz: The options available depend on $L(x)$, the length of x. If $L(x) = $ NFFT, then the time is 1 s, and we can accept the defaults and the command to produce the power spectrum. For example, when the digitization rate is 256 points per second, the command would be

    `mlab.psd(x,256,256)`

    We can increase the resolution using the `pad_to` option. For example, for a sample initially digitized at 256 Hz, we can calculate the power spectrum for a 512-Hz digitization rate using the command

    `mlab.psd(x,256,512,pad_to=512)`

    Finally, we can average the time series by taking x to be X_n, an integer multiple of NFFT. This procedure is useful when data are noisy and when we want to use the `noverlap` option in addition to windowing to minimize power leakage. Thus, the command takes the form

    `mlab.psd(X_n,256,256,noverlap=128)`

    It should be noted that the optimal choice for `noverlap` is one-half of NFFT, namely half the length of the time series, including zero padding.
2. Frequencies less than 1 Hz: It is necessary first to determine the minimal length of time series that is required. Suppose we wanted to look for periodic components of order 0.1 Hz. A 0.1-Hz sinusoidal rhythm will have one cycle every 10 seconds. Thus the length of the time series must be at least 20 seconds. However, we will clearly get a much better looking power spectrum for longer time series, so perhaps we should use five times longer, e.g., 100 s. Thus the command becomes

    `mlab.psd(x,6400,64)`

    It is important to note that the characterization of the frequency content of biological time series in the less than 1 Hz frequency range is made problematic because of the effects of nonstationarity. Thus, the length of the time series and often the particular segment of the time series to be analyzed must be chosen with great care. In fact, it is in the analysis of such time series that many of the options available in `mlab.psd()` are most useful.

8.5 What Have We Learned?

1. What is the fundamental conclusion of Fourier's theorem?

   All experimentally measured time series can be represented mathematically as a sum of sines and cosines with different frequencies, phase shifts, and amplitudes.

2. What are important examples of a Fourier integral transform pair?

   Examples of Fourier integral transform pairs include a time series and its Fourier integral transform, the power spectrum and the autocorrelation function (see Section 14.6), the probability density function and the characteristic function (see Section 14.2.3), and the momentum and position wave functions in quantum mechanics.

3. How can a sine wave be obtained from a periodic square-wave time series?

   Low-pass filtering a periodic square-wave time series yields a sine wave.

4. What is the power spectrum?

   The power spectrum describes how the energy (or variance) of a signal is distributed with respect to frequency.

5. What are the effects of the length of a time series on the power spectrum?

   In general, the longer the time series, the better the resolution of the frequency components. However, most biological signals are nonstationary, and hence using too long a time series can introduce artifacts. A practical solution is to pad with zeros.

6. Why it is important to use a windowing function for the calculation of a power spectrum, and which windowing functions are recommended for routine laboratory use?

   The procedure of extracting a segment of finite length from an infinitely long signal introduces power leakage and ripples into the power spectrum. Data windowing functions are used to minimize these effects. It is useful to keep in mind that using any windowing function is better than using none at all. The Hanning windowing function is the default for `matplotlib.mlab.psd()`.

7. What is the effect of an abrupt change in a time series on its power spectrum?

   The effect of an abrupt change in a time series is to introduce high-frequency components into the power spectrum (Gibbs phenomenon). An often overlooked but very important way that an abrupt change can be introduced into an experimentally collected time series is through the initial and final points not being the same. Recall that the Fourier transform assumes that the signal is periodic.

8. What is the fundamental uncertainty that underlies the representation of trajectories in the time and frequency domains?

   The uncertainty arises because of the reciprocal relationship between time and frequency. Consequently, it is impossible to concentrate both the time series $g(t)$ and its Fourier transform $G(f)$.

9. What factors determine the resolution of a power spectrum?

> The reciprocal relationship between time and frequency imposes an unavoidable limitation on the resolution of a power spectrum. A second factor is the length of the time series used for determining the power spectrum. For a stationary signal, the longer the time series, the better the resolution.

10. How can the power spectrum be determined from the Fourier transform of the output of a dynamical system?

> If $X(f)$ is the Fourier transform of the output of a dynamical system, then the power spectrum is given by $W(f) = X(f)X^*(f)$, where $X^*(f)$ denotes the complex conjugate of $X(f)$.

11. Under what conditions can a continuous time series be faithfully represented by a discrete one?

> A continuous signal can be faithfully represented by a discrete one if it is low-pass filtered to ensure that there are no frequencies present higher than the Nyquist frequency and if the digitization rate is at least twice the corner (cutoff) frequency of the low-pass filter.

12. The Shannon–Nyquist theorem stresses the importance of low-pass filtering. What are the important constraints on this low-pass filtering process?

> The time series must be appropriately low-pass filtered before it is digitized.

13. Why is calculating a convolution using the FFT typically faster than using, for example, np.convolve?

> Computer multiplication of two Fourier transforms is typically much faster than the evaluation of a sum used by np.convolve.

8.6 Exercises for Practice and Insight

1. There are many ways that the Fourier series for a periodic square wave can be written.

   a. Consider a square wave defined as the $2L$-periodic function

   $$g(t) = \begin{cases} 0 & \text{if } -L \leq t < 0, \\ 1 & \text{if } 0 \leq t < L, \end{cases}$$

   and $g(t) = g(t+2L)$. Show that

   $$g(t) = \frac{1}{2} + \sum_{k=1}^{\infty} \frac{2}{(2k-1)\pi} \sin\left[\frac{(2k-1)\pi t}{L}\right].$$

   b. Of course, the Fourier series can be written in terms of frequencies. Show that

$$g(t) = \frac{1}{2} + \frac{4}{\pi} \sum_{k=1}^{\infty} \frac{\sin(2\pi(2k-1)ft}{(2k-1)}.$$

c. A periodic square wave can also be considered an even function (Figure 8.5b). Show that in this case,

$$f(t) = \frac{1}{2} + \frac{2}{\pi} \sum_{k=1}^{\infty} \frac{(-1)^{k-1}}{2k-1} \cos((2k-1)\omega_0 t).$$

The take-home message from this exercise is that one must check a Fourier series representation obtained from a mathematical table or on the Internet carefully to ensure that one has understood the situation that is being described.

2. Write a computer program to show that (8.31) produces the results shown in Figure 8.5.

3. An investigator wants to use a 25-Hz periodic square wave to make a 25-Hz sine wave for an experiment. The investigator proposes to use an RC circuit to perform the required low-pass filtering. What should be the value of the corner frequency, namely the product RC, where R is the resistance and C is the capacitance? Write a computer program to confirm your prediction. If you do so, you will see that the filtered square wave looks either like a square wave with rounded edges or a sawtooth wave, depending on the value of RC. However, if you pass the filtered signal through the same RC circuit a second time, the result looks much more sinusoidal. Explain this observation.

4. Using the Fourier transform of $G(\sin 2\pi f_0 t)$ given by (8.44), calculate the power spectrum $W(f)$. What is the one-sided power spectrum and what is the two-sided power spectrum?

5. Let $F(f \leq X)$ be the cumulative variance spectrum giving the variance arising from all frequencies less than X. Using dimensional analysis, show that

$$\frac{dF(f \leq X)}{df} = k_e f F(f \leq X),$$

where $k_e = \log(10)$.

6. Determine the power spectrum for the limit cycle generated by the van der Pol oscillator (Section 4.5.1). Hint: It is useful to choose the time step of the integration to be 2^n.

7. A good way to learn how to calculate power spectra is to practice doing it. Ask your professor to show you how to collect a time series of something that you yourself can measure, such as the electrical activity generated by electrical fish, the electrical activity of muscle contraction, a pendulum swinging, postural sway collected using a force platform, or finger tremors. Once you have collected your time series, use `matplotlib.mlab.psd()` to calculate the power spectrum.

Chapter 9
Feedback and Control Systems

One of the most important regulatory mechanisms is feedback [24, 37]. Virtually, every physiological variable is associated with a feedback control loop. Indeed, some would even say that every such variable has multiple feedback loops associated with it [35, 277]. Examples arise in the nervous system (e.g., pupil light reflex, recurrent inhibition, and stretch reflex), protein synthesis and gene regulation, endocrine systems, respiration, blood cell production, and the control of blood pressure. Of course, the importance of feedback is not limited to physiology. Feedback mechanisms regulate population size, control climate, and even help us operate our automobiles.

Feedback control mechanisms have a looped structure in which the output is fed back as input to influence future output. In contrast, in the dynamical systems we have considered up to now that the output has depended only on the input. Such systems are referred to as *open-loop* systems to distinguish them from the *closed-loop* architecture of a feedback mechanism. Figure 9.1 compares open-loop and closed-loop controls. The output of a closed-loop controller is fed back to a comparator ($\otimes$) that computes the difference between the input and output. This difference is called the *error signal*. We are most often interested in feedback controllers that keep the error signal as small as possible. For example, the input might be the temperature we want our room to be at, and then the feedback controller would be a thermostat, which regulates a furnace to ensure that the room achieves and maintains the desired temperature. Such feedback controllers are called negative feedback controllers: if the temperature is not where we want it to be, the controller works to minimize the temperature error signal.

The dynamics of feedback control can be analyzed either in the time domain using ordinary differential equations or in the frequency domain using the Laplace and Fourier transforms. The important point is that when the dynamical system is linear, the description of the dynamics in the time and frequency domains must be the same. Most often investigators choose the appropriate analysis based on convenience. For example, at the benchtop it is typically much easier to investigate a biological system operating in a linear range using methods based on the Fourier

© Springer Nature Switzerland AG 2021
J. Milton and T. Ohira, *Mathematics as a Laboratory Tool*,
https://doi.org/10.1007/978-3-030-69579-8_9

transform and thus take advantage of the FFT. On the other hand, for systems not operating in a linear range, such as systems which produce sustained oscillations, it is typically easier to use techniques developed for the analysis of nonlinear ordinary differential equations.

Three issues common to control problems are of particular interest to biologists. First, since the various components of a feedback loop are spatially separated, it is necessary to take into account the effects of time delays, which account for the time it takes to detect an error and then act on it. Second, as we shall soon see, time-delayed feedback control loops have a propensity to generate oscillations. Engineers aim to design controllers that eliminate, or at least minimize, the effects of oscillations. Biological control does not shun oscillations. Nonetheless, the concept that oscillations in biology can interfere with orderly control is reinforced by the observations that certain diseases (see Section 13.2), appropriately called *dynamical diseases*, are characterized either by the appearance of oscillations that are not present in healthy individuals or by changes in the nature of normally occurring oscillations, such as the heartbeat and gait cycle [522, 523, 571]. Finally, it is necessary to understand the effects of random perturbations, referred to as *noise*. Noise interferes with the accurate determination of the error signal, but paradoxically, it can sometimes have stabilizing effects [501, 580].

In this chapter, we introduce the study of feedback control in biological systems and introduce the role played by time delays in control. In the following chapter, we expand our discussion of time delays and then in the chapters which follow we examine the properties of oscillations (Chapters 11 and 13) and finally, the effects of noise (Chapters 14–16).

The questions we will answer in this chapter are

1. What is the difference between open-loop and closed-loop controls?
2. What is meant by the term "negative feedback?"
3. How does one determine in the laboratory whether a biological feedback control mechanism is operating in a linear range?
4. What properties of a first-order DDE with proportional negative feedback determine its stability?
5. How does a Andronov–Hopf bifurcation occur?
6. Why are Andronov–Hopf bifurcations important?
7. What is the relationship between the period of an oscillation generated by a time-delayed negative feedback controller and the time delay?
8. What observation suggests that biological feedback control mechanisms operate is a manner similar to the feedback mechanisms used by engineers?

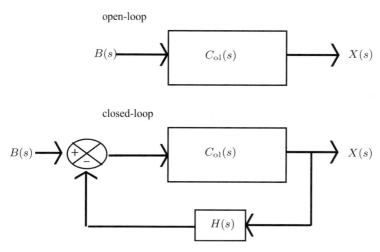

Fig. 9.1 A comparison of the block diagram for an open-loop control mechanism (top) and a closed-loop negative feedback control mechanism (bottom), illustrating how an experimentally determined open-loop transfer function $C_{ol}(s)$ is incorporated into the corresponding closed-loop feedback controller.

9.1 Feedback Control: frequency domain approaches

The *block diagram* is a useful and widely employed graphical shorthand for the analysis of the feedback control of complex linear dynamical systems. Figure 9.1 compares the block diagrams for a system with open-loop control and a system with closed-loop control. In constructing a block diagram, the components of the system most relevant for control are represented by boxes, referred to as *blocks* (see below), which are connected by lines that show the relationships between them.

It is the properties of the integral transforms we discussed in Chapters 7 and 8 that make block diagrams so useful. By convention, block diagrams are discussed using the Laplace transform; however, one could equally well use the Fourier transform. Recall from Section 7.3 that the transfer function $C(s)$ for a system is given by

$$C(s) = \frac{X(s)}{B(s)}, \tag{9.1}$$

where $B(s)$ is the Laplace transform of the input to the system (often referred to as the referenced input), and $X(s)$ is the Laplace transform of the output of the system. The goal of the block diagram is to calculate $C(s)$ given the transfer functions $C_i(s)$ of each of the n blocks, where $i = 1, 2, \ldots, n$. Each block is labeled by its transfer function. Examples of blocks include the transfer function of a time delay, the transfer function for a photoreceptor (7.49), the signal transduction in the yeast Hog1 cascade, the transfer function for the closed-loop control of the pupil light reflex (9.18), and so on. Depending on the system under consideration, the transfer functions of the blocks may be known analytically or have been determined experimentally using techniques such as the Bode plot (Section 7.5).

Block diagram descriptions of biological control systems can become very complex. For example, Guyton's[1] famous block diagram for the control of circulation contains over 350 blocks [434]. Thus, it is useful to simplify the block diagram as much as possible so that the important relationships between the blocks can be more easily seen. There are two rules that are used to simplify a block diagram. The first is that the transfer function of n blocks arranged in series is equal to the product of the transfer function of each of the blocks. This can easily be understood from (9.1).

Suppose that the open-loop control system shown in the top of Figure 9.1 consists of n blocks arranged in series. Then the output of the first block is $B(s)C_1(s)$, the output of the second block is $B(s)C_1(s)C_2(s)$, and the output of the nth block is $B(s)\Pi_{n=1}^n C_i(s) = X(s)$. Thus $C_{ol}(s) = \Pi_{n=1}^n C_i(s)$. For example, Stark considered that the block diagram for the crayfish photoreceptor system consists of the block diagram for the time delay in series with the block diagram that describes the process by which the photoreceptor converts light intensity into a neural spike train. Thus (7.49) is the product of two transfer functions.

The second rule concerns the nature of the transfer function for a closed loop. Here we determine this for the case of negative feedback control. Exercise 4 at the end of the chapter determines the transfer function for a positive feedback control loop. The bottom of Figure 9.1 shows the block diagram for closed-loop negative feedback control. For closed-loop control, the system contains two additional blocks that are not seen for the same system under open-loop control: the comparator, denoted by $\otimes$, whose output is the error signal $E(s)$, and a block for the feedback controller $H(s)$.

In biological applications, it is often the case that the transfer function for the the block that is controlled, or *plant* as the engineers call it, can be measured under open-loop conditions. We have therefore used the same label $C_{ol}(s)$ in Figure 9.1b that we used in Figure 9.1a. Here, we show how the closed-loop transfer function $C_{cl}(s)$ for the system can be determined if $C_{ol}(s)$ is known. Referring to the bottom of Figure 9.1, we can write

$$X(s) = E(s)C_{ol}(s) \tag{9.2}$$

and

$$E(s) = B(s) - X(s)H(s). \tag{9.3}$$

If we combine (9.2) and (9.3), we get

$$X(s) = [B(s) - X(s)H(s)]C_{ol}(s),$$

or

$$C_{ol}(s)B(s) = X(s)[1 + C_{ol}(s)H(s)].$$

In other words,

---

[1] Arthur Clifton Guyton (1919–2003), American physiologist.

$$C_{cl}(s) = \frac{\text{output}}{\text{input}} = \frac{X(s)}{B(s)} = \frac{C_{ol}(s)}{1 + H(s)C_{ol}(s)}. \qquad (9.4)$$

We see that in a closed-loop system, the change in the output for a given input is less than that observed for the related open-loop system. By choosing $H(s)$ appropriately, the error can be made small, and the effects of an external disturbance can be largely canceled.

The effectiveness of a negative feedback controller is limited by its tendency to generate oscillations. To illustrate this point, let us assume that

$$C_{ol}(s) = \frac{1}{s^2 + 3s + 2}.$$

As we showed in Section 7.3, the characteristic equation is obtained by setting the denominator of $C_{ol}(s)$ equal to zero,

$$s^2 + 3s + 2 = 0,$$

which corresponds to the ordinary differential equation

$$\frac{d^2x}{dt^2} + 3\frac{dx}{dt} + 2x = 0.$$

In this case, the fixed point $x^* = 0$ is a stable node ($\lambda_1 = -1, \lambda = -2$).

Now suppose we close the loop and assume that $H(s) = 1$ (this case is referred to as *unitary feedback*). Now we have

$$C_{cl}(s) = \frac{C_{ol}(s)}{1 + C_{ol}(s)} = \frac{1}{s^2 + 3s + 3}.$$

In this case, the characteristic equation corresponds to the differential equation

$$\frac{d^2x}{dt^2} + 3\frac{dx}{dt} + 3x = 0,$$

and we see that the fixed point is a stable spiral point ($\lambda_{1,2} = -1.5 \pm 3j$). In contrast to a stable node, a stable spiral point exhibits a damped oscillation in response to a perturbation. This propensity to generate oscillations becomes even greater when the feedback loop is time-delayed, since $H(s)$ becomes $e^{-\tau s}H(s)$. In fact, it is possible that the feedback controller can become unstable. Clearly, there is a trade-off between the accuracy of a feedback controller and its propensity to generate oscillation and instability.

9.2 "Linear" Feedback

Given the importance of measuring $C_{ol}(s)$ in order to determine $C_{cl}(s)$ for biological systems, investigators have employed a combination of surgical resections and drugs that inhibit unwanted inputs into the system in order to isolate the system under consideration. For example, $C_{ol}(s)$ for the crayfish photoreceptor in Section 7.4.1 was measured under "open-loop" conditions by surgically isolating the sixth abdominal ganglion from all inputs except light. However, these approaches are not generally feasible and hence determining $C_{cl}(s)$ by measuring $C_{ol}(s)$ is not likely to be an approach of wide applicability in biology.

A moment's reflection suggests a more encouraging possibility. What is the operating range of the feedback controller? It is possible that certain feedback controllers in biology are tuned to operate very close to the fixed point of the dynamical system. The concept of homeostasis in physiology [57, 103], namely the relative constancy of the body's internal environment, suggests that this will generally be true for a healthy individual. If we accept this postulate, then the dynamics sufficiently close to the fixed point will be described by a linear differential equation, allowing us to define a transfer function $C_{\ell}(s)$. This assumption will hold only when the fixed point of the dynamical system is stable. Of course, $C_{\ell}(s)$ determined in this linear range is not necessarily the same as $C_{cl}(s)$ determined for the whole operating range (hence the difference in notation). However, this is not necessarily a limitation if our goal is to understand the input–output response for inputs that cause small deviations from the fixed point.

How can we determine experimentally whether a feedback controller is operating in a linear range? One way is to use the principle of superposition (Tool 1) introduced in Chapter 1. However, since we do not know a priori the extent of the linear range, this approach can be time-consuming to implement in the laboratory. Consequently, investigators typically use other approaches to estimate linearity.

The simplest method is to inject a sinusoidal input and see whether the output is also sinusoidal. In general, the frequency of a feedback system depends on its linear properties, whereas the shape of the waveform depends on its nonlinear properties [792]. Thus if we observed that both the input and output are sinusoidal, then we can infer that, at least to a first approximation, the feedback system is operating in a linear range. This observation can be used to justify the use of Bode plot analysis. An example is the hyperosmolar response of the Hog1 signaling cascade we discussed in Sections 7.4.2 and 8.1.1. Consequently, the use of integral transforms dominates biomedical engineering approaches to living organisms [874], and they have been used, for example, to investigate the properties of the thalamus [892] and even the electroencephalogram (EEG) generated by the human brain [648]. Moreover, the possibility of determining transfer functions for individual biochemical reactions may make it possible to understand the behaviors of large metabolic networks [21]. Recently, an "optogenetic approach" has been developed to deliver a customized signaling input at the cellular level in situ [828, 829]. The fact that this method is light-based suggests that it may ultimately be possible to determine Bode plots for cellular and biochemical feedback mechanisms using sinusoidally modulated light intensities much as Stark did in his classic experiments (see, in particular, the online supplementary material provided for [828]).

A frequently used test for linearity is the coherence function

$$\gamma_{xy}^2(f) = \frac{|S_{xy}(f)|^2}{S_{xx}(f)S_{yy}(f)}, \tag{9.5}$$

where the quantities $S_{xx}(f)$ and $S_{yy}(f)$ are, respectively, the autospectral density functions of the input $x(t)$ and the output $y(t)$, and $S_{xy}(f)$ is the cross-spectral density function between $x(t)$ and $y(t)$. In Section 14.8.3, we show how these spectral density functions can be readily estimated from experimental data using the FFT (see also `cohere_demo.py` on the `matplotlib` website). For a linear noise-free dynamical system that receives a single input, $\gamma_{xy}^2(f) = 1$. In practice, $\gamma_{xy}^2(f) > 0.8$ is usually accepted as evidence of linearity [874]. Low coherence, e.g., $\gamma_{xy}^2(f) < 0.8$, does not necessarily mean that the dynamical system is nonlinear. Common causes of low coherence are measurement noise, failure to isolate the inputs such that the measured output reflects multiple inputs, and the possibility that there is no relationship between the input and output. In order to conclude that low coherence reflects nonlinearity, it is necessary to systematically rule out these alternative interpretations.

9.2.1 Proportional–Integral–Derivative (PID) Controllers

PID controllers are very important in industrial applications and are becoming recognized as important in biological applications as well. The meaning of PID can readily be seen from the transfer function

$$H(s) = K_p + \frac{K_I}{s} + K_D s = \frac{K_D s^2 + K_p s + K_I}{s}, \tag{9.6}$$

where K_p is the proportional gain, K_I is the integral gain, and K_D is the derivative gain. When there are no time delays present, PID controllers often do not use derivative action and hence are better termed PI controllers[2]. It has been estimated that over 90% of all industrial control problems are solved by PID control [24]. Biological examples of PI control arise in the context of calcium homeostasis [196] and adaptation such as occurs in the hyperosmolar response of the Hog1 signaling cascade (Section 7.4.2). An example of PD control occurs in the response of cone photoreceptors located in the retina [880]. Often overlooked is the fact that when time delays are present, derivative action may be necessary to maintain stability [373, 800]. Indeed, numerical simulations indicate that for balance control with delayed feedback the contribution due to the integral term is negligible [465].

---

[2] Common usage is to apply the term PID to the whole class of controllers that includes PID, PI, and PD controls.

How does a PID controller work? The tracking error $E(s)$ is the difference between the desired input $R(s)$ and the actual output $Y(s)$. This error signal is sent to the PID controller, and the controller computes both the derivative and the integral of this error signal. The signal $X(s)$ just past the controller is

$$Y(s) = K_P E + K_I \int E\, dt + K_D \frac{dE}{dt},$$

where E is the magnitude of the error. This signal will be sent to the system, and the new output will be obtained. This new output will be sent back to the sensor to compute the new error signal. The controller takes this new error signal and computes its derivative and its integral again. This process goes on and on.

Example 9.1 The transfer function for the yeast hyperosmolar response is characterized by the presence of a derivative term s in the numerator. This type of transfer function is most suggestive of a PI controller. To see how this is possible, take

$$C_{ol}(s) = \frac{k}{k+s}$$

and

$$H(s) = K_p + \frac{K_I}{s}.$$

From (9.4), we see that for PI negative feedback control, we have

$$C_{cl}(s) = \frac{C_{ol}(s)}{1 + H(s)C_{ol}(s)} = \frac{\dfrac{k}{k+s}}{1 + \left(\dfrac{K_p s + K_I}{s}\right)\left(\dfrac{k}{k+s}\right)}$$

$$= \frac{\dfrac{k}{k+s}}{\dfrac{s(k+s) + k(K_p s + K_I k)}{s(k+s)}} = \frac{ks}{s^2 + k(1 + K_p)s + K_I k}.$$

This result has the same form as the transfer function we obtained using Bode analysis in Section 7.6.   ◇

A detailed discussion of PID controllers is beyond the scope of this introductory textbook (for a very readable introduction, see [24]). It is useful to know a few properties of PID controllers. For a pure proportional feedback controller, there always will be a steady-state error. This error can be decreased by increasing the proportional gain K_p; however, this increases the oscillatory component. There are two better ways to reduce the steady-state error to zero. First, we can add feedforward control. Namely, the disturbance is measured before it enters the system, and corrective actions are taken before the disturbance has influenced the system. Biologists, particularly neuroscientists, tend to emphasize the role of feedforward control. However, the design of an effective feedforward controller requires exact knowledge of the process dynamics, which is unlikely to be available. Second, we can add an integral action. Since the use of integral control does not require information about process dynamics, the addition of integral control provides a more robust way to reduce the steady-state error to zero. Finally, the addition of a derivative action is used to dampen oscillations that might arise as a consequence of PI control.

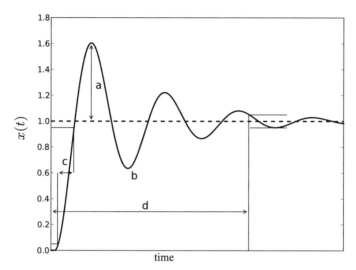

Fig. 9.2 Parameters describing the response of a feedback controller to a unit step. (a) Maximum overshoot expressed as a percentage of the step size. (b) Number of clearly defined oscillations before the response settles down to the steady state. (c) Rise time, defined as the time to increase from 5 % to 95 % of the step size. (d) Settling time, defined as the time until the output remains within ±5 % of the steady-state value. (e) Steady-state error defined as the error that remains after the settling time.

Typically, a "trial and error" approach is used to optimize the performance of a PID controller for a given application. Thus it is useful to understand how to tune a PID controller for a given task and to appreciate the empirical measurements used by engineers to assess the goodness of a feedback controller. The parameters are most often assessed in response to a step input and are shown in Figure 9.2 (see caption for definitions). Table 9.1 shows the effects of changing K_P, K_I, and K_D on each of these parameters in a PID controller.

9.3 Feedback control: time-domain approaches

Historically, biomathematicians have modeled biological control using ordinary differential equations, for example,

Table 9.1 Effect of increasing separately each component of a PID controller on its response to a perturbation.

Response	K_P	K_I	K_D
Rise time	Decrease	Decrease	Small change
Overshoot	Increase	Increase	Decrease
Settling time	Small change	Increase	Decrease
Steady-state error	Decrease	Eliminate	Small change

$$\frac{dx}{dt} + kx = f(x) \tag{9.7}$$

where k is a constant. The *plant*, or system to be controlled, corresponds to the left-hand size of (9.7). The error signal, $e(t)$, is equal to the difference between $x(t)$ and the desired outcome, $x_{des}(t)$, namely $e(t) = x(t) - x_{des}(t)$. Typically $x_{des}(t)$ is chosen to correspond to the fixed point, x^*, of (9.7) where x^* is determined by setting $dx/dt = 0$ and assuming that $x = x^*$. Thus the determination of the stability of feedback control in the time domain corresponds to the linear stability analysis of the fixed point. The linearized equation for the error is

$$\frac{de}{dt} + ke = K_P e$$

where K_p is the proportional gain which is equal to the slope of $f(x)$ evaluated at x^*. For negative feedback, $K_p < 0$.

9.3.1 Gene regulatory systems

It is well documented in cell cultures that some proteins show periodic increases in their activity during cell division, and these increases reflect periodic changes in the rate of protein synthesis. Regulating these activities requires feedback control, and indeed, many gene-regulating mechanisms take the form of the feedback controllers [283, 602, 618, 835]. An example of negative feedback control of gene expression is shown schematically in Figure 9.3. The metabolite (m) feeds back to control the rate of mRNA synthesis. The negative feedback arises because the higher the concentration of m, the lower the concentration of mRNA. In other words, some enzymes repress their own synthesis. Less commonly, the feedback can be positive. Such a situation arises in the setting of inducible enzymes: the substrate induces the formation of the enzyme in order to metabolize it. Molecular biologists refer to this type of regulatory mechanism as an *operon*, such as the *lac* operon and the *tryptophan* operon. It should be noted that the synthesis of many proteins is not controlled by feedback controllers: such proteins are *constitutive*, meaning that they are continually synthesized.

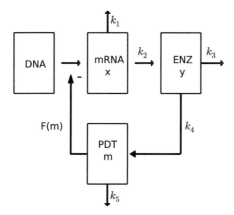

Fig. 9.3 Schematic control system for the production of an enzyme (y) according to the Goodwin model. The enzyme combines with the substrate to produce a product m that represses the transcription of DNA to mRNA (x), the template for making the enzyme.

Brian Goodwin[3] [283] proposed a model to describe the negative feedback control of mRNA synthesis and to demonstrate that oscillations could arise from the interactions between mRNA (x), enzyme (y), and enzyme product (m). He made the following assumptions:

1. The concentration of enzyme y increases at a rate proportional to the concentration of x and is removed at a constant rate.
2. The metabolite concentration $[m]$ increases at a rate proportional to the concentration of enzyme and is removed at a rate proportional to its own concentration.
3. The synthesis mRNA is under the negative feedback control of $[m]$, and it is removed at a constant rate.

Using the law of mass action, we obtain the third-order system of differential equations

$$\frac{d[x]}{dt} = F([m]) - \alpha, \quad \frac{d[y]}{dt} = A[x] - \beta, \quad \frac{d[m]}{dt} = C[y] - \gamma[m], \qquad (9.8)$$

where $F([m])$ describes negative feedback.

Goodwin took a particularly simple choice of $F([m])$, namely

$$F([m]) \approx \frac{1}{[m]},$$

and so we have

$$\frac{d[x]}{dt} = \frac{K}{[m]} - \alpha, \quad \frac{d[y]}{dt} = A[x] - \beta, \quad \frac{d[m]}{dt} = C[y] - \gamma[m], \qquad (9.9)$$

---

[3] Brian Carey Goodwin (1931–2009), Canadian mathematician and biologist.

where A, K, α, β are positive constants. The synthesis of mRNA and enzyme takes minutes, but the binding of the enzyme product to the DNA takes place much more quickly (certainly less than a microsecond). Thus we could assume that

$$\frac{d[m]}{dt} \approx 0,$$

so that

$$[m] = \frac{C}{\gamma}[y] = \theta[y]. \tag{9.10}$$

The reader should note that we have used the steady-state approximation to change $[m]$ from a variable to a parameter. Thus we can rewrite the Goodwin equations as

$$\frac{d[x]}{dt} = \frac{K}{\theta[y]} - \alpha, \quad \frac{d[y]}{dt} = A[x] - \beta. \tag{9.11}$$

Since the fixed point is a center, (9.11) generates periodic oscillations.

Extensions to (9.8) often take the form of more realistic choices of the negative feedback term $F([m])$ [130, 286, 320, 618], for example

$$F([m]) = \frac{A}{B + [m]^n}.$$

In addition, it is necessary to account explicitly for the time it takes to synthesize the proteins. In this case, (9.8) takes the form of a delay differential equation [528, 529, 736, 906, 907]. By incorporating these time delays into mathematical models, it has been possible to obtain essentially quantitative agreement between model and experimental observations for the *lac* operon [906, 907] and the *tryptophan* operon [736]. These models have also been able to account for the experimentally observed bistability that occurs in the *lac* operon [647, 907].

9.4 Production–Destruction

In the 1970s, a group of mathematical physiologists, led by Michael Mackey[4], Uwe an der Heiden, and Leon Glass[5], recognized that many models in physiology have the form [17, 275, 522]

$$\frac{dx}{dt} = \text{production} - \text{destruction} = P(x(t - \tau)) - xD(x(t - \tau)).$$

The usefulness of the production–destruction framework is that it is often easy to identify the biological mechanisms that cause a variable to increase ("production") and those that cause it to decrease ("destruction"). There are three physiologically

---

[4] Michael C. Mackey (b. 1943), American-born Canadian mathematical physiologist and physicist.

[5] Leon Glass (b. 1943), American-born Canadian mathematical physiologist and physicist.

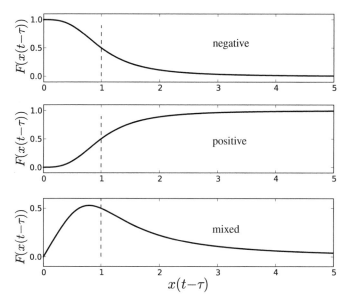

Fig. 9.4 Three common types of feedback observed in physiological systems: (a) negative, (b) positive, and (c) "mixed," given respectively by (9.12), (9.13), and (9.14). The vertical dashed lines show $x = K$. Parameters: $K = 1, n = 3$.

relevant choices of the feedback (Figure 9.4). Of course, there are many different ways to write a mathematical expression that duplicates the functional forms for these different types of feedback. Mathematicians tend to favor expressions that include exponentials or the hyperbolic tangent. Mathematical physiologists favor forms based on the Hill function (see Section 3.2.1). The feedbacks shown in Figure 9.4 correspond, respectively, to

$$F_N(x) = \frac{K^n}{K^n + x^n} \quad \text{(negative feedback, Figure 9.4a)}, \tag{9.12}$$

$$F_P(x) = \frac{x^n}{K^n + x^n} \quad \text{(positive feedback, Figure 9.4b)}, \tag{9.13}$$

$$F_M(x) = \frac{xK^n}{K^n + x^n} \quad \text{(mixed feedback, Figure 9.4c)}. \tag{9.14}$$

These forms of feedback have a number of advantages, including the fact that they are specified by only two parameters, n and K.

Feedback functions having these forms have been studied in a number of contexts. Examples include the propagation of activity in random networks of excitable elements [710, 833], the Naka–Rushton[6] equation in vision research [622, 880], enzyme kinetics discussed in Chapter 3, poverty traps [71], ice albedo effects [88],

---

[6] Ken-Ichi Naka (d. 2006), Japanese-American retinal physiologist; William Albert Hugh Rushton (1901–1980), English physiologist.

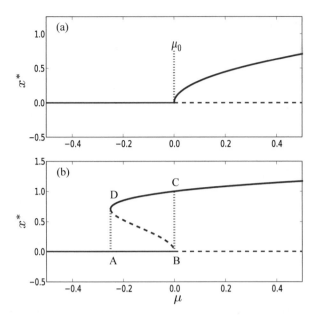

Fig. 9.5 Schematic representation of (a) a supercritical Andronov–Hopf bifurcation and (b) a subcritical Andronov–Hopf bifurcation. The solid lines represent the stable solutions, and the dashed lines the unstable ones. Both of these bifurcations are symmetric about the $x^* = 0$ axis; however, the lower branch is not shown in order to simplify the figure.

and the occurrence of crises in climate and population models [741]. Here we discuss examples that incorporate negative and positive feedback in physiological models (see also Exercises 9.9–9.9). We discuss an example with negative feedback in Section 9.5, an example with positive feedback in Section 10.5.3, and an example with mixed feedback in Section 13.2.

Oscillations arise in production–destruction models. Most frequently these oscillations arise via an Andronov–Hopf bifurcation[7]. Andronov–Hopf bifurcations arise when a conjugate pair of complex eigenvalues cross the imaginary axis into the right plane at the same time. These bifurcations can be of two types: supercritical and subcritical. In the case of a supercritical Andronov–Hopf bifurcation (Figure 9.5a) when the fixed point loses stability, a small amplitude, stable limit cycle is created. The amplitude is zero at oscillation onset, but since the frequency is determined by the imaginary part of the eigenvalues it is not equal to 0. Since the amplitude of the oscillation grows progressively for $\mu > \mu_0$, this bifurcation is often referred to as a "soft bifurcation."

In the case of a subcritical Andronov–Hopf bifurcation (Figure 9.5b), when the fixed point loses stability a small amplitude, unstable limit cycle is created. How-

---

[7] Aleksandra Andronov (1901–1952), Russian physicist, worked extensively on the stability of dynamical systems. The crater Andronov on the Moon is named after him. Eberhard Hopf (1902-1983), Austrian-Hungarian-American mathematician, was a pioneer in bifurcation theory.

ever, this type of bifurcation is typically associated with a saddle-node, or fold limit cycle, bifurcation which creates a large amplitude limit cycle "out of the clear blue sky." Thus when the system trajectory diverges from the unstable fixed point, it "jumps" to this larger amplitude stable limit cycle. Since the amplitude of the stable limit cycle is fixed at oscillation onset, this type of bifurcation is sometimes referred to as a "hard bifurcation." Note that the subcritical Andronov–Hopf bifurcation is associated with bistability since there can be a coexistence of a stable fixed point and a stable limit cycle.

The cardinal features of the Andronov–Hopf bifurcations can be understood in terms of prototypical second-order differential equations. These equations take the form

$$\frac{dr}{dt} = \mu r - r^3 \tag{9.15}$$

$$\frac{d\theta}{dt} = \omega + br^2$$

for a supercritical Andronov–Hopf bifurcation and

$$\frac{dr}{dt} = \mu r + r^3 - r^5 \tag{9.16}$$

$$\frac{d\theta}{dt} = \omega + br^2$$

for a subcritical Andronov–Hopf bifurcation, where μ controls the stability of the fixed point, ω is the angular frequency ($\omega = 2\pi f$) at oscillation onset, and b describes the strength of the dependence of frequency on amplitude for the large amplitude oscillations. It is important to note that these equations are written in terms of the polar coordinates. This is because when written in this way, the mathematical analysis is simplified. However, for numerical investigations it is necessary to change to the Cartesian coordinates x, y using the relations $x = r\cos\theta$, $y = r\sin\theta$ and $r^2 = x^2 + y^2$. This procedure is illustrated in Exercises 4 and 5.

The *cardinal properties of a supercritical Andronov–Hopf bifurcation* are (Figure 9.5a) [276] as follows:

1. The amplitude of the limit cycle increases in proportion to $\sqrt{\mu - \mu_0}$, where μ_0 is the value of the control parameter at which the fixed point becomes unstable.
2. Since the amplitude of the oscillation grows gradually from 0, this type of bifurcation is sometimes referred to as a "soft bifurcation" [276].
3. The frequency of the limit cycle is approximately $\omega = \Im(\lambda)$.

The *cardinal properties of a subcritical Andronov–Hopf bifurcation* are (Figure 9.5b) [276] these:

1. Sudden appearance of large-amplitude oscillation as μ crosses a stability boundary at μ_0. For this reason, this type of Andronov–Hopf bifurcation is sometimes called a "hard excitation."
2. Little further change in amplitude and frequency as μ increases beyond μ_0.

3. Bistability between a fixed point and a limit-cycle attractor occurs when μ is tuned to the region outlined by ABCD in Figure 9.5b. Single-pulse annihilation occurs when a perturbation is large enough to displace the dynamical system from the basin of attraction of the limit cycle into the basin of attraction associated with the stable fixed point.
4. Hysteresis is present: This refers to the fact that the point of oscillation onset as μ increases from A to B is not the same as the point of oscillation offset when C goes to D.

In their textbook *From Clocks to Chaos: The Rhythms of Life*, Leon Glass and Michael C. Mackey systematically drew attention to a large number of phenomena in physiological systems that suggest the occurrence of an Andronov–Hopf bifurcation [276]. For example, the decrease in the amplitude of contractions of the uterus in a female with dysmenorrhea following administration of a nonsteroidal anti-inflammatory drug [747] closely resembles the changes in amplitude expected to occur for a supercritical Andronov–Hopf bifurcation [276]. Similarly, the contractions of the uterus during pharmacologically induced labor build up in a manner suggestive of a supercritical Andronov–Hopf bifurcation [382]. Both subcritical and supercritical Andronov–Hopf bifurcations occur in dynamical systems that contain time-delayed feedback-control mechanisms, such as the pupil light reflex (PLR) (see below). They can also occur in certain periodic chemical reactions, such as the chlorine dioxide–iodine–malonic acid reaction [468, 469]. Subcritical Andronov–Hopf bifurcations readily arise in neurons (see Chapter 11). Other examples include certain periodic chemical reactions, such as the oxidation of nicotinamide adenine dinucleotide (NADH) catalyzed by horseradish peroxidase [167], aeroelastic flutter and other vibrations of airplane wings [177], and instabilities of fluid flows [178].

At first sight, it seems straightforward to determine whether a limit cycle has arisen via a supercritical or subcritical Andronov–Hopf bifurcation. For example, the AUTO computer software package makes it possible to determine the nature of the bifurcation if the equations describing the dynamical system are known (see Section 11.5). In the laboratory, however, this determination can be quite difficult [567]. In particular, the identity of the critical control parameter may not be known, and even if it is known, it may be difficult to vary it experimentally. The effects of the critical slowing-down phenomena and the interplay between noise and time delays are important confounding factors that limit the experimentalist's ability to pinpoint the parameter value for which the bifurcation occurred (see Section 5.2.1). These problems can be circumvented when the feedback can be clamped.

9.5 Time-delayed feedback: Pupil Light Reflex (PLR)

The pupil light reflex (PLR) was the first human neural feedback control mechanism to be studied from the point of view of transfer function analysis [793]. The advantage offered by the PLR stems from the ease with which the feedback loop can be noninvasively opened. It is this property of the PLR that earned it the title *paradigm of human neural control* [794]. The PLR can also be studied in the time domain. This makes it possible to directly compare the role of time-domain and frequency-domain approaches for studying biological control.

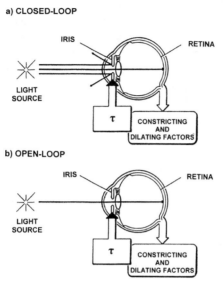

Fig. 9.6 Diagram of the pupil light reflex to illustrate the difference between (a) closed-loop and (b) open-loop ("Maxwellian" view) conditions for illuminating the retina. Under closed-loop conditions, the iris acts like the aperture of a camera and thereby controls the illuminance of the retina. Under open-loop conditions, changes in the size of the pupil do not affect the illuminance of the retina [797].

Pupillary size reflects a balance between constricting and dilating mechanisms [494]. Pupil constriction is caused by contraction of the circularly arranged pupillary constrictor muscle, which is innervated by parasympathetic fibers. The motor nucleus for this muscle is the Edinger–Westphal[8] nucleus, located in the oculomotor complex of the midbrain. There are two main neural mechanisms for pupil dilation: a mechanism that involves contraction of the radially arranged pupillary dilator muscle innervated by sympathetic fibers[9] (traditionally referred to as "active" reflex dilation) and a mechanism that operates by inhibition of the activity of the Edinger–Westphal nucleus (traditionally referred to as "passive" reflex dilation).

The PLR does not control pupil size, but acts to regulate the retinal luminous flux J_r (measured in lumens),

$$J_r = IA, \tag{9.17}$$

---

[8] Ludwig Edinger (1855–1918), German anatomist and neurologist; Carl Friedrich Otto Westphal (1833–1890), German neurologist and psychiatrist.

[9] In the human iris, the dilator muscle is a poorly developed myoepithelial cell and likely does not exert much force. In contrast, the pigeon pupillary dilator is striated, and hence pigeon pupillary movements are faster than those of humans [494].

where I is the illuminance (lumens $\cdot$ mm^{-2}) and A is the pupil area (mm^2). The iris performs this function by acting like the aperture in a camera (Figure 9.6a). When light strikes the retina, the pupil constricts, and hence both A and J_r decrease (negative feedback). If we are a little more attentive, we notice that pupil size does not immediately change in response to a change in illumination: there is typically a delay, or latency, of $\approx$ 180–400 ms before changes in pupil area can be detected. Thus the PLR is an example of a time-delayed negative feedback control mechanism.

The reason why the feedback loop can be readily opened stems from the fact that even with the brightest light, pupil diameter does not get much smaller than $\approx$ 1 mm. Suppose we focus a light beam whose diameter is smaller than 1 mm onto the center of the pupil (Figure 9.6b): although pupil size can get smaller or larger, the retinal light flux remains constant. In other words, the feedback loop has been "opened" [797]. We can measure the transfer function for the open PLR and then use it to calculate the closed-loop transfer function.

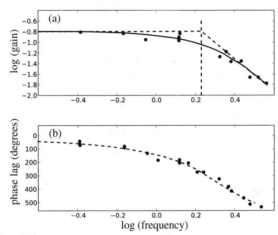

Fig. 9.7 Bode plot for the pupil light reflex. the overall shape of the amplitude and phase plot suggests that the PLR acts as a simple lag. At the breakpoint, there is a difference between the data and the linear approximation to the simple lag of $\approx$ 9 dB, indicating that it is a third-order lag. For low frequencies, the phase decreases linearly with slope $\approx$ 0.2 s. Finally, the amplitude as f approaches 0 gives a gain of 0.16. Putting all this together, we obtain the result given by (9.18). Figure reproduced from [793] with permission.

9.5.1 PLR: frequency-domain analysis

Using the above open-loop conditions, Stark and colleagues were able to measure the open-loop transfer function C_{ol} for the pupil light reflex [793, 797]. Applying the same approach that was used to measure C_{ol} for the crayfish photoreceptor, Stark illuminated the retina under "open-loop" conditions with a light beam whose intensity varied sinusoidally with varying frequency and measured the amplitude and phase shift of the oscillations that occurred in pupil size. In this way, he obtained the Bode plot shown in Figure 9.7, suggesting that

$$C_{ol}(s) = \frac{0.16e^{-0.2s}}{(1+0.1s)^3}. \tag{9.18}$$

Using (9.4) with $H(s) = 1$ we obtain the closed-loop transfer function

$$C_{cl}(s) = \frac{C_{ol}(s)}{1+C_{ol}(s)}. $$

9.5.2 PRL: time-domain analysis

The $e^{-0.2s}$ term in the numerator of $C_{cl}(s)$ means that there is a time delay, τ, of ≈ 200ms between when a change in retinal illuminance occurs and when the resultant changes in the pupil area occur (see Section 7.5.4). This time delay can also be determined by measuring the change in pupil size in response to a brief light pulse (Figure 9.8a, c). A delay of ≈ 300 ms is equal to the difference between the time that the light pulse starts and the time that the pupil first begins to constrict (grow smaller). For healthy people, $\tau \approx 180$–400 ms, where the shortest latencies are measured for the highest light intensities [494]. Stark used a very high light intensity for determining $C_{cl}(s)$, which is why he obtained $\tau \approx 200$ ms. The observation that the pupil area gets transiently smaller in response to the light pulse implies that the PLR is a time-delayed negative feedback control mechanism.

The importance of τ can be demonstrated by carefully focusing a narrow light beam at the pupillary margin. If correctly placed, the light beam causes pupil size to undergo oscillations with period ≈ 900 ms [796, 804]. Initially, the pupil constricts, since the retina is exposed to light; however, at some point, the iris blocks the light from reaching the retina, and the pupil subsequently dilates (the retina is in the dark), thus exposing again the retina to light. The cycle repeats. The oscillation in pupil size is made possible by the time delay between the change in retinal luminous flux and the resultant changes in the pupil area. In the clinical literature, this technique is referred to as "edge-light" pupil cycling [586, 804]. In the control literature, it is referred to as "high-gain" pupil cycling [794, 796]. The latter terminology stems from the observation that when a slit-like light beam is focused on the pupillary margin, the change in the retinal light flux for a change in pupil area is large since the retina is either illuminated or it is not.

The presence of a time delay means that what happened in the past has consequences for what happens in the present. Mathematical models of these physiological systems take the form of a delay differential equation (DDE) of the form, for example,

$$\frac{dx}{dt} + k_1 x(t) = F(x(t-\tau)), \tag{9.19}$$

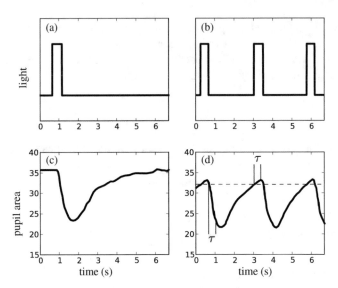

Fig. 9.8 The time delay in the pupillary light reflex is the time between (a) the onset of the light pulse and (c) the onset of the transient decrease in the pupil area. In this example, $\tau \approx 300$ ms. Pupil cycling can be induced using the measured pupil area to control the retinal illumination (see Section 9.6 for more details). By comparing (b) and (d), it can be seen that the retina is illuminated only when the pupil area exceeds an area threshold (horizontal line in (d)). There is a time delay of ≈ 300 ms for the onset of constriction and a similar delay for the onset of dilation. Since the illuminance is less than that used by Stark, the delay is longer.

where F describes the feedback, τ is the time delay, and $x(t), x(t-\tau)$ are, respectively, the values of the state variable x at times $t, t-\tau$. We will discuss the analysis of DDEs in more detail in the following chapter. The point we make here is that many of the tools we have already used for the analysis of ordinary differential equation can be used to analyze DDEs. In other words, let's see how far we can get using what we already know!

It has been shown that the description of the PLR based on the closed-loop transfer function $C_{cl}(s)$ is equivalent to the DDE [499]

$$\frac{dA}{dt} + kA = F_N(A(t-\tau)),\qquad(9.20)$$

where A is the pupil area, $k > 0$ describes the rate of pupillary movements, and F_N describes negative feedback, provided that the fixed point is stable. Since the PLR functions as a negative feedback control mechanism, a convenient way to describe negative feedback mathematically is to use (9.12) [500]:

$$F_N(A) = \frac{K^n}{K^n + A(t-\tau)^n},\qquad(9.21)$$

where n, K are constants. Subsequently, investigators extended this model to a second-order DDE by including a variable that represented the membrane potential of the retinal ganglion cells [78, 79]. This refinement to the model was required in order to account for the frequency-dependent shift in average pupil size that occurs when the retina is illuminated with a sinusoidally modulated light source.

The advantage of (9.20) is that we can use it to gain insight into the behavior of the PLR once the fixed point becomes unstable. The fixed point A^* can be determined graphically. Defining $u = A - A^*$, we obtain the linearized DDE

$$\frac{du}{dt} + ku(t) = du(t - \tau),\tag{9.22}$$

where $d = (dx/dt)\big|_{A=A^*} < 0$ is the slope of F evaluated at the fixed point. We note that we have used Tool 7 to determine the parameter d.

The characteristic equation is

$$\lambda + k = de^{-\lambda \tau}.\tag{9.23}$$

Here we have used the "usual Ansatz" (Tool 4). By writing $\lambda = \gamma + j\omega$ and setting $\gamma = 0$, we have

$$\frac{k}{d} + j\frac{\omega}{d} = e^{-j\omega\tau} = \cos\omega\tau - j\sin\omega\tau,\tag{9.24}$$

where we have used Euler's formula (Tool 8).

A graphical approach can be used to obtain insight into (9.24) [519] (Figure 9.9). The left-hand side of (9.24) is a straight line, and the right-hand side is a circle in the complex plane [519]. The roots of (9.24) depend on the interplay between τ and the gain G, where

$$G := \left(\frac{k}{d}\right)^{-1}.$$

Since $k > 0$ and $d < 0$, we have $G < 0$, and hence we see that negative gain corresponds to negative feedback. When $d = k$, we have $G = -1$. When G is to the left of $G = -1$ (low-gain region in Figure 9.9), there is no intersection between the straight line and the circle shown in Figure 9.9, and hence no solution of (9.24). This means that there can be no change in stability caused by changes in d, k. In other words, the fixed point is stable for all τ. On the other hand, for choices of d, k that place G in the high-gain region of Figure 9.9, there can be a nonempty intersection, and hence it is possible for the fixed point to become unstable.

We can solve for the values of ω and τ for which the fixed point becomes unstable, respectively, ω_{sb} and τ_{sb}. From (9.24), we have, on equating the real and imaginary parts,

$$\cos\omega_{sb}\tau_{sb} = \frac{k}{d},\tag{9.25}$$

$$\sin\omega_{sb}\tau_{sb} = -\frac{\omega_{sb}}{d}.\tag{9.26}$$

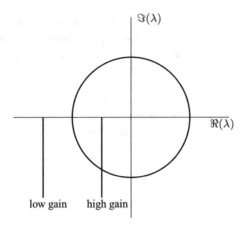

Fig. 9.9 Graphical representation of the stability of (9.24) as a function of d and k.

Now we can square both sides of these equations and then add them, to obtain

$$\frac{k^2}{d^2} + \frac{\omega_{sb}^2}{d^2} = \frac{k^2 + \omega_{sb}^2}{d^2} = 1,$$

and thence

$$\omega_{sb} = \sqrt{d^2 - k^2}. \tag{9.27}$$

We can obtain τ_{sb} by dividing (9.26) by (9.25):

$$\tan \omega_{sb} \tau_{sb} = -\frac{\omega_{sb}}{k}, \tag{9.28}$$

or

$$\tau_{sb} = \frac{1}{\omega_{sb}} \tan^{-1}\left(-\frac{\omega_{sb}}{k}\right).$$

Edge-light pupil cycling corresponds to a high-gain feedback loop [584, 793, 796]. This is because for a narrow light beam placed on the pupillary margin, the change in the retinal light flux for a change in the pupil area is large: the retina is either illuminated or not. We can directly compare the predicted period of the high-gain oscillations for (9.20) and the experimentally measured pupil cycle time.

Referring to the black dot ($\bullet$) in Figure 9.10, we see that (9.28) implies that

$$\frac{\pi}{2} < \omega_{sb} \tau_{sb} < \pi,$$

or

$$\frac{1}{4\tau} < f_{sb} < \frac{1}{2\tau},$$

where we have used the definition $\omega = 2\pi f$ and dropped the subscripts. It is convenient to rewrite this last expression in terms of the period T, namely $f = 1/T$,

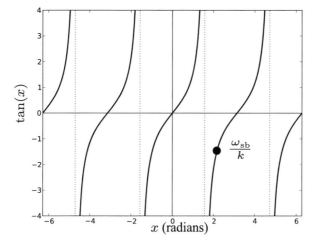

Fig. 9.10 Graphical representation of the tangent. The vertical dotted lines are placed at, from left to right, $x = -3\pi/2$, $x = -\pi/2$, $x = \pi/2$ and $x = 3\pi/2$.

$$4\tau > T > 2\tau. \tag{9.29}$$

Equation (9.29) represents a useful rule of thumb for investigating biological oscillations in the laboratory. A first step in explaining the appearance of an oscillation is to identify the time delays. If the observed period is longer than 2τ, it is possible that the oscillation is generated by a delayed feedback control mechanism. If, in addition, the observed period satisfies (9.29), then it is possible that the biological oscillation has arisen because of instability in a negative feedback control mechanism. For example, the measured PLR delay is ≈ 300 ms for typical light intensities, and the observed pupil cycle time is ≈ 900 ms. These experimental observations are in excellent agreement with (9.29):

$$1200 \text{ ms} > T > 600 \text{ ms}.$$

Example 9.2 If you hold your arm still out in front of you, you will notice the presence of very small amplitude oscillations referred to as physiological tremor. Its frequency is 9–12 Hz and hence the period is ≈ 80–125ms. It has been suggested that this tremor is related to the stretch reflex [488] for which the time delay of the short latency reflex is 20-45ms [412].◊

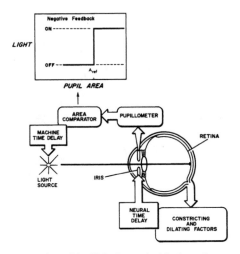

Fig. 9.11 Schematic representation of the PLR clamped with piecewise constant negative feedback (PCNF) (inset). Figure reproduced from [584] with permission.

9.6 Clamping the PLR

Lawrence Stark[10] was the first to demonstrate the feasibility of controlling a human neural feedback control mechanism by manipulating parameters in an external feedback mechanism [792].   He referred to this technique as "environmental clamping." This terminology draws an analogy with the development of voltage [364] and dynamic [757–759] clamps, used to regulate the membrane potential of neurons. Since the PLR feedback loop can be easily "opened," it is possible to close the loop using an external feedback that controls retinal illuminance I as a function of the pupil area A [498, 584, 714, 792]. Experimentally, this is accomplished using a pupillometer to measure A and then using A to adjust the intensity of the illuminating light source (see Figures 9.8b, d, and 9.11). The feedback in the clamped PLR is constructed electronically [498, 584, 714]. Figure 9.11 shows the PRL clamped with piecewise constant negative feedback (PCNF). In this case, the retina is illuminated only if A is greater than a threshold A_{ref} (Figure 9.8b, d). This form of clamping mimics edge-light pupil cycling.

The panels on the left-hand side of Figure 9.12 show the dynamics of the PLR clamped with PCNF as a function of A_{ref}. For all choices of A_{ref} that are less than the maximum area of the pupil, A oscillates. It is useful to examine closely the interplay between the retinal illuminance and the changes in A. Choose the initial point of the oscillation to be the smallest value A_{min} of A. Since $A_{min} < A_{ref}$, the light is OFF and the pupil responds by dilating. When $A = A_{ref}$, the light is turned ON; however, the pupil keeps dilating for a time τ_d, which is the time delay for the dilating pathway of the PLR. At time τ_d after the light went ON, A attains its largest

---

[10] Lawrence Stark (1926–2004), American neurologist, a recognized authority in the use of engineering analysis to characterize neurological control mechanisms.

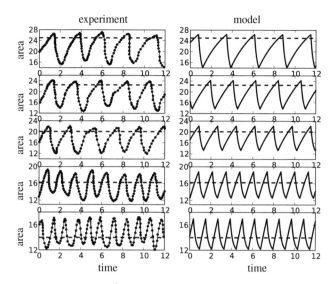

Fig. 9.12 Changes in pupil area (mm^2) as a function of time when the PLR is clamped with PCNF. The value of A_{ref} (dashed lines) was (top to bottom) 25.0 mm^2, 22.5 mm^2, 20.1 mm^2, 16.2 mm^2, and 14.0 mm^2. The observed changes in pupil area (left) are compared to those predicted from (9.30) (right). The values of the parameters used in the simulations are $k_c = 4.46\ \mathrm{s}^{-1}$, $k_d = 0.42\ \mathrm{s}^{-1}$, $A_{\mathrm{ON}} = 11.8$ mm^2, $A_{\mathrm{OFF}} = 34.0$ mm^2, and $\tau = 0.38$ s. Note that the time delay is the neural time delay plus a machine time delay. Figure reproduced from [584] with permission.

value, $A_{\max}$. Subsequently, the pupil begins to constrict. When $A = A_{\mathrm{ref}}$, the light is turned OFF. However, the pupil continues to constrict for a time τ_{sb}, the time delay in the constricting pathway, until it reaches $A_{\min}$, and the cycle repeats.

A mathematical model that captures these dynamics is

$$\frac{dA}{dt} = \begin{cases} A_{\mathrm{ON}} - k_c A & \text{if} A(t-\tau) < A_{\mathrm{ref}}, \\ A_{\mathrm{OFF}} - k_d A & \text{otherwise}, \end{cases} \tag{9.30}$$

where $A_{\mathrm{ON}}, A_{\mathrm{OFF}}$ are the asymptotic values attained by the pupil when the light is respectively ON or OFF indefinitely, and k_c, k_d are the pupillary rate constants for constriction and dilation. The advantage of this model is that the solution can be constructed by piecing together exponential functions. The solution of (9.30) once the limit cycle oscillation has been established is

$$A(t) = \begin{cases} A_{\mathrm{ON}} + [A(t_0) - A_{\mathrm{ON}}] e^{k_c(t-t_0)}, \\ A_{\mathrm{OFF}} + [A(t_0) - A_{\mathrm{OFF}}] e^{k_d(t-t_0)}. \end{cases} \tag{9.31}$$

As can be seen in Figure 9.12, there is remarkable agreement between the predictions of (9.31) and the observed behaviors of A. The period of the oscillation can be determined from (9.31), and it is in good agreement with that observed experimentally (see Exercise 9.9).

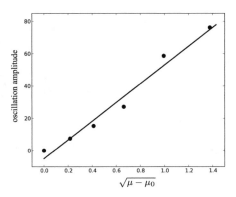

Fig. 9.13 A plot of the amplitude of pupil cycling as the gain in the feedback is increased, showing the occurrence of a supercritical Andronov–Hopf bifurcation.

Example 9.3 The Andronov–Hopf bifurcation that arises in a mathematical model of the pupil light reflex (PLR) (see equation (9.20)) is supercritical [500]. The clamped PLR can be used to investigate experimentally the nature of oscillation onset when reflex gain becomes sufficiently high [501]. For this purpose, the feedback is constructed to make I a linear function of A, and μ is the slope of this linear relationship: increasing the slope is equivalent to increasing the feedback gain. At sufficiently high μ, regular oscillations occur in the pupil area. Figure 9.13 shows that the amplitude of the oscillations is proportional to $\sqrt{\mu - \mu_0}$. This observation supports the prediction that the onset of the oscillation in the pupil area occurs via a supercritical Andronov–Hopf bifurcation. ◇

9.7 The future

Developments in engineering and biomathematics have made it possible to construct *cyborgs*, namely living organisms that are part protoplasm, part machine [581].

The basic idea is that the measured physiological variable x is controlled by an implanted electronic circuit that calculates the required feedback and then outputs an appropriate stimulus. In this way, it becomes possible to develop nonpharmacological treatment strategies that not only regulate physiological variables but also restore functions that have been lost. Examples include the construction of hybrid neural circuits [238, 542, 757–759, 900], cochlear implants to aid the hearing-impaired [487, 917], implantable devices to control cardiac arrhythmias [385] and abort seizures [335, 508, 574, 661], and the development of an artificial pancreas to treat patients with diabetes [132, 305]. Over the last decade, these approaches have been extended to restore mobility to patients who have lost their limbs [321, 451] and to allow patients with locked-in syndrome to communicate [887]. Indeed, it has even been possible to construct *insect bots* [834], making it theoretically possible to use such organisms to gather intelligence.

9.8 What Have We Learned?

1. What is the difference between open-loop and closed-loop control?

    The output of a closed-loop controller depends on both the input and the output. In contrast, the output of an open-loop controller depends only on the input. For example, patients with hypertension typically take medication to control blood pressure according to a schedule that does not depend on the measured blood pressure at the time they take their pills. This situation represents open-loop control of blood pressure. Closed-loop control of blood pressure would require that the blood pressure be monitored and that the measured blood pressure be used to adjust the dose of the blood pressure pill.

2. What is meant by the term "negative feedback?"

    Negative feedback is feedback that acts to minimize the error.

3. How does one determine in the laboratory whether a biological feedback control mechanism is operating in a linear range?

    See Section 9.2.

4. What properties of a first-order DDE with proportional negative feedback determine its stability?

    The magnitude of the gain and the time delay.

5. How does a Andronov–Hopf bifurcation occur?

    An Andronov-Hopf bifurcation occurs when a pair of complex eigenvalues cross the complex plane imaginary axis at the same time.

6. Why are Andronov–Hopf bifurcations important?

    An Andronov-Hopf bifurcation is one of the bifurcations that can create or destroy a limit-cycle oscillation. Our interest in an Andronov–Hopf bifurcation stems from the observation that it is a typical bifurcation that generates oscillations in feedback control mechanisms described by DDEs.

7. What is the relationship between the period of an oscillation generated by a time-delayed negative feedback controller and the time delay?

In general, the period of an oscillation generated by a time-delayed feedback control mechanism is greater than twice the delay. If the period is expected to be between two and four times the delay, then it is possible that the oscillation is generated by a time-delayed negative feedback control mechanism. In other words, if we observe an oscillation with a period of 1 s in a biological preparation, it could have arisen because of the presence of a time-delayed control mechanism with a delay between 0.25 s and 0.5 s. Although this rule of thumb is strictly true only for the simplest case of a first-order time-delayed negative feedback control mechanism, it turns out in practice to be quite useful.

8. What observation suggests that biological feedback control mechanisms operate in a similar manner to the feedback mechanisms used by engineers?

> The transfer function measured for biological feedback control mechanisms is the same as that measured for man-made feedback control mechanisms such as the PID controller. For example, the transfer function for the yeast hyperosmolar response is the same as that for a PI controller.

9.9 Exercises for Practice and Insight

1. For positive feedback $E(s) = B(s) + X(s)H(s)$, show that

$$C_{cl}(s) = \frac{C_{ol}(s)}{1 - H(s)C_{ol}(s)}.$$

2. The Laplace transform of a delay is typically written using a step function. Thus if $f(t) = H(t - \tau)f(t - \tau)$, then $F(s) = e^{-\tau s}F(s)$. Why? Hint: See Section 6.7.
3. Equation (9.31) can be used to determine the period of pupil cycling when the pupil light reflex is clamped with piecewise constant negative feedback, and it also suggests a graphical method to determine the parameters. All of these can be determined by choosing the appropriate initial conditions.

   a. Draw the solution when pupil cycling occurs and identify the portions of the period that have length τ.
   b. Show that the period T is given by

   $$T = 2\tau + t_1 + t_2$$
   $$= 2\tau + k_c^{-1} \ln \left[\frac{A_{max} - A_{ON}}{A_{ref} - A_{ON}} \right] + k_d^{-1} \ln \left[\frac{A_{min} - A_{OFF}}{A_{ref} - A_{OFF}} \right].$$

   Hint: Assume that the dynamics are already on the limit cycle.
   c. The value of A_{max} can be determined by choosing $A(t_0) = A_{ref}$. Show that k_d and A_{OFF} can be determined, respectively, from the slope and intercept of a plot of A_{max} versus A_{ref}.
   d. The value of A_{min} can be determined by choosing $A(t_0) = A_{ref}$. Show that k_c and A_{ON} can be determined, respectively, from the slope and intercept of a plot of A_{min} versus A_{ref}.

4. Consider the system

$$\frac{dr}{dt} = r(1 - r^2)$$
$$\frac{d\theta}{dt} = 1.$$

Show that

a. Use the graphical technique we used for fixed-point bifurcations in Chapter 5 to show that there are two fixed points one of which is stable.

b. Now show that in Cartesian coordinates, this system becomes

$$\frac{dx}{dt} = x - x^3 - xy^2 - y$$
$$\frac{dy}{dt} = y - y*x^2 - y^3 + x.$$

(Hint: $\dot{x} = \dot{r}\cos\theta - r\dot{\theta}\sin\theta$ and $\dot{y} = \dot{r}\sin\theta + r\dot{\theta}\cos\theta$ where the dot notation indicates the first derivative.)

c. Write a computer program to integrate this system and show that it generates a circular limit cycle with radius 1.

d. Suppose you want to generate a limit cycle with radius R where R is an arbitrary positive number. How would you modify your computer program?

5. The prototypical equation for an Andronov-Hopf bifurcation is given by (9.15).

a. Take $b = 0$ and $\omega = 1$. Show that in Cartesian coordinates, this system of equations becomes

$$\frac{dx}{dt} = \mu x + x^3 + x^2 y - x^5 - 2x^3 y^2 - xy^4 - y$$
$$\frac{dy}{dt} = \mu y + yx^2 + y^3 - yx^4 - 2x^2 y^3 - y^5 - x.$$

b. Write a computer program to integrate this equation and show that there can be bistability, in this case the coexistence between a fixed stable point and a stable limit cycle.

c. Now take $b \neq 0$. What is the effect of the parameter b?

6. Tyson[11] (1979) considered a generalization of the Goodwin model and obtained the following equations for the feedback inhibition of enzyme synthesis

---

[11] John J. Tyson (b. 1947), American biologist.

$$\frac{dx}{dt} = \frac{1}{y^n + 1} - Kx, \quad \frac{dy}{dt} = x - Ky,$$

where n, K are positive constants [835].

a. What are the fixed points?
b. What are the linearized equations about each fixed point?
c. What is the stability of each fixed point?
d. Write a computer program to numerically integrate these differential equations and verify your conclusions.

7. It has been suggested that bistable gene regulatory circuits can act as toggle switches [258]. A possible realization arises in a situation of two repressor genes whose products x and y inhibit the production of each other. The mathematical model takes the form

$$\frac{dx}{dt} = k_x F_N(y) - x, \quad \frac{dy}{dt} = k_y F_N(x) - y, \tag{9.32}$$

where k_x, k_y are, respectively, the production rates for x and y, F_N is negative feedback given by (9.12) with $K = 1$, and the values of the parameter n that controls the gain in F_N are not necessarily the same for x and y.

a. Under what conditions does there exist a single positive fixed point, and what is its stability?
b. Under what conditions do three fixed points exist, and what is the stability of each of them?
c. What happens to the size of the region of bistability as the parameter n increases?
d. Write a computer program to demonstrate the existence of bistability in (9.32).
e. Are there parameter ranges for which a limit-cycle oscillation arises (compare your observations with those in the next exercise).

8. It is not known how cellular organisms can produce oscillations, such as the circadian rhythm, whose period exceeds the cell division time. One possibility is that the rhythm is generated by a transcriptional network [194]. Long-period oscillations can be generated by extending the two-gene repressor system studied in the previous exercise to the three-gene repressor system referred to as the *repressilator* [92, 194, 456]:

$$\frac{dx}{dt} = k_x F_N(y) - x, \quad \frac{dy}{dt} = k_y F_N(z) - y, \quad \frac{dz}{dt} = k_z F_N(x) - z, \tag{9.33}$$

where F_N is given by (9.12). Consider the special case in which the synthetic rates of all the repressor genes are identical, namely $k = k_x = k_y = k_z$, and set $K = 1$.

a. Write a computer program to integrate (9.33). Verify that for $n \geq 2$ in F_N, an oscillation arises if k is large enough.

b. What happens to the shape of the phase-plane portrait as k becomes very large? It has been suggested that the formation of this starlike structure in the phase plane having three sharp turns points to a new route for the formation of limit cycles that is presently under investigation [456].

c. The fixed-point stability analysis of (9.33) is time-consuming and must be done with care. However, if you want a challenge, see how much you can accomplish and then check the analysis given in [92, 194, 456] for hints.

9. The Wilson–Cowan[12][13] equations describe the dynamics of large populations of inhibitory and excitatory neurons [880, 881]. A simplification of these equations to two neurons yields the Wilson–Cowan oscillator [880]. This neural circuit is composed of an excitatory neuron and an inhibitory neuron. The excitatory neuron receives two inputs: one self-excitatory, and the other a recurrent inhibitory input. The mathematical model takes the form

$$T_x \frac{dx}{dt} = -x + k_y F_P(1.6x - y + I), \quad T_y \frac{dy}{dt} = -y + k_x F_P(1.5x), \tag{9.34}$$

where x, y are the respective firing rates of the excitatory and inhibitory neurons, T_x and T_y are their respective time constants, I is an external input, and F_P is given by (9.13). Take the following parameter choices [880]: $T_x = 5$, $T_y = 10$, $n = 2$, $k_x = k_y = 100$, and $K = 30$.

a. Take $I = 0$. Show that there is only one fixed point and evaluate its stability.

b. Write a computer program to integrate (9.34). Slowly increase I. Are there values of I for which a limit-cycle oscillation arises?

c. In order to account for the conduction time delay τ between the excitatory and inhibitory neurons, (9.34) becomes

$$T_x \frac{dx}{dt} = -x + k_y F_P(1.6x - y(t - \tau) + I),$$

$$T_y \frac{dy}{dt} = -y + k_x F_P(1.5x).$$

What is the effect of τ on the dynamics?

---

[12] Hugh R. Wilson (b. 1943), Canadian-American computational neuroscientist.

[13] Jack D. Cowan (b. 1933), Scottish-born computational neuroscientist.

Chapter 10
Time delays

People who have been stuck in traffic are familiar with the negative impact that slow human reaction times have on performance [659, 813]. Thus, it is somewhat surprising that it is only in the last 20–30 years that the study of delay differential equations has begun to attract attention in biology. One of the main reasons that DDEs have become so successful as models of biological control stems from the fact that many of the details of the mechanisms which produce the delay are hidden from the control process. In other words, all of this biology is replaced by a time delay which can be measured experimentally.

An important reason why DDEs were not a staple in the education of life scientists is that there are few textbooks written in a style accessible to undergraduate students. Fortunately, this is changing [208, 276, 519, 775, 800]. The availability of accessible textbooks has become possible because of the hard work of mathematicians and engineers over the last 50 years. The major hurdle that had to be overcome was to understand the role of time delays in feedback control [15, 208, 309, 436, 448, 519, 775, 800]. The presence of a time delay can often result in counterintuitive phenomena that provide interesting insights into biological control [578, 591, 641, 642].

Time delays in biological systems typically reflect a combination of transmission times and processing times. Transmission times arise because the different components of the feedback mechanism are spatially distributed, i.e., sensors that detect the error and effectors that act to minimize the error are not located in the same place. The time required for information to travel between these various parts of the feedback controller must be taken into account. Biological examples of transmission delays include the time required for hormonal signals to travel from their site of production to target organs either by diffusion or by passage through the circulation, and the time required for a nerve impulse to travel along an axon and across a synapse. Time delays can also be due to processing times that arise because it takes time to act upon an error signal. Examples include the time it takes to synthesize a protein ($\approx$ 1–2 min), for blood cells to mature in the bone marrow before their release into circulation ($\approx$ 7 days), and in animal population models, gestation

© Springer Nature Switzerland AG 2021
J. Milton and T. Ohira, *Mathematics as a Laboratory Tool*,
https://doi.org/10.1007/978-3-030-69579-8_10

times (weeks to months). Often in biology, the contribution of the processing times to the time delay, τ, is much larger than the contribution due to transmission. For example, in the pupil light reflex considered in the last chapter, only about 20ms of the 300ms delay can be accounted for by axonal conduction times. Similarly, in the glucose–insulin feedback loop, the delay is of the order of minutes compared to a circulation time of a few seconds.

Historically, the use of DDEs caused great concern because of their perceived challenge to the validity of Newtonian dynamics [519, 858]. In a Newtonian dynamical system, the future dynamics can be completely predicted by specifying the initial conditions at one instant in time. However, when time delays are present, what happened in the past affects what happens in the present! This debate was largely resolved in the context of feedback control by recognizing the importance of *hidden variables*, namely variables that are not sensed by the control mechanism. A famous example is a pendulum clock in which the descending weight is enclosed in a case. If a person cannot see the weight, then the only way to predict when the clock next needs to be wound is to know when it was last wound [519]. Thus, for example, the fact that the propagation of an action potential along an axon can be well modeled using a partial differential equation may not be useful for constructing a model of neural feedback. This is because there are no sensors in the organism that can detect a nerve impulse in progress. All that can be done is to introduce a delay between when an action potential was generated by one neuron and when the resultant postsynaptic potential subsequently appears in another neuron.

In Section 9.5.2, we introduced a first-order DDE as a model for the pupil light reflex. This model took the general form

$$\dot{x}(t) + kx(t) = F(x(t), x(t - \tau)) \tag{10.1}$$

where $\dot{x} := dx/dt$. When dealing with DDEs, it is most convenient to use the "dot convention for the derivative" since it is necessary to distinguish between those changes in the variables which occur in the present from those for which the changes occur in the past. Biological systems are typically quite sensitive to sudden changes. Important examples arise in the neural control of movement and balance. Indeed, the human nervous system can detect changes in the position, speed, and even the acceleration of the body and its joints. Thus, we can anticipate that F will often contain terms related to $x(t - \tau)$, $\dot{x}(t - \tau)$, and possibly even $\ddot{x}(t - \tau)$.

The questions we will answer in this chapter are

1. Why do time delays frequently arise in biology?
2. What is a state-dependent delay?
3. What is a distributed delay?
4. What is the difference between a lag and a time delay?
5. What is the order of a delay differential equation (DDE), and what is its dimension?
6. A DDE is sometimes referred to as a functional differential equation. Why?

7. What is a retarded functional differential equation (RFDE)?
8. What is a neutral functional differential equation (NFDE)?
9. What are the different types of state-dependent feedback and what factors limit its stability?
10. What is predictive feedback and what factors limit its stability?
11. How does a jump discontinuity arise in a DDE?
12. How do the effects of a jump discontinuity differ for a RFDE and a NFDE?
13. What is the D-subdivision method for determining the stability of a DDE?
14. How can the natural frequency, ω_n, of an inverted pendulum be determined experimentally?
15. What is the difference in stability between a hung-down pendulum and an inverted pendulum?
16. What is postural sway?
17. Under what conditions can an inverted pendulum be stabilized by vertically vibrating the pivot point? Why does this not account for the stabilizing effects of vibration and postural sway?
18. What is a sensory dead zone and how does its presence affect the equations of motion for the feedback control of an inverted pendulum?
19. What computer packages are currently available to integrate DDEs?
20. What is the method of steps for integrating a DDE?
21. What is the semi-discretization method for integrating DDEs?
22. Can the same integration methods for ODEs be used to integrate a DDE?

10.1 Classification of DDEs

There are two classifications that are used to discuss equations such as (10.1). The first classification compares the highest derivative in the plant (left-hand side of (10.1)) to that contained in the feedback (right-hand side). When the highest derivative in the feedback is less than that in the plant, the DDE is referred to as a *retarded functional differential equation* (RFDE). When the highest derivative in the feedback equals that in the feedback, the DDE is referred to as a *neutral functional differential equation* (NFDE). The distinction between RFDEs and NFDEs has important consequences for stability. For a RFDE, instability can arise when only a finite number of eigenvalues have a positive real part. In contrast for a NFDE, instability is associated with an infinite number of eigenvalues with a positive real part.

Although our focus is on RFDEs, NFDEs occasionally arise in biological systems. Perhaps the earliest application of a neutral DDE to biology is the neutral delayed logistic equation for population growth

$$\dot{x}(t) = r(1 - (x(t-\tau) + k_2\dot{x}(t-\tau))/K)x(t)$$

where r is a growth rate constant and K is a constant equal to the carrying capacity of the environment. This equation was developed to explain the effects on the growth rate of consumption of a resource which takes time τ to recover [448]. It is interesting to note that present- day mathematical biologists prefer to model this phenomenon using a DDE with a state-dependent delay (see below). Neutral delays also arise in the setting of the control of movement when acceleration terms are included in the feedback [371, 492, 770].

A second classification considers the nature of the feedback. In order to classify the different kinds of feedback that arise in second-order DDEs, it is useful to note that all forms of time-delayed feedback control can be considered as a type of predictive control in which observations made in the past are used to predict corrective actions taken in the present. From this perspective, two general classes of delayed feedback controllers can be distinguished (Fig. 10.1). *State-dependent feedback* uses information obtained at time $t - \tau$ to predict the corrections made at time t. The type of information available for feedback control depends on the existence of biological sensors to detect the magnitude and time-dependent changes in the controlled variable. Thus, the feedback can be proportional (P)

$$F_\mathrm{P} = k_\mathrm{p}x(t - \tau),\tag{10.2}$$

proportional–derivative (PD)

$$F_\mathrm{PD} = k_\mathrm{p}x(t - \tau) + k_\mathrm{d}\dot{x}(t - \tau),\tag{10.3}$$

or proportional–derivative–accelerative (PDA)

$$F_\mathrm{PDA} = k_\mathrm{p}x(t - \tau) + k_\mathrm{d}\dot{x}(t - \tau) + k_\mathrm{a}\ddot{x}(t - \tau),\tag{10.4}$$

where $k_\mathrm{d}, k_\mathrm{d}, k_\mathrm{a}$ are, respectively, the proportional, derivative, and accelerative gains.

In contrast, *predictive feedback* (PF) uses all of the information available up to $t - \tau$ to develop an internal model upon which to predict corrective actions taken at time t [445, 591]. The corresponding control force is

$$F_\mathrm{PF} = k_\mathrm{p}e_\mathrm{p}(t)\tag{10.5}$$

where $e_\mathrm{p}(t)$ is the prediction of the actual error, which is determined based on an internal model, namely

$$e_\mathrm{p}(t) = \exp{(k\tau)}\,e_\mathrm{p}(t - \tau) + \int_{t-\tau}^{t} \exp{(ks)}\,F(s+t)\mathrm{d}s,\tag{10.6}$$

where the second term on the right represents the model-dependent prediction. Optimal control can be obtained by solving the equations over the delay period [430, 536]. When the parameters of the internal model precisely match those of the actual system, then the prediction gives the exact state. In this case the feedback eliminates, or compensates, for the delay in the feedback loop.

Our focus is on dynamical systems which possess a single constant delay. There have been some studies of DDEs with 2 or more delays [14, 45]. Often there can be a distribution of delays. For example, in a nerve there is typically a distribution of

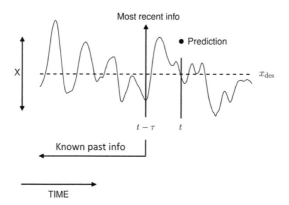

Fig. 10.1 Schematic representation of prediction schemes for time-delayed feedback. The exact motion is given by the solid line and the prediction at time t by the •. State-dependent controllers use the most reliable available data, i.e., $x, \dot{x}, \ddot{x}$, measured at $t - \tau$ to predict the correction (•) made at time t. Model predictive controllers use information obtained from the known time history from $-\infty$ to $t - \tau$ to develop an inverse model to predict the correction made at time t.

axonal diameters which implies that there is a distribution of conduction velocities and hence transmission delays. Thus, (10.1) becomes

$$\dot{x}(t) - k_1 x(t) = k_2 \int_{-\infty}^{t} x(s) G(t-s) ds = k_2 \int_0^{\infty} x(t-s) G(s) ds$$

where $G(\tau)$ is a kernel which describes the distribution of delays and k_1, k_2 are constants [519]. For special choices of $G(\tau)$, it has been shown that a distributed delay has a stabilizing effect on negative feedback control in both neural and ecological dynamical systems [210, 212, 560]. However, it must be kept in mind that $G(\tau)$ has not been well characterized experimentally and hence the relative contributions of a distribution of transmission times versus a distribution of processing times have not yet been investigated. We can consider that (10.1) represents the situation when a distribution of delays is replaced by its mean value

$$\overline{\tau} = \int_0^{\infty} G(u) u \, du.$$

In other words, the distribution of delays, $G(\tau)$, is replaced by a delta distribution, $G_\delta(\tau) = \delta(\tau - \overline{\tau})$ which is equal to zero except when $\tau = \overline{\tau}$.

An important observation is that τ can be state-dependent. Time delays for insect and animal and even blood cell maturation are a function of population density [7, 527]. Thus, (10.1) becomes

$$\dot{x}(t) + x(t) = k' x(t - \tau(x(t))), \tag{10.7}$$

where k' is a constant and the notation $\tau(x(t))$ indicates that the delay is a function of the error. State-dependent delays also arise in a number of engineering applications including milling, cooling systems, and in a spatially extended networks [39, 376].

State-dependent delays also arise very frequently in the nervous system. As we discussed in Section 9.5.2, the time delay in the pupil light reflex is a function of the illuminance: the shortest τ are measured for the most intense light stimulus. A much more common manifestation of a neural state-dependent delay is the fact that central processing times increase as the complexity of the voluntary task increases [342]. This is referred to as Hick's law in the psychology literature. Finally, for certain human motor, balance and computational tasks there can be a central (or psychological) refractory time of about $200 - 400$ms [841, 853]. This refractory time arises because the response to the second stimulus cannot be acted upon until the response to the first stimulus is completed.

10.2 Estimating time delays

A variety of methods have been used to measure time delays in biology. In some cases, time delays have been measured directly as the time between when a stimulus is applied and when the response is first detected (see Figure 9.8). Examples include gestation and maturation times, delays in reflexes such as the pupil light reflex, and the times to make a protein. In engineering application, time delays can be measured using cross-correlation techniques as described in Section 14.8.2. However, this approach has been considered to be less reliable than using the response to a perturbation in biological applications [842].

Example 10.1 For motor tasks dependent on visual input, the time delay can be measured from the responses following a visual blackout. An example arises in the determination of τ for pole balancing at the fingertip [591]. In this experiment, subjects were required to balance a pole on the surface of a table tennis racket while wearing glasses equipped with liquid crystal optical beam shutters (Figure 10.2). These shutters have the property that they are transparent but become opaque when a voltage is applied to them. The purpose of the table tennis racket is to minimize sensory inputs from cutaneous mechanoreceptors located in the fingertip. Reflective markers are attached to each end of the pole so that the pole's movements can be monitored using motion capture cameras. Following a 500ms long visual blackout, the first corrective movement is identified from the change in the velocity of the bottom marker (Figure 10.2(top)). This delay determined in this way is equal to ≈ 223ms and is equal to that obtained from the response to a brief mechanical perturbation applied to the balance pole [553]. Averaging techniques (see Section 14.4.4) were used to minimize the effects of those changes in the velocity of the bottom reflective marker which are uncorrelated with the visual blackout. Visual blackouts are particularly well suited for estimating τ in virtual tasks which involve the interaction between a human and a task displayed on a computer screen (see, for example, [405]). A naturally occurring visual blackout is an eye blink (duration $\approx 150 - 225$ ms [457, 860]). $\diamondsuit$

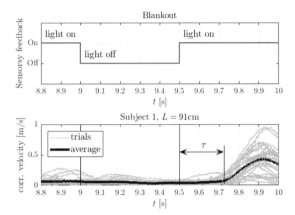

Fig. 10.2 Measuring time delay for pole balancing using a visual blank out. (a) After 9s of pole balancing ("light on"), the optical shutter becomes opaque ("light off") for 500ms, then again transparent ("light on"). (b) The averaged velocity of the movements of the bottom of the pole (thick dark line) is determined from 25 poles balancing blank out trials (light lines). See text for more discussion).

In the nervous system, transmission delays are most commonly estimated indirectly from measurements of the axonal conduction velocity. Such estimations of τ are limited by the fact that the distance between two interacting neurons is seldom known with precision. Conduction velocities, v, can be measured using electrophysiological techniques. Practical problems arise when the diameter of the axons becomes smaller than that of the recording electrode [565]. Conduction velocities can also be estimated indirectly from the diameter, d, of the axon and its myelination: for myelinated axons $v_a \approx \sqrt{d}$ and for an unmyelinated axon, $v_a \approx dg\sqrt{-\ln g_m}$, where $0 < g_m < 1$ is the ratio of axon diameter to overall fiber diameter [865]. The dependence of v_a on g_m raises the possibility that transmission times are adapted during learning [105, 224, 789]. In the central nervous system, g_m is determined by patterns of action potentials and glial–neuron interactions [224, 807, 912]. Moreover, it has been suggested that the self-organization of transmission delays may be an important, but under-recognized, mechanism for learning in both living [211] and artificial [891] neural networks.

10.3 Delays versus lags

The distinction between a lag and a time delay often causes confusion. For example, in the input line of MATLAB's dde23, delays as referred to as lags (see Section 10.9.3)! However, in control theory it is important to be precise in the use of these terms. Control systems with lags can be modeled with ODEs but if delays are present, the correct model takes the form of a DDE. Moreover, it is possible that both lags and delays may be present [491].

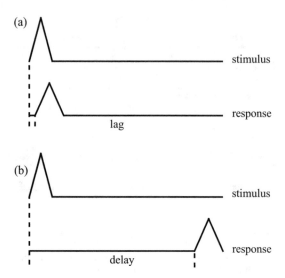

Fig. 10.3 Comparison between (a) a lag and (b) a time delay. See text for discussion.

It is easiest to appreciate the difference between a lag and a delay using our black-box analogy. In the case of a lag (Figure 10.3a), the time course of the stimulus overlaps that of the response. Indeed, for a pure lag the response begins at the onset of the stimulus. A consequence is that the shape of the stimulus becomes spread out over time. In other words, the lag produces a change in the shape of the input signal. An example is the response of a linear dynamical system to a step input discussed in Section 6.1 (see Figure 6.2). In contrast, in the presence of a delay, the input stimulus is exactly postponed by an amount equal to the delay. Importantly, there is no significant change in the shape of the stimulus. In the typical case where the time course of the stimulus does not overlap that of the response (Figure 10.3b), a time delay must be used.

Historically investigators were interested in approximating the effects of a delay by a lag. This interest was motivated by the fact that computer algorithms for the numerical integration of DDEs were not widely available and mathematical methods for the analysis of stability of DDEs were not widely appreciated. It is possible that the effects of a delay can be approximated by a lag in situations in which the delay is so small that the time course of the stimulus and response significantly overlap (Figure 10.3a). Many of these approximations took the form of a linear chain of ODEs [519], for example,

$$x_1 \xrightarrow{k_1} x_2 \xrightarrow{k_2} x_3 \longrightarrow \cdots x_{n-1} \xrightarrow{k_n} x_n, , \tag{10.8}$$

or

$$\frac{dx_2}{dt} = k_1 x_1 - k_2 x_2$$

$$\frac{dx_3}{dt} = k_2 x_2 - k_3 x_3$$

$$\cdots$$

$$\frac{dx_n}{dt} = k_{n-1} x_{n-1} - k_n x_n$$

$$\frac{dx_{n+1}}{dt} = k_n x_n,$$

the case where $n = 2$ was discussed in Section 3.4.2. The corresponding approach in the frequency domain is referred to as a Padé approximation [50]. These approaches work best when the time delay is small. However, it should be kept in mind that even an infinitesimally small time delay can be destabilizing [448].

A second approximation is to use a Taylor series expansion (Tool 7) to approximate terms of the form $x(t - \tau)$, namely

$$x(t - \tau) = x(t) - \tau \dot{x}(t) + \frac{1}{2}\tau^2 \ddot{x}(t) + \cdots + (-1)^n \frac{1}{n!} 2^n x^n(t) + R(t, \tau, n)$$

where

$$R(t, \tau, n) = (-1)^{n+1} \frac{1}{(n+1)!} \tau^{n+1} x^{n+1}(t - \theta \tau)$$

and $\theta \in [0, 1]$. However, as emphasized by Tamas Insperger[1], this approximation is completely without mathematical justification [369]. In particular, if the order of the Taylor series expansion, n, exceeds the order of the leading derivative in the DDE by 2, then the approximated system becomes unstable independently of the system's parameters.

At present there is little reason for using these approximations. However, it is possible that the computational load can be reduced in certain applications by replacing a small delay with a lag. In these cases, it is prudent to first determine the stability of the DDE model and then use these results to justify replacing the delay with a lag.

10.4 Jump discontinuities

In order to solve a DDE such as (10.1), it is not sufficient to specify an initial value $x(t_0) = X_0$ (which, by the way, is what we would do to solve $dx/dt = -kx$). Instead we must specify an initial function $\phi(t_0)$, namely all of the values of $x(t)$ that lie on the interval $t \in [t_0 - \tau, t_0]$, i.e., $\phi(t) \in [t_0 - \tau, t_0]$. This is the way that the dynamical system remembers its past. Since the number of points on an interval is infinite, (10.1) is also referred to as an infinite-dimensional differential equation or as a functional differential equation.

---

[1] Tamas Insperger (PhD, 2002), Hungarian theoretical mechanical engineer.

An important issue is that, in general, the right and left derivatives evaluated at $t = t_0$ are not the same. For example, if

$$\dot{x}(t) = F(x(t), x(t-\tau)) = x(t-\tau)$$

with $\phi(s) = 1$ for $s \in [t_0 - \tau, t_0]$, the derivative from the left at $x = 0$ is equal to 0 and that from the right is equal to 1. This means that even if F and ϕ have continuous derivatives, there will necessarily always be a jump discontinuity in the first derivatives at t_0. This is because it is impossible to satisfy simultaneously the conditions $x(t_0) = \phi(t_0)$ and $\dot{x}(t_0^+) = \dot{\phi}(t_0^-)$.

The effects of this jump discontinuity depend on whether the DDE is a RFDE or a NFDE. In the case of a RFDE, the initial first discontinuity at t_0 is propagated as a discontinuity in the second derivative at time $t_0 + \tau$, as a discontinuity in the third-degree derivative at time $t_0 + 2\tau$ and, more generally, as a discontinuity in the $(n+1)$st derivative at time $t_0 + n\tau$. In other words, the solution is progressively smoothed as a function of time as the initial derivative discontinuity is propagated successively to higher-order derivatives. In contrast, for a NFDE the effect of the jump discontinuities are always present though they may diminish as a function of time.

10.5 Stability analysis: First-order DDEs

We discuss the *D-subdivision method* to investigate the stability of DDEs[2] [373, 436, 800]. The D-subdivision method consists of four steps: 1) make the usual ansatz that $x(t) \approx \exp(\lambda t)$, 2) obtain the characteristic equation $D(\lambda)$, 3) take $\lambda = \gamma + j\omega$ and substitute into the equation $D(\lambda) = 0$, and 4) decompose this equation into real and imaginary parts and complete the analysis. Readers should quickly realize that this is essentially the same approach that we used to analyze the stability of the model for the pupil light reflex discussed in Section 9.5!

10.5.1 Wright's Equation

The delay version of our familiar first-order differential equation is

$$\frac{dx}{dt} = kx(t-\tau), \tag{10.9}$$

where k is a parameter. This equation is referred to as *Wright's*[3] *equation* [182, 373, 775, 890].

---

[2] The stability and solution of a first-order DDE can be obtained using the Lambert W function. See Exercise 10.11 for details.

[3] Edward Maitland Wright (1906–2005), English mathematician.

Figure 10.4 shows the solutions of (10.9) when $k = -1$ for different choices of τ. It is convenient to divide the initial function $\Phi(t)$ into the value at $t = 0$ plus a part that describes $\Phi(t)$ on the half-open interval $[-\tau, 0)$. This is the way, for example, that the initial function is represented in XPPAUT computer programs for integrating DDEs [202] (see below). Thus we assume that initially, the system is at rest, and then at $t = 0$, a brief input of unit magnitude is applied, i.e., $\Phi(0) = 1$ and $\Phi(t) = 0, t \in [-\tau, 0)$. When τ is small, the solutions look very much like those we obtained for $\tau = 0$. However, as τ gets bigger, we have first damped oscillatory solutions, and then, when $\tau > \pi/2$, we have instability. Clearly, the presence of the time delay can have major effects.

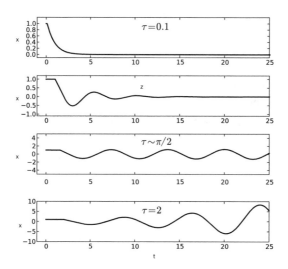

Fig. 10.4 Effect of changing τ on the behavior of (10.9) when $k = -1$. The initial function is $\Phi(0) = 1$ and $\Phi(t) = 0, t \in [-\tau, 0)$.

When $\dot{x}(t) = 0$, the fixed point of (10.9) is $x^* = 0$. We assume that $x(t) \approx \exp(\lambda t)$, where λ is an eigenvalue, to obtain the characteristic equation

$$\lambda = ke^{-\lambda \tau}. \tag{10.10}$$

Equation (10.10) is a transcendental equation when λ is complex, and hence in general, there is an infinite number of solutions. We can easily verify this using Euler's formula (Tool 8) (see Exercise 9.9).

When $\tau = 0$, (10.10) has only one root,

$$\lambda = k.$$

Stability implies $k < 0$. When $\tau > 0$, then λ can be either real or imaginary, depending on the choice of τ.

Real Eigenvalues

In order to determine the real roots λ_{real} of (10.10) when $\tau > 0$, it is useful to rewrite (10.10) as

$$k = \lambda_{real} e^{\lambda_{real} \tau}. \tag{10.11}$$

Figure 10.5a shows a plot of k versus λ_{real} for $\tau = 1$ (solid line). The minimum value of k occurs when $\lambda_{real} = -1/\tau$, for which $k = -(e\tau)^{-1}$. When $k > 0$, then $\lambda_{real} > 0$, and hence the fixed point is unstable for all $\tau > 0$. For $0 > k > -(e\tau)^{-1}$, there are two real and negative eigenvalues λ_{real}. Thus the fixed point is a stable node.

Complex Eigenvalues

When $k = -(e\tau)^{-1}$, the two real eigenvalues merge, and when $-(e\tau)^{-1} > k$, an infinite number of pairs of complex eigenvalues emerges, where $\lambda_{complex} = \gamma \pm j\omega$. The condition for stability is that the real part of all of the $\lambda_{complex}$ must be negative: instability occurs when the real part of at least one $\lambda_{complex}$ becomes positive. Thus, the boundary that separates the stable from the unstable solutions occurs at $\gamma = 0$.

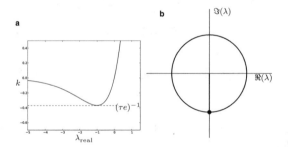

Fig. 10.5 (a) Plot of k as a function of λ_{real} when $\tau = 1$. (b) Graphical stability analysis of (10.9) using (10.12). The right-hand side of (10.12) coincides with the negative portion of the imaginary axis. The value (ω_{sb}, τ_{sb}) is indicated by the dot ($\bullet$).

We can determine the stability boundary by setting $\gamma = 0$, so that (10.10) becomes

$$e^{-j\omega\tau} = \frac{j\omega}{k}. \tag{10.12}$$

This representation facilitates the use of graphical techniques to investigate stability [519]. The left-hand side represents a circle in the complex plane, and the right-hand side represents the negative part of the imaginary axis (Figure 10.5b).

A change in stability occurs when the imaginary axis intersects the unit circle. What are the values of the frequency ω_{sb} and delay τ_{sb} at this intersection point? In order to determine these values, we use Euler's formula (Tool 8) to rewrite (10.12) as

$$\frac{j\omega_{sb}}{k} = -\left[\cos \omega_{sb}\tau_{sb} - j\sin \omega_{sb}\tau_{sb}\right]. \tag{10.13}$$

For (10.13) to hold, it is necessary that the real and imaginary parts of the two sides of the equation equal each other. Thus

$$\cos \omega_{sb}\tau_{sb} = 0, \tag{10.14}$$

$$\sin \omega_{sb}\tau_{sb} = \frac{\omega_{sb}}{k}. \tag{10.15}$$

The general solution of (10.14) is $\omega_{sb}\tau_{sb} = (2n+1)\pi/2$, where $n = 0, \pm 1, \pm 2, \dots$. From Figure 10.5b, we see that $n = -1$, and hence

$$\omega_{sb}\tau_{sb} = -\frac{\pi}{2}. \tag{10.16}$$

Recall from Section 7.5.4 that in the frequency domain, a delay introduces a phase shift proportional to $-\omega\tau$. Thus (10.16) indicates that at the point of instability, there will be a phase shift of $\pi/2$. We can square both sides of (10.14)–(10.15) and add to get

$$\cos^2 \omega_{sb}\tau_{sb} + \sin^2 \omega_{sb}\tau_{sb} = \frac{\omega_{sb}^2}{k^2} = 1 \tag{10.17}$$

and hence obtain

$$\omega_{sb} = k. \tag{10.18}$$

By combining (10.16) and (10.18), we obtain

$$\tau_{sb} = \left|\frac{\pi}{2k}\right|.$$

If we take $k = -1.0$, we predict that instability will occur when $\tau > \pi/2$, which is what we observed in Figure 10.4.

Finally, we note that

$$\left|\frac{1}{ek}\right| < \tau_{sb} < \left|\frac{\pi}{2k}\right|.$$

In other words, the real part of the complex eigenvalues λ_{sb} for these values of τ must be negative. Hence the fixed point is a stable spiral point. Thus, our stability analysis accounts for all of the observations in Figure 10.4.

10.5.2 Hayes equation

The production–destruction models discussed in Section 9.4 often have the general form

$$\dot{x}(t) = F(x(t), x(t-\tau)).$$

The linearized version about the fixed point x^* is

$$\frac{du}{dt} = Au(t) + Bu(t-\tau), \tag{10.19}$$

where

$$u = x - x^*, \quad A = \frac{dF}{dx}\bigg|_{x=x^*}, \quad B = \frac{dF}{dx(t-\tau)}\bigg|_{x=x^*}.$$

We refer to (10.19) as the *Hayes equation* [182, 332]. Equation 9.19 is an example of a Hayes equation with negative feedback.

Following the D-subdivision method, we have

$$D(\lambda) = \lambda - a - be^{-k\tau}. \tag{10.20}$$

Setting $D(\lambda) = 0$, taking $\lambda = \gamma + j\omega$, and decomposing into real and imaginary parts yields

$$Re : \gamma - A - Be^{-\gamma\tau}\cos(\omega\tau) = 0,$$
$$Im : \omega + Be^{\gamma\tau}\sin(\omega\tau) = 0.$$

The boundaries of the stable region can be determined as a parametric function of ω. When $\omega = 0$, we have

$$B = -A. \tag{10.21}$$

When $\omega\tau \neq k\pi$, we have

$$A = \frac{\omega\cos(\omega\tau)}{\sin(\omega\tau)} \tag{10.22}$$

$$B = \frac{-\omega}{\sin(\omega\tau)}.$$

Figure 10.6 shows the stability diagram for (10.19) when $\tau = 1$. For choices of all choices of A, B outside of this region, the solutions of (10.19) are unstable. The number of unstable characteristic exponents in these regions as a function of A, B can be determined [800].

In 1950, Hayes [332] gave a complete treatment for the conditions that ensure that the fixed point of (10.19) is stable. The general forms obtained by Hayes are easiest to appreciate if the characteristic equation is written as

$$\frac{A}{B} + j\frac{\omega}{B} = e^{-j\omega\tau}. \tag{10.23}$$

The stability conditions are

$$|A| > |B|, \tag{10.24}$$

or

$$|A| < |B| \quad \text{and} \quad \tau < \frac{\cos^{-1}(-A/B)}{(B^2 - A^2)^{1/2}}, \tag{10.25}$$

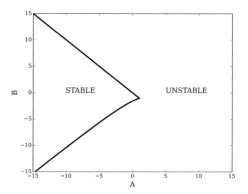

Fig. 10.6 Stability diagram for (10.19). The upper boundary of the stable region is given by (10.21) and the lower boundary by (10.22). For the number of unstable exponents in the unstable region, see [373].

where

$$0 \leq \cos^{-1}\left(-\frac{A}{B}\right) \leq \pi.$$

Alternatively, these conditions can be written as (see Exercise)

$$A\tau < 1, \quad A\tau < -B\tau < \sqrt{(A\tau)^2 + a_1^2}, \tag{10.26}$$

where $0 < a_1 < \pi$ is the root of the equation

$$a\cot(a) = A\tau,$$

and $a_1 = \pi/2$ when $A\tau = 0$.

We can get some insight into Hayes' stability conditions by setting $\lambda = jf$ and rewriting (10.23) as

$$j\omega = A + Be^{j\omega\tau} \quad = A + B[\cos\omega\tau - j\sin\omega\tau] = A + B\cos\omega\tau - jB\sin\omega\tau.$$

Equating the real and imaginary parts, we obtain

$$-A = B\cos\omega\tau, \quad \omega = -B\sin\omega\tau.$$

We can solve for τ and ω by squaring both equations and add to obtain

$$\frac{A^2 + \omega^2}{B^2} = \cos^2\omega\tau + \sin^2\omega\tau = 1.$$

Since

$$\omega\tau = \cos^{-1}\left(-\frac{A}{B}\right),$$

we have

$$\tau = \frac{\cos^{-1}(-A/B)}{\sqrt{B^2 - A^2}},$$

which is exactly what Hayes obtained.

10.5.3 Positive Feedback: Cheyne–Stokes Respiration

The human breathing rhythm is typically quite regular and generally attracts little attention. However, respiratory dysrhythmias occur, and they attract great interest [119, 120], particularly when they are associated with intervals during which breathing ceases, referred to as apneas. By far the most common of these dysrhythmias is Cheyne–Stokes[4] respiration (CSR) [119]. CSR is a rhythmic breathing pattern in which intervals of apnea alternate with a crescendo–diminuendo pattern of hyperpnea (Figure 10.7). It is currently believed that CSR is related to instability in the feedback control of respiration [119]. The observation that CSR can be induced in anesthetized dogs by artificially prolonging the circulation time [304] supports the relevance of a time-delayed feedback controller. Here, we discuss a model for CSR developed by Michael Mackey and Leon Glass [275, 276, 522].

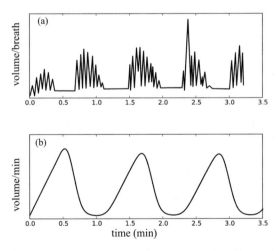

Fig. 10.7 (a) Ventilation as a function of time in a patient with Cheyne–Stokes respiration. (b) Solution of (10.28) when S^* is increased to a range in which the fixed point is unstable. Figure adapted from [523] with permission.

---

[4] John Cheyne (1777–1836), British physician; William Stokes (1804–1878), Irish physician.

Carbon dioxide (CO_2) is eliminated from the body by ventilation. Under constant conditions, CO_2 is produced by the body at a constant rate R_b ($\text{mm Hg} \cdot \text{min}^{-1}$). There is a feedback control mechanism that regulates the arterial CO_2 partial pressure. The sensor that detects arterial CO_2 is located in the brainstem; the "bellows," i.e., your lungs, are located in the chest. The time delay arises because of the blood transit time from the brainstem (where ventilation is determined by chemoreceptors and by the respiratory oscillator) to the lungs (where CO_2 elimination takes place). It is observed that the ventilation V ($\text{liters} \cdot \text{min}^{-1}$) is a monotonically increasing function of arterial CO_2 at a time τ (min) in the past (positive feedback). Taking these considerations into account, we obtain the following model [275, 276, 522]:

$$\frac{dx}{dt} = \text{production} - x \cdot \text{destruction} = R_b - k_v x F_p(x(t - \tau)), \tag{10.27}$$

where x (mm Hg) denotes the arterial pCO_2, i.e., the partial pressure of CO_2, and the rate of ventilation is

$$F_P(x(t - \tau)) := V(x(t - \tau)) = \frac{V_m x(t - \tau)^n}{K^n + x(t - \tau)^n},$$

where we have used (9.13). This choice of the feedback function takes into consideration the fact that there is a finite limit to how much V can be increased: V cannot exceed V_m. By setting $K = x$, we see that K is the pCO_2 at which V equals one-half V_m. Thus, the control of arterial pCO_2 by ventilation rate can be described by

$$\frac{dx}{dt} = R_b - k_v x V(x(t - \tau)) = R_b - k_v x \frac{V_m x(t - \tau)^n}{K^n + x(t - \tau))^n}. \tag{10.28}$$

The fixed point, x^*, of (10.28) is

$$x^* = \frac{R_b}{k_v V^*},$$

where V^* is F_P evaluated at $dx/dt = 0$. The linearized equation becomes

$$\frac{du}{dt} = \frac{R_b}{x^*} u(t) - \frac{R_b S^*}{V^*} u(t - \tau), \tag{10.29}$$

where S^* is $dF_P/dx(t - \tau)$ evaluated at the fixed point, and $u = x - x^*$.

Using the stability conditions given by (10.26), we have

$$\frac{R_b S^* \tau}{V^*} < \sqrt{\left(\frac{R_b \tau}{x^*}\right)^2 + a_1^2},$$

where a_1 is determined by solving

$$a_1 \cot(a_1) = -\frac{R_b \tau}{x^*}.$$

The parameters x^*, R_b, V^*, S^*, τ can be estimated from experimental data for healthy humans [275, 276, 522] and are, respectively, 40 mm Hg, 6 mm Hg $\cdot$ min^{-1}, 7 L $\cdot$ min^{-1}, 4 L $\cdot$ min^{-1} mm Hg, and 0.25 min. We see that

$$\left(\frac{R_b \tau}{x^*}\right)^2 \approx 0$$

and

$$a_1 \sim \frac{\pi}{2}.$$

Hence, the stability condition for the fixed point for ventilation becomes

$$\frac{R_b S^* \tau}{V^*} < \frac{\pi}{2}, \tag{10.30}$$

or

$$S^* < \frac{\pi V^*}{2 R_b \tau}. \tag{10.31}$$

As with the pupil light reflex, instability results in the appearance of a stable limit cycle via a supercritical Hopf bifurcation. Equation (10.31) predicts that there are four possible ways that the steady-state ventilation rate can become unstable [275, 522]: increase S^*, increase τ, increase R_b, and decrease V^*. All of these predictions are borne out by clinical and experimental observations. For example, Cheyne–Stokes respiration arises when the circulatory time is increased either in the setting of congestive heart failure or in dogs with arterial extensions [119]. In both cases, τ is increased. Cheyne–Stokes respiration is frequently observed in morbidly obese people, consistent with the prediction that increasing R_b can destabilize the fixed point. Increases in S^* occur in patients with certain brainstem strokes, and decreased V^* occurs in terminal illness. Thus, this simple model provides a framework that encompasses most, if not all, of the situations in which CSR occurs. However, the appearance of CSR in anesthetized dogs required that the circulation time be prolonged to minutes [304]. In other words, this model may be too simplified. Subsequent extension to a "minimal model" leads to a DDE with 15 parameters for which it has not yet been possible to obtain analytical insight [104, 120].

10.6 Stability analysis: Second-order DDEs

Second-order DDEs most commonly arise in the setting of the feedback control of the position and motions of mechanical systems. Here, our focus in on the second-order DDE with PD feedback

$$\ddot{x}(t) + \omega_n^2 x(t) = F(x(t - \tau), \dot{x}(t - \tau)) \tag{10.32}$$

where ω_n^2 is the natural frequency and x is the state variable. This equation is frequently discussed in the control and dynamics literature [208, 371, 460, 770, 800]. Here, we illustrate the properties of (10.32) through their application to problems related to human balancing tasks [460, 591–593, 599, 801].

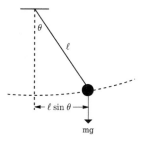

Fig. 10.8 The plane pendulum. Although the gravitational force, mg, acts in the vertical direction, the component of this force which influences the motion of the mass acts perpendicular to the rod.

10.6.1 Planar pendulum

The plant, i.e., the left-hand side of (10.32), describes the dynamics of a pendulum. The simple planar pendulum consists of a mass, m, attached to a rigid, massless rod of length ℓ (Figure 10.8). The purpose of the rod is to restrain the movements of the mass to the plane and, in particular, confine its movements to arcs of circles with radius ℓ. The component of the gravitational force, F_g, that influences the motion of the mass is

$$F_g = -mg \sin \theta$$

where g is the gravitational constant and θ is the vertical displacement angle such that $\theta = 0$ corresponds to the "hung-down" position and $\theta = \pi$ to the inverted (upright) position. The equation of motion for the pendulum is obtained by equating the torque about the support axis to the product of the angular acceleration and the moment of inertia, I, about the same axis. Consequently

$$I\frac{d^2\theta}{dt^2} = \ell F_g.$$

The moment of inertia of a point mass located at a distance ℓ from the axis of rotation is $m\ell^2$. Thus, we obtain

$$\frac{d^2\theta}{dt^2} + \omega_n^2 \sin\theta = 0 \tag{10.33}$$

where the natural frequency is

$$\omega_n = \sqrt{\frac{g}{\ell}}.$$

We can analyze the stability of (10.33) using the techniques discussed in Section 4.3. First, we rewrite (10.33) as a system of first-order differential equations

$$\frac{d\theta}{dt} = v$$

$$\frac{dv}{dt} = -\omega_n^2 \sin\theta$$

where v is the angular velocity. The fixed points are $(n\pi, 0)$, where $n = 0, \pm 1, \pm 2, \cdots$.

The fixed points $(0,0), (2\pi, 0), \cdots, (2n\pi, 0)$ for $n = 1, 2, 3, \cdots$ correspond to the pendulum in the "hung-down" position. The linearized equation of motion is

$$\frac{d^2\theta}{dt^2} + \omega_n^2 \theta(t) = 0. \tag{10.34}$$

The fixed point is center. Thus, the solutions consist of sinusoidal oscillations with period, T_p, where

$$T_p = 2\pi \sqrt{\frac{\ell}{g}}. \tag{10.35}$$

For an excellent discussion of the dynamics of the "hung-down" pendulum, see [32].

The fixed points $((2n+1)\pi, 0)$, $n = 1, 2, 3, \cdots$ correspond to the pendulum in the inverted position. The linearized equation of motion is

$$\frac{d^2\theta}{dt^2} - \omega_n^2 \theta(t) = 0.$$

The fixed point is a saddle. It is the stabilization of the inverted pendulum with delayed feedback described by (10.32) that is relevant for investigations into human balance problems.

10.6.2 Stability analysis

Our starting point is to consider whether the inverted position of the pendulum can be stabilized using PD feedback, namely

$$\frac{d^2\theta}{dt^2} - k\theta(t) = -k_p\theta(t-\tau) - k_d\theta(t-\tau), \tag{10.36}$$

where k is a constant which will, in general, depend on the details of the model for the inverted pendulum. We analyze stability using the D-subdivision method [373, 436, 800]. Making the usual Ansatz $\theta(t) \approx \exp(\lambda t)$, we obtain the characteristic equation

$$D(\lambda) = \lambda^2 + k + k_p e^{-\lambda\tau} + k_d \lambda e^{-\lambda\tau}.$$

By taking $\lambda = \gamma + j\omega$, $\omega \geq 0$, setting $D(\lambda) = 0$, and decomposing into real and imaginary parts, we obtain

$$Re : \gamma^2 - \omega^2 + k + k_p e^{-\gamma\tau}\cos(\omega\tau) + k_d\gamma e^{-\gamma\tau}\cos(\omega\tau) + k_d\omega e^{-\gamma\tau}\sin(\omega\tau) = 0$$

$$Im : 2\gamma\omega - k_p e^{-\gamma\tau}\sin(\omega\tau) + k_d\omega e^{-\gamma\tau}\cos(\omega\tau) - k_d\gamma e^{-\gamma\tau}\sin(\omega\tau) = 0.$$

When $\gamma = 0$, we can simplify these relations to

$$\text{if } \omega = 0: \quad k_p = -k, \quad k_d \in R \tag{10.37}$$

$$\text{if } \omega \neq 0: \quad k_p = (\omega^2 - k)\cos(\omega\tau), \quad k_d = \frac{\omega^2 - k}{\omega}\sin(\omega\tau). \tag{10.38}$$

Figure 10.9 shows that the inverted pendulum will be stable provided that k_p and k_d are located within the D-shaped stability region.

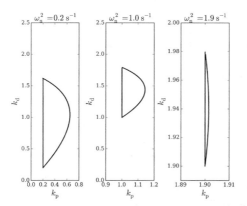

Fig. 10.9 Stability diagram as a function of k_p and k_d when $\tau = 1$. The upright position is stable provided that (k_p, k_d) are chosen within the D-shaped stability region. As ω_n^2 increases, the size of the D-shaped stability region decreases (note the change in the scales) until it disappears when $\omega_n^2 = 2$. For the number of unstable exponents in the unstable regions, see [373].

As ω_n^2 decreases, the size of the stability domain decreases until it finally disappears when $\omega_n^2 = 2$. This can be shown by analyzing the tangent of the parameter curve (10.38) at $\omega_n = 0$. It can be shown that [367, 374]

$$\lim_{\omega \to 0} \frac{dk_d}{dk_p} = \lim_{\omega \to 0} \frac{\frac{dk_d}{d\omega}}{\frac{dk_p}{d\omega}} = \frac{6\tau + k\tau^3}{3k\tau^2 + 6}.$$

The tangent is vertical if $6 + 3k\tau^2 = 0$ and hence

$$k_{\text{crit}} = -\frac{2}{\tau^2}.$$

If $k < k_{\text{crit}}$, then the fixed point of (10.36) is unstable for all k_p and k_d. When $\tau = 1$, $k_{\text{crit}} = -2$.

It is quite surprising that it is possible to stabilize an inverted pendulum with delayed feedback. Traditionally, delays are considered to be destabilizing in the sense that they often induce oscillatory dynamics (see Figure 10.4). Thus, the stabilization of an inverted pendulum has become an important benchmark for comparing control strategies developed for engineering applications [606]: the more robust the control strategy, the shorter the ℓ that can be stabilized.

We can obtain some insights into the role that a time delay plays in stabilizing an inverted pendulum by balancing a wooden dowel at our fingertip (Figure 14.1). Wooden dowels can be bought at a typical hardware store usually of 3–4 foot lengths. We recommend a dowel with a diameter of 0.25 inches (0.625 mm) which is about the diameter of a pencil. The advantage of a wooden dowel is that we can cut it into different lengths and thus directly test our ability to balance a pole on our fingertip as a function of ℓ. When the dowel is the length of a typical wooden pencil (≈ 0.2m), we cannot balance it at our fingertip: typically, the dowel quickly falls in less than 2 seconds despite our desperate attempts to keep it balanced by moving our hand. However, as the dowel becomes longer (say 0.6-1.0m), it becomes easier to balance it for longer times at our fingertip. Indeed, an expert stick balancer can balance a 0.3m dowel at their fingertip for over 15 minutes while seated [591]! The reason that longer sticks are easier to balance than shorter ones is because our reaction times are finite. Once the dowel becomes sufficiently long, its rate of movement becomes slow relative to the time required by our sensorimotor nervous system to make a corrective movement [553, 591]). We expect that to balance a dowel of length ℓ, the time delay must be less than a critical value proportional to $\sqrt{\ell}$ (see below).

Pole balancers will soon discover that no matter how much they practice the pole always eventually falls, especially if the pole balancing is attempted from a seated position. Even for the one with the most expertise, the pole can fall within a few seconds for some trials, while remaining balanced for minutes on others. This behavior is consistent with the presence of a sensory dead zone (see below). Consequently, even though the balancer perceives that the initial position of the pole is always the same, in fact, it isn't!

It is often more useful in laboratory investigations to think of these stability results in terms of a parameter that can be measured and varied, namely τ and ℓ.

The estimates of τ_{crit} and ℓ_{crit} depend, of course, on the details of the balance control problem. From this perspective, the fixed point of (10.36) is stable provided that $\tau \leq \tau_{\text{crit}}$. In the case of an inverted pendulum stabilized by time-delayed PD feedback, we have $k = -\omega_n^2$ and hence

$$\tau_{\text{crit}} = \frac{\sqrt{2}}{\omega_n^2} = \frac{T_p}{\pi\sqrt{2}} \tag{10.39}$$

where T_p is given by (10.35). This estimate of τ_{crit} is determined for a pendulum with one degree of freedom. An important point is that (10.39) is also valid for an nth-degree-of-freedom model for an inverted pendulum provided that the time period T_p is measured for small oscillations of the same mechanical structure hanging at its stable hung-down position [801]. This provides a useful "trick" for measuring τ_{crit} and ω_n experimentally.

The critical stick length is

$$\ell_{crit} = \frac{1}{2}g\tau^2. \tag{10.40}$$

For a given τ, stability requires that $\ell \geq \ell_{crit}$.

It is more realistic to model the inverted pendulum as a rod rather than as a point mass. For a rod, the moment of inertia is $m^2/12$, the center of mass is located at $\ell/2$, and we have $\omega_n^2 = 6g/\ell$ and hence

$$\ell_{crit} = 3g\tau^2.$$

Thus, we see that τ_{crit} is proportional to $\sqrt{\ell}$ as we observed for (10.40).

For all choices of feedback controller, F, it is possible to estimate ℓ_{crit}. Different feedback controllers are characterized by different ℓ_{crit}'s; however, for many types of controllers, ℓ_{cri} must be determined numerically. Thus, if we know τ and can vary ℓ, it should be possible to identify the nature of the feedback controller that is involved in stick balancing at the fingertip [591]. The following example illustrates how this procedure can be used to estimate the control strategy likely used by expert stick balancers.

Example 10.2 As we have seen, the time delay for pole balancing at the fingertip is ≈ 0.223s (Figure 10.2). The vertical dashed lines in Figure 10.10a show ℓ_{crit} for $\tau = 0.223$s for five different feedback control strategies. The feedback for PD, PDA, PF are given, respectively, by (10.3), (10.4), and (10.5). An act and wait (AAW) controller is a special case of a time-varying controller in which the feedback term is periodically switched ON and OFF [368, 375]. The intermittent controller (IP) makes use of sampled feedback of the state variable but uses a special system-matched hold rather than a zero-order hold [261, 262]. The feedback control for AAW and IP are given in [370]. The •'s in Figure 15.2 show the measured balanced times for five expert pole balancers. Balance trials were considered successful if the pole remained balanced for 4 min. for at least 1 out of 5 consecutive balancing trials. Thus, expert pole balancers appear to use an internal model that compensates for the time delay by predicting the sensory consequences of the pole's movements. This same type of control strategy is also thought to be involved in the control of other skilled movements by the nervous system [419, 754]. $\diamond$

Since the fingertip is continually moving during pole balancing, a more accurate model takes the form of a pendulum balanced on a moving cart (Figure 10.11c and d). In order to simplify the modeling, the subject is required to sit in a chair with their back placed against the back of the chair. The purpose of this seated position is to eliminate movements of the trunk and body which could influence the

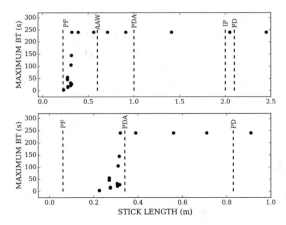

Fig. 10.10 Comparison of the maximum balance times obtained for five consecutive balance trials as a function of ℓ for expert pole balancers. (a) The dashed vertical lines show the calculated ℓ_{crit} for AAW, IP, PF, PD, and PDA feedback controllers determined in [370] for 1% sensory uncertainty. (b) The dashed vertical lines for PD, PDA, and PF are calculated numerically for a pendulum–cart model of stick balancing when there is a $1°$ sensory dead zone [591]. Under these conditions, ℓ_{crit} for AAW and IP are much larger than determined for PD feedback and hence are not shown.

movements of the pole. Expert pole balancers hold the position of their wrist and fingers rigid and hence the movements of the arm occur at the elbow and shoulder. The consequence is that the mass of the cart is equal to the mass of the arm ($\approx$ 1.4kg). The effects of increasing the mass of the cart are to decrease the length of the pole that can be balanced by half [591].

The reader should quickly realize that in this situation there are actually two control problems: 1) stabilize the upright position of the pole, and 2) keep the bottom of the pole within arm's reach. Thus, we have a system of four differential equations (see Exercise 10.11). The feedback controller can be identified by estimating ℓ_{crit}; however, due to the complexity of the model, this estimation must be obtained using numerical integration. Figure 15.2b shows that ℓ_{crit} appears to agree best with that estimated for PDA feedback. However, the human visual nervous system is not very sensitive for detecting changes in acceleration [180]. This uncertainty is not included in the model, but would certainly shift the estimates to the right [370]. Thus, it is most likely that the nervous system is using predictive feedback to balance a pole at the fingertip. This observation is consistent with the concept that the nervous system develops an internal model to predict the sensory consequence of a movement in order to compensate for the effects of the time delay [591].

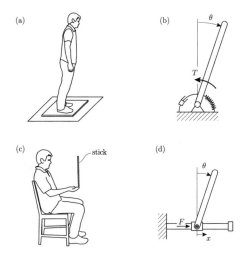

Fig. 10.11 (a) Postural sway as a pinned pendulum where $\mathscr{Q}$ is to torque; (b) pole balancing at the fingertip as a Pendulum–cart model where F is a force. Figure reproduced from [589] with permission.

10.7 Human balancing

Falls are a leading cause of accidental death and morbidity in the elderly. The first "baby boomer" turned 65 in 2011. Each day since then, 10,000 people become 65 years old. In the over 65's age group, nearly one-third of elders experience a fall each year and this frequency increases as elders continue to age. Thus, it is not hard to imagine that aging Western societies are facing an epidemic of falling. The medical costs alone are expected to be staggering. These concerns provide the impetus for current research activity into human balance control. Two experimental paradigms have been extensively studied: 1) postural sway during quiet standing (Figure 10.11a) and 2) pole balancing at the fingertip (Figure 10.11b).

Postural sway refers to the movements made by the body to maintain balance during quiet standing (Figure 10.12). It is typically monitored by having a subject stand quietly with eyes closed on a force platform. Balance is maintained by applying torque at the ankle joint. The force platform measures the center of pressure (COP) which is the weighted average of all of the downward forces acting on the force platform through the soles of the feet. The COP depends primarily on stance width and the motor control of the position of the ankle. Postural sway results from changes in the relative positions of the COP and the center of mass, COM, during quiet standing [885]. In 1D, as the COP moves to the right of the COM, the COM moves left, and vice versa. For an adult human standing quietly, the COM lies approximately at the level of the second sacral vertebrae ($\approx 55\%$ of their height [667]). The biomechanical condition for stable balancing is that the COM must be located within the base of support defined by the area under and between the feet.

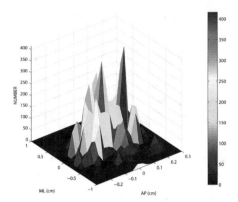

Fig. 10.12 Two-dimensional histogram constructed from the anterior–posterior (AP) and medial–lateral (ML) components of the COP while a healthy subject stands quietly on a force platform for 2 minutes. The sampling frequency was 200 Hz. See also Figure 16.7.

In a quiet room with eyes closed, the COP oscillations are of the order of 0.5 deg for healthy individuals with no history of falling. The bandwidth of the fluctuations is 0-3 Hz with a mean frequency of 0.9-1.3Hz[502, 886, 896]. The complex nature of the fluctuations in COP has attracted a great deal of interest. There have been three lines of investigation. One possibility is that the fluctuations represent the dynamics of a stochastic time-delayed dynamical system [142, 580, 636]. A second possibility is that the dynamics represent a sampled feedback control system [263, 592]. This possibility is supported by experimental observations demonstrating that corrective movements for balance control are exerted by the nervous system intermittently [502]. This observation has also been made for stick balancing on the fingertip [94] and other human tracking tasks [628]. Possible mechanisms for generating a sampled data system include 1) central control mechanisms related to the central refractory time [263, 502], 2) the presence of a sensory dead zone to produce a switching-type feedback [580, 592], and 3) a time-delayed dynamical system tuned at the edge of stability [94, 591].

A third possibility is based on the observation that fluctuations in COP in either the anterior–posterior direction or the medial–lateral direction can, at times, appear to contain two components [914]: a "slow" nonoscillatory component (referred to as "rambling") and a "fast" oscillatory component (referred to as "trembling"). The "slow" component corresponds to an exponential decay back to equilibrium that is perturbed stochastically. The "fast" component is related to the time-delayed feedback control involved in the stabilization of an inverted pendulum. It has been suggested that the "slow" component is present inside the feedback loop of the "fast" component [403].

These three possibilities are not mutually exclusive. The common feature of each of them is an inverted pendulum stabilized by state-dependent and time-delayed feedbacks [371, 492, 548, 599, 646] (Fig. 10.8). The governing equation takes the form

$$\ddot{\theta}(t) - \omega_n^2 \theta(t) = \mathcal{T}(t - \tau), \tag{10.41}$$

where $T(t - \tau)$ is the delayed torque. However, all time-invariant nonlinear feedback controllers that can be written in the form $f(t) = h(\theta(t - \tau), \dot{\theta}(t - \tau))$ can be reduced to PD feedback after linearization if the function h is smooth in both of its arguments. Consequently, most present-day models for human postural sway during quiet standing take the form of (10.36) [403, 482, 548, 592, 646, 671, 801].

10.7.1 The vibration paradox

An important insight into the control of human balance can be obtained from investigations into the stabilizing effects of vibration. It is known that the inverted pendulum can be stabilized by moving the pivot point vertically either periodically [4, 5, 474, 805] or noisily [66]. These stabilizing effects are typically interpreted in the context of the Mathieu equation

$$\ddot{\theta}(t) - (\omega_n^2 + \beta \cos(2\pi f_V t))\theta(t) = 0. \tag{10.42}$$

The complete mathematical analysis of (10.42) requires the use of mathematical techniques, such as Floquet theory of linear periodic dynamical systems, which are beyond our scope. However, two conclusions from these analyses are relevant for our purposes. First, the condition for stability of the upright position is that the vibration frequency, f_V, exceeds

$$f_V > \frac{\sqrt{2g\ell}}{2\pi a} \tag{10.43}$$

where a is the peak-to-peak amplitude. Second, the term $\cos(2\pi f_V t)$ means that the downward acceleration of the pole can periodically exceed that of gravity. This is possible only if the pole is firmly attached to the vibrating pivot point. It is particularly important to remember this point should the reader decide to verify this phenomenon by attaching the bottom of the pole to a jigsaw (as suggested in [474])!

What would be expected to happen if we vertically vibrated the fingertip during pole balancing? On the basis of the above discussion, we would expect no improvement. Surprisingly, vertical vibration at the fingertip using a whole-body vibration benefits stick balancing [595]. In this experiment, the vibrating platform (Physioplate) generated a vertical vibration at the sole of the feet with an amplitude of 0.9–2.3mm as the vibration frequency was varied, respectively, from 15 to 30Hz. This resulted in a 0.1–0.3mm amplitude vibration at the fingertip. The observation that the amplitude of the vibration of the fingertip is smaller than that of the soles of the feet is a manifestation of the filtering properties of the human body for vibration. Consequently, in this experiment subjects self-selected the degree of flexion of their knees for comfort. The effect of the vertical vibration was to increase the mean balance time 2.1–2.6-fold. No improvement in balancing time was observed when subjects stood on a vibrating platform which did not produce a vertical displacement of the feet. These observations are surprising for two reasons. First, the downward acceleration cannot exceed the gravitational acceleration, g, since the pole is not physically attached to the fingertip. Second, the vibration frequency that benefits pole balancing (50 Hz) is nearly 100 times smaller than that predicted by (10.43) (when $\ell = 0.55$m and $a = 0.001$m, $f_V = 5525$Hz).

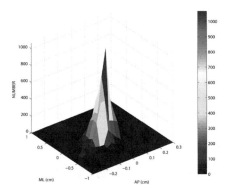

Fig. 10.13 Effect of bilateral Achilles tendon vibration on postural sway for a subject standing quietly with eyes closed on a force platform. Postural sway in the absence of vibration for the same subject is shown in Figure 10.12. Both Achilles tendons were vibrated synchronously using Bruel–Kjaer Type 4810 minishakers mounted on a movable trolley that was powered by a Bruel–Kjaer Type 2706 power supply driven by a wave generator (HP model 33120A) at 40 Hz. The peak-to-peak amplitude of the vibration was 0.18 mm.

Figure 10.13 shows that the stabilizing effects of vibration are also observed in human postural sway with bilateral Achilles tendon vibration while a subject stands with eyes closed [573]. It is clear that vibration limits the range of the fluctuations in COP in both the AP and ML directions. This observation suggests that balance has been stabilized. Similar effects have been observed using vibrating insoles [697, 698]. Indeed, it has been suggested that a shoe with a vibrating insole may be useful for decreasing the risk of falling in the elderly [489]. The vibration amplitudes are very small (≤ 0.2 mm) compared to the mass of the body (on average 70 kg). Thus, it is unlikely that the beneficial effects of vibration are due to vertical displacements of the body. It is more likely that these effects are a manifestation of stochastic resonance [609, 610]. This effect of noisy stimuli acts through the enhancement of the response of neural sensory receptors.

There have been two approaches to attempt to resolve this paradox. The first approach was to evaluate the stability of a Mathieu-type equation with time-delayed PD type feedback [372],

$$\ddot{\theta}(t) + (\omega_n + \beta \cos 2\pi f_V t)\theta(t) = -k_p \theta(t - \tau) - k_q \dot{\theta}(t - \tau). \qquad (10.44)$$

It was found that (10.44) predicts that for a given τ, ℓ_{crit} can be shorter than predicted by (10.42). This effect can occur even if the maximum acceleration of the stick's base does not exceed g. However, these beneficial effects of vibration required a considerably larger vibration amplitude than used for human stick balancing and postural sway.

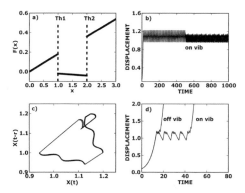

Fig. 10.14 (a) Graphical representation of $F(x(t-\tau))$ used in eqrefeq:quail. (b) When $Th_2 \gg Th_1$, periodic forcing decreases both the peak-to-peak and mean amplitude of the oscillations. (c) A complex attractor exists under these conditions; however, its exact nature has not yet been characterized. (d) When $Th_2 > Th_1$, the system can only be transiently confined and eventually escapes to infinity. Turning on the periodic forcing approximately doubles the survival time. Figure reproduced from [579] with permission.

The second approach has been to interpret the effects of vibration in the context of a "drift and act" hypothesis [579, 580, 595, 596]. This mechanism proposes that the basin of attraction for the stabilized upright position is small enough so that escape ("falls") are possible. Inside the basin of attraction trajectories "drift" freely. Corrective actions are taken only when the trajectories leave the basin of attraction. Consequently, any strategy that decreases the amplitude of the fluctuations in θ will have a stabilizing effect since they decrease the probability that the trajectory escapes the basin of attraction.

The main features of the observations can be qualitatively captured by a simple model which incorporates an unstable equilibrium point, a time-delayed switch-type controller, and parametric periodic excitation. This model takes the form

$$\frac{dx}{dt} = F(x(t-\tau))x(t) + kx(t)\sin 2\pi ft. \tag{10.45}$$

$F(x(t-\tau))$ is given in Figure 10.14a. Equation (13.9) describes a "drift and act" controller: corrective actions ("act") are taken only when $Th_2 > x(t-\tau) > Th_2$. When $Th_2 \gg Th_1$, there is a range of parameters for which a complex periodic attractor exists (Figure 10.14b and c). As shown in Figure 10.14, (13.9) accounts for three properties produced in human postural sway with ~ 0.1mm amplitude, bilateral Achilles tendon vibration while a subject stands with eyes closed [573]: 1) a decrease in the extent of the COP fluctuations in the horizontal plane (maximal at

~ 40Hz); 2) a small shift in the centroid of the COP fluctuations without a change in postural alignment; and 3) an increase in the path length of the COP trajectories, maximal at $\sim 40 - 70$ Hz.

10.8 Intermittent control

The effects of vibration on human balance control draw attention to the importance of a sensory dead zone. Sensory dead zones exist both for postural sway and pole balancing at the fingertip. The fluctuations in the ankle joint angles during postural sway are less than tenths of a degree and hence are much smaller that the thresholds for the both the vestibular and visual systems. Thus it is currently thought that the principle sensory input for postural sway is proprioception [228, 503]. This sensory dead zone is of the order of $0.05 - 0.08°$ [228, 589]. At one time, it was thought that the stiffness in the Achilles tendon by itself was sufficient to maintain balance during quiet standing [886]. However, recent measurements have shown that although Achilles tendon stiffness contributes to balance control, active feedback control by the nervous system is required [503].

Dead zones are also present for pole balancing at the fingertip. The fluctuations in θ during pole balancing are more than an order of magnitude larger than the variations in joint angle observed during postural sway. The observation that an initially balanced pole falls quickly after eye closure suggests that vision provides the more important sensory input. A sensory dead zone arises because it is very difficult for the nervous system to estimate the vertical displacement angle of the pole in the anterior–posterior direction. Indeed, an expert pole balancer capable of balancing a 0.3m pole at their fingertip for over 4 minutes can balance the same pole for less than a few seconds when one eye is patched. Hikers are very familiar with this problem: from a distance, the next mountain to climb looks vertical; however, as the hiker nears the mountain, it becomes easier to appreciate that the mountain has a slope. This sensory dead zone is of the order of 1-3° during pole balancing [591].

In the context of control, a dead zone represents a strong, small-scale nonlinearity. While this nonlinearity has no effect on large-scale stabilization in the linear system, it may lead to complex dynamics on the small scale including limit-cycle oscillations and even microchaos (see Section 13.4). Although the presence of a sensory dead zone might be considered to be a nuisance, it has advantages in situations in which noisy perturbations are also present. In this situation, there exists the fundamental problem of distinguishing between those fluctuations that need to be acted upon by the controller and those that do not [580, 918]. This is because, by definition, there is a finite probability that an initial deviation away from a set point will be counterbalanced by one toward the set point just by chance. Too quick a response by a controller to a given deviation can lead to the phenomenon of "over control" leading to destabilization, particularly when time delays are appreciable. On the other hand, waiting too long runs the risk that the control may be applied too late to be effective. Thus, methods based on continuous feedback control (e.g.,

[703, 897]) are not only anticipated to be very difficult to implement by the nervous system, but are also unlikely to be effective.

The presence of a sensory dead zone means that the feedback is either ON or OFF depending on whether the controlled variable is greater than or less than the sensory threshold Φ. Thus, (10.36) is replaced by

$$\ddot{\theta}(t) - \omega_n^2 \theta(t) = \begin{cases} 0 & \text{if } \theta(t-\tau) < \Pi, \\ f(\theta(t-\tau), \dot{\theta}(t-\tau), \cdots) & \text{otherwise}. \end{cases} \quad (10.46)$$

Switch-like controllers are well known to engineers and have the property that they are optimal when the control is bounded [231]. In addition, they act as a "noise gate" to reduce the effects of noise. A familiar example of discontinuous control arises in the thermostatic control of room temperature.

The presence of a sensory dead zone emphasizes that time-delayed feedback control is also constrained by sensorimotor uncertainties in the control process. These uncertainties arise because of, for example, limitations in the ability of sensory receptors to estimate state variables and errors in motor control realization. Although these uncertainties cannot be controlled, they nonetheless place limitations that help identify those control strategies that would be expected to work best in an unpredictable environment. The sensory uncertainties are particularly important for predictive feedback since they limit the ability of such models to predict the sensory consequences of movements [370, 405].

Recent attention in balance research has focused on the possibility that balance control is exerted intermittently much like a rocket is maneuvered to a space station dock by applying intermittent jets of force. Brief, intermittent corrective movements are observed experimentally during stick balancing at the fingertip [94, 95] and during quiet standing in the contractions of the soleus and gastrocnemius muscles [502]. It has been suggested that the ballistic corrective movements observed during postural balance [94, 95, 502] reflect "chattering." the dynamical signature of a switch-like controller [69, 70]. These observations have spawned research into two general types of intermittent feedback controllers: 1) event-driven intermittent feedback and 2) clock-driven intermittent feedback.

10.8.1 Event-driven intermittent control

For event-driven intermittent control, the switching feedback can be the consequence of a peripherally located sensory dead zone or can be related to motor planning by the brain that takes advantage of the properties of a saddle node.

10.8.1.1 Switching feedback: sensory dead zones

The principal sensory input during quiet standing with eyes closed is from proprioceptive sensors in the ankle joint [228, 876], namely the Golgi tendon organs and muscle spindles [503]. In view of the above considerations, a possible model for postural stability during quiet standing that emphasizes the role of sensory dead zones takes the form

$$J_A \ddot{\theta}(t) - \omega_n^2 \theta(t) = \mathscr{Q}_p(t) + \mathscr{Q}_d(t), \tag{10.47}$$

where $\mathscr{Q}_p, \mathscr{Q}_d$ are the components of the ankle torque related to the angular position and the angular velocity and

$$\mathscr{Q}_p = \begin{cases} 0 & \text{if } |\theta(t-\tau)| < \Phi_{pos}, \\ -K_p \theta(t-\tau) & \text{otherwise}, \end{cases}$$

$$\mathscr{Q}_d = \begin{cases} 0 & \text{if } |\dot{\theta}(t-\tau)| < \Phi_{vel}, \\ -K_d \dot{\theta}(t-\tau) & \text{otherwise}, \end{cases}$$

where K_p, K_d are, respectively, the proportional and derivative control gains and Φ_{pos}, Φ_{vel} are the sensory dead zones for the body's angular position and angular velocity. The moment of inertia of the body, J_A, is taken with respect to the normal line through the pivot point shown in Figure 10.11a. The passive stiffness, K_{pass}, is chosen to be insufficient to completely resist falling against gravity. It should be noted that at one time it was thought that postural stability during quiet standing was maintained completely by stiffness in the Achilles tendon [886].

The model described by (10.47) reproduces oscillations in the postural sway angle of 0.5 deg as shown in Figure 16.7. The dynamics of this model are "microchaotic" when the frequency-dependent encoding of force is taken into account (see Example 13.1).

10.8.1.2 Switching feedback: saddle node

A novel suggestion concerning the role played by switch-like controllers in the stabilization of an inverted pendulum has been advanced which is based on the properties of a saddle point [13, 70]. The fixed point for an inverted pendulum in the absence of feedback is a saddle point. Figure 10.15a shows a phase representation of a saddle point. Each trajectory in the phase plane represents the solution for a different choice of the initial condition $(\theta(t_0), \dot{\theta}(t_0))$. If we look carefully at each trajectory, we can see that even though the fixed point is unstable, there are intervals during which the trajectories approach the fixed point and other intervals where they diverge from the fixed point. There are only two exceptions to this rule: both correspond to a very special choice of the initial conditions and hence for all practical purposes can be ignored.

Figure 10.15b shows a switch-like control strategy that takes advantage of the phase plane properties of a saddle point. By rotating the axis of the phase plane

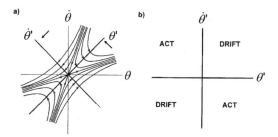

Fig. 10.15 (a) Phase plane representation of a saddle point. The axis is rotated. (b) The quadrants labeled "ACT" are when the control is switched ON, otherwise the control is OFF ("DRIFT"). Figure reproduced from [579] with permission.

from $\dot{\theta}$ versus θ to $\dot{\theta}'$ versus θ' (Figure 10.15a), we can see that the phase plane can be divided into regions where the trajectories approach the saddle point and other regions where the trajectories diverge away. Thus, the investigators turned the control OFF ("drift") when the trajectories were in regions that naturally moved toward the fixed point and activated control ("act") in the regions where the trajectories would naturally move away from the fixed point (Figure 10.15b). Since this switch-like controller depends on both $\theta(t)$ and $\dot{\theta}(t)$, it is a PD-type controller, albeit a more complicated one than the PD controllers typically used in engineering applications.

There were three surprising observations. First, it was possible to determine an act mechanism that resulted in robust control. Second, this strategy worked in the presence of a delay and hence for a saddle point that exists in an infinite-dimensional space. Finally, in the presence of additive noise, this control strategy results in the presence of power-law scaling regions similar to those observed for human postural sway [13].

10.8.2 Switching feedback: clock driven

The development of virtual balancing tasks that involve the interplay between a person and a computer makes it easy to manipulate parameters related to balance control, such as time delay and feedback gain, by making changes in a computer program [93, 405, 504, 553, 572, 673]. Ian Loram[5] and colleagues used this approach to compare the effectiveness of continuous versus intermittent feedback control for maintaining balance [504]. This virtual balancing task involves using a joystick to control the position of a target on the computer screen that is programmed to move as an unstable load. Subjects moved the joystick either while maintaining continuous manual contact with it ("continuous control") or by gently tapping it with a ruler ("intermittent control"). Surprisingly, joystick movements produced by tap-

---

[5] Ian Loram (PhD 2003), English professor of biomechanics.

ping at an optimal frequency of ≈ 2 Hz provided better control than continuous joystick control.

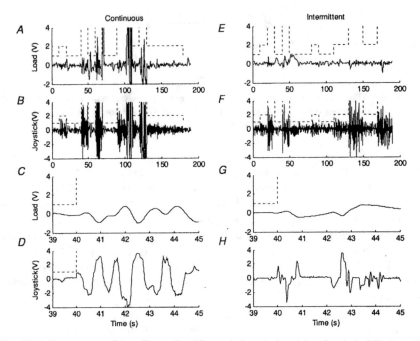

Fig. 10.16 Comparison of the effects of sudden variations in joystick gain (dashed line) on the control of joystick position for a continuous and intermittent control strategy (solid lines). The subject was asked to maintain the joystick position at the center of the screen ($V = 0$), where v is the voltage. The top four panels (A, B, E, F) show the effects of multiple sudden changes in joystick gain, and the bottom four panels (C, D, G, H) show the effect of a single change in joystick gain. Figure reproduced from [504] with permission.

Figure 10.16 compares the effectiveness of continuous and intermittent control strategies in the face of sudden unexpected changes in joystick gain. The joystick gain can be altered by changing the voltage per unit distance of the cursor movement seen on the computer screen. The top four panels (A, B, E, F) show the effects of multiple unexpected changes in joystick gain, and the bottom four panels (C, D, G, H) show the response to a single sudden change in joystick gain. In response to a single change in joystick gain, uncorrected oscillations occur for continuous control. In contrast, for intermittent control, the fluctuations in load and joystick position are smaller overall, and no clear oscillatory component is seen. The benefits of intermittent control are even more dramatic when multiple unexpected changes in joystick gain are introduced (panels A, B, E, F). If we associate fixed-point instability with the appearance of an oscillation, then the intermittent control strategy is more stable. An unexpected observation was that the feedback time delay was smaller for the intermittent control strategy.

10.9 Numerical Integration

Numerical algorithms to integrate DDEs with constant delays are now available in a variety of numerical packages including XPPAUT, Maple, MATLAB, and Mathematica. These computer programs have opened the door to allow investigators to appreciate the richness of the dynamics exhibited by DDEs. Nonetheless, it is important to keep in mind that numerical algorithms can only approximate the stable solutions of differential equations. Of course, the presence of unstable solutions can often be deduced from transient behaviors. However, obtaining details of the structure of the unstable solutions requires the use of other methods. A well-studied example is the presence of canards in the differential equations that describe excitable cells [292].

Two general approaches have been used to numerically integrate DDEs [47]. The first approach relies on using the same methods used to integrate ODEs. An example is the use of the Runge–Kutta method employed in XPPAUT. The second approach uses numerical techniques specifically developed to integrate DDEs such as the method of steps (MATLAB's dde23 [755], Maple's desolve and Mathematica's NDSolve), and the semi-discretization method [373]. In our discussion, we illustrate each method with a program that integrates (10.36) with $k_1 = 0$ and F given by (10.3). These sample programs serve as useful templates for other applications.

10.9.1 XPPAUT

Numerical integrators for differential equations generally assume that at least the first few derivatives are continuous. These numerical methods often behave badly at points where jump discontinuities occur. Nonetheless, numerical methods developed for ODEs, such as the Euler and Runge–Kutta algorithms, often perform very well for the integration of DDEs. However, it is always prudent to repeat the numerical integration using smaller time steps in order to establish that the effects of the jump discontinuities present in DDEs are not producing significant numerical artifacts.

XPPAUT [202] is freely downloadable from the Internet and can be readily installed on either PCs or Macs (and Linux systems as well). This free, stand-alone package can perform numerical integration using a variety of fixed-step numerical methods, including integrating discrete time systems, ordinary and delay differential equations, stochastic differential equations, and partial differential equations. This program can determine the fixed points and evaluate their stability. In addition, XPPAUT is particularly well suited for the investigation of the dynamics of excitable systems composed of neurons and cardiac cells. It is not uncommon that XPPAUT programs can be located on the Internet or in the Supplemental materials for published papers located on journal websites.

XPPAUT is run through a graphical user interface which is used not just to visualize results, but also to modify parameters, initial conditions, and even the numerical integration method. The main benefits of this program are its flexibility and the ease with which different simulations can be compared. Information on how to download the package as well as documentation and tutorials are available at www.math.pitt.edu/ bard/xpp/xpp.html.

Computer programs are stored with the file extension *.ode. These files can be made using a text editor, for example, Notepad or Wordpad on a PC or TextEdit on a Mac.

All XPPAUT files have the form

```
# pendulum.ode
# this program integrates the equations for
# an inverted pendulum stabilized by
# time-delayed negative feedback
dx/dt=y
dy/dt=a*x-k1*delay(x,tau)-k2*delay(y,tau)
init x=0.1,y=0.1
par a=0.5,k1=0.6,k2=1.0,tau=1
@ TOTAL=100,dt=0.01,delay=10,xlo=0,xhi=100,ylo=-2,yhi=2,maxstor=100000
done
```

Useful hints:

Since the XPPAUT package is nearly 30 years old, a few comments are useful to help readers use it to integrate DDEs. It should be noted that the default integrator for DDEs in XPPAUT is the Runge–Kutta method.

1. The term $x(t - \tau)$ in a DDE corresponds to delay(x,tau) in XPPAUT.

2. The parameter tau refers to the value of the time delay used in the simulation. The value of delay refers to the largest value of the delay that you might use. Thus, the value of delay must be greater than or equal to the tau value. It should be noted that the value of delay reserves in memory the amount of space that will be required to store the initial function. This observation should be kept in mind in choosing the largest delay that you will use.

3. XPPAUT constructs the initial function $\Phi(s)$ in two parts: the value of X(s) at $t(0)$ (IC menu) and a second part that is a function defined on the interval $[-\tau, 0)$ (Delay menu).

4. The functional part of the initial function can be introduced in two ways. First, we can use the Delay tab in the XPPAUT window. This is most useful when $\Phi(s)$ is either a constant or is described by a reserved function such as exp(). Second, $\Phi(s)$ can be constructed in the form of tables which are read into the program.

5. TOTAL refers to the total time for the integration and dt is the step size for the numerical integration. Thus if TOTAL=20 and dt=0.05, there will be $20 \times 20 = 400$ time steps. The parameters xlo,xhi,ylo,yhi refer, respectively, to the dimensions of the x-axis and y-axis of the figure displayed by XPPAUT and not to the variables that appear in the differential equation. The parameter maxstor sets the total number of time steps that will be kept in memory (default value is 5000).

6. Likely the most common mistake made using XPPAUT is to add spaces to the entered equation to "make it look better." Unfortunately, the computer is not impressed! Thus do not add spaces when you type in the equation. A good rule of thumb is that when in doubt, no spaces!

7. The values of the initial conditions and parameters can readily be changed using the on-screen menus provided by XPPAUT. In other words, it really does not matter which values you type into the *.ode program since you can easily change them later.

8. It is sometimes useful to define an auxiliary variable by adding the line

   ```
   aux v=delay(x,tau)
   ```

   to the program. This line defines the auxiliary variable, v, as the delayed value of the variable, x. This auxiliary variable can be used to construct a phase plane representation of the dynamics by using Viewaxes. For example, we can replace t by v in the Viewaxes window.

10.9.2 Method of steps (dde23, desolve, NDSolve)

When dealing with time-delayed dynamical systems, it is often useful to draw a timeline such as shown in Figure 10.17. The symbols $\hat{X}_i, \phi_i$ refer to, respectively, the solutions and initial conditions defined on an interval of length τ. The subscript for $\hat{X}$ refers to the step number for $t > 0$ in units of τ and the subscript for ϕ refers to the time at the beginning of the interval. Thus in order to obtain the first step solution on $[0, \tau]$, $\hat{X}_1$, it is necessary to know the initial condition on the interval $[-\tau, 0]$, $\phi_{-\tau}$. Similarly, to obtain the solution for the second step, $\hat{X}_2$, defined on the interval $[\tau, 2\tau]$, we need to know, ϕ_0, namely the initial condition defined on the interval $[0, \tau]$. It should be noted that ϕ_0 equals $\hat{X}_1$, and so on. Thus, we see that the solution of a DDE with constant τ's can be constructed as a sequence of steps of length τ. This method is widely used by mathematicians to establish the existence and uniqueness of solutions of a DDE. It is also employed in two software packages for integrating DDEs with constant delays, dde23 (Matlab), desolve (Maple), and NDSolve (Mathematica).

To illustrate this method, consider (10.9) with $k = 1$ and $\tau = 1$

$$\dot{x}(t) = x(t-1). \tag{10.48}$$

In general, we can calculate $\hat{X}_{t_i}$ if we know ϕ_{t_i-1} by solving

$$\int_{\phi_{t_i-1}}^{\hat{X}_{t_i}} dx = -\int_{t_i}^{t} \phi_{i-1}(s-1)ds.$$

Hence

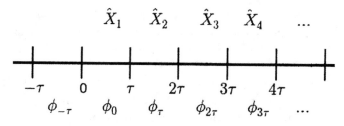

Fig. 10.17 Illustration of method of steps.

$$\hat{X}_{t_i} = \phi_{t_i} = \phi_{i-1} - \int_{t_i}^{t} \phi_{i-1}(s-1)ds.$$

We can apply these equations to solve (10.48) step by step. On the interval $[0,1]$, the solution is

$$\hat{X}_1 = 1 - \int_0^t 1 ds = 1 - t$$

and on the interval $[1,2]$

$$\hat{X}_2 = 0 - \int_1^t [1-(s-1)]ds = -2t + \frac{t^2}{2} + \frac{3}{2},$$

and on the interval $[2,3]$

$$\hat{X}_3 = -\frac{1}{2} - \int_2^t [-2(s-1) + \frac{1}{2}(s-1)^2 + \frac{3}{2}]ds$$
$$= \frac{5}{3} + (t-1)^2 - \frac{1}{6}(t-1)^3 - \frac{3}{2}t,$$

and so on.

10.9.3 dde23 (matlab)

MATLAB's dde23() makes it possible to solve DDEs with constant delays [755]. It is assumed that the system of DDE takes the form

$$\dot{y}(t) = f(t, y(t), y(t-\tau_1), y(t-\tau_2), \cdots, y(t-\tau_k))$$

where the τ_k are constant delays and are solved on $a \leq t \leq b$ with a given history $y(t) = S(t)$ for $t \leq a$. This function is included in version 6.5 and later of MATLAB.
The use of dde23 is similar to that of ode23. The function call takes the form

```
sol = dde23(ddefile, lags, history, tspan)
```

where `ddefile` is an m-file that gives the equation to be integrated, `lags` is a vector given the values of the time delays, `history` gives the initial function, and `tspan` is a vector that gives the time interval for the integration. Keep in mind that the use of `dde23()` requires the creation of two m-files: one m-file describe the DDE and the second m-file performs the integration. Finally, it should be noted that `dde23` does not assume that terms of the form $x(t - \tau_j)$ appear in the equations.

We illustrate the use of `dde23` by integrating (10.36). There are two programs (the program names are arbitrary). The first program DRHS.m defines (10.36), and the second program `first_pend`.m calls the first program and performs the numerical integration.

DRHS.m

```
function yp = DRHS(t,y,Z)
global k
ylag1 = Z(:,1);
ylag2 = Z(:,2);
yp=[
y(2);
k(1)*y(1)-k(2)*ylag1(1)-k(3)*ylag2(2);
];
end
```

delay_pend.m

```
function delay_pend()
global k
global tau
% parameters
k=[0.5,0.6,1.0]';
% delays
ylag1=1.0;
ylag2=1.0;
% initial conditions
y10=0.1;
y20=0.1;
yi=[y10,y20]';
% start/end values of t
t0=0;
t1=100;
interval=[t0,t1];
sol=dde23('DRHS',[1,1],yi,interval)
ysol=sol.y';
yx=ysol(:,1);
yv=ysol(:,2);
t=sol.x';
plot(t,yv,'k-')
end
```

10.9.4 Semi-discretization

The semi-discretization method for DDEs shares the same philosophy as the finite element method does for PDEs. In PDEs, the finite element method discretizes the spatial coordinates while the time coordinates remain continuous. This approach has been particularly useful to integrate the PDEs that arise in the analysis of solid bodies and for problems in computational fluid mechanics. In the case of DDEs, the delayed terms and the time-periodic coefficients are discretized (such as in discrete time sampling), while all of the other terms in the differential equations are continuous. The implementation of this semi-discretization strategy is complicated by the need to keep track of past effects encompassed by the interval $[t - \tau, t]$ on the present time dynamics. For a complete discussion of the semi-discretization approach for DDEs, see [367, 373, 374].

Here, we briefly outline the semi-discretization approach for DDEs and illustrate it with an application to (10.1). Semi-discretization methods arise naturally in the control of movement and balance by the nervous system. In this case, the continuous time coordinate is associated with the dynamics of the muscle-skeletal system (described using Newtonian mechanics) and the discrete-time coordinate is associated with the neuro-muscular feedback controller. It is quite likely that the sampled spike time codes are neither continuous nor discrete but lie somewhere in between [374]. A practical advantage of the semi-discretization approach is that the code can be programmed into whatever computer language that the user knows best. Thus, it is not necessary for the user to learn how to use additional software packages.

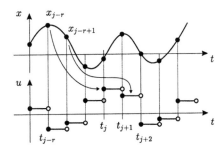

Fig. 10.18 Discrete-time sampling effect.

In order to appreciate that the semi-discretized version of (10.1) lies in between its discrete time and continuous versions, we first rewrite the time sampled versions of (10.1) when $\tau = 0$ as

$$\dot{x}(t) + k(t) = F(x(t_i)) \tag{10.49}$$

where $t_i = i\Delta t$ and Δt is the sample period of the feedback control. The semi-discretized version of (10.49) is

$$\dot{x}(t) + k(t) = F(x(t_{i-r})) \tag{10.50}$$

where r is a positive integer referred to as the discrete delay. The control force is determined using discrete delayed values of the angular position, and angular velocity and is kept piecewise constant over each sampling period $[t_i, t_{i+1})$. Consequently, we have both a feedback delay of $r\Delta t$ and a zero-order hold. Thus on each sampling interval we have a finite-dimensional representation which can be solved as an ODE. This concept is illustrated in Figure 10.18.

We can rewrite (10.50) as

$$\dot{x}(t) + k(t) = F(x(t - \rho(t))) \tag{10.51}$$

where (see Figure 10.19)

$$\rho(t) = r\Delta t + t + \text{Int}\left(\frac{t}{\Delta t}\right)$$

and the average delay $\bar{\tau}$ is

$$\bar{\tau} = \frac{1}{\Delta t}\int_0^{\Delta t} \rho(t)dt = \left(r + \frac{1}{2}\right)\Delta t.$$

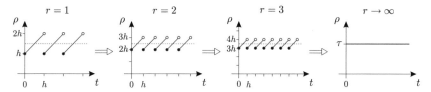

Fig. 10.19 Discrete-time sampling effect as time-periodic delay. Figure reproduced from [373] with permission.

Approximation of the term $x(t - \tau)$ by term $x(t_{i-r})$ with piecewise constant argument over the interval $t \in [t_i, t_{i+1})$ actually corresponds to a perturbation in the delay, since the term $x(t_{i-r})$ can also be written as $x(t - \rho(t))$, where

$$\rho(t) = t - t_i + rh, \qquad t \in [t_i, t_{i+1}), \tag{10.52}$$

or

$$\rho(t) = t - h\,\text{Int}\left(\frac{t}{h}\right) \tag{10.53}$$

in continuous time representation, where the function Int rounds toward zero. Thus ρ is the sawtooth-like time-periodic time delay shown in Figure 10.19. The relation between (10.50) and (10.51) can be established by increasing the discrete delay r while at the same time decreasing the sampling period Δt such that the average delay remains constant. Obviously, if $r = 0$ then (10.50) is equal to (10.51). In the limit case where $r \to \infty$ and $\Delta t \to 0$ such that $(r + \frac{1}{2})\Delta t = \tau$, it can be shown that the solution of (10.50) approaches that of (10.1) [367, 373].

The MATLAB code which uses the semi-discretization method to generate the dynamics of an inverted pendulum is the following.

```
% create time history for
% x"(t) - a x(t) = -p x(t-tau) -d x'(t-tau)
% by (zeroth-order) semidiscretizaton

clear

tau = 0.5;                 % delay
a = 1;                     % system parameter
p = 2;                     % proportional gain
d = 2;                     % derivative gain
tmax = 20;                 % duration of simulation
r = 20;                    % delay resolution
h = tau/r;                 % discretization step
x0 = ones(r+2,1);          % initial condition
A = [[0, 1];[a, 0]];       % system matrix
B = [[0];[1]];             % input matrix
K = [-p, -d];              % matrix for control gains

Phi = zeros(r+2,r+2);
dv = ones(r+1,1);
dv(1:2) = 0;
Phi = Phi + diag(dv,-1);
Phi(3,1) = 1;
P = expm(A*h);
R = (expm(A*h)-eye(2))*inv(A)*B;
Phi(1:2,1:2) = P;
Phi(1:2,r+2) = R;
Phi(3,1:2) = K;

tv = zeros(tmax/h,1);
xv = zeros(tmax/h,1);
for i = 1:tmax/h
    x1 = Phi*x0;
    tv(i) = i*h;
    xv(i) = x1(1);
    x0 = x1;
end

figure
hold on
plot(tv,xv)
box on
xlabel('t')
ylabel('x')
```

10.10 What have we learned?

1. Why do time delays frequently arise in biology?

   Time delays arise because typically the sensors that detect the error are not located in the same place as the effectors that act to minimize the error. Time delays also arise because of processing times related to the time it takes to act upon a stimulus.

2. What is a state-dependent delay?

   The magnitude of the delay depends on the state variable(s).

3. What is a distributed delay?

   In many cases there is not a single delay present but many different delays. In this case we can represent the many different delays by a continuous distribution of delays.

4. What is the difference between a lag and a time delay?

   In the case of a lag the time course for the stimulus to a system overlaps the system's response. In contrast for a delay the time courses for the stimulus and the response do not overlap.

5. What is the order of a delay differential equation (DDE), and what is its dimension?

   The order of the delay is equal to the highest derivative. The dimension of the DDE is always infinite because the initial condition is given by a function of length τ.

6. A DDE is sometimes referred to as a functional differential equation. Why?

   This description of a DDE emphasizes that the initial condition is a function of length τ.

7. What is a retarded functional differential equation (RFDE)?

   A DDE in which the highest derivative in the feedback is less than the highest derivative present in the plant (i.e., the process that is to be controlled).

8. What is a neutral functional differential equation (NFDE)?

   A DDE in which the highest derivative in the feedback is equal to the highest derivative present in the plant.

9. What is state-dependent feedback and what factors limit its stability?

   State-dependent control uses measurements of the state variable (position, velocity and acceleration) made at time $t - \tau$ to predict corrective movements made at time t. This type of control is limited by the magnitude of the time delay.

10. What is predictive feedback and what factors limit its stability?

   Predictive feedback uses all information that is available up to $t - \tau$ in order to predict actions taken at time t. In neural control this type of feedback is associated with the development of an internal which predicts the sensory consequences of the movement. This results in a compensation of the time delay. This type of control is limited by sensory uncertainties [370, 405].

11. How does a jump discontinuity arise in a DDE?

   A jump discontinuity necessarily arises in a DDE since the left and right derivatives
   evaluated at $t = t_0$ are necessarily not the same.

12. How do the effects of a jump discontinuity differ for a RFDE and a NFDE?

   The effects of the initial jump discontinuity for a RFDE are progressively smoothed
   as a function of time as the initial derivative discontinuity is propagated successively
   to higher derivatives. In the case of NFDE the effects of the jump discontinuities are
   always present though they may diminish as a function of time.

13. What is the D-subdivision method for determining the stability of a DDE?

   The D-subdivision method is a method to determine the local stability of a fixed point of
   a DDE. It consists of four steps: 1) make the usual ansatz that $x(t) \approx \exp(\lambda t)$, 2) obtain
   the characteristic equation $D(\lambda)$, 3) take $\lambda = \gamma + j\omega$ and substitute into the equation
   $D(\lambda) = 0$, and 4) decompose this equation into real and imaginary parts and complete
   the analysis.

14. How can the natural frequency, ω_n, of an inverted pendulum be determined
    experimentally?

   First, the time period, T_p, is measured for the small oscillations for the mechanical structure hanging at its stable Hung-down position and then ω_n is calculated using (10.39).

15. What is the difference in stability of the fixed point for a hung-down pendulum
    and an inverted pendulum?

   The fixed point for a hung-down pendulum is a center (if a term of the form $k\dot{\theta}(t)$ is
   present then it is a stable spiral point). In contrast, for an inverted pendulum the fixed
   point is a saddle point.

16. What is postural sway?

   Postural sway refers to the movements made by the body to maintain balance during
   quiet standing. It is typically measured by having a subject stand quietly on a force
   platform with eyes closed.

17. Under what conditions can an inverted pendulum be stabilized by vertically
    vibrating the pivot point? Why does this not account for the stabilizing effects
    of vibration and postural sway?

   The requirement for stability is that the pivot point be attached physically to the vibrating stimulus in order that the downward acceleration exceeds that of gravity. This cannot
   explain the stabilizing effects of vibration on human balance since the body is not physically attached to the vibrating platform.

18. What is a sensory dead zone and how does its presence affect the equation of
    motion for the feedback control of an inverted pendulum?

   A sensory deadzone arises because sensory receptors cannot detect changes which are
   below a certain threshold. This means that the feedback is either "on" or "off" depending
   on whether the controlled variable is greater than or less than the threshold.

19. What computer packages are currently available to integrate DDEs?

> Computer programs that can integrate a DDE can be found in XPPAUT, Matlab, Maple and Mathematica.

20. What is the method of steps for integrating a DDE?

> The solution of the DDE is calculated in steps of length τ. For example, the initial function on the interval $[t_0 - \tau, t_0]$ is used to calculate the solution on the interval $[t_0, t_0 + \tau]$.

21. What is the semi-discretization method for integrating DDEs?

> This numerical method discretizes the delayed terms and time-periodic coefficients while all of the other terms in the differential equation are continuous. This method is particularly useful in situations in which the feedback is time sampled.

22. Can the same integration methods for ODEs be used to integrate a DDE?

> In principle, programs designed to integrate ODEs such as the Euler and Runge–Kutta method (XPPAUT) should behave badly when integrating DDEs. However, in practice they often do quick well. When these methods are used it is prudent to repeat the integration with a smaller time step in order to establish that the effects of a jump continuity are not causing a significant numerical artifact.

10.11 Exercises for Practice and Insight

1. Show that the stability conditions given by (10.24)–(10.25) can be rewritten as (10.26).
2. Show that (10.35) is the period of the oscillation generated by (10.34).
3. The Pulfrich[6] phenomenon demonstrates the effects of delay on visual perception [490, 779]. This phenomenon can be readily illustrated by attaching a string to a paper clip to make a pendulum. Allow the paper clip to move back and forth in the frontal plane. Cover one eye with a filter that reduces the light intensity but still makes it possible to see the paper clip. The paper clip now appears to move in an elliptical orbit. This is the Pulfrich phenomenon.

   a. Why does the paper clip seem to move along an elliptical path? Is this path traversed clockwise or counterclockwise?
   b. Compare what happens when you cover your left eye with the filter and when you cover your right eye.

4. Write a computer program to integrate (9.20) using the initial parameter choices: $\tau = 0.3$, $K = 50$, $n = 10$, $k = 3.21$. The stability of the PLR depends very much on the choice of these parameters. Choose the parameter values so that the fixed point is stable, and construct a Bode plot by measuring the response to a sinu-

---

[6] Carl Pulfrich (1858–1927), German physicist.

soidal input. What are the differences and similarities between these simulations and the experimental observations in Figure 9.7?

5. Curiously, the induction of autoimmune hemolytic anemia (AIH) in rabbits is sometimes marked by a steady depression of hemoglobin levels, and at other times by sustained oscillations in hemoglobin levels, and at other times by sustained oscillations in hemoglobin concentration and reticulocyte levels with a period of 16–17 days [660]. In 1979, Michael Mackey developed the following model to explain these observations [520]:

$$\frac{dE}{dt} = k_1 F_N(E(t-\tau)) - k_2 E(t), \tag{10.54}$$

where

$$F_N(E(t-\tau)) = \frac{K^n}{K^n + E^n(t-\tau)}.$$

The parameters for this model obtained from the literature are $\tau = 5.7$ days, $k_1 = 7.62 \times 10^{10}$ cells$\cdot$kg$\cdot$day^{-1}, $k_2 = 2.31 \times 10^{-2}$ days^{-1}, $K = 2.47 \times 10^{11}$ cells$\cdot$kg^{-1}, $E^*_{norm} = 3.3 \times 10^{11}$, where E^*_{norm} is the normal circulating number of erythrocytes, and $n = 7.6$.

a. Write a computer program to integrate equation (10.54).

b. You will find that the program becomes unstable whenever you run it. This is because the numbers are fairly large.

c. Rewrite (10.54) by defining the dimensionless variable

$$\hat{E} = \frac{E}{E^*_{norm}}.$$

Now what happens when you run the program?

d. In autoimmune hemolytic anemia, the peripheral destruction rate k_2 is increased. What happens as k_2 is increased from 0.02 in steps up to, say, 0.5? For each value of k_2, estimate steady-state values of $\hat{E}$. If there is an oscillation, estimate the minimum and maximum values of the oscillation. Plot these values as a function of k_2 along the y-axis. The graph shows a Andronov–Hopf bifurcation followed by a reverse Andronov–Hopf bifurcation as k_2 increases.

e. Can you identify the precise value of k_2 that marks the onset of the Andronov–Hopf bifurcation and the reverse Andronov–Hopf bifurcation using numerical simulations? Why not? (Hint: see Section 5.4.1) Calculate the value of β for which the oscillation arises and the value at which the oscillation disappears using local stability analysis.

6. Consider that the inverted pendulum is stabilized by PDA feedback, i.e.,

$$\ddot{\theta}(t) - \omega_n^2 \theta(t) = -k_p \theta(t-\tau) - k_d \dot{\theta}(t-\tau) - k_a \ddot{\theta}(t-\tau)$$

where k_p, k_d, k_a are, respectively, the proportional, derivative, and accelerative gains. Show that

$$\ell_{\text{crit}} = \frac{3g\tau^2}{2}.$$

7. A simple model for switching feedback control of balance was initially proposed by Eurich and Milton [209, 374, 442]. In its simplest and dimensionless form, this model is

$$\dot{x}(t) = x(t) + f(x(t - \tau)),\tag{10.55}$$

where the control input is a function of the delayed state $x(t - \tau)$ of the form

$$f(x(t - \tau)) = \begin{cases} C & \text{if } x(t - \tau) < -1, \\ 0 & \text{if } -1 \le x(t - \tau) \le 1, \\ -C & \text{if } x(t - \tau) > 1, \end{cases}$$

where $C \ge 1$ is a constant and $\tau \le \ln C$. Despite its simplicity, this model incorporates the three essential features of human balance control, namely an unstable upright position in the absence of feedback, stabilizing time-delayed feedback, and a sensory dead zone. Although (10.55) is a simplified model of the Newtonian mechanics described by (10.36), it still presents the main feature of the inverted pendulum, namely instability occurs as an exponential fall (and not oscillatory). The solution of the linearized open-loop system $\ddot{\theta}(t) - \omega_n^2 \theta(t) = 0$ is given as $\theta(t) = c_1 e^{\omega_n t} + c_2 e^{-\omega_n t}$ with c_1 and c_2 being constants depending on the initial conditions. The feedback control operates by allowing the system to drift for small displacements (open-loop control when $|x| \le \Pi$) with stabilizing negative feedback (closed-loop control) only becoming active once θ exceeds the sensory threshold, i.e., $|x| > \Pi$.

The dynamical behaviors of (10.55) are shown in Figure 10.20. There are no stable fixed-point solutions; however, there are three types of limit-cycle solutions possible (O1, O2, and O3). These solutions and their stability depend on two parameters, τ and C. Here we take $\Pi = 1$ for simplicity.

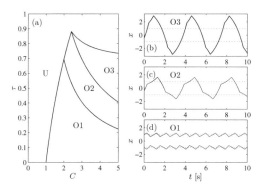

Fig. 10.20 Left panel: Steady-state behaviors of (10.55) as a function of C and τ. Right panel: Oscillatory solutions that arise in regions O1, O2, and O3. Copyright ©2015 Society for Industrial and Applied Mathematics. Reprinted with permission. All rights reserved.

- Show that the region in parameter space which encloses these solutions is bounded by

$$\tau \le \tau_a(C) = \ln C \tag{10.56}$$

and

$$\tau \le \tau_d = \ln \frac{2C^2}{C^2 - 1} \tag{10.57}$$

(Hint: Choose the initial conditions $x(t_0) = 1$ and $0 \le x(s) \le 1$ for $s \in [t_0 - \tau, t_0]$.)

- The stable limit cycle O1 encircles $+1$ for $x > 0$ and coexists with a stable limit cycle which encircles -1 for $x < 0$. Show that the region in parameter space for which these solutions coexist is bounded by

$$\tau < \tau_a(C) \quad \text{and} \quad 0 < \tau < \tau_b(C) = \ln \frac{C}{C - 1}. \tag{10.58}$$

- The regions O2 and O3 contain two qualitatively different limit cycles. Show that the regions in parameter space for which these solutions exist are, respectively,

$$\tau < \tau_a(C) \quad \text{and} \quad \tau_b(C) < \tau < \tau_c = \ln \frac{C + 1}{C - 1} \tag{10.59}$$

and

$$\tau_c(C) < \tau < \tau_d(C). \tag{10.60}$$

8. A pendulum–cart model for pole balancing at the fingertip takes the form[7]

$$\frac{1}{3} m\ell^2 \ddot{\theta}(t) + \frac{1}{2} m\ell \ddot{x}(t) \cos \theta(t) - \frac{1}{2} mg\ell \sin \theta(t) = 0 \tag{10.62}$$

$$\frac{1}{2} m\ell \ddot{\theta}(t) \cos \theta(t) + (m + m_0)\ddot{x}(t) - \frac{1}{2} m\ell \dot{\theta}^2(t) \sin \theta(t) = f(t - \tau)$$

where θ is the vertical displacement angle of the pole, m, m_0 are, respectively, the mass of the pole and the cart, and f describes the control force. The displacement of the fingertip, x, is measured from the typical starting point for pole balancing located at $\approx L/2$ in front of the subject, where L is the total length of the arm. A derivation of a pendulum–cart model can be found on Wikipedia.

a. Show that if $f(t - \tau) = 0$, (10.62) can be reduced to

$$\ddot{\theta}(t) - \omega_n^2 \theta(t) = 0,$$

---

[7] Using the matrix notation introduced in a course on linear algebra, we can rewrite this equation as

$$\begin{pmatrix} \frac{1}{3} m\ell^2 & \frac{1}{2} m\ell \cos \theta \\ \frac{1}{2} m\ell \cos \theta & m + m_0 \end{pmatrix} \begin{pmatrix} \ddot{\theta} \\ \ddot{x} \end{pmatrix} + \begin{pmatrix} -\frac{1}{2} mg\ell \sin \theta \\ -\frac{1}{2} m\ell \dot{\theta}^2 \sin \theta \end{pmatrix} = \begin{pmatrix} 0 \\ f(t) \end{pmatrix}. \tag{10.61}$$

This is a much easier way to understand and hence is an example of how taking a course in linear algebra is so useful!

where $\omega_n = \sqrt{6g/cl}$ and $c = 4 - 3m/(m+m_0)$. (Hint: Eliminate the cyclic coordinate x and linearize around the upper fixed point.)

b. Show that when (10.62) is linearized about the upper equilibrium point, we obtain[8]

$$\frac{1}{3}m\ell^2\ddot{\theta}(t) + \frac{1}{2}m\ell\ddot{x}(t) - \frac{1}{2}mg\ell\theta(t) = 0$$

$$\frac{1}{2}m\ell\ddot{\theta}(t) + (m+M_0)\ddot{x}(t) = f(t-\tau).$$

c. Show that when $\Phi_p = 0$ and $\Phi_d = 0$, (10.47) reduces to (10.36) where

$$\omega_n = \sqrt{\frac{mg\ell - K_{pass}}{J_A}},$$

$$k_p = \frac{K_p}{J_A},$$

$$k_d = \frac{K_d}{J_A}.$$

d. The Lambert W function describes the branches, W_k, of the inverse relation of the function $f(z) = ze^z$ [145, 466]. There are two branches where z is real, denoted by W_0 and W_{-1}. For the remaining branches, k is a complex number. The W function can be used to solve first-order time delay equations [701, 768]. If we have $\dot{x}(t) = ax(t) + bx(t-\tau)$ then the characteristic equation is $\lambda - a - be^{-\lambda\tau}$.

   i. Show that in terms of $W(x)$, the roots of the characteristic equation are

   $$\lambda_k = \frac{1}{\tau}W_k(\tau be^{-\lambda\tau}) + a$$

   where $k = 0, \pm 1, \pm 2, \cdots$. Each of the infinitely many roots of the characteristic equation corresponds to one of the branches of W_k. It has been proven that the solution which corresponds to $k = 0$ always has the largest real part and, therefore, is the dominant mode of the solution of this delay differential equation [768].

   ii. Draw the real branches of the $W(x)$. (Hint: It is easiest to first determine the graph for $y = xe^x$ and then determine $x = W(y)$.)

   iii. Show that

   $$\frac{d}{dx}W(x) = \frac{W(x)}{x(1+W(x))}.$$

   iv. Show that

---

[8] Using matrix notation that can be written as

$$\begin{pmatrix} \frac{1}{3}m\ell^2 & \frac{1}{2}m\ell \\ \frac{1}{2}m\ell & m+m_0 \end{pmatrix} \begin{pmatrix} \ddot{\theta} \\ \ddot{x} \end{pmatrix} + \begin{pmatrix} -\frac{1}{2}mg\ell & 0 \\ 0 & 0 \end{pmatrix} \begin{pmatrix} \theta \\ x \end{pmatrix} = \begin{pmatrix} 0 \\ f(t) \end{pmatrix}. \tag{10.63}$$

$$\int W(x)dx = x\left(W(x) - 1 + \frac{1}{W(x)} \right) + C.$$

v. It is possible to use $W(x)$ to compute the solutions of first-order delay differential equations using commands already embedded in MATLAB [902]
   and Mathematica [701]. In this way, compute the solutions of (10.9) and
   compare your results to those obtained in Section 10.5.1.

Chapter 11
Oscillations

Oscillations fascinate biologists: cells cycle, hearts beat, lungs pump, wings flap, animals sleep and then wake. It is the widespread occurrence of oscillations that most clearly distinguishes biological control from the control of manmade devices. Engineers care about minimizing the effects of unwanted oscillations, but Mother Nature seems content to exploit their benefits. Indeed, oscillations make it possible to coordinate the activities of large populations of cells, such as tissues and organ systems, through the tendency of individual oscillators to synchronize their activities with other oscillators [683, 811, 822].

Arguably the most important biological oscillations are those associated with excitable cells. Without excitable cells, hearts would not beat, wings would not flap, and brains would not think. Excitability is intimately associated with the presence of a cubic nonlinearity in the relevant mathematical models [67, 226, 227, 621, 844, 845]. Here we derive the famous Hodgkin–Huxley (HH) equations for a neuron to demonstrate that the molecular engines for excitability reside in the membrane ion channels. The same principles used to understand the RC (resistor and capacitor) circuit (Section 7.1) are used to investigate how a neuron generates an action potential [294, 343]. Indeed, it has become possible using these principles to develop a solid state silicon chip that mimics the behavior of living neurons [3].

Neuroscientists use the spiking patterns of neuron in order to classify them. Figure 11.1 shows a selection of the spiking patterns that have been identified for neurons in the brain. As an experienced neuro-physiologist inserts an electrode into the brain they listen to the electrical activity detected by the electrode on a loud speaker. They use the spiking patterns they hear to both identify the region of the brain that the electrode is in and eventually to locate the neuron that they wish to investigate more carefully. On the other hand, neurosurgeons record the spiking patterns of neurons in the basal ganglia to guide the insertion of an electrode to help treat patients with Parkinson's disease using deep brain stimulation [363].

The transition between a resting state and a continuous spiking state in a neuron corresponds to a bifurcation in a mathematical model for neural dynamics. Intuitively we would expect that the nature of the bifurcation depends in some way on

© Springer Nature Switzerland AG 2021
J. Milton and T. Ohira, *Mathematics as a Laboratory Tool*,
https://doi.org/10.1007/978-3-030-69579-8_11

the types and numbers of ion channels in the neuronal membrane. However, neurons which possess different ion channels exhibit very similar spike initiation mechanisms [691]. This observation suggests that the nature of neural oscillations is not related to the presence of specific combinations of ion channels, but more likely reflect the action of robust underlying principles. In other words, neural dynamical behaviors are degenerate in the sense that many different combinations of ion channels can work together to generate the same dynamics. Up to now we have emphasized that the important property of models of excitable cells is the presence of a cubic nonlinearity. Thus we need to address the molecular basis for the cubic nonlinearity.

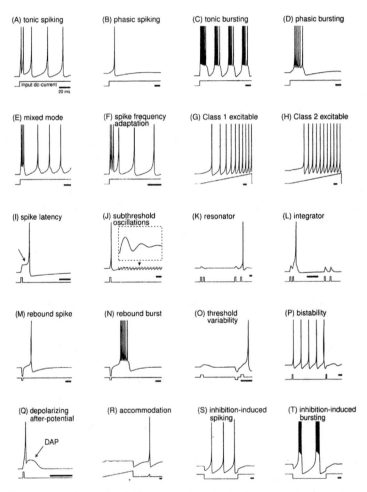

Fig. 11.1 Examples of the different spiking behaviors observed for cortical neurons. Figure reproduced from izhikevich.org with permission.

The questions we will answer in this chapter are

1. Why is a limit cycle a more robust explanation for a biological oscillator than a center?
2. What can be said about the nature of the fixed point that is surrounded by a limit cycle?
3. How can a limit cycle occur when the fixed point is stable?
4. An ion channel is often modeled by an equivalent RC circuit. Why are the RC circuits arranged in parallel rather than in series?
5. What is the key difference between the resistance in an electrical circuit and the conductance of an ion channel?
6. How does the capacitance affect the flow of charges across the neuronal membrane?
7. What is the relationship between the current, I, and the charge, Q?
8. What is a voltage clamp?
9. What is the typical shape of the fast and slow nullclines for neurons?
10. What does the phrase "conductance-based model" for a neuron mean?
11. Although the Hodgkin–Huxley equation was developed to model to squid giant axon, it does not accurately describe the dynamics of this axon. Explain.
12. What is the difference between an ionotropic and a metabotropic ion channel?
13. What is the mathematical requirement for excitability?
14. What is the mathematical interpretation of Type 1, Type 2, and Type 3 excitability?
15. What does the term "co-dimension 1" mean?
16. The resting potential for a neuron integrator behaves like what kind of fixed point?
17. The resting potential for a neuron resonator behaves like what kind of fixed point?
18. What types of bifurcations involve a saddle node?
19. What are the types of bifurcations of equilibria that produce limit-cycle oscillations in excitable systems?
20. What are the types of bifurcations of limit cycles that eliminate the oscillations in excitable systems?
21. What experimental measurements are useful for determining the type of bifurcation that produces and eliminates a limit-cycle oscillation?
22. How many co-dimension 1 bursting neurons are possible?
23. What is AUTO?
24. How is the computational load associated with numerically integrating a differential equation measured?

11.1 Hodgkin–Huxley Neuron Model

Hodgkin and Huxley were awarded the 1963 Nobel Prize in physiology or medicine for establishing the electrophysiological basis for action potential generation by a neuron and the subsequent propagation of the action potential along the axon

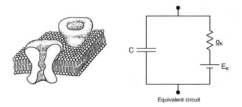

Fig. 11.2 (a) Schematic representation of sodium ion channel inserted into lipid bilayer. (b) Electrically equivalent RC circuit. Figure reproduced from [295] with permission.

[344–348]. Here we combine experimental observations with elementary principles of electricity to derive the Hodgkin–Huxley (HH) equation. Our presentation is from the modern perspective of ion channels, a concept that was unknown to Hodgkin and Huxley.

The study of the electrophysiology of excitable cells owes much to two organisms: the squid and the green alga *Nitella*. Both organisms contain a "giant axon": the squid giant axon is about 1 mm in diameter; the *Nitella* cell is about 2–3 mm in diameter. In contrast, many neurons in the brain have cell body (soma) diameters of ≈ 10 microns, and axon diameters can be less than 1 micron. The large diameters of the squid axon and *Nitella* cell enabled biologists to perform the key experiments and collect the precise data that formed the basis of the HH model.

There is convincing experimental evidence that the neuron's action potential is determined solely by properties of its membrane. In particular, the cytoplasm from a squid giant axon can be rolled out and replaced with a solution that has the same ionic composition as cytoplasm [348]. It is observed that this preparation can generate an action potential that very closely resembles that produced by the intact axon.

Neuronal excitation and signaling involve movement of ions through membrane-bound ion channels (Figure 11.2). The important ions are sodium (Na^+), potassium (K^+), chloride (Cl^-), and calcium (Ca^{++}). Ion channels are what is "excitable" in the term "excitable cell." Thus each ion channel may be regarded as an excitable molecular structure that responds to a specific stimulus, e.g., a change in membrane potential, the binding of a neurotransmitter, or the mechanical deformation of the membrane. The channel's response, or *gating*, is one of opening or closing. Each type of ion channel exhibits selective permeability, and hence when it opens, only a very restricted class of ions can pass through. The ions flow passively down their electrochemical activity gradient at a high rate ($> 10^6$ ions per second). Since ions carry charge, we anticipate that it will be possible to learn a great deal about properties of ion channels by applying the laws of physics that govern the flow of charge.

There are three principles from physics that lie at the basis of the HH equation: Ohm's[1] law, capacitance, and Nernst[2] potential.

---

[1] Georg Simon Ohm (1789–1854), German physicist and mathematician.

[2] Walther Hermann Nernst (1864–1941), German physical chemist and physicist.

Ohm's Law

Conductance (in siemens) g is given by

$$g = \frac{I}{V}, \tag{11.1}$$

where V is the voltage difference (volts) across a conductor and I is the current flow (amperes). If we note that

$$g = \frac{1}{R},$$

where R is the resistance (ohms), then we can rewrite (11.1) in the form of Ohm's law that is most likely to be familiar to the reader, namely

$$V = IR.$$

However, (11.1) is much better for our purposes here. It is useful to keep in mind that in physics, R is typically a constant and hence a parameter, whereas the conductance of the ion channel in a neural membrane is not constant and hence is a variable.

What is the resistance across the membrane for an excitable cell? Biologists have found that the resistance is quite variable, with a range of ≈ 10–$10^6 \ \Omega \cdot cm^{-1}$ [343]. These values lie between those observed for an ionic solution ($\approx 20 \ \Omega \cdot cm^{-1}$) and those observed for pure lipid bilayers ($\approx 10^{15} \ \Omega \cdot cm^{-1}$). These observations together with measurements of the properties of membrane noise [850] led to the conclusions that neural membranes contain ion channels and the number of open ion channels varies.

The reason that Ohm's law plays a central role in membrane physics is that each ion channel behaves like an elementary conductor that spans the lipid bilayer (Figure 11.2a). The total electrical conductance of a membrane is the sum of all these elementary conductances arranged in *parallel*, each of which is associated with an individual ion channel. Thus the conductance provides a measure of how many ion channels are open, how many ions are available to pass through them, and how easily the ions pass.

Membrane as Capacitor

There is a potential difference between the inside and outside of a resting excitable cell. For a typical resting neuron, this potential difference is ≈ -70 mV, and for a skeletal muscle cell, it is ≈ -90 mV. The negative sign indicates that the inside of the neuron is electrically negative with respect to the outside. However, the cell membrane is very thin, e.g., 2.5–10 nanometers. An electrical capacitor is formed when two conductors are separated by a narrow gap. Capacitance C is a measure of how much charge Q must be transferred from one conductor to another to set up a given potential V, namely

$$C = \frac{Q}{V}.$$ (11.2)

The capacitance of the neural membrane is ≈ 1 μF per square centimeter, where F is the abbreviation for farad. A 1-farad capacitor will charge to 1 V when 1 C (coulomb) of charge is in one conductor and -1 C is in the other. This is an enormous capacitance, and in fact, 1μF for the neural membrane is rather large. The high electrical capacitance of the neural membrane is a direct consequence of its small width.

The capacitance places a lower limit on the number of ions (charges) that must move and how rapidly they must move to produce a given voltage signal. In general, capacitance slows the voltage response to any current by a characteristic time t_c that depends on the product of the capacitance and the effective resistance (see below).

Hodgkin and Huxley made the assumption that for each ion type X, the membrane behaves linearly, that is,

$$I_X = g_X(V - E_X),$$ (11.3)

where g_X and E_X are, respectively, the conductance and equilibrium (Nernst) potential of the ion channel for in X. This *linear membrane hypothesis* is not exactly the same as Ohm's law, since the current flow becomes zero at E_X, not at zero volts. However, just like Ohm's law, (11.3) is an empirical law that holds for many, but not all, ion channels. The sign convention means that when $V < E_X$, there will be an inward flow of ion X, and when $V > E_X$, the flow of X is outward.

The total capacitance current per unit neuronal membrane surface area is

$$I = \frac{dQ}{dt} = C\frac{dV}{dt}$$

The total current into the neuron must sum to zero (Kirchhoff's current law). Thus

$$0 = I + I_x = C\frac{dV}{dt} + g_X(V - E_x).$$ (11.4)

where g_X and E_X are, respectively, the conductance and equilibrium (Nernst) potential of the channel for ion X. On re-arranging we obtain

$$\frac{dV}{dt} = -\frac{V}{t_c} + \frac{E_X}{t_c},$$ (11.5)

where $t_c = C/g_X$ is the membrane time constant for ion X. Thus we see that

$$V(t) = E_X - (E_X - V(0))e^{-t/t_c}.$$

Thus if we suddenly perturb the membrane potential away from E_X, the membrane potential will return to E_X exponentially fast with time constant t_c. A shorthand way to depict this description of an ion channel's dynamics is the RC circuit shown in Figure 11.2b.

Nernst Potential

Table 11.1 shows that the concentrations of Na^+, K^+, Cl^-, and Ca^{++} for a resting neuron are not the same on both sides of the excitable membrane. In particular, the concentration of K^+ ions is greater inside the neuron, and the concentration of Na^+ is greater outside.

Table 11.1 Free ionic concentrations and calculated room temperature equilibrium potentials for a model resting neuron.

Ion	$[\cdot]_{out}$ (mM)	$[\cdot]_{in}$ (mM)	$[\cdot]_{out}/[\cdot]_{in}$	Equilibrium Potential (mV)
Na^+	142	10	14.2	+67
K^+	4	90	0.044	−78
Ca^{++}	2.5	10^{-4}	25000	+134
Cl^{-1}	103	4	25.75	−82

Consider a hypothetical membrane that contains only K^+ channels, and distribute the K^+ ions in accordance with the values in Table 11.1. What happens when the ion channel is opened? Obviously, K^+ ions will flow "outside," since we know that at equilibrium, there can be no concentration gradients unless another force is acting on the system. However, the important property of the ion channel is that it is permeable only to K^+ ions. This means that the negative counter ions cannot also flow with K^+ through the ion channel. Consequently, as a result of the movement of K^+ ions, a very small positive charge builds up outside of the cell, and a very small negative charge builds up on the inside. The more K^+ ions that pass through the ion channel, the larger the charge difference between the two sides. However, like charges repel. At some point, the force due to the concentration gradient will be exactly balanced by the force due to the electrochemical gradient. This point is referred to as the *Nernst equilibrium*

$$E_X = -\frac{58.1}{n} \ln \frac{[X]_{in}}{[X]_{out}} \text{ (mV)}, \tag{11.6}$$

where n is the valence of the ion, and we have calculated the Nernst potential E_X at temperature 298 K.

Examining Table 11.1, we see that the resting potential across the neural membrane (≈ -70 mV) is very close to the equilibrium potential of K^+ and Cl^-. During the formation of an action potential, the concentration of Cl^- ions inside and outside the neuron does not change. Thus the resting membrane potential and its changes during action potential generation must be related, at least in some way, to flows of K^+ across the neural membrane. In contrast, in excitable cells in the plant kingdom, the membrane potential is controlled by movement of Cl^- ions [396].

At this point, we can guess that the HH equation that governs the changes in the membrane potential of the neuron will look something like

$$C\frac{dV}{dt} = -\sum_{i=1}^{n} I_i \approx -(I_{Na} + I_K + I_L + I_{ext}),$$

where I_{Na} is the current related to the Na channels, I_K is the current related to the K channels, I_L is the "leak" current, and I_{ext} is an external current, supplied, for example, by means of a stimulating electrode. Moreover, we can actually do a little bit better and use our notion of conductances to obtain

$$C\frac{dV}{dt} = -[g_{Na}(t)(V - E_{Na}) + g_K(t)(V - E_K) + g_L(t)(V - E_L) + I_{ext}]. \quad (11.7)$$

Based on these observations, Hodgkin and Huxley formulated the following hypothesis for the formation of an action potential. At rest, the permeability of the membrane potential to K^+ ions is about 100 times its permeability to Na^+ ions. This is why the membrane potential at rest is close to the equilibrium potential of K^+. When an action potential is triggered, the permeability of the membrane to Na^+ suddenly increases dramatically. This causes Na^+ ions to diffuse down their concentration gradient toward E_{Na}, and hence the membrane potential of the neuron becomes positive. In order to re-establish the resting membrane potential, the K^+ channels open more, and gradually both channels close, returning to their resting values with different time constants. The characteristic shape of the action potential arises because the changes in the Na^+ ion flows occur on a faster time scale than those for the K^+ ion flows. The fact that the membrane potential transiently hyperpolarizes, i.e., becomes transiently more negative than the resting potential, is evidence that the K^+ channels become transiently "more open" after the action potential has formed.

Evidence to support this hypothesis was obtained by measuring the current flows associated with each type of ion channel during action potential generation. In order to do this, Hodgkin and Huxley made use of a technique known as the *voltage clamp* [139]. Due to the large diameter of the squid axon, it was possible to thread two thin electrodes inside the axon. One of the electrodes was used to measure the potential, and the other was used to inject current. Using electronic feedback circuitry similar to that described in Section 9.2.1, current is injected so that a predetermined fixed voltage is maintained across the membrane (hence the term voltage clamp). The injected current is the mirror image of the current generated by ion flows across the neural membrane at that potential. The individual currents were identified using a combination of techniques including ion substitution, specific ion channel blockers, and clamping the membrane potential at different values.

In order to translate these experimental observations into a mathematical model, Hodgkin and Huxley assumed that

$$g_X(t) \approx \overline{g}_X n(t),$$

where $n(t)$ is a gating variable that controls the opening and closing of the ion channel, and $\overline{g}_X$ is the channel conductance when the channel is open. They used "educated curve-fitting" techniques to obtain $n(t)$, which means that they began with a simple model for ion channel dynamics that could be solved exactly and then made simple modifications to this basic model to account for deviations between the predicted and observed dynamics. The advantage of this approach over that of simply guessing the functional forms is that it is often easier to use experimental observations to guide subsequent changes to the model.

Their first step was to assume that the dynamics of the opening and closing of the ion channel are governed by the law of mass action

$$C \underset{\beta_n}{\overset{\alpha_n}{\rightleftharpoons}} O,$$

where C is the number of closed channels, and O is the number of open channels. The rate constants for channel opening and closing are, respectively, α_n and β_n. These rate constants were observed to depend on the membrane potential V, but for constant V, they are time-independent. Let n be the fraction of K^+ channels that are open, and $1 - n$ the fraction of the gates that are closed. Thus for constant V, we can write

$$\frac{dn}{dt} = \alpha_n(1-n) - \beta_n n = \alpha_n - n(\alpha_n + \beta_n). \tag{11.8}$$

Hodgkin and Huxley used the voltage clamp to hold V constant and determine α_n, β_n. By repeating the same measurements for different values of V, they were able to determine $\alpha_n(V)$ and $\beta_n(V)$. To evaluate the rate constants from experimental measurements, it is convenient to rewrite (11.8) as

$$\frac{dn}{dt} = \frac{n_\infty - n}{t_v},$$

where

$$n_\infty := \frac{\alpha_n}{\alpha_n + \beta_n}, \quad t_v := \frac{1}{\alpha_n + \beta_n}.$$

Thus $\alpha_n = n_\infty/t_v$ and β_n can be determined.

Proceeding in this way, they determined that the dynamics of the K^+ channel could be described in terms of one gating variable, namely

$$g_K(t) = \bar{g}_K n^4 (V - E_K), \quad \frac{dn}{dt} = \alpha_n(V)(1-n) - \beta_n(V)n.$$

The term n^4 accounts for the measured lag in the K^+ channel opening [295].

The same approach was applied to determining the contribution of the Na^+ current. This current is more complex than the K^+ current, since two gating variables are required to describe its dynamics: an activation gating variable m and an inactivation gating variable h. The equations that describe the Na^+ current are

$$g_{Na} = \bar{g}_{Na} m^3 h (V - E_{Na}),$$

$$\frac{dm}{dt} = \alpha_m(V)(1-m) - \beta_m(V)m,$$

$$\frac{dh}{dt} = \alpha_h(V)(1-h) - \beta_h(V)h.$$

Thus the HH equation is a system of equations: one differential equation is of the form of (11.7), and the others take into account the openings and closings of the various types of ion channels. The modern-day form[3] of the HH equation is [295]

$$C\frac{dV}{dt} = -\left[\bar{g}_{Na} m^3 h(V - E_{Na}) + \bar{g}_k n^4(V - E_K) + \bar{g}_L(V - E_L) + I_{ext}\right],$$

$$\frac{dm}{dt} = \alpha_m(V)(1-m) - \beta_m(V)m,$$

$$\frac{dh}{dt} = \alpha_h(V)(1-h) - \beta_h(V)h, \qquad (11.9)$$

$$\frac{dn}{dt} = \alpha_n(V)(1-n) - \beta_n(V)n,$$

where

$$\alpha_m(V) = 0.1(V+35)/(1 - \exp(-(V+35)/10)),$$
$$\beta_m(V) = 4\exp(-(V+60)/18),$$
$$\alpha_h(V) = 0.07\exp(-V+60)/20),$$
$$\beta_h(V) = 1/(\exp(-(V+30)/10)+1),$$
$$\alpha_n(V) = 0.01(V+50)/(1 - \exp(-(V+50)/10)),$$
$$\beta_n(V) = 0.125\exp(-(V+60)/80).$$

Using (11.9), Hodgkin and Huxley were able to predict the current flows that occur during an action potential, the experimentally observed absolute and relative refractory periods, and the conduction velocity of the squid giant axon [346]. This last finding is truly amazing, since one would normally expect that the prediction of a conduction velocity would require a model expressed in terms of a partial differential equation (PDE), namely a differential equation with the two independent variables time and space. That this was possible can again be attributed to the use of the voltage clamp technique, which mitigates effects related to the intracellular spread of current. Mathematically, this experimental trick makes it possible to replace a PDE by an ODE. Surprisingly (11.9) does not accurately describe the dynamics of a squid giant axon to a sustained physiologically relevant, suprathreshold depolarizing current. In particular, (11.9) predicts repetitive spiking, whereas the squid giant axon typically produces a single spike and then is quiescent (see also section 11.3.3). It has been shown that a modification of a single parameter description of the K^+

---

[3] These equations are different from those proposed by Hodgkin and Huxley because V has been interpreted as the transmembrane potential with the sign convention that the inside of a resting neuron is negative with respect to the outside.

channel (11.9) is sufficient to explain the observed behavior of the squid giant axon [134, 136]. Repetitive spiking can be produced by placing the squid giant axon in a low Ca^{++} medium [303] or a solution with a higher NaCl concentration that normal sea water [546] or using much higher, non-physiological sustained depolarizing currents which produce irreversible changes in the axon [115].

Hodgkin and Huxley did not have access to high-speed digital computers. Thus they were unable to appreciate the rich nature of the dynamics that (11.9) possessed, including bistability, limit-cycle oscillations, and bursting behaviors. It was the subsequent efforts of mathematicians that brought these dynamics to light. The first important step was made independently by Fitzhugh [226, 227] and the Nagumo group [621], who realized that the time scales for the gating parameters m, n, h were not the same. In particular, the time scale for m is much faster than the others, so it is reasonable to assume that it is sufficiently fast that it relaxes immediately, and hence $dm/dt = 0$. Moreover, if we set $h = h_0$ to a constant, then the resulting two-variable FHN equations discussed in Section 5.1.1 retain many of the features observed experimentally.[4] In other words, the dynamics observed in the FHN equation are also seen in the HH equation [8].

11.2 Ion channels

Approximately 25 years after the work of Hodgkin and Huxley, a powerful technique for studying the dynamics of single ion channels, called the *patch clamp*, became available [312]. By combining the patch and voltage clamps, the dynamical changes in conductance for each type of ion channel can be precisely measured typically over the range $-100\,mV$ to $+50\,mV$ (for a careful discussion of this procedure, see [295]). Thus it became possible to interpret Hodgkin and Huxley's original concept of gating in terms of the molecular dynamics of ion channel opening and closing. When the voltage clamp is turned on, the ion channel opens after a variable delay and tends to remain open until the voltage clamp is turned off. Since the opening time is variable, it is necessary to average multiple trials in order to obtain the current measured by Hodgkin and Huxley using the voltage clamp. This ensemble average can be constructed either by averaging multiple trials of opening recorded from a single ion channel or by averaging single trials measured from a very large number of ion channels of the same type. Surprisingly, the ensemble averages obtained using these two approaches are the same. This observation is a manifestation of the *ergodic hypothesis* (see also Section 16.1).

The above observations indicate that (11.9) represents the "averaged" version of a stochastic dynamical system. The stochastic underpinnings of the HH equation may be important to keep in mind when neuronal behaviors are examined in the so-called *channelopathies*, a term that refers to certain clinical disorders, including

---

[4] The mathematical steps involved in reducing the HH equation to the FHN equation are well described in [422].

certain types of epilepsy, that are related to decreases in the number of specific ion channels and abnormalities in their function [427, 597, 690]. Presumably, as the number of ion channels in the neuronal membrane decreases, there comes a point at which the above "ergodic hypothesis" is no longer valid, and hence (11.9) must be replaced by an appropriately formulated stochastic differential equation.

Ion channels such as those described by Hodgkin and Huxley are referred to as ionotropic ion channels. An important point from the modeling point of view is that the ion channel and the receptor which activates the ion channel reside in the same protein complex. Neuro-physiologists have followed the HH recipe and made all of the necessary measurements to determine the parameters that describe a large range of channel dynamics. Presently, the exact functional forms of the gating equations are known for many types of ion channels (for convenient summaries of the parameters for different ion channels see [207, 491]). For example, the HH-type model that describes the behavior of the leech pacemaker neuron takes into account eight different types of ion channels, resulting in a system of 14 nonlinear differential equations [160, 765]. In order to avoid confusion, it is best to refer to (11.9) as the HH model that is applicable to the squid giant axon and use the term "HH-type models" to refer to other applications.

There is a second class of ion channels which are opened indirectly through a cascade of intracellular events. This type of ion channel is referred to as metabotropic. An example is the gamma-aminobutyric acid, Type B ($GABA_B$) ion channel. Here the binding of a GABA molecule to the membrane receptor activates an intracellular complex called a G-protein which, in turn, activates a K^+ channel. The mathematical model for the dynamics of a metabotropic ion channel differs from that of an ionotropic ion channel since it is necessary to account for the time delay between receptor binding and ion channel activation. A consequence is that metabotropic ion channels are nonlinear, slow to activate and their effects can be long lasting. If this time delay is considered to be small compared to the characteristic time scale, then the delay can be approximated as a lag (see Section 10.3). This assumption leads to the following model [207]

$$I_{GABA_B} = \overline{g}_{GABA} \frac{s^n}{K_d + s^n}(V - E_K)$$
$$\frac{dr}{dt} = a_r[T](1 - r) - b_r r$$
$$\frac{ds}{dt} = k_1 r - k_2 s$$

where a_r, b_r, k_1, k_2 are constants, $[T]$ is the concentration of $GABA_B$ in the synaptic cleft, r denotes the $GABA_B$ receptors, and s is the fraction of open channels. The lag is produced by the cascade formed by the two ODEs. Obviously replacing the lag with a discrete time delays would reduce the number of parameters that need to be estimated. However, to our knowledge, this has not yet been attempted.

The remarkable insight of mathematical neuroscientists is that the spiking dynamics of neurons can often be understood by considering differential equations contain-

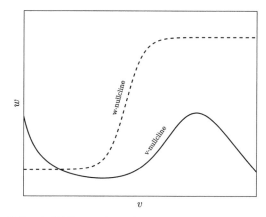

Fig. 11.3 The v-nullcline (solid line) and w-nullcline (dashed line) for a resting model neuron. Note that the fixed point where the v and w nullclines intersect is near the left knee of the v-nullcline. These nullclines were calculated using the Morris–Lecar model for a neuron which exhibits Type 1 excitability. This model is described in Exercise 10 at the end of this chapter.

ing just two variables: a membrane potential variable, v, and a recovery variable, w [721]. Since there are only two variables, we can use phase-plane analysis to obtain a qualitative description of the dynamical behavior (Figure 11.3). For neurons the v-nullcline, namely the nullcline obtained by setting $dv/dt = 0$, is a cubic nonlinearity (the "fast" variable). The difference between various types of neurons resides in the shape of the w-nullcline (the "slow" variable). In the case of the Fitzhugh–Nagumo model for a neuron, the w-nullcline is a straight line. More generally, the w-nullcline takes the form of a sigmoidal nonlinearity. The resting state of the neuron is described by the intersection of the nullcline near the "left knee." Hence the different types of excitability relate to the different ways the nullcline intersects to create fixed points. Here we illustrate the usefulness of this approach by discussing three problems related to neurons: 1) excitability, 2) the onset of oscillations, and 3) reduced neuron models for large-scale computations.

11.3 Excitability

Neuroscientists often introduce the concept of excitability from the point of view of the spiking threshold. A small "subthreshold" stimulus evokes a small, graded post-synaptic potential, whereas a "super-threshold" input causes an all or none action potential whose magnitude is at least an order of magnitude larger than the post-synaptic potential. However, there are fundamental problems associated with this approach. For example, as readily can be seen with the Fitzhugh–Nagumo model of the neuron, the spiking threshold is not constant but itself depends on the magnitude of the input current.

Modern-day computational approaches to excitable systems emphasize a more geometrical definition of excitability [207, 382]. First, it is assumed that at rest the neuronal system has a stable fixed point. This system is excitable if there exists a trajectory that starts in a small neighborhood of the stable fixed point, leaves the neighborhood of the fixed point, and then eventually returns to the fixed point.

A. L. Hodgkin (1948) observed that there were three modes of onset of oscillation in excitable cells [344, 345]. *Type I excitability* describes oscillation onset in which the spiking can occur arbitrarily slowly. *Type II excitability* describes an oscillation onset characterized by abrupt spiking whose frequency could not be made slower than a threshold frequency. *Type III excitability* describes neurons which were incapable of repetitive spiking. In mathematical models for neural dynamics, neurons are excitable because "they are tuned" close to a bifurcation from a rest state to a limit-cycle attractor [382]. Thus we can anticipate that it will be possible to obtain a dynamical explanation for Hodgkin's observations.

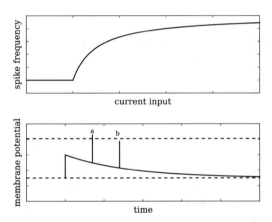

Fig. 11.4 (Top). The spike frequency as a function of the input current for a neuron which exhibits Type 1 excitability. (Bottom). The response of the neuron to subthreshold inputs. It is assumed that the first subthreshold input arrives at $t = 0$ and the next at either $t = a$ or $t = b$.

11.3.1 Type 1 excitability

Figure 11.4 summarizes two important properties of a neuron that exhibits Type 1 excitability. First there is a continuous relationship between the current, I, input to the neuron of the output spike frequency, f. Note that f can be arbitrarily low (in other words, the time interval between two neural action potentials (spikes) can be arbitrarily long). Second, if the input to the neuron is subthreshold, the membrane potential returns to its resting values exponentially. From the point of view of dynamical systems theory, this means that the resting membrane potential responds to small perturbations as if it were a stable node. Thus the eigenvalues, λ, in the linearized equations of proposed mathematical models are real valued and negative.

The observation that the fixed point for a Type 1 neuron behaves like a stable node has important implications for how a Type 1 neuron responds to an input consisting of a periodic train of neural spikes (more properly one might say a periodic train of excitatory postsynaptic potentials). If the second pulse arrives right after a subthreshold input at $t = a$, then the summed membrane potential may be sufficient to cross the spiking threshold and the neuron generates an action potential ("fires"). However, if the same subthreshold input arrives at $t = b$ after a longer time interval, then the membrane potential may not exceed the spiking threshold. In other words, neurons which exhibit Type 1 excitability perform a temporal integration of the incoming pulse trains. Consequently these neurons are referred to as *integrators*. Integrator neurons prefer high-frequency input: the higher the frequency, the sooner they fire. Decreasing the frequency of the input periodic spike train delays or even terminates their firing.

Many cortical neurons function as integrators, for example, most cortical pyramidal neurons, including regular spiking, intrinsically bursting and chattering [382]. Specific examples include neurons in layer 5 of neocortex, neurons in cat sensorimotor cortex and rat barrel sensory cortex (a useful source is [861]). Type 1 neurons also occur in the areas of brain including the CA1 pyramidal neurons in the hippocampus. However, it is important to keep in mind that some neurons are able to behave as Type 1 neurons to certain inputs and as Type 2 neurons to other inputs [861].

A possible phase-plane interpretation of Type 1 excitability is shown in Figure 11.5a. Mathematical models which reproduce this phase plane are referred to as belonging to Class 1[5]. When there is no injected current there are three fixed points. The left-most fixed point is stable and the other two are unstable. The effect of injecting a current is to shift the v-nullcline upwards. Once the injected current is sufficiently large, the two fixed points on the left coalesce to make a saddle node. Thus, the limit cycle arises from a saddle-node bifurcation. It should be noted that each time the limit-cycle trajectory comes close to the "ghost" of the saddle node, the speed of moving on the trajectory slows (see Section 5.1). Since the rightmost fixed point is unstable, there is no bistability.

The properties of Type 1 neurons can be understood in terms of a saddle node on invariant circle (SNIC) bifurcation. The limit cycle, or circle, corresponds to a homoclinic trajectory that originates at the saddle node and then terminates on the

---

[5] The convention is to refer to mathematical models that account for Hodgkin's Type 1, Type 2, and Type 3 excitability as belonging, respectively, to Class 1, Class 2, and Class 3.

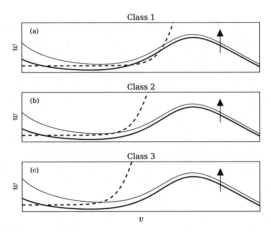

Fig. 11.5 Phase-plane representations of (a) Class 1, (b) Class 2, and (c) Class 3 models for neuronal excitability. The w-nullcline (dashed line) and the v-nullcline (solid lines) are computed for the Morris–Lecar model described in Exercise 9 at the end of this chapter. The advantage of this model is that the three types of excitability can be reproduced by varying just one parameter β_w [691]: $\beta_w = -21mV$ (Class 1), $\beta_w = -13mV$ (Class 2), and $\beta_w = 0mV$ (Class 3).

same saddle node[6]. The important point is that the saddle node is located on the invariant circle. In other words, the limit cycle passes exactly through the point in phase space where the saddle node existed. The magnitude of μ does not affect the amplitude of the limit cycle (hence the amplitude of the spikes is fixed). The time it takes to complete one cycle on the invariant circle is infinite. Thus the spiking frequency at oscillation onset is zero. As μ approaches μ_c from above, there occurs the bottleneck phenomena we discussed in Section 5.1. Hence the frequency grows as $\sqrt{\mu - \mu_0}$ (alternatively the period decreases as $1/\sqrt{\mu - \mu_0}$). The amplitude of the limit does not change very much when $\mu > \mu_c$. There is no bistability because the stable fixed point is destroyed at $\mu = \mu_c$. From a neurophysiological point of view, this type of bifurcation becomes possible when the net current at peri-threshold potentials is inward (depolarizing) at steady state [691].

11.3.2 Type 2 excitability

Two key properties of Type 2 neurons are that the onset of the oscillation occurs with finite nonzero frequency and the response to small stimuli results in an oscillatory return of the membrane potential to its resting value (Figure 11.6 (Top)). Thus from

---

[6] A homoclinic trajectory starts and ends at the same fixed point. Such trajectories are rare and if present indicates that the dynamical system can undergo a bifurcation that either creates or destroys a limit cycle.

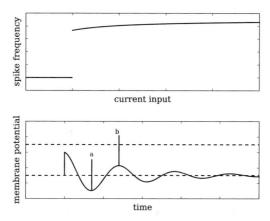

Fig. 11.6 (Top). The spike frequency as a function of the input current for a neuron which exhibits Type 2 excitability. (Bottom). The response of the neuron to subthreshold inputs. It is assumed that the first subthreshold input arrives at $t = 0$ and the next at either $t = a$ or $t = b$.

a dynamical point of view, the resting membrane potential responds to a perturbation as would be expected for a stable spiral point. This means that the eigenvalues, λ, in the linearized equations of proposed mathematical models are complex with negative real parts. Note that in contrast to Type 1 neurons we have avoided the use of the word subthreshold. This is because for Type 1 neurons the threshold is well defined (in the above discussion it is $\mu = 0$), but it is not as well defined for Type 2 neurons [382].

The observation that the membrane potential of Type 2 neurons exhibits an oscillatory return to its resting state has implications for how the neuron responds to a periodic train of inputs (Figure 11.6 (Bottom)). In contrast to a Type I neuron, the effect of the next pulse in the input after a subthreshold input on the output of a Type 2 neuron depends on its timing relative to the period of the damped oscillatory response. If the timing between the two input pulses is near half the period (e.g., $t = a$), then the pulses effectively cancel each other. On the other hand, if the timing between the two input pulses is near the period (e.g., $t = b$), then the pulses add up. Thus such neurons prefer inputs that have a certain resonant frequency that is equal to the frequency of the subthreshold oscillation. Such neurons are referred to as *resonators* [378]. Resonators include most inhibitory neurons as well as neurons in the mesencephalic V region of the brainstem and stellate neurons in entorhinal cortex [382]. Again we note that certain neurons can exhibit Type 1 or Type 2 behaviors depending on the nature of their inputs [382].

Type 2 excitability corresponds to the phase plane shown as Class 2 in Figure 11.5b. In the absence of an injected current there is one stable fixed point. When a current is injected the v-nullcline moves upward and the fixed point moves to the right. As long as the fixed point remains sufficiently to the left of the local minimum of the v-nullcline, the fixed point remains stable. However, once the fixed point moves sufficiently far to the right of the local minimum in the v-nullcline, it becomes unstable and a limit cycle appears. It cannot be due to a SNIC bifurcation since the fixed point when $\mu < \mu_c$ is a spiral point.

Bifurcations that produce oscillations when a conjugate pair of complex eigenvalues cross into the right half of the complex plane are referred to as Andronov–Hopf bifurcations. As we saw in Section 9.4, there are two types of Andronov–Hopf bifurcation: supercritical and subcritical. Type 2 excitability is associated with Class 2 models that exhibit a subcritical Andronov–Hopf bifurcation. From a neurophysiological point of view, this bifurcation becomes possible when the inward current at spike initiation is faster than the outward current despite the net depolarizing current being outward at steady state [691].

It is tempting to identify Type 1 excitability with a supercritical Andronov–Hopf bifurcation. However, we have just seen that this interpretation is wrong. Let's see why our intuition is incorrect. Since the fixed point for a Andronov–Kopf bifurcation when $\mu < \mu_c$ are complex eigenvalues, we know that $\lambda = \gamma \pm j\omega$ where $\omega = 2\pi f$. When $\mu = \mu_c$ we have $\gamma = 0$ but $\omega \neq 0$. A fixed point where $\lambda = \pm j\omega$ is a center. This means that at the bifurcation point there exists an oscillation whose frequency is not zero, hence we cannot have Type 1 excitability. This is true whether we have a supercritical or a subcritical Andronov–Hopf bifurcation. Thus another distinction between the two types of Andronov–Hopf bifurcations is related to the stability of the oscillation which arises when $\mu = \mu_c$. If this oscillation is stable, we have a supercritical Andronov–Hopf bifurcation. If it is unstable we have a subcritical Andronov–Hopf bifurcation. In this case, there often exists a second limit cycle which is stable and has a much larger amplitude.

11.3.3 Type 3 excitability

Type 3 neurons are not capable of generating repetitive spiking. A well-studied example of Type 3 excitability is the squid giant axon [136]. Type 3 excitability corresponds to the phase plane shown as Class 3 in Figure 11.5c. In the absence of an injected current, there is one stable fixed point. When a current is injected the v-nullcline moves upward and the fixed point moves to the right. As long as the fixed point remains sufficiently to the left of the local minimum of the v-nullcline, the fixed point remains stable. However, even though the injected current gets larger, the fixed point never appears to the right of the local minimum in the v-nullcline. From a neurophysiological point of view, Class 3 behaviors arise because a fast-activating inward current overpowers the slow-activating outward current during the stimulus transient. This occurs even though the slow-acting outward current dominates during constant stimulation [691].

11.4 Creating and destroying limit cycles

From a mathematical point of view, considerations of how an oscillation starts and how it is destroyed place stringent requirements on the nature of the underlying mechanisms. Bifurcations that create limit cycles are referred to as "global bifurcations" since they involve larger regions of the phase plane than just the neighborhood of a fixed point. Thus the linearization procedure that we used to classify fixed point bifurcations is of no use when it comes to understanding the mechanisms that generate limit cycles. Nonetheless over the last 45 years, mathematical neuroscientists have made tremendous progress toward identifying these mechanisms. In this chapter, we consider the co-dimension 1 bifurcations that create and destroy oscillations in neurons. The term *co-dimension 1* means that changes in a single parameter, μ, are sufficient to cause the bifurcation which occurs when $\mu = \mu_c$. Typically μ is the current input to the neuron. Surprisingly there are only four co-dimension 1 bifurcations that create an oscillation and only four than can destroy it [379, 382]. This means that there are only 16 possible co-dimension 1 bursting neurons.

A detailed mathematical analyses of these bifurcations is beyond the scope of our discussion which focuses on the laboratory investigation of dynamical systems. The important point for our discussion is the realization that the nature of the onset and offset bifurcations can be identified experimentally by asking four questions: 1) Is bistability present? 2) Is there a DC shift in potential[7] 3) Does the spiking frequency change? and 4) Does the amplitude of the spikes change? Below we summarize the relevant experimentally observations for each bifurcation.

11.4.1 Oscillation onset bifurcations

The four co-dimension 1 bifurcations from a fixed point to a limit cycle are 1) saddle node (fold limit cycle) bifurcation, 2) saddle node on invariant circle (SNIC) bifurcation, 3) supercritical Andronov–Hopf bifurcation, and 4) subcritical Andronov–Hopf bifurcation. Figure 11.7 shows examples of the changes in membrane potential associated with each of these bifurcations.

We have already discussed oscillation onset via a SNIC or Andronov–Hopf bifurcation. The saddle-node bifurcation refers to situation in which a stable node coexists with a saddle point and a stable stable limit cycle. As μ increases the node and saddle coalesce to make a saddle node which is annihilated leaving behind the stable limit cycle.

Laboratory identifiers of the saddle-node bifurcation are

1. When $\mu < \mu_c$ small perturbations are damped out exponentially since the stable fixed point is a node.

---

[7] DC shifts refer to very slow changes in membrane potential that are best measured using DC amplifiers.

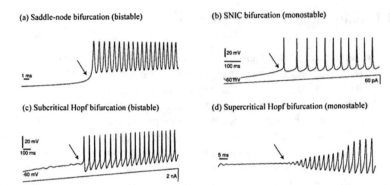

(a) Saddle-node bifurcation (bistable)

(b) SNIC bifurcation (monostable)

(c) Subcritical Hopf bifurcation (bistable)

(d) Supercritical Hopf bifurcation (monostable)

Fig. 11.7 Time courses of the membrane potential for a neuron undergoing four types of bifurcations which produce an oscillation. The two bifurcations on the left are associated with bistability. The current ramp is used in the case of the subcritical Hopf bifurcation (lower left) and the SNIC bifurcation (upper right) to minimize stimulus artifacts [382].

2. When $\mu < \mu_c$ there is bistability since a stable fixed point coexists with a stable limit cycle.
3. There is a DC baseline shift before the bifurcation.
4. The frequency of the limit cycle is nonzero at onset.
5. The amplitude of the limit cycle is fixed and nonzero at onset and does not change for $\mu > \mu_c$.

A saddle-node bifurcation has been observed experimentally for the onset of electrographic seizures [394].

Laboratory identifiers of the saddle node on invariant circle (SNIC) bifurcation are

1. When $\mu < \mu_c$ small perturbations are damped out exponentially since the stable fixed point is a node.
2. When $\mu < \mu_c$ there is no bistability.
3. A DC baseline shift is not present.
4. The frequency of the limit cycle at oscillation onset is 0 and grows as $\sqrt{\mu - \mu_c}$ when $\mu > \mu_c$.
5. The amplitude of the limit cycle is fixed and nonzero at onset and does not change for $\mu > \mu_c$.

A SNIC bifurcation has been observed for pyramidal neurons in the primary visual cortex of rats [382].

Laboratory identifiers of the supercritical Andronov–Hopf bifurcation are

1. When $\mu < \mu_c$ small perturbations produced an oscillatory return to baseline since the stable fixed point is a spiral.
2. There is no DC baseline shift before oscillation onset.

3. The frequency of the oscillation is fixed and does not change for $\mu > \mu_0$.
4. The amplitude of the limit cycle at oscillation onset is 0 and grows as $\sqrt{\mu - \mu_c}$ when $\mu > \mu_c$.

A supercritical Andronov–Hopf bifurcation has been observed for brainstem mesencephalic V neurons [382] as well as in a variety of physiological feedback control mechanisms [276] including pupil light reflex (see Section 9.4).

Laboratory identifiers of the subcritical Andronov–Hopf bifurcation are

1. When $\mu < \mu_c$ small perturbations produce an oscillatory return to baseline since the stable fixed point is a spiral.
2. There can be bistability. Other properties of this bifurcation are discussed in Section 9.4.
3. There is no baseline shift before oscillation onset.
4. The spiking frequency is fixed and does not change when $\mu > \mu_c$.
5. There is a sudden appearance of large-amplitude oscillation as μ crosses a stability boundary at μ_0. There is little further change in the amplitude of the oscullauo as μ increases beyond μ_0.

A subcritical Andronov–Hopf bifurcation has been observed for brainstem mesencephalic V neurons [382] as well as in a variety of physiological feedback control mechanisms [276].

11.4.2 Oscillation destroying bifurcations

There are four co-dimension 1 bifurcations that destroy an oscillation and produce a fixed point: 1) SNIC bifurcation, 2) supercritical Andronov–Hopf bifurcation, 3) fold limit cycle bifurcation, and 4) saddle homoclinic orbit bifurcation. Figure 11.8 shows examples of the changes in membrane potential associated with each of these bifurcations.

We have already discussed two of these bifurcations in the context of oscillation onset. In the case of oscillation offset, a SNIC bifurcation has been observed for pyramidal neurons in layer 5 of rat visual cortex [382], a supercritical Andronov–Hopf bifurcation for excitation block in pyramidal neurons located in rat visual cortex [382] and several physiological feedback control mechanisms [276], a fold limit cycle bifurcation for neurons in rat brainstem mesencephalic V neurons [382] and a saddle homoclinic orbit bifurcatiis accounts for the termination of many types of seizure-like events (SLEs) [394].

The fold limit cycle bifurcation is the easiest to understand since it behaves much like the saddle-node bifurcation for fixed points. The main difference is that the role of the stable and unstable fixed points is replaced by a stable and unstable limit cycle. Thus at $\mu = \mu_c$ two limit cycles appear "out of the clear blue sky": one is stable and the other is unstable. This bifurcation is associated with a co-existent stable fixed point which is usually a stable spiral point. An example of the type

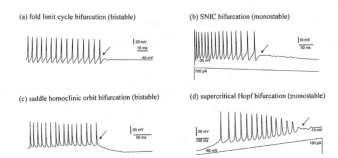

Fig. 11.8 Time courses of the membrane potential for a neuron undergoing four types of bifurcations which destroy an oscillation. The two bifurcations on the left are associated with bistability. The current ramp is used in the case of the subcritical Andronov–Hopf bifurcation (upper right) and the SNIC bifurcation (lower right) to minimize stimulus artifacts [382].

of bifurcation arises in the formation of a stable limit cycle in association with a subcritical Andronov–Hopf bifurcation.

The saddle homoclinic orbit bifurcation closely resembles the SNIC bifurcation. The important difference between these two bifurcations is that in the form the fixed point is a saddle point and in the latter it is a saddle node. Whereas in the SNIC bifurcation the saddle node either disappears or bifurcates into two points depending on the direction of the changes in μ, in the case of the saddle homoclinic orbit bifurcation the saddle point persists. As $\mu \to \mu_c$ the limit cycle becomes a homoclinic orbit equilibrium and the period becomes infinite. After the bifurcation the cycle no longer exists.

Laboratory identifiers for a fold limit cycle bifurcation

1. There is bistability.
2. There is a DC shift in membrane potential.
3. The spike frequency remains for $\mu < \mu_c$, but is zero when $\mu = \mu_c$
4. The amplitude of the spikes can be variable for $\mu < \mu_c$.

Laboratory identifiers for a saddle homoclinic orbit bifurcation: The laboratory identifiers are

1. There is bistability.
2. There is a DC shift.
3. As $\mu \to \mu_c$ from below, the frequency $\frac{1}{\ln|\mu - \mu_c|} \to 0$.
4. Amplitude remains fixed for $\mu = \mu_c$.

This bifurcation has been observed for neurons in the pre-Boltzinger complex of rats [382] and for the offset of electrographic seizures [394].

Example 11.1 Viktor Jirsa[8] and his co-workers [124, 191, 394, 395] used experimental observations to derive an empirical model for the onset and offset of epileptic seizure-like events (SLEs) in mouse, zebrafish, and human brains. They observed that the onset of SLEs was characterized by the abrupt appearance of fast electrical discharges. Since the amplitude and frequency of the discharges did not smoothly ramp up from baseline, the bifurcation for SLE onset must be either a subcritical Andronov–Hopf bifurcation or a saddle-node (fold limit cycle) bifurcation. The presence or absence of a DC baseline shift can be used to distinguish between these possibilities. Experimentally they observed a systematic DC shift from baseline. Consequently they concluded that the onset of a SLE occurs via a fold limit cycle bifurcation.

In order to determine the nature of the bifurcation for SLE offset they noted that a DC shift back toward baseline occurs as the SLE stops. This observation rules out a SNIC or supercritical Andronov–Hopf bifurcation as possibilities. The saddle-node (fold limit cycle) bifurcation maintains a constant spike frequency to seizure offset. However, it is typically observed that spike frequency slows toward seizure offset. Thus, they concluded that seizure offset occurs via a saddle homoclinic bifurcation. Remarkably they were able to experimentally confirm the prediction that the interspike intervals scale logarithmically as seizure offset approaches. They constructed a five variable mathematical model called the "Epileptor" which undergoes a saddle-node bifurcation for oscillation onset and a saddle homoclinic bifurcation for oscillation offset. Subsequently they used this model to monitor the spread of epileptic activity from an epileptic focus in human patients with focal onset seizures [395]. However, it is important to keep in mind that despite the elegance of the derivation of this model, it is a phenomenological model that makes the assumption that both seizure onset and offset are co-dimension 1 bifurcations. This assumption remains to be established. ◇

11.5 AUTO

AUTO is a powerful software package which calculates the bifurcation diagram for ordinary differential equations [175, 202]. The software package XPPAUT introduced in Section 10.9.1 provides a convenient and seamless access to AUTO [202]. It should be kept in mind that the calculation of bifurcation diagrams is a nontrivial task and many pitfalls exist. A software package, DDE-BIFTOOL, is available for determining the bifurcation diagram for delay differential equation [199], but is not discussed further here.

We illustrate how to use AUTO-XPPAUT by determining the bifurcation for the generic form of the saddle-node bifurcation described by

$$\dot{x} = \mu + x^2. \tag{11.10}$$

---

[8] Viktor K. Jirsa (b. 1968) is a German physicist and neuroscientist.

For more practice see Exercises 5–9 at the end of the chapter. The steps for determining the bifurcation diagram shown in Figure 5.2 are:

1. Write an .ode program to integrate (11.10).

2. It is very important to ensure that the solution has settled onto either a fixed-point or a limit cycle before using AUTO to continue the solution as a parameter changes. In the case of (11.10), we know that a fixed-point solution $(x^* < 0)$ exists for $\mu < 0$. Pick the initial condition $x = -1$ and run XPPAUT. The most common cause of failure using AUTO is to use this program when the solution is not already on a branch. With this in mind we extend the integration by clicking on Initialconds Last. This XPPAUT command extends the integration by an amount defined by TOTAL, from the step at which the previous integration finished. Depending on the differential equation, the choice of initial condition, and the parameter values it may be finished to click on Initialconds Last several times.

3. The AUTO window is activated by clicking on Auto which is located under File. The most important functions for our purposes are Parameter, Axes, Numerics, Run, and Grab. A complete description of all of the commands in AUTO is provided in the online user's guides cited above. The circle in the lower left of the window summarizes the number of stable and unstable eigenvalues for a given choice of the bifurcation parameter(s): the number of crosses inside the circle corresponds to the number of stable eigenvalues (i.e., eigenvalues with negative real part) and the number outside the circle corresponds to the number of unstable eigenvalues (the number of eigenvalues with positive real part). How many crosses do you see for (11.10) and are they stable or unstable? How many crosses did you expect to see and why?

4. Click on Parameters. It is possible to declare up to 5 bifurcation parameters in this box. By default, XPPAUT chooses the first five parameters in the *.ode program. In our case there should be just one bifurcation parameter declared. Is this correct?

5. Click on Axes and choose HiLo. Most of this box will already be filled out correctly. We see that the variable X will be plotted on the Y-axis and we are able to choose the range to vary the bifurcation parameter. Note that $X_{min}, X_{max}, Y_{min}, Y_{max}$ refer to the dimensions of the bifurcation diagram and not the variables and thus X_{min}, X_{max} refers to the range over which μ is varied. We suggest that you choose $X_{min} = -3$ and $X_{max} = 1$. The values Y_{min}, Y_{max} refer to the maximum and minimum values of a periodic orbit. Since (11.10) does not possess a periodic orbit, these values are irrelevant for our purposes (see Exercise 2). Once you have completed the menu press OK.

6. Click on Numerics. For our purposes, we can accept all of the default values except those for Par Min and Par Max. Choose these values so that μ is varied from Par Min = -3 to Par Max =1, then push OK.

7. Click on `Run Steady state`. A thick black line will appear that extends over the range that we varied μ. The fact that the line is thin means that the fixed-point is stable (a thin black line indicates an unstable fixed-point). If you do not see a line then the most likely cause is that you were not sufficiently close to the stable fixed-point solution when you started. Go back to Step 2.

8. Click on `Grab`. You will see a cross appear on the thick black line. You can use the right and left arrow keys to move back and forth along this line. As you move along this line in this way what happens to the eigenvalues (see circle in lower left hand corner). Also if you look in the window below the bifurcation diagram you can see more information pertaining to the current location of the movable cross in the diagram, such as the value of μ and, in the case of a limit circle, its period. Note that there are also crosses with numbers attached which do not move as you use the arrow keys. The points are special points since it is possible to continue the solution from them. Using the `Tab` key, you can move between the special points. When you are on one of these special points, the bottom window provides information the point type. In our case you will see `EP` which means end point. If you see `MX` then this means you made a mistake and it is back to step 2 that you go!

9. In order to complete the bifurcation diagram, it is necessary to continue to the solution from one of the numbered special points. Use the `Tab` to position the cross on the special point labeled 1. Press `Enter`. Click on `Numerics` and change the sign of `Ds`. Note that when we change the sign of `Ds` we reverse the starting direction for AUTO. Now click on `Run` and you should see the bifurcation diagram shown in Figure 5.2.

Example 11.2 AUTO can be used to characterize the bifurcations in (11.9) (see Exercise 8). As shown in Figure 11.9a, (11.9) exhibits two different Hopf bifurcations when the magnitude of the stimulating current is the bifurcation parameter μ. The bifurcation that occurs at the lower value of μ is a subcritical Andronov–Hopf bifurcation, whereas the bifurcation at higher μ is a supercritical Andronov–Hopf bifurcation. In addition to the two Andronov–Hopf bifurcations, there are two-fold limit cycle bifurcations. Can you identify where they are? Figure 11.9b shows that if the magnitude of the stimulating current is adjusted to be in the range of the subcritical Hopf bifurcation, then a brief pulse can abort the spiking pattern. $\diamondsuit$

11.6 Reduced Hodgkin–Huxley Models

Many functions of the nervous system, such as cognition, consciousness, and the skilled performance of complex voluntary movements, cannot be ascribed to the behaviors of single neurons, but likely emerge from the properties of large populations of neurons.

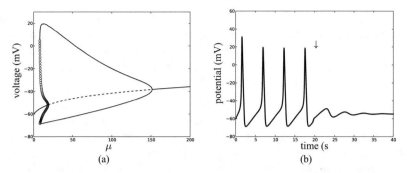

(a)   (b)

Fig. 11.9 (a). Bifurcation diagram for the Hodgkin–Huxley equation for a neuron determined using AUTO. The bifurcation parameter μ is the magnitude of the stimulating current. See Exercise 8 to see how this diagram was constructed. (b) Stopping a regular spiking Hodgkin–Huxley neuron by applying a single 10-mA pulse for 0.4 s ($\downarrow$).

How many neurons do we need to scale up to the basic functional units of the cortex? The neocortex is composed of vertically oriented *cortical hypercolumns*, each containing $\approx 10^3 - 10^4$ neurons [446, 612]. About 10^6-10^7 columns make up the neocortex. These cortical columns are thought to represent the functional unit of information processing in the awake brain [355, 446, 467]. From a neurophysiological perspective, a column represents a population of neurons having a common input that collectively perform a function. However, the neurons within the column can exhibit different firing patterns. Thus, to answer our question, we need at the very least to include $10^3 - 10^4$ neurons just to model a single cortical column.

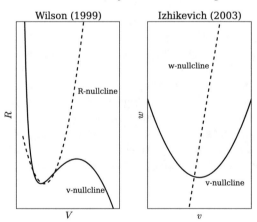

Fig. 11.10 (a) Nullclines for the reduced neuron model developed by Hugh R. Wilson [879, 880]. (b) Nullclines for the reduced neuron model developed by Eugene Izhikevich [380, 381].

A reasonable approach to describing the dynamics of a neural population of this size or larger is to use computer simulations. Since we are attempting to model a very large and complex system, we are going to have to be concerned with the efficiency of our algorithms, implementations, and computer hardware. A measure of the computational load associated with a given computer implementation is the number of floating-point operations (addition, subtraction, multiplication, etc.) performed per second. "Floating-point operation(s) per second" is abbreviated FLOPS. The computationally most efficient model of a neuron to implement in a computer program is the integrate-and-fire neuron model [381] (≈ 5 MFLOPS, where M stands for "mega," i.e., 1 MFLOPS is 10^6 FLOPS). However, an integrate-and-fire model can generate only tonic spiking at a constant frequency. In other words, although this neuron model is computationally efficient, it is too simple. At the other end of the spectrum, we have the HH-type equations for a neuron, which require ≈ 1200 MFLOPS [381]. The use of such models increases the computational load by a factor of 240, which is prohibitive if our goal is to construct neural networks having 10^5 to 10^{11} neurons.

The above observations motivated neuroscientists to search for simple models of neurons that could generate a wide range of spiking patterns but at the same time possess a small computational load [380, 381, 879, 880]. These approaches were motivated by two observations: 1) the typical shapes of the "fast" and "slow" nullclines (Figure 11.3), and 2) the observation that the resting membrane potential of neurons typically corresponds to the fixed point located near the left knee of the cubic nonlinearity. These observations motivated two approaches. The first attempt was made by Hugh R. Wilson and approximated the w-nullcline as a parabola while maintaining the cubic nature of the v-nullcline (Figure 11.10b). The second approach championed by Eugene Izhikevich[9] approximates the v-nullcline as a quadratic function and the w-nullcline as a linear function (Figure 11.10a).

11.6.1 Wilson-type reduced neurons

Wilson's model takes advantage of the observation that Na^+ and K^+ are the primary contributors to the action potential for neo-cortical neurons [879, 880]. Thus (11.9) can be reduced to

$$C\frac{dV}{dt} = -m_\infty(V)(V - E_{Na}) - g_K R(V - E_K) + I \qquad (11.11)$$
$$\frac{dR}{dt} = \frac{1}{\tau_R}(-R + R_\infty(V))$$

where

---

[9] Eugene Izhikevich (b. 1967), Russian-born American mathematical neuroscientist.

$$m_\infty = 17.8 + 47.6V + 33.8V^2$$
$$R_\infty = 1.24 + 3.7V + 3.2V^2.$$

Wilson made the following assumptions:

1. Na^+ channel activation is assumed to be sufficiently fast so that the concentration of Na^+ can be described by its equilibrium value, m_∞.
2. The dynamics of the Na^+ channels have no inactivation variable. It should be noted that human and mammalian neo-cortical neurons contain Na^+ channels that do not activate. Consequently, the Wilson model is designed to account for the neural dynamics of neocortex.
3. The membrane time constant, τ_R, is independent of V.
4. There is no leakage current. On closer inspection, it can be seen that the leakage current in Wilson's model is actually absorbed in the Na+ and K^+ currents. This assumption can be justified by assuming that E_L lies between E_{Na} and E_K.

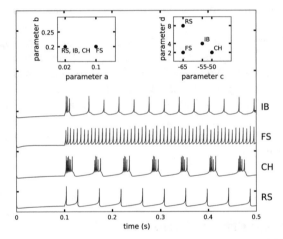

Fig. 11.11 Four neuronal firing types generated by the reduced neuronal model for the Izhikevich-type reduced neurons [380, 381]: regular spiking (RS), fast spiking (FS), intrinsic bursting (IB), and chattering (CH). A single voltage trace for each neuron type with a constant current input is shown in the lower four panels; the choices a, b, c, d of the four parameters for each spiking type are shown in the insets. This figure was prepared by the late John D. Hunter.

The V-nullcline is the cubic nonlinearity

$$R = \frac{-m_\infty(V)(V - E_{Na}) + I}{g_K(V - V_K)}$$

and the R nullcline is the parabola

$$R = R_\infty := 1.24 + 3.7V + 3.2V^2.$$

The resultant model is an improvement over the FHN model for a neo-cortical neuron. In order to describe the dynamics of bursting neurons, Wilson added a Ca^{++} current and a slow Ca^{++}-mediated K^+ hyperpolarizing current. This model is an improvement over the Hindmarsh–Rose model for a bursting neuron.

11.6.2 Izhikevich-type reduced neurons

Izhikevich [380, 381] showed that all of the spiking patterns shown in Figure 11.11 could be generated by the model

$$\frac{dv}{dt} = 0.04v^2 + 5v + 140 - w + I_{ext}, \quad \frac{dw}{dt} = a(bv - w), \qquad (11.12)$$

together with the after-spiking reset condition

$$\text{if } v \geq 30 \text{ mV}, \quad \text{then } \begin{cases} v \leftarrow c, \\ w \leftarrow w + d, \end{cases}$$

where v, w are variables, a, b, c, d are parameters, and I_{ext} is an external current. In fact all of the spiking patterns shown in Figure 11.1 can be generated by this model (for parameter choices see [381, 382]). The computational load for the Izhikevich model neurons is only ≈ 13 MFLOPS [381]. Currently, large-scale models of neural networks employ this model [383, 854]. This type of reduced model emphasizes the subthreshold dynamics of the neuron and places less emphasis on getting the shape of the action potential correct. This is because it retains detailed information about the left knee while simplifying the vector field outside of this neighborhood. In contrast the Wilson model does a better job describing the subthreshold dynamics and shape of the action potentials. However, the Wilson models are about 10 times less computationally efficient than the Izhikevich model. Using Izhikevich-type models it was possible to simulate the dynamics of a realistic, large-scale corticothalamic population of neurons [383].

11.7 What Have We Learned?

1. Why is a limit cycle a more robust explanation for a biological oscillator than a center?

   A limit-cycle oscillation is much more robust to the effects of perturbations than the oscillation associated with a center. It also occurs in the dissipative dynamical systems studied experimentally.

2. What can be said about the nature of the fixed point that is surrounded by a limit cycle?

We would anticipate that the fixed point would be a spiral point. Intuitively, we would expect it to be an unstable spiral point, as we observed for the van der Pol oscillator. However, as the next item shows, it could be a stable spiral point.

3. How can a limit cycle occur when the fixed point is stable?

It is possible for a limit cycle to surround a stable spiral point. For this to occur, there needs to be some form of instability once perturbations away from the fixed point become large enough. This instability typically takes the form of an unstable separatrix (likely associated with a saddle point): for initial conditions inside the separatrix, trajectories spiral toward the fixed point, while outside, they spiral toward the limit cycle. This occurs, for example, in the case of a subcritical Andronov-Hopf bifurcation for a neuron.

4. An ion channel is often modeled by an equivalent RC circuit. Why are the RC circuits arranged in parallel rather than in series?

The RC circuits are arranged in parallel since each ion channel is separate from another ion channel.

5. What is the key difference between the resistance in an electrical circuit and the conductance of an ion channel?

The resistance in an electrical circuit is a parameter, but the conductance of an ion channel is a variable.

6. How does the capacitance affect the flow of charges across the neuronal membrane?

The capacitance places a lower limit on the number of ions (charges) that must move and how rapidly they must move to produce a given voltage signal. Thus it slows the voltage response to any current by a characteristic time t_c that depends on the product of the capacitance and the effective resistance (see below).

7. What is the relationship between the current, I, and the charge, Q?

$I = dQ/dt$.

8. What is a voltage clamp?

Due to the large diameter of the squid axon, it was possible for Hodgkin and Huxley to thread two thin electrodes inside the axon. One of the electrodes was used to measure the potential, and the other was used to inject current. The "clamp" corresponds to the electronic feedback that controls the injected current to maintain a predetermined fixed voltage across the membrane. The injected current is the mirror image of the current generated by ion flows across the neural membrane at that potential.

9. What is the typical shape of the fast and slow nullclines for neurons?

The "fast" nullcline is a cubic and the "slow" nullcline is a sigmoid.

10. What does the phrase "conductance-based model for a neuron" mean?

The system of nonlinear differential equations is developed to describe the electrical characteristics of the excitable cell membrane and its ion channels.

11. Although the Hodgkin–Huxley equation was developed to model to squid giant axon, it does not accurately describe the dynamics of this axon. Explain.

> (11.9) predicts repetitive spiking, whereas the squid giant axon typically produces a single spike and then is quiescent. It has been shown that a modification of a single parameter description of the K^+ channel (11.9) is sufficient to explain the observed behavior of the squid giant axon [134, 136]. Repetitive spiking can be produced by placing the squid giant axon in a low Ca^{++} medium [303] or a solution with a higher NaCl concentration that normal seawater [546] or using much higher, non-physiological sustained depolarizing currents which produce irreversible changes in the axon [115].

12. What is the difference between an ionotropic and a metabotropic ion channel?

> For an ionotropic ion channel the ion channel and the receptor which activates the ion channel are located in the same protein complex. In the case of a metabotropic ion channel the ion channel is opened indirectly through a cascade of intracellular events. Thus the ion channel and its receptor are not located in the same place.

13. What is the mathematical requirement for excitability?

> In mathematical models for neural dynamics, neurons are excitable because "they are tuned" close to a bifurcation from a rest state to a limit cycle attractor.

14. What is the mathematical interpretation of Type 1, Type 2, and Type 3 excitability?

> The three types of excitability correspond to three different ways that the v and w nullclines intersect.

15. What does the term "co-dimension 1" mean?

> A co-dimension 1 dynamical system is a system in which changes in a single parameter are sufficient to cause a bifurcation.

16. What types of co-dimension 1 bifurcations involve a saddle node?

> The co-dimension 1 bifurcations that involve a saddle node are 1) saddle-node (fold limit cycle) bifurcation, 2) saddle-node on invariant circle (SNIC) bifurcation and 3) saddle-node homoclinic orbit bifurcation.

17. The resting potential for a neuron integrator behaves like what kind of fixed point?

> A stable node.

18. The resting potential for a neuron resonator behaves like what kind of fixed point?

> A stable spiral point.

19. What are the types of bifurcations of equilibria that produce limit-cycle oscillations in excitable systems?

> The four co-dimension 1 bifurcations that produce an oscillation are: 1) saddle-node (fold limit cycle) bifurcation, 2) SNIC bifurcation, 3) subcritical Hopf bifurcation and 4) supercritical Hopf bifurcation.

20. What are the types of bifurcations of limit cycles that eliminate the oscillations in excitable systems?

The four co-dimension 1 bifurcations that can destroy an oscillations are: 1) saddle node (fold limit cycle) bifurcation, 2) SNIC bifurcation, 3) saddle homoclinic orbit bifurcation, and 4) supercritical Hopf bifurcation.

21. What experimental measurements are useful for determining the type of bifurcation that produces and eliminates a limit-cycle oscillation?

The experimental observations useful for identifying the type of bifurcation which leads to oscillation onset are 1) response of the resting membrane potential to low amplitude, depolarizing impulses, 2) the presence or absence of bistability, 3) the amplitude and frequency of the oscillation when $\mu = \mu_0$, and 4) the dependence of the oscillation amplitude and frequency when $\mu > \mu_0$.

22. How many co-dimension 1 bursting neurons are possible?

There are 4 co-dimension 1 bifurcations that create an oscillation and 4 that destroy an oscillation. Thus there are 16 possible co-dimension 1 bursting neurons.

23. What is AUTO?

A computer program that can be used to determine the bifurcation diagram for an ordinary differenial equation.

24. How is the computational load associated with numerically integrating a differential equation measured?

A simple measure of computational load is to determine the number of floating-point operations required to perform the calculation. The term FLOPS refers to the number of floating-point operations (addition, subtraction, multiplication, etc.) performed per second.

11.8 Exercises for Practice and Insight

1. Write a computer program to investigate the properties of the Fitzhugh–Nagumo equations (See (5.2) in Section 5.1).

   a. Take $a = 0.25$, $I_a = 0$, and $I_{ext} = 0$. Determine b and γ such that there is one stable critical point. Pick two initial conditions to demonstrate that the system is excitable for this parameter choice.
   b. Take $a = 0.25$ and $I_a = 0$. Determine b and γ such that there are three critical points, two of which are stable and one unstable. Verify your predictions by choosing two initial conditions to demonstrate the bistability.
   c. Now assume that $I_{ext} > 0$. Choose b and γ such that a limit cycle exists. Prepare the phase diagram and plot both the limit-cycle solution and the nullclines. Label the "fast" and "slow" parts of the limit-cycle trajectory.

2. An investigator observes that she can stop an oscillation by the application of a single carefully chosen perturbation. Draw an example of a phase-plane diagram that would be consistent with this observation.

3. A very useful model for understanding limit cycles is the Poincaré oscillator

$$\frac{dr}{dt} = r(k_1 - r), \quad \frac{d\theta}{dt} = k_2.$$

Here (r, θ) are polar coordinates (recall that a limit cycle describes a closed trajectory in the phase plane, e.g., a circle). These equations describe a system moving with constant angular velocity $d\theta/dt = k_2$ along the closed trajectory. The Poincaré oscillator is of fundamental importance, because it is possible to determine the dynamics both analytically and qualitatively (in contrast to, for example, the van der Pol oscillator, which can be determined only qualitatively). Determine the dynamics of the Poincaré oscillator. What happens when $k_1 < r$? when $k_1 > r$? when $k_1 = r$? Hint: Since the two equations are not coupled, they can be considered separately.

4. An example of a dynamical system that possesses a supercritical Andronov–Hopf bifurcation is the Poincaré oscillator [810]

$$\frac{dr}{dt} = \mu r - r^3, \quad \frac{d\theta}{dt} = \omega + k_1 r^2,$$

where r, θ are polar coordinates, μ controls the stability of the fixed point at the origin, ω gives the frequency of oscillations, and k_1 gives the dependence of frequency on amplitude. Confirm that at instability, a pair of complex eigenvalues crosses the imaginary axis from left to right. Hint: Rewrite this equation in Cartesian coordinates by defining $x = r\cos\phi$ and $y = r\sin\phi$.

5. Use XPPAUT to determine the bifurcation diagram for the transcritical bifurcation whose generic equation is

$$\frac{dx}{dt} = \mu x - x^2.$$

6. Use XPPAUT to determine the bifurcation diagram for the supercritical bifurcation whose generic equation is

$$\frac{dx}{dt} = \mu x - x^3.$$

7. Use XPPAUT to determine the bifurcation diagram for the subcritical bifurcation whose generic equation is

$$\frac{dx}{dt} = \mu x + x^3.$$

8. In this exercise, we show how to use AUTO to determine the bifurcation diagram for the Hodgkin–Huxley equation for a neuron. The steps to make this bifurcation diagram are

a. Download hh.ode from https://cnd.mcgill.ca/ebook/index.html. Note that this equation has more parameters that can fit in the shown parameter box shown in XPPAUT. You can maneuver up and down the parameter list by using the ∨, ∨∨, ∧, ∧∧ buttons located on the parameter box.

b. Run XPPAUT to integrate this equation when the parameter `curbias=0`. Use the `Initialconds Last` button several times to ensure that the HH equation has settled onto the initial conditions of v, m, h, n (we obtained, respectively, $-59.9996379, 0.052955, 0.5959941, 0.317732$).

c. After starting `AUTO` set the following parameters:

- In the `Parameter` window set the first parameter to be `curbias` (the menu will show `blockna` which is the first parameter declared in hh.ode).

- In the `Axes` window click on `Hi-Lo`. The y-axis should be V and the x-axis (called `MainParm`) should be `curbias`. Set $X_{min} = 0$ and $X_{max} = 200$ so that `curbias` ranges from 0 to 200. Set $Y_{min} = -80$ and $Y_{max} = 20$ so that the amplitude of the Limit-cycle oscillation ranges from -80 to 20.

- In the `Numerics` menu set `Par min = 0` and `Par max = 200` (this sets the range over which the bifurcation parameter `curbias` will be varied). Set `Nmax=500` to limit the number of points that will be computed along a given branch of the bifurcation diagram to 500. In addition set `Npr=500` and `Norm Max =150`.

- Click on `Run` and then `Steady state`. You will see a solid with 4 special points.

- Click on `Grab` and use the `Tab` to move between them. From the bottom screen, we see that special points 1 and 4 are end points (EP) and special points 2 and 3 are labeled as Hopf bifurcation points (HB). Note that the line between special points 2 and 3 is thin meaning that the fixed-point that exists for this range of `curbias` values is unstable. Confirm this by looking at the circle diagrams in the lower left hand corner of the AUTO screen.

- We can complete the bifurcation diagram by either starting at special point 2 and increasing `curbias`, or by starting at special point 3 and decreasing `curbias`. Let's start at special point 2. Use the `Tab` key to place the cross at this point and then press `Enter`. Note that the `Run` menu changes. Click on `Periodic` and the bifurcation diagram unfolds before your eyes!

- XPPAUT (and mathematicians) typically use lines to indicate the stability of fixed-points (remember that thick lines correspond to stable fixed-points and thin lines to unstable ones). Limit cycles are indicated by circles placed which are positioned to indicate the maximum and minimum values of the oscillation: filled circles correspond to stable limit cycle oscillations and open circles to unstable limit-cycle oscillations.

9. Consider the Fitzhugh–Nagumo equation

$$\frac{dv}{dt} = v - \frac{v^3}{3} - w + I_{ext}$$

$$\frac{dw}{dt} = \frac{v + a - b * w}{\tau}$$

where $a = 0.7$, $b = 0.8$ and $\tau = 12.5$. Note that this way of writing the Fitzhugh–Nagumo equation is slightly different from what we used in the previous problem. Write a XPPAUT program to integrate this equations and use AUTO to determine the bifurcation diagram. The bifurcation parameter is I_{ext}. Use the initial conditions $v(0) = 0.1$ and $w(0) = 0.0$ and we recommend the choices: $TOTAL = 500$, $dt = 0.005$ and $maxstor = 5000000$.

10. The Morris–Lecar model [607] is convenient for investigating excitability [691]. It can be written as

$$CdV/dt = I_{stim} - \overline{g}_{fast}m_\infty(V)(V - E_{Na}) - \overline{g}_{slow}w(V - E_K) - g_{leak}(V - E_{leak})$$

$$dw/dt = \phi_w \left[\frac{w_\infty(V) - w}{\tau_w} \right]$$

where

$$m_\infty(V) = 0.5 \left[1 + \tanh\left(\frac{V - \beta_m}{\gamma_m} \right) \right]$$

$$w_\infty(V) = 0.5 \left[1 + \tanh\left(\frac{V - \beta_w}{\gamma_w} \right) \right]$$

$$\tau_w(V) = 1/\cosh\left(\frac{V - \beta_w}{2\gamma_w} \right)$$

and $E_{Na} = 50$ mV, $E_K = -100$mV, $E_{leak} = -70$mV, $\overline{g}_{fast} = 20$ mS/cm^2, $\overline{g}_{slow} = 20$ mS/cm^2, $\overline{g}_{leak} = 2$ mS/cm^2, $\phi_w = 0.15$, $C = 2\mu$F/cm^2, $\beta_m = -1.2$ mV, $\gamma_m 18$mV and $\gamma_w = 10$mV. The advantage of using the Morris–Lecar model in this form with these parameter values is that the three classes of excitability can be readily studied by changing the value of only one parameter, namely β_w (see legend to Figure 11.5).

a. Determine the equations for the V and w nullclines.
b. Write a computer program to reproduce panels a), b), and c) in Figure 11.5.

Chapter 12
Characterizing and Manipulating Oscillations

Oscillations are characterized by their amplitudes and frequency content. As we saw in Section 8.3.3, determination of the power spectrum is a powerful technique to describe oscillations. In chronobiology, circadian rhythms are fit to a sinusoid with period 24 hours using cosinor analysis [146]. However, in many applications we are not satisfied with merely describing the oscillations. Here we focus on how the oscillation responds to perturbations.

A variety of techniques are available to characterize the responses of oscillators to perturbations [276, 882]. These techniques are very much in the spirit of our "black box" approach to stability, in that we apply perturbations (inputs) to a limit-cycle oscillator (black box) and see what happens (output). Three of these techniques are particularly useful for laboratory investigations: the determination of the Poincaré section, phase-resetting curves, and phase-locking diagrams. All of these techniques exploit the generic properties of stable limit-cycle oscillators. Here the term "generic" refers to properties that all limit-cycle oscillators possess. Thus a basic knowledge of these techniques is very useful in designing experiments to investigate the properties of biological oscillators. We take advantage of research related to falling, human balance control and excitable cells to illustrate these applications.

The questions we will answer in this chapter are

1. What is the importance of a Poincaré section?
2. What is the phase-resetting curve and why is it important?
3. What is the difference between phase-locking between two oscillators and synchronization between two oscillators?
4. What is an Arnold tongue diagram and why is it important?
5. Why does the green alga *Nitella* generate an action potential?
6. What is the optimal stimulus for generating a spike in a Hodgkin–Huxley neuron?
7. What is dynamic clamping?

© Springer Nature Switzerland AG 2021
J. Milton and T. Ohira, *Mathematics as a Laboratory Tool*,
https://doi.org/10.1007/978-3-030-69579-8_12

8. What is the "curse of dimensionality?"

12.1 Poincaré Sections

In the phase plane, a limit cycle corresponds to a closed curve, such as a circle. Suppose we could carefully place a section, referred to as the Poincaré section, directly in line with the trajectory of the limit cycle. Then the limit cycle would impact the section once per period T (Figure 12.1). If we use a single variable X to represent some property of the limit cycle, say the amplitude, then we can calculate its value as it hits the Poincaré section the next time, at $t + T$, using a discrete time map

$$X_{t+T} = f(X_t), \tag{12.1}$$

where $f(X_t)$ is often called the *first return map*.

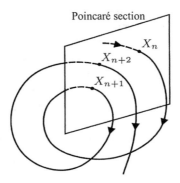

Fig. 12.1 Schematic representation of a Poincaré section for a dynamical system. The two trajectory crossings that are seen in the lower part of the figure do not represent a violation of uniqueness. Since the limit cycle is by assumption subjected to perturbations, it is necessary to set the time axis perpendicular to the page of the book to show when the perturbations occur. Thus the appearance of a crossing is simply an artifact produced by plotting a three-dimensional dynamical system in two dimensions.

If the limit cycle were not subject to random perturbations, it would always hit the section at the same point in Figure 12.1. In other words, a fixed point of (12.1) corresponds to the limit cycle in the 2-dimensional phase plane. However, since random perturbations are always present, the values of X_t for different returns to the Poincaré section typically differ from one another. These fluctuations can be used to investigate the stability of the underlying dynamical system. In mathematical models, it can be quite difficult to place the Poincaré section appropriately. However, the first return map can often be readily measured in a laboratory setting. Much-studied examples include self-paced walking over a level walking surface (see below) and periodic forcing of excitable cells (Figure 12.13).

12.1.1 Gait Stride Variability

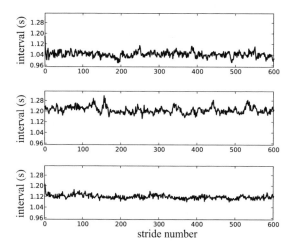

Fig. 12.2 Gait stride variability for self-paced walking for three healthy subjects. The subjects made two laps (800 m) on the outside lanes of an athletic track. Data were collected with the assistance of Adam Coleman (Claremont McKenna College), Joey Haber (Pitzer College), and Christopher Ruth (Pitzer College).

Self-paced walking refers to walking that occurs automatically without much thought. Of course, we can voluntarily choose to walk by purposely controlling the movements of our limbs (see, for example, Monty Python's Ministry of Silly Walks on YouTube). That is not what we consider here. Self-paced walking for healthy, active individuals is remarkably rhythmic. Indeed, the coefficient of variation (CV), namely the standard deviation divided by the mean, is typically $< 2\%$. This observation suggests that the gait cycle behaves like a limit-cycle oscillator. However, it is not presently known to what extent this rhythm is controlled actively by a central pattern generator [98, 428] or passively by the biomechanical properties of the musculoskeletal system [141, 452]. In either case, the limit cycle interpretation for self-paced walking implies that the times at which the feet contact the ground provide a measure of the first return map for the gait cycle.

The first return map for self-paced gait can be measured by taping a force-sensitive electrode (FSR) to the insole of the shoe [327]. An FSR is a conductive-polymer-layer sensor that changes its resistance when loaded (FSRs are commercially available and inexpensive). Thus changes in voltage between successive contacts of the foot with the ground can be used to determine the stride interval time. A *stride* is defined as successive contacts of the same foot with the ground; hence two steps equal one stride.

Figure 12.2 shows the gait stride intervals measured while healthy, active subjects walk around a running track. Despite the fact that the CV is small, there are occasional larger fluctuations in the stride interval. It has been observed that the CV for gait strides is increased in the elderly, particularly in elders with a history of falling [325], and in patients with degenerative diseases of the basal ganglia, such as Parkinson's disease [324]. Readers can confirm this observation by downloading gait stride data from PhysioNet. However, the first return map for gait strides typically appears as a scatter of points with, to the eye, no clear temporal structure. Consequently, the dynamics of gait stride variability tend to be discussed in terms of the mathematical methods developed in Chapters 15 and 16.

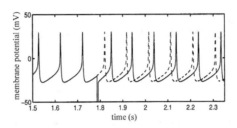

Fig. 12.3 Effect of a single −15-nA, 5-ms hyperpolarizing square-wave pulse on a repetitively spiking, slowly adapting motoneuron of the sea slug *Aplysia*. A 3-nA depolarizing current was injected into the motoneuron to cause it to fire repetitively. Figure reproduced from [238] with permission.

12.2 Phase Resetting

Many biological oscillators exhibit a remarkably stereotyped behavior in their response to single brief perturbations [276, 617, 882]. Figure 12.3 illustrates this phenomenon for a periodically spiking *Aplysia* motoneuron that receives a brief hyperpolarizing current pulse [238]. We see that the interspike interval (ISI) when the pulse was delivered is longer than that of the unperturbed ISI. However, the subsequent ISIs are the same length as that observed prior to the perturbation. This observation has been repeated in many contexts, including the effects of electrical pulses on periodically spiking excitable cells [296, 386, 684], the effect of light pulses on the emergence of *Drosophila* pupae [882], mechanical perturbations of essential tremor [799], the effects of stumbles on the gait cycles of cockroaches [392] and humans [645], and the effects of shift work [156] and jet lag [155, 163] on the human sleep–wake cycle.

The following are some common features of experiments of this type [276]:

1. A single stimulus may lead to either a lengthening (positive resetting) or shortening (negative resetting) of the perturbed cycle length. This depends on both the magnitude of the stimulus and the point during the cycle at which the stimulus was delivered.
2. Following a perturbation, the rhythm is generally reestablished quickly, often by the following cycle. The reestablished rhythm has the same amplitude and

frequency as the rhythm before the stimulus was applied. The lone exception occurs when a single pulse abolishes the rhythm.

3. Although the rhythm is reestablished following a perturbation, subsequent maxima occur at different times from what would have been observed in the absence of the perturbation (compare the solid and dotted lines in Figure 12.3).

4. The graph of the length of the interval that received no stimulus as a function of the point during the cycle at which the stimulus was delivered may be continuous or show apparent discontinuities.

The study of the responses of biological limit cycles to brief pulses was pioneered by the late Art Winfree[1] [882–884] and championed subsequently by his colleagues Leon Glass [273, 276] and Steven Strogatz[2] [810, 811]. These applications to biological systems are based on over 100 years of mathematical analyses, including pioneering work by Poincaré and Arnold.[3] In principle, these methods can be applied to any situation in which the existence of a limit cycle is suspected. The basic concept is that in the phase plane, a limit cycle is represented by a closed curve. As the dynamical system evolves in time, the dynamics revolve around the closed curve in the phase plane. Since every closed curve is topologically equivalent to a circle, it is convenient to use the concept of *phase*, denoted by ϕ. Pick an arbitrary time t_0 on the limit cycle and assign it to $\phi = 0$. Then the phase ϕ_p for every other time t_p on the cycle can be determined as

$$\phi_p := \frac{t_p - t_0}{T} \quad (\text{mod } 1), \tag{12.2}$$

where T is the period of the unperturbed oscillation. Thus phase is a number between 0 and 1, which indicates a unique location on the cycle.

On the limit cycle (circle), the phase is constantly increasing with time, and we can write the associated differential equation

$$\frac{d\phi}{dt} = \frac{1}{T}. \tag{12.3}$$

The solutions are of the form

$$\phi(t) = \frac{1}{T}t + \phi(0). \tag{12.4}$$

Different solutions differ only in their respective values of $\phi(0)$ and are said to be out of phase with one another. Now suppose that we pick an arbitrary point (x_0, y_0) in the phase plane located in the basin of attraction for the limit cycle.[4] We can determine ϕ by waiting until the trajectory returns to the circle (steady-state

---

[1] Arthur Taylor Winfree (1942–2002), American theoretical biologist.

[2] Steven Henry Strogatz (b. 1959), American mathematician

[3] Vladimir Igorevich Arnold (1937–2010), Russian mathematician.

[4] Points that are not in the basin of attraction for the limit cycle form the set of *phaseless points*. For example, for the supercritical Hopf bifurcation, the basin of attraction is everything except the unstable fixed point. In this case, the set of phaseless points contains one point. For the subcritical

solution) and then use (12.2). In this way, we can assign a phase to every point in the basin of attraction for the limit cycle. We expect that points very close together will have phases that are very close together. Thus we can construct contour lines in the phase plane, referred to as *isochrons*, which group together points with the same phase (Figure 12.4).

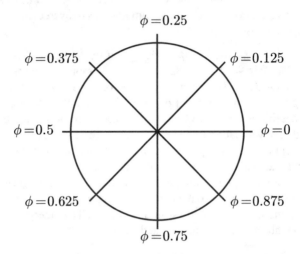

Fig. 12.4 Schematic representation of the isochrons for a simple Poincaré-type limit-cycle oscillator [276]. In general, the arrangement of isochrons is much more complex [882].

What happens when a limit-cycle oscillation is momentarily perturbed by a single brief perturbation? Designate by $\phi_{initial}$ the phase of the limit cycle at which the perturbation was delivered. Suppose we consider that the perturbation does not take us out of the basin of attraction for the limit cycle. Then, after some time, we expect that the trajectory will again be on the cycle, or very close to it. Does this mean that nothing happens in the long term? Not quite. The phase of the solution may have changed. More precisely, if the point in the phase space to which the system is perturbed has been assigned an effective phase, then when the trajectory returns to the cycle, we know what its phase $\phi_{final}(0)$ will be. We can say that the phase has been reset from $\phi_{initial}$ to ϕ_{final}. At this point, we will dispense with the $\phi(0)$ notation and consider the instantaneous phase ϕ of the oscillator, and hence the perturbation causes a change from ϕ to ϕ'.

If the perturbation occurs when the oscillator is at phase ϕ and takes almost no time, we can say that the phase has been instantly reset to a new phase ϕ'. There are two ways to determine ϕ'. First, if we can visualize the phase plane, then we can read ϕ' from the isochrons (Figure 12.4). However, in the laboratory, we typically see only one variable of the oscillator as a function of time, and hence we do not have access to the whole phase plane. Thus we must use another method to determine ϕ':

Hopf bifurcation, the basin of attraction is everything outside of the separatrix that separates the limit cycle from the stable fixed point.

we wait until we have the next clear instance of $\phi = 0$ and then work backward to determine the phase right after the perturbation. Note that if we consider discrete cycles to be between any two $\phi = 0$ events, then we can determine the cycle length for any cycle. Normally, this is just the period T of the oscillator, but when the oscillator is perturbed, at least one cycle will have a different length (Figure 12.3).

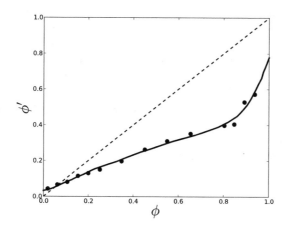

Fig. 12.5 Phase-transition curve determined for a slowly adapting motoneuron of the sea slug *Aplysia*. A 3-nA depolarizing current was injected into the motoneuron to cause it to fire repetitively. Single hyperpolarizing (-15 nA) 5-ms square-wave current pulses were injected at different phases ϕ of the cycle, and the new phase ϕ' was determined as described in the text. Figure redrawn from [238] with permission.

Let T' be the length of the perturbed cycle. The time from the beginning of the perturbed cycle to the perturbation is $T\phi$, and hence $T' = T\phi + T_r$. The time T_r remaining in the perturbed cycle is proportional to the phase $1 - \phi'$ remaining to be traversed, where ϕ' is the new phase. Hence

$$T' = T\phi + T(1 - \phi') = T(1 + \phi - \phi').\tag{12.5}$$

We can measure ϕ, T, T', and we want to determine ϕ'. So

$$\phi' = 1 + \phi - \frac{T'}{T}.\tag{12.6}$$

Since phase is defined between 0 and 1, we can rewrite (12.6) as

$$\phi' = \phi - \frac{T'}{T} \quad (\bmod 1).\tag{12.7}$$

A plot of new phase ϕ' versus old phase ϕ is called the *phase-transition curve* (PTC). Figure 12.5 shows the experimentally determined PTC for an *Aplysia* motoneuron.

Many authors prefer using a *phase-resetting curve* (PRC), namely a plot of $\Delta\phi = \phi - \phi'$ versus ϕ.

Experimentally, two types of PRCs have been identified. Type 1 PRCs are those in which perturbations cause only phase advances (i.e., $T' < T$), whereas for Type 2 PRCs, both phase advances and delays (i.e., $T' > T$) can be produced. Figure 12.5 shows an example of Type 2 phase resetting. These differences in resetting reflect differences in the mechanisms for spike initiation [204, 344]. Moreover, the type of PRC characterizes the ability of the oscillator to synchronize with other oscillators [102, 250, 679]. Although it is tempting to related Type 1 and Type 2 PRCs to, respectively, Type 1 and Type 2 excitability [204], this relationship is not necessarily valid [205]. Indeed, the same neuron can exhibit either a Type 1 or a Type 2 PRC depending on its inputs [861].

The most important two facts about phase-resetting curves are that 1) they can be measured experimentally and 2) completely describe the response of the oscillator to perturbations. However, there are three ways in which the phase-resetting theory can breakdown in experimental applications [409]. First, the period of the oscillation may not be constant. A varying period results in uncertainty in the estimates of the new and old phase. It is commonly observed that PRCs determined from invertebrate neurons [238, 684, 693, 771] are much less "noisy" than those measured for mammalian neurons [340, 533, 634, 743].

Second, the time to return to the limit cycle following a perturbation may not be sufficiently fast. In the laboratory, this can be tested by delivering stimuli repeatedly at fixed phase [238, 411, 476] to obtain the fixed phase-resetting curve, FPRC. The observation that the PRC and FRPC are not the same implies that the phase resetting depends on the past stimulation history of the neuron. In other words, the perturbed trajectory has not been able to get back to the limit cycle in a sufficiently short period of time before the next perturbation arrives. An alternate approach is to quantify the effects of perturbations on subsequent limit cycles, namely, compare the effect of the stimulus on the first-order PRC (effect on first limit cycle), second-order PRC (effect on second period of the limit cycle with the perturbation) [154, 658, 715].

A final concern arises when an abrupt change occurs in the PRC when the perturbation is applied at a particular phase. Abrupt transitions have frequently been observed for cardiac cells [296, 297, 385, 386]. A long-standing question is whether or not these abrupt changes represent discontinuities in the PRC [276]. The PRC is expected to be continuous for stimulus amplitudes that do not lead to a transition outside the basin of attraction for the limit cycle. Recent theoretical studies indicate that abrupt transitions between advance and delay in the PRC are related to the "all-or-none" dynamics of excitable cells due to the fast positive feedback dynamics of the sodium current which generates the action potential upstroke [408]. Surprisingly this all-or-none transition is so sensitive that even the stochastic gating of a single ion channel is sufficient to convert an "all" to a "none" response and vice versa!

12.2.1 Stumbling

The risk of falling is increased when a walker stumbles over, for example, a rock or tree root. The human stumbling reaction is phase-dependent [632, 645, 742]. In other words, the strategy used to recover from the stumble depends on the phase of the gait cycle when the perturbation was delivered. The importance of determining the PRC for stumbling stems from the fact that oscillators with the same PRC respond similarly to the same perturbations. Thus it becomes possible, for example, to construct a walking robot that has the same PRC as a human and use the responses of the robot to perturbations to devise strategies to improve recovery from stumbles in humans [645].

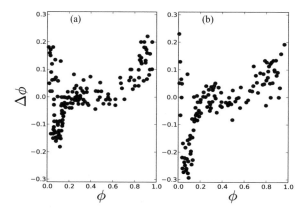

Fig. 12.6 Phase-resetting curves (PRC) for human gait in response to a mechanical perturbation. The mechanical perturbation was a brief tug applied to the shank of the right leg. The PRC on the left is the transient PRC, and on the right is the steady-state PRC. Figure redrawn from [645] with permission.

The phase-resetting curve for walking can be measured in the laboratory using mechanical perturbations (Figure 12.6). In these experiments, the perturbation is delivered by pulling the shank of one leg briefly backward as the subject walks on a treadmill. These perturbations are applied in a manner such that their timing cannot be predicted by the walker. By delivering these perturbations at different phases of the gait cycle, the stumbling PRC can be constructed from the data obtained. In Figure 12.6, the zero phase of the gait cycle is defined as the time at which maximal flexion of the right knee occurred. As can be seen, the gait phase was delayed when the perturbation was applied early in the gait cycle (early swing phase) and was advanced when the perturbation was applied late (late swing phase).

An important question for the use of the PRC in the description of stumbling while walking is whether the limit cycle following the first stumble recovers completely before the next stumble occurs. This question can be investigated by determining whether the PRC obtained using a single perturbation (the transient PRC) is the same as that obtained when the perturbation is applied repeatedly at the same

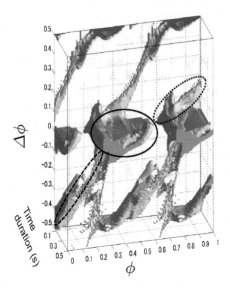

Fig. 12.7 The three-dimensional phase-resetting curve (PRC) for the response of human gait to mechanical perturbations. Since the transient and steady-state PRC are not the same (Figure 12.6), $\Delta\phi$ depends not only on ϕ, but also on the duration of the perturbation. Figure reproduced from [645] with permission.

phase of every gait cycle (the steady-state PRC) [418]. In the case of stumbling, the transient PRC (Figure 12.6a) is not exactly the same as the steady-state PRC (Figure 12.6b). Thus the PRC depends on both the duration of the perturbation and the phase at which it was delivered (Figure 12.7). Interestingly, the transient and steady-state PRC are more similar when the perturbations are applied in late swing phase than in early swing phase, implying that the effects of the duration of the perturbation are also phase-dependent.

12.3 Interacting Oscillators

The characterization of the phenomena that arise from the interactions between two or more oscillators has become a dominant theme in biology [683, 811, 822]. The surprising observation is that under appropriate conditions, such phenomena are quite stereotypical, and in particular, do not depend on the details of the mechanisms that generate the oscillation. Before embarking on this demonstration, we need to familiarize ourselves with the terminology of this field. Consider two oscillators A and B. The term *periodic forcing* means that oscillator A affects oscillator B, but not vice versa. When there are reciprocal interactions between oscillators A and B, they are said to be *coupled*. The term *weak* means that the effects of the interaction between the oscillators are manifested only by changes in the phase of the oscillator.

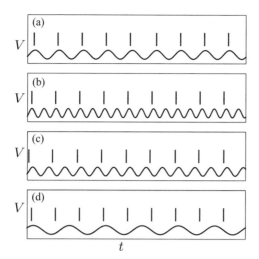

Fig. 12.8 Examples of $n : m$ phase-locking for a periodically forced integrate-and-fire (IF) neuron (see Sections 1.3.2 and 6.4.1): (a) 1 : 1, (b) 1 : 2, (c) 3 : 2, and (d) 2 : 3. The stimulus is presented by the sinusoidally varying solid line, and the action potentials generated by the IF neuron by the solid vertical lines. The model is $dV/dt = k_1 V + k_2 \sin(2\pi ft)$, where $k_1 = 1.6094$ s^{-1} is the integration time constant, $k_2 = 0.05$ V$\cdot$s^{-1} is the strength of the stimulus, and f is the stimulus frequency. For simplicity, we adjusted the membrane potential V so that the resting value is 0 and the spiking threshold is 1. Whenever $V = 1$, V is immediately reset to 0, and an action potential is recorded. When $k_2 = 0$ V$\cdot$s^{-1}, the IF neuron periodically generated action potentials with frequency, $f_0 \approx 0.8985$ Hz. The resting membrane potential is 0, and the reset condition is $V = 0$ when $V \geq 1$. We performed the numerical simulations using the global flags option in XPPAUT [202] with a step size of 0.0001 [761].

If more than the phase is affected, say the amplitude and period, then the interaction is said to be *strong*.

12.3.1 Periodic Forcing: Continuous stimulation

Figure 12.8 illustrates examples of stable phase-locking patterns observed when a hypothetical neuron periodically spiking with frequency f_0 is stimulated with a sinusoidally varying current of frequency f. The phase-locked patterns are labeled $n : m$, where n is the number of action potentials (cycles of spontaneous rhythm), and m is the number of stimulus cycles per phase-locked cycle. Thus the 2 : 3 pattern shown in Figure 12.8c has two neural action potentials for every three cycles of the phase-locked rhythm. The phrase "stable phase-locked" means that following a single additional perturbation, the same $n : m$ pattern is re-established.

It is observed that the phase-locked patterns depend on both the ratio f/f_0 and the amplitude a_s of the stimulus. In the frequency–amplitude domain, the stable phase-

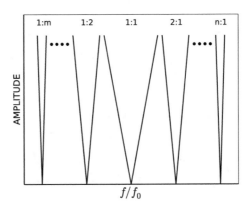

Fig. 12.9 Schematic representation of an Arnold tongue diagram of a hypothetical biological oscillator with frequency f_0 that is stimulated sinusoidally at frequency f.

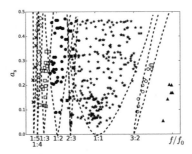

Fig. 12.10 Arnold tongue diagram for an *Aplysia* interneuron subjected to periodic sinusoidal stimulation with amplitude a_s (*y*-axis) and frequency f/f_0 (*x*-axis). A depolarizing current was injected into the motoneuron to cause it to fire repetitively at frequency f_0. A total of $1,292$ combinations of a_s, f were used to construct this figure; 645 of these combinations resulted in an obvious phase-locked pattern. Only the phase-locked patterns are shown. These patterns were determined from inspection of the time series. The dashed lines have been drawn to enclose data with the same phase-locking pattern so as to emphasize the underlying Arnold tongue structure. Code: $\times = 1:5$, $\diamondsuit = 1:4$, $\square = 1:3$, $\bullet = 1:2$, $\blacksquare = 2:3$, $\star = 1:1$, $\circ = 3:2$, $\blacktriangle = 2:1$. The solid lines have been drawn to enclose data with the same phase-locking pattern so as to emphasize the underlying Arnold tongue structure. Figure redrawn from [359] with permission.

locked patterns are arranged in an orderly fashion (Figure 12.9). The wedge-shaped regions observed for each $n:m$ are referred to as *tongues*. The diagram is referred to as an *Arnold tongue diagram* in honor of Vladimir Arnold, the mathematician who extensively investigated this phenomenon.

Figure 12.10 shows an example of an Arnold tongue diagram measured for a periodically spiking *Aplysia* motoneuron [359]. Only low-integer phase-locked patterns are observed. Since the $1:1$ tongue occurs even in the limit of vanishingly small a_s [274, 433], it is most robust in the face of noisy perturbations. Consequently, it is the largest Arnold tongue. The dashed lines that enclose each tongue have been estimated experimentally from examination of the time series. It is important to note that $\approx 50\%$ of the choices of a_s and f/f_0 used in this study did not result in a pattern that could be described as $n : m$, where n and m are constant. Similar observations have been obtained for a wide variety of neural preparations, including the squid giant axon [544, 545], invertebrate pacemaker neurons [329], neurons in the rat cortex [743], synchronized neurons in hippocampal slices [330], and even the green alga *Nitella* [331].

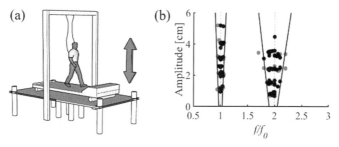

Fig. 12.11 (a) Schematic representation of a subject walking on a treadmill with a vertically oscillating platform. (b) Arnold tongue diagram showing for subjects walking on a vertically oscillating platform. The Arnold tongues are delinated by solid lines. For values of f, f_0 and platform oscillation amplitude outside of the Arnold tongue no significant entrainment is observed. Figure supplied by Jeff Nessler, California State University, San Marcos.

Example 12.1 The therapeutic use of synchronization strategies for the rehabilitation is based on the assumption that interactions between two oscillators are highly stereotyped and do not depend on the mechanisms that generate the oscillations. This assumption has motivated the use of sensory motor synchronization as a therapy for aiding the rehabilitation of patients with gait disorders. Examples include the side-by-side walking of therapist and patient [631] and entraining a person to walk to a periodic auditory cue provided by a metronome [824]. A novel approach is to have a person walk on a treadmill that is vertically oscillating at a normal gait frequency (Figure 12.11a) [633]. The Arnold tongue diagram (Figure 12.11b) shows that a healthy subject can phase lock to the oscillating platform in either a 1 : 1 or a 2 : 1 pattern. The 1 : 1 phase-locking pattern corresponds to one right heel strike per period of the treadmill motion, and the 2 : 1 corresponds to two right heel strikes per period of the treadmill motion. A curious observation was that the phase-locked patterns existed only transiently. Intermittent phase-locking pattern arises when the coupling strength is not strong enough to maintain synchronization. It is frequently observed in the synchronization patterns of spiking in neural networks [731].   ◇

12.4 Periodic forcing: Pulsatile stimulation

An experimental paradigm that is much more amenable to mathematical analysis is obtained by replacing the sinusoidally varying stimulus by a periodic pulsatile stimulus [40, 43, 276, 296]. To see why this is so, it is necessary to assume that the effects of the pulsatile stimulus are weak and that following each stimulus, the biological oscillator completely recovers before the next stimulus arrives. Together, these assumptions imply that the phase-transition curve $PTC(\phi, a_s)$ determined from a single-pulse experiment as described in the preceding section is the same as that determined from a periodic pulsatile stimulation experiment. In this case, the effects of periodic stimulation can be predicted by the one-dimensional map

$$\phi_{n+1} = PTC(\phi, a_s) + \frac{t_s}{T} \pmod 1, \tag{12.8}$$

where t_s/T is the phase at which the stimulus with amplitude a_s is delivered. The advantage of this formalism is that it makes it possible to interpret the experimental observations within the context of a mathematical model for which considerable mathematical insight has been obtained. Indeed, this why in 1981, Michael Guevara, Leon Glass, and Alvin Shrier[5] were able to provide the first convincing demonstration that chaotic dynamical behaviors can arise in a living cell, namely spontaneously beating cardiac cells in culture [296]. In order to understand the basis for these claims, it is necessary to explore the properties of (12.8) in more detail.

Equation 12.8 is an example of a *circle map*: a point on a circle is moved to another point on the circle. There are two important mathematical concepts involved in the analysis of circle maps: invertibility and rotation number ρ. A circle map is invertible if there is a $1 : 1$ relationship between ϕ and ϕ'. A $1 : 1$ relationship means that for each ϕ, there is only one possible ϕ', and vice versa. The rotation number, ρ, is the average number of cycles covered per stimulus [207, 273, 276]. It is calculated as

$$\rho = \lim_{n \to \infty} \frac{1}{T} \frac{\phi_n}{n}. \tag{12.9}$$

where T is the period of the oscillator. An important practical point is that (12.9) refers to the accumulated phase, and hence we do not take the modulus in evaluating this expression [202]. In terms of ρ, an $n : m$ Arnold tongue includes all solutions having the same value of ρ.

The best way to determine the Arnold tongue diagram is not by computer-aided inspection of time series as we did for Figure 12.10, but to use (12.8) together with (12.9). In particular, we measure $PTC(\phi, a_s)$ and then use (12.9) to calculate ρ as a function of t_s/T for a given value of a_s. This corresponds to a horizontal slice through the Arnold tongue diagram. Repeating this procedure for different choices of a_s enables the Arnold tongue diagram to be constructed one horizontal slice at a time. Note that the nature of the solution characterized by a given value of ρ is best visualized by constructing a cobweb diagram as discussed in Chapter 13.

---

[5] Alvin Shrier (b. 1949), Canadian physiologist and biophysicist.

Example 12.2 The standard circle map is described by the equation

$$\phi_{n+1} = [\phi_n + b + k_1 \sin(2\pi\phi_n)] \pmod 1, \tag{12.10}$$

where $b := n/m$. This map is invertible, provided that $k_1 < 0.16$. Comparing with (12.8), we have $b = t_s/T$ and $\text{PTC}(\phi, a_s) = \phi + k_1 \sin(2\pi\phi_n)$. Figure 12.12 shows a plot of ρ versus b when $a_s = 0.15$. This diagram is often referred to as the *devil's staircase*. ◊

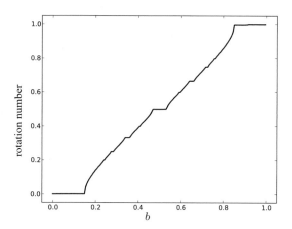

Fig. 12.12 The devil's staircase, a plot of the rotation number ρ as a function of the parameter b for (12.10).

Example 12.3 The concept of an Arnold tongue diagram observed for pulsatile stimulation has been particularly important for understanding the development of certain cardiac arrhythmias [276, 278, 296]. The pacemaker for the normal cardiac rhythm is the sinoatrial (SA) node, located in the left atrium. The SA node sends periodic pulses that perturb the spontaneously beating ventricles. Normally, for each pulse delivered by the SA node, the ventricles contract once. Pathological cardiac arrhythmias arise when this 1 : 1 relationship is lost. Many clinically observed cardiac arrhythmias can be reproduced by altering a_s and f/f_0 in spontaneously beating embryonic chick cardiac cells in culture (where electrical stimulation plays the role of the SA node, and the cardiac cells the role of the ventricle). All of the observed patterns in these experiments could be predicted from the measured PTC for the beating heart cells. ◊

Mathematicians have been able to show that Arnold tongues will be observed experimentally only for small integer values of n and m [276]. For higher values of n, m, the tongues will be so thin that it would be impossible to observe the phase-locking pattern. Moreover, mathematicians have been able to understand that two other types of solutions can exist in the regions between the tongues. Both of these solutions represent complex aperiodic phase-locking patterns. One type of aperiodic phase-locking is termed *quasiperiodic*. Suppose we have a time series of the form

$$x(t) = a \sin k_1 t + b \sin k_2 t .$$

What is its period? A period exists if and only if k_1 and k_2 are commensurate numbers. If k_1 and k_2 are incommensurate, then we have a *quasiperiodic* solution. Quasiperiodic phase-locked solutions exhibit variations in n and m that never repeat. Such solutions are possible, even though the circle map is invertible.

Second, there can be a *chaotic* pattern (see also Chapter 13). Chaotic solutions can arise when the circle map is not invertible; for example, a circle map can fail to be invertible because a bistable solution occurs as the amplitude increases. A chaotic phase-locking pattern resembles a quasiperiodic solution in that both n and m vary in an aperiodic manner. The key difference between a quasiperiodic and a chaotic phase-locking pattern becomes apparent when the solutions for two different initial conditions are compared. If for two initial conditions that are close together, the solutions remain close together, then we have a quasiperiodic phase-locking pattern. On the other hand, if the two solutions diverge from each other exponentially (*sensitivity to initial conditions*), then the phase-locking pattern is chaotic.

The existence of these two types of solutions explains why in the experiment that produced Figure 12.10, only about half of the solutions could be classified as $n : m$.

12.5 An illustrative biological example

An elegant experiment that illustrates the use of the techniques we have described up to this point involves the periodic stimulation of *Nitella*, a green alga that was used as a surrogate for the squid giant axon in the early days of electrophysiology.[6] The main difference between *Nitella* and the squid giant axon is that in *Nitella*, a Cl^- ion channel is the dominant determinant of membrane potential [396]. The experimental design used by Hatsuo Hayashi[7] and colleagues was as follows. The ionic composition of the medium surrounding a *Nitella* cell was adjusted so that it produced action potentials periodically. The frequency of periodic spiking can be changed by applying a constant-current input to the cell. To this constant stimulating current, the investigators added a periodically varying component. Thus the total stimulus received by the *Nitella* cell had two parts:

---

[6] Of course, the reader might wonder why *Nitella* produces an action potential. It does so because a major dynamical behavior of *Nitella* is *cytoplasmic streaming* [766, 864, 895]. An action potential stops cytoplasmic streaming. Hence when a *Nitella* cell is partially transected by, let us say the teeth of a hungry moose, the occurrence of an action potential stops cytoplasm from flowing out of the cell. In other words, action potential formation is part of a damage control mechanism.

[7] Hatsuo Hayashi (b. 1947), Japanese neuroscientist.

1. A constant part I_0. By adjusting I_0, we can change the frequency of the action potentials produced by the cell. Denote the frequency of the action potentials in the absence of a periodic input by f_0.
2. $\sin 2\pi f_i t$. This component of the stimulating current describes the external periodic forcing. Its frequency is denoted by f_i, and its amplitude by I.

The experiment involved determining the Arnold tongue diagram by varying the parameters I and f_i/f_0. Our concern here is to show how for each I and f_i/f_0, the investigators were able to classify the observed behaviors.

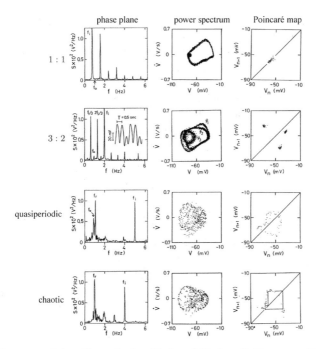

Fig. 12.13 Responses of the self-sustained oscillation of *Nitella* cells (frequency f_0) to sinusoidal stimulation (frequency f_i). The 1 : 1 pattern corresponds to $f_i/f_0 = 0.8$ with $f_i = 0.8$ Hz, the 3 : 2 pattern to $f_i/f_0 = 2.0$ with $f_i = 2.0$ Hz, the quasiperiodic pattern to $f_i/f_0 = 5.3$ with $f_i = 4.1$ Hz, and the chaotic pattern to $f_i/f_0 = 4.1$ with $f_i = 4.0$ Hz. This figure represents a montage constructed with permission from several figures in [331]

They observed four qualitatively different types of dynamics:

Entrained Oscillations

Entrainment means that the *Nitella* produces an action potential for each period of the external forcing (Figure 12.13. first row). From the phase plane and the Poincaré map, we see that the *Nitella* is generating action potentials periodically. What is the period? From the power spectrum, we see that the period is not f_0, but f_i (the other peaks are integer multiples, or harmonics, of f_i). Since there is one action potential for every cycle of stimulus, we have a 1 : 1 phase-locking pattern.

Harmonic Oscillations

The presence of distinct peaks in the power spectrum indicates that there is a periodic pattern (Figure 12.13, second row). This is confirmed by the observation that the phase plane is a closed curve; however, this closed curve is much more complicated than that observed for 1 : 1 phase-locking. By looking at the data, the investigators concluded that Ψ_1 and Ψ_2 represented action potentials and Ψ_3 was associated with a subthreshold response to stimulation. Thus we have two *Nitella* action potentials per phase-locked cycle period, and hence $m = 2$. From the power spectrum, we see that there are three major peaks related to f_i, namely f_i, $f_i/3$, and $2f_i/3$. Thus there are three cycles of stimulus for each phase-locked cycle, and hence $n = 3$ (the inset makes this point more clearly). Thus we have a 3 : 2 phase-locking pattern. This is confirmed by looking at the Poincaré section, where we see three points: each one corresponds to a different return to the section.

Quasiperiodic Oscillations

The power spectrum shows the presence of periodic components (Figure 12.13, third row). Curiously, there is a periodic component f_r that is not simply related to either f_0 or f_i. By "not simply related" we mean that f_r is not a harmonic or subharmonic of either f_0 or f_i. The phase plane is strikingly different, since we do not see a closed curve, but rather a "cloud of points"; the trajectory seems to be irregularly filling up a region of the phase plane. However, before we decide what kind of solution we have obtained, let us look at the last case.

Chaotic Oscillations

In some sense, the results in the third and fourth rows of Figure 12.13 look qualitatively the same. However, there is an important difference, namely that the return map shown in the last row admits a period-3 solution. At the time that this study was conducted, the presence of a period-3 solution was considered to be strong evidence of chaos ("period three implies chaos" [483]). Since that time, better methods have been developed to detect the presence of chaos. Thus the behavior shown in the third row indicates a quasiperiodic solution rather than a chaotic one.

12.6 Coupled Oscillators

The observation that identical weakly interacting oscillators tend to coordinate their activities lies at the basis of phenomena ranging from firefly flashing [203, 598, 810] to the binding of sensory stimuli by the cortex [197], the generation of epileptic seizures [574], and the tendency for menstrual cycles to occur at the same time in women living in close quarters in university residences and military barracks [316, 550].

Even couples walking side by side tend to synchronize their gaits [630]. Christiaan Huygens[8] described an early practical application of synchronization to adjust clocks to keep time correctly.

Compared to the richness of the dynamical behaviors observed for the periodic forcing of oscillators, only three types of behaviors are seen for coupled oscillators. There can be in-phase synchronization (phase difference between oscillators is 0) and out-of-phase synchronization (phase difference is π). The third behavior, which has been recently emphasized, is the occurrence of oscillator and amplitude death [206, 921, 922]. In other words, the population of coupled oscillators as a whole shows no oscillatory behavior at all! Such solutions occur for strong coupling [206] and are particularly likely to occur when the coupling is time-delayed [921].

Example 12.4 A much popularized example concerns the synchronized flashing of populations of male fireflies in some parts of southeast Asia [87, 203, 598, 810]. The male fireflies of this species flashes to attract females, and at dusk, they congregate in trees. Although they do not begin to flash synchronously, eventually they come to flash at the same rate and in unison. We conclude that this must be a result of the fireflies influencing one another. Suppose we have two oscillators running at slightly different natural frequencies. Then trivial differential equations for the uncoupled oscillators can be written as

$$\frac{d\phi_1}{dt} = \omega_1, \quad \frac{d\phi_2}{dt} = \omega_2,$$

where ϕ_1, ϕ_2 are the respective phases of the oscillators with angular frequencies ω_1, ω_2. Introducing coupling of the form described above, we obtain

$$\frac{d\phi_1}{dt} = \omega_1 + k_1 \sin(\phi_2 - \phi_1), \quad \frac{d\phi_2}{dt} = \omega_2 + k_2 \sin(\phi_1 - \phi_2).$$

The form of these equations allows us to write a single differential equation in a new variable representing the difference between the two variables: $\varphi := \phi_1 - \phi_2$. Then the equation is

$$\frac{d\varphi}{dt} = \omega - K \sin(\varphi), \tag{12.11}$$

---

[8] Christiaan Huygens (1629–1695), Dutch mathematician, astronomer, and physicist.

where $\omega = \omega_1 - \omega_2$ and $K = k_1 + k_2$. These are two stable solutions (see Exercise 6). One of these solutions corresponds to the in-phase solution, and the other to the out-of-phase solution.   ◇

12.7 Dynamic clamping

Dynamical clamping of neurons is a first step toward developing neuromorphic circuits to repair diseased neural circuits with implants that can adapt to biofeedback. In these circuits the membrane potential of the neuron is monitored and the "clamping" box plays the role of, for example, an interneuron (Figure 12.14). Variations of this technique have been used to study the dynamics of recurrent inhibitory [404, 851] and excitatory [542] loops involving crayfish slowly adapting stretch receptors, to form artificial electrical [757] and chemical [758, 759] synapses, to form simple neural circuits [429] and to study rhythmogenesis in networks of coupled oscillators [898].

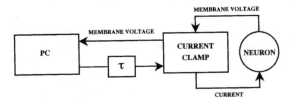

Fig. 12.14 Schematic representation of a dynamically clamped neuron. The membrane potential of a neuron is monitored by a computer via a voltage/current clamp and a DA/AD converter. Each time the computer detects a spike, at a time τ later it directs the clamp to inject either an excitatory (recurrent excitatory loop) or inhibitory (recurrent inhibitory loop) current pulse into the neuron. Figure reproduced from [238] with permission.

An example of a delayed clamped recurrent loop is a recurrent inhibitory neural loop composed of a periodically spiking *Aplysia* motoneuron which is reciprocally connected to a computer which plays the role of the inhibitory interneuron [238] (see also Section 13.5). In response to an action potential generated by the neuron at time t, the computer delivers a brief hyperpolarizing current to the neuron at $t + \tau$ [236, 237, 515]. The PRC plays an essential role in the analysis of the dynamical behaviors generated by this loop. Since the experimentally determined PRC (Figure 12.5) summarizes the response of the Aplysia motoneuron to a hyperpolarizing pulse as a function of phase the pulse is delivered, it is possible to develop a mathematical model for dynamics of this delayed inhibitory loop. In other words, it is not necessary to develop a Hodgkin–Huxley type model of the neuron in order to predict the behavior of the delayed recurrent loop. Thus quantitative comparisons of prediction to observation are possible [238]. The results demonstrated that multistable spiking patterns arise once τ becomes sufficiently large.

Many of the feedback loops in the nervous system are pulse-coupled. In other words, the interaction between neurons is characterized by pulse-like interactions [398]. This pulse-coupling is a direct consequence of the fact that when neurons are physically separated by more than a few millimeters, interactions between them are in the form of discrete synaptic potentials driven by action potentials. The critical parameters for describing the dynamics of such feedback loops are those related to the timing of the action potentials. Consequently, all of the physiological processes involved in the transmission of information around the neural loop can be incorporated into τ. Thus we can expect that similar results will be seen if the feedback loop is replaced by a recurrent inhibitory loop in the olfactory cortex or hippocampus, peripheral sensory feedback to a central pattern generator or feedback from brain stem and subcortical structures to an epileptic cortical region, and so on.

12.8 Switching bistable states

In a bistable system the introduction of a carefully timed stimuli with the appropriate strength, duration and frequency can cause a switch between the two coexistent attractors. This observation has sparked interest in the development of electroceutical devices to treat disease. Examples include the use of stimuli to abort epileptic seizures and terminate certain cardiac arrhythmias. However, it can be quite difficult to optimize electroceutical dosing protocols on a trial and error basis [106, 410, 879]. In addition to timing and strength of stimuli clinical concerns include minimizing tissue damage, propagation of energy to neighboring regions and the need to maximize battery life in implantable devices [114]. Thus, depending on the application, it is often necessary to decide between using stimuli that optimize peak energy versus those that optimize energy consumption.

A starting point is to determine the optimal stimulus that causes a switch between attractors in mathematical models known to exhibit bistability. Thus a great deal of effort has been devoted to determining the optimal stimulus that can cause a spiking Hodgkin-Huxley [112, 113, 135, 233, 303] or Fitzhugh–Nagumo [112] neuron to stop spiking. We remind the reader that the onset of repetitive spiking in both neuron models is due to a subcritical Andronov–Hopf bifurcation. In these situations, a variety of analytical methods can be used (for a review see [114]). In these studies, the optimal stimulus is the one which minimizes the energy of the stimulus. The energy of a signal is proportional to the variance and hence from a practical point of view the optimal stimulus corresponds to the one with the smallest L^2-norm.

The optimal stimulus for eliciting a spike in a Hodgkin-Huxley neuron is shown in Figure 12.15 (solid line). Curiously the analytically and experimentally determined stimulus is a smoothly varying biphasic current waveform having a relatively long and shallow hyperpolarizing phase followed by a depolarizing phase of briefer duration [135]. The hyperpolarizing phase removes a small degree of the resting level of Na^+ channel inactivation. This result together with the subsequent depolarizing phase provides a signal that is energetically more efficient for eliciting spikes

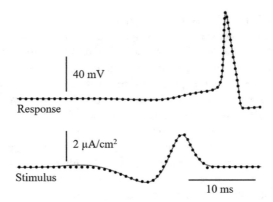

Fig. 12.15 Optimal stimuli for eliciting an action potential for the Hodgkin-Huxley equation. The solid lines show the action potential (top) for the optimal stimulus (bottom) determined using model-dependent methods. The •'s show the optimal stimulus and response determined using the extrema-based, stochastic hill climbing model-independent method. In both cases, the stimulus is the one that minimizes the L^2-norm. Figure prepared by Joshua Chang, Dell Medical School, University of Texas, Austin.

than rectangular current pulses. Although this result at first may appear surprising it is consistent with the observation that brief inhibition can promote spiking in a Hodgkin–Huxley neuron [174].

It must be kept in mind that a Andronov–Hopf subcritical bifurcation is not the only bifurcation in neuronal models associated with bistability. A saddle node on an invariant circle (SNIC) bifurcation is also associated with bistability and can arise, for example, in the Morris–Lecar neuron. Whereas in the Andronov–Hopf subcritical bifurcation the fixed point is enclosed by the limit cycle, in the case of the SNIC bifurcation the fixed point lies outside of the area in phase space enclosed by the limit cycle. In this situation a carefully timed uniphasic, square-wave pulse is an optimal stimulus.

In general, the mathematical equations for the bistable system are not known. Thus it is necessary to identify the optimal stimulus using model-independent methods. One such method is the extrema-based, stochastic hill climbing approach recently developed by Chang and Paydarfar [114]. This approach starts with a randomly generated stimulus, searches a few of the neighboring stimuli, picks the best one and then uses this as the starting point for the next iteration. It continues to do this over and over again, always choosing the best seen stimulus as the starting point for the next iteration. Figure 12.15 (•) shows the optimal stimulus for eliciting a spike in a Hodgkin–Huxley neuron. It is identical to that obtained using analytical methods (Figure 12.15 (solid line)).

A fundamental problem with model-independent method for determining optimal stimuli is the "curse of dimensionality," namely as the number of dimensions increases linearly, the search space increases exponentially [48]. Since each point in the search for an optimal stimulus becomes a unique dimension, a 50ms stimulus sampled at 0.1ms becomes a 500-dimension problem! Thus it is necessary to combine these approaches with techniques that reduce the dimension such as using extrema as key feature points [113].

12.9 What Have We Learned?

1. Why is a limit cycle a more robust explanation for a biological oscillator than a center?

   A limit cycle oscillation is much more robust to the effects of perturbations than the oscillation associated with a center. It also occurs in the dissipative dynamical systems studied experimentally.

2. What can be said about the nature of the fixed point that is surrounded by a limit cycle?

   We would anticipate that the fixed point would be a spiral point. Intuitively, we would expect it to be an unstable spiral point, as we observed for the van der Pol oscillator. However, as the next item shows, it could be a stable spiral point.

3. How can a limit cycle occur when the fixed point is stable?

   It is possible for a limit cycle to surround a stable spiral point. For this to occur, there needs to be some form of instability once perturbations away from the fixed point become large enough. This instability typically takes the form of an unstable separatrix (likely associated with a saddle point): for initial conditions inside the separatrix, trajectories spiral toward the fixed point, while outside, they spiral toward the limit cycle. This occurs, for example, in the case of a subcritical Andronov-Hopf bifurcation for a neuron.

4. What is the importance of a Poincaré section?

   The Poincaré section provides a direct connection between a discrete time map and a continuous dynamical trajectories.

5. What is the phase-resetting curve and why is it important?

   The PRC characterizes the responses of a biological oscillator to perturbations and hence characterizes the ability of the oscillator to synchronize with other oscillators [102, 250, 679]. The fact that the PRC is experimentally measured often facilitates comparisons between prediction and observation.

6. What is the difference between phase-locking between two oscillators and synchronization between two oscillators?

   Both phase-locking and synchronization refer to the interaction between two oscillators. In the case of phase-locking, oscillator A can influence oscillator B, but not vice versa, whereas synchronization refers to the mutual interaction between two oscillators having the same frequency.

7. What is an Arnold tongue diagram and why is it important?

   An Arnold tongue diagram summarizes the phase-locking patterns of a periodically forced oscillator as a function of the ratio between the frequencies of the two oscillators and the amplitude of the forcing stimulus. The periodic forcing of an oscillator arises frequently in biology [276]. In particular, the Arnold tongue diagram has been useful for understanding the etiology of certain cardiac arrhythmias.

8. Why does the green alga *Nitella* generate an action potential?

   Action potentials are often used by organisms to synchronize the activity of a spatially extended dynamical system. In the case of *Nitella*, which has very large cells, the generation of an action potential is used to stop cytoplasmic transport when the cell is partially damaged. In the case of neuromuscular control, an action potential is used to coordinate the contraction of muscle cells.

9. What is the optimal stimulus for generating a spike in a Hodgkin–Huxley neuron?

   The analytically and experimentally determined stimulus is a smoothly varying biphasic current waveform having a relatively long and shallow hyper-polarizing phase followed by a depolarizing phase of briefer duration. The hyper-polarizing phase removes a small degree of the resting level of Na+ channel inactivation. This result together with the subsequent depolarizing phase provides a signal that is energetically more efficient for eliciting spikes than rectangular current pulses.

10. What is dynamic clamping?

    A closed-loop circuit in which the membrane potential of the neuron is monitored by a computer ("clamping box") which, in turn, delivers a stimulus to the neuron. For example, if the dynamically clamped neuron is designed to mimic a recurrent inhibitory loop the delivered stimulus would be an inhibitory pulse.

11. What is the "curse of dimensionality?"

    As the number of dimensions increases linearly, the search space increases exponentially.

12.10 Exercises for Practice and Insight

1. Determine the fixed points of (12.11) and evaluate their stability.
2. Suppose a neuron periodically spikes with period $T_0 = 0.75$. Sketch the relationship between the phase ϕ_i at which a spike occurs and the phase ϕ_{i+1} at which the next spike occurs.
3. Write a computer program and determine the phase-resetting curves for the following neuron models:

   a. The Fitzhugh–Nagumo model.
   b. The Hodgkin–Huxley model.
   c. The Morris–Lecar model.

   Are the PRCs Type 1 or Type 2?

Chapter 13
Beyond Limit Cycles

Dynamical approaches to biological systems emphasize the qualitative nature of the time-dependent behaviors that can be observed as parameter values are changed. If the important parameters can be identified, then they can be manipulated experimentally. For example, in feedback control, the stability of the fixed point depends on the interplay between the delay τ and the feedback gain. It is surprising that experimentalists did not begin earlier than they did to investigate systematically the behaviors of biological systems as parameters were changed. One barrier was that inexpensive personal computers did not become available until the 1970s. The second barrier was that experimentalists did not have catalogs that documented the types of behaviors that could occur (see, for example, [291, 810]). It can be very difficult at the benchtop to interpret complex time series unless there is some sense of the nature of the phenomena that one is looking for.

The successes of biomathematicians in understanding the behavior of fixed points and limit-cycle oscillators naturally raised the question, "Is this all there is?" The answer turned out to be negative. This realization influenced how scientists view the world. For nearly 100 years, the concept of *homeostasis*, namely the constancy of the physiological environment, dominated the thinking of physiologists [57, 103]. The necessity for mechanisms to enable homeostasis seemed self-evident: "normal" cell function requires a constant composition of the fluid that bathes the cells. Mathematically, homeostasis was translated into feedback controllers having stable fixed points, while controllers with unstable fixed points were considered to be nonphysiological. In retrospect, this perceived need for constancy seems puzzling, given the fact that many cellular and physiological processes in the body are periodic, including the cell cycle, the heartbeat, and the endocrine cycles.

This orderly concept of how living organisms work was challenged following the advent of affordable personal digital computers. It is not clear whether the seeds of doubt first began in the thoughts of mathematicians [558], computer scientists [836], or climate physicists [506]. However, the person who most clearly brought the issue to the forefront of thinking biologists was Robert M. May[1]

---

[1] Robert McCredie May, Baron May of Oxford (b. 1936), Australian theoretical ecologist.

© Springer Nature Switzerland AG 2021
J. Milton and T. Ohira, *Mathematics as a Laboratory Tool*,
https://doi.org/10.1007/978-3-030-69579-8_13

(now Sir Robert May) [547]. Within a year, two investigators, Leon Glass and Michael C. Mackey, had placed the issue squarely at the feet of physiologists [522]: Simple deterministic systems can generate complex dynamics, said to be *chaotic*, which can be indistinguishable from those generated by a random process. Do fluctuations, for example, in serum sodium ion concentrations, reflect random fluctuations about a constant fixed point as predicted by homeostasis, or the chaotic dynamics of a physiological control mechanism? In this chapter, we discuss the events that led to the realization that biological systems can generate chaotic dynamics.

The questions we will answer in this chapter are

1. What are four types of problems whose dynamics are described by a one-dimensional map?
2. What is the condition for stability of a continuous-time dynamical system and of a discrete-time dynamical system?
3. What is a dynamical disease?
4. What are the conditions that establish the existence of chaotic dynamics in a mathematical model?
5. What are the properties of a chaotic attractor?
6. What are examples of biological experimental paradigms that have produced chaotic dynamics?
7. What are the effects of digital control on the properties of a continuous-time dynamical system?
8. What is microchaos?
9. With respect to the mechanisms that produce microchaos, what do the terms "small scale" and "large scale" mean?
10. What is transient microchaos?
11. What is a neuronal motif?
12. What is multistability?
13. How has multistability been demonstrated in a 2-neuron delayed inhibitory recurrent loop?
14. What is a delay-induced transient oscillation?

13.1 Chaos and Parameters

The focus of May's 1976 paper was the behaviors exhibited by one-dimensional maps used to model population growth,

$$x_{t+1} = f(x_t),$$ (13.1)

where x is a suitable estimate of population size measured at times t and $t + 1$, and f is in general a nonlinear function [547]. In particular, May was interested in the deceptively simple-looking quadratic map

$$x_{t+1} = \mu x_t(1 - x_t),\tag{13.2}$$

where $0 < \mu \le 4$ is a parameter.[2] This equation describes the growth of a population with, for example, a once per year breeding season whose growth is limited by competition (the x_t^2 term). It is convenient to express population size as a dimensionless number equal to the number of organisms per unit area divided by the carrying capacity, namely, the maximum population size that can be supported per unit area. Since there is a single maximum, (13.2) is sometimes called a one-humped map. Provided that $0 < \mu \le 4$, the growth of the population described by (13.2) will always be confined to the closed interval $[0, 1]$. In mathematical terms, this choice of $f(x_t)$ maps the interval $[0, 1]$ to itself.

What is the behavior of (13.2) as a function of μ? We answer this question by discussing the different ways that the dynamics of discrete-time dynamical systems can be evaluated. The most straightforward way is simply to integrate (13.2) numerically for different choices of μ.

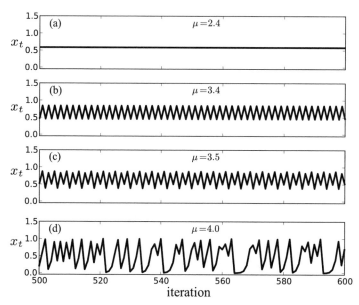

Fig. 13.1 Dynamics of (13.2) as the parameter μ is changed. For each value of μ, an initial value was picked randomly on the interval $(0, 1)$. The first 500 iterates were discarded, and the next 100 iterates were plotted.

---

[2] As in Chapter 5, we use the symbol μ to designate the bifurcation parameter.

Figure 13.1 shows examples of the solutions of (13.2) for different values of μ. When the fixed point becomes unstable, namely when $\mu > 3$ (see below), oscillatory solutions arise. The first oscillation to appear as μ increases is a period-2 oscillation (Figure 13.1b). Then comes a period-4 oscillation (Figure 13.1c) and then a succession of oscillations with period 2^n, until finally, these regular oscillations are replaced by highly irregular ones (Figure 13.1d).

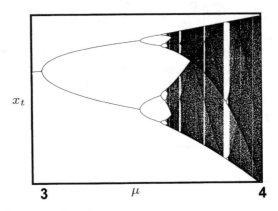

Fig. 13.2 Bifurcation diagram for (13.2).

A convenient way to summarize these observations is to construct a bifurcation diagram (Figure 13.2). The bifurcation diagram for (13.2) is generated by first choosing an initial value of μ and an initial value of x_t (compare this procedure to that discussed in Section 5.2.1). Then (13.2) is iterated: the first 1000, say, iterations are discarded as transients, and then the values of x_t are plotted for the next 1000 iterations. Once completed, we increment μ and repeat the process. In this bifurcation diagram, a stable fixed point appears as a single point, a period-two oscillation as two distinct points, a period-four oscillation as four distinct points, and so on.

It is worthwhile to examine Figure 13.2 carefully. First, the qualitative changes in dynamics occur abruptly. Consequently, it is very easy to determine the value of μ that marks the onset of the period-two oscillation, the period-four oscillation, and so on. Second, the range of μ values in which each successive period-doubled oscillation occurs becomes smaller as the length of the period increases. If we call Δ_{μ^n} the range of values of μ over which a stable cycle of period n is present, then it was shown by Feigenbaum[3] [221] using a pocket calculator that

$$\lim_{n \to \infty} \frac{\Delta_{\mu^n}}{\Delta_{\mu^{2n}}} = 4.6692016\ldots.$$

This ratio turns out to be independent of the precise analytic form of the map as long as the map has a single maximum. For this reason, the number $4.6692016\ldots$

---

[3] Mitchell Jay Feigenbaum (b. 1944), American mathematical physicist.

is called the *Feigenbaum number*. As μ continues to increase in the range $3.57\ldots <$ $\mu < 4$, stable periodic orbits appear having odd periods along with solutions having more complex behaviors. For a complete description of the dynamics of (13.2), see [171].

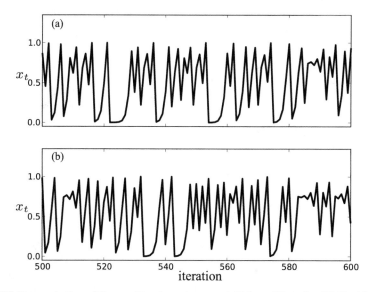

Fig. 13.3 Demonstration of the sensitive dependence on initial conditions for (13.2) with $\mu = 4$. The difference in the choice of initial conditions was 0.00000001.

A particularly complex and interesting solution arises when $\mu = 4$. In this case, aperiodic fluctuations arise that exhibit a sensitive dependence on the choice of initial condition (see also Section 13.3.1 below). In Figure 13.3, the difference in the choice of initial conditions was only 0.00000001. These solutions are referred to as *chaotic*, and they captured the imagination of a generation of scientists. One of the authors of this book (JM) can recall as a university student going to parties at which the focus of attention was on a small programmable pocket calculator that was generating a chaotic solution for (13.2). It was simply hard to believe that a simple completely deterministic equation could generate a solution that was indistinguishable from noise (see also Section 14.4.2). The demonstration that "period three implies chaos" [483] ($\mu \approx 3.8$ in Figure 13.2) prompted many investigators to design experimental paradigms capable of generating chaotic dynamics (for example, see [166, 296, 330, 331, 377, 498, 544, 743]).

The second numerical method for investigating (13.2) is to construct a so-called *cobweb* diagram (Figure 13.4). The first step is to plot $f(x_t)$ as a function of x_t. Next, we plot the line $x_t = x_{t+1}$ (dashed line in Figure 13.4). We can now use a simple iterative graphical procedure to visualize the nature of the solutions generated by (13.2). To illustrate, take $\mu = 2.4$ and choose $x_0 = 0.14$ (Figure 13.4a). In order to obtain x_1, we simply draw a vertical line (green) until it hits the curve described by

(13.2) (solid black line). In order to obtain x_2, we draw a horizontal line (red) to the line $x_t = x_{t+1}$ (dashed black line), and from this point, a vertical line (green) to the curve described by (13.2) (solid black line). And so on, and so on. Figure 13.4 shows the cobweb diagrams for the same values of μ used to generate the solutions shown in Figure 13.1. As is illustrated in Exercise 2, an important application of the cobweb diagram technique arises in the context of the Arnold tongue diagrams discussed in the previous chapter.

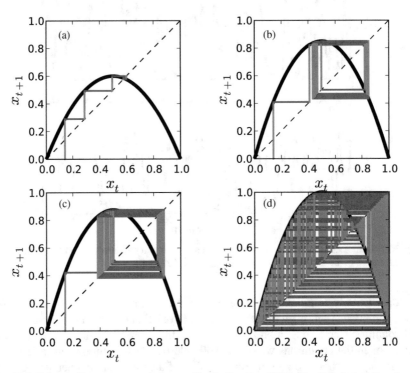

Fig. 13.4 Cobweb diagram approach for graphically iterating (13.2). The thin dashed line shows $x_t = x_{t+1}$. The alternating vertical (green) and horizontal (red) lines illustrate the iterative process for generating the cobweb diagram. The values of μ are the same as used in Figure 13.1, namely (a) 2.4, (b) 3.4, (c) 3.5, (d) 4.0.

Finally, we can use the same tools that we used for the analysis of stability of fixed points of differential equations (Chapter 4) to determine the stability of the fixed points of (13.1) as a function of μ. Here we illustrate this procedure by evaluating the stability of the fixed points of (13.2). The fixed points x^* of (13.2) are the values of x_t for which $x_t = x_{t+1}$ and thus are solutions of the equation

$$x^* [(\mu - 1) - \mu x^*] = 0 .$$

There are two different critical points: $x^* = 0$ and $x^* = \frac{\mu-1}{\mu}$. Linearizing about the critical point $x^* = 0$ yields the equation

$$x_{t+1} = \mu x_t .$$

The criterion for stability is

$$\left| \frac{df(x_t)}{dx_t} \right|_{x_t=x^*} < 1 . \tag{13.3}$$

Thus the $x^* = 0$ fixed point is stable, provided that

$$|\mu| < 1 .$$

Similarly, we can show that the $x^* = \frac{\mu-1}{\mu}$ fixed point is stable, provided that

$$|2 - \mu| < 1 ,$$

or equivalently,

$$1 < \mu < 3 .$$

A one-dimensional map can be thought of as a singular perturbation of a one-dimensional delay differential equation (DDE). Namely, if we have

$$\varepsilon \frac{dx}{dt} + k_1 x = f(x(t - 1)) ,$$

then the map is obtained by letting ε approach 0. Since chaotic dynamics arise in (13.2), it is natural to ask whether such dynamics can also arise in suitably constructed DDEs. As we shall see in the next section, the answer to this speculation is a resounding yes.

13.2 Dynamical Diseases and Parameters

A number of diseases in human beings and other animals are characterized by either the appearance of an oscillation in a variable that does not normally oscillate or an alteration in the qualitative properties of an oscillation (Table 13.1). By analogy with Figure 13.1, is it possible that these diseases arise because a parameter (μ in this figure) has been tuned into a parameter range associated with abnormal dynamics? In other words, are certain diseases *dynamical* [44, 275, 522, 523, 571, 574, 577]? If so, then treatment could be effected by first identifying the control parameter and then shifting it back into the parameter range associated with healthy dynamics.

As an illustration of this concept, Mackey and Glass [522] examined certain hematological diseases in which the number of circulating white blood cells (WBCs)

Table 13.1 Regular and irregular dynamics in human health and disease [523, 571].

Field	Regularly recurring	Irregularly occurring
Cardiac rhythms	Sinus rhythm	Atrial fibrillation
	Ventricular bigeminy	Ventricular fibrillation
Eye movements	Nystagmus	Opsoclonus
Hematology	Periodic hematopoiesis	Periodic CML
	Autoimmune hemolytic anemia	Cyclical thrombocytopenia
Locomotion	Gait, marching	Cerebellar ataxia
Movement disorders	Tremors	Choreoathetosis
	Hiccups	Myoclonus
Myology	Fibrillations	Fasciculations
	Myotonic discharges	
	Myokymia	
Pupils	Pupil cycle time	Hippus
Respiration	Periodic breathing	Ataxic breathing
	Cheyne–Stokes	Cluster breathing

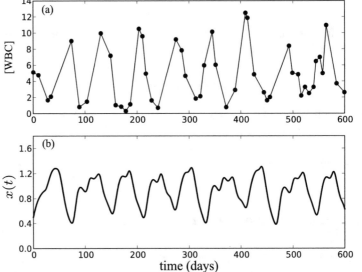

Fig. 13.5 A comparison of the changes in WBC counts for (a) a patient with periodic CML and (b) a solution of (13.4). Figure reproduced from [523] with permission.

undergoes spontaneous irregular oscillations (Figure 13.5a). This phenomenon occurs in gray collies and certain forms of leukemia in humans, for example, periodic chronic myelogenous leukemia (CML). The presence of unsuspected spontaneous oscillations greatly complicates treatment: Typically, chemotherapeutic agents are given when WBC levels are very high. In the case of a patient with unsuspected periodic CML, the decrease in WBC due to chemotherapy could coincide with a decrease in WBC due to the underlying dynamical disease. The combination of these two effects would produce very low WBC levels and thus subject the patient (collie or human) to serious infections.

Their model took the form of the now famous *Mackey–Glass equation*

$$\frac{dx}{dt} + k_1 x(t) = k_2 x(t-\tau) \frac{K^\mu}{K^\mu + x(t-\tau)^\mu}, \tag{13.4}$$

where k_1, k_2, K are constants, μ is the bifurcation parameter, τ is the time delay, and x is the circulating WBC density (namely, the number of WBCs per unit blood volume). The right-hand side takes into account the observation that the control of x exhibits elements of both negative and positive feedback. When x is low, a hormone called granulopoietin is produced that stimulates WBC production (positive feedback). However, when x becomes high, WBC production rates fall toward zero (negative feedback). In other words, the feedback is *mixed* (Section 9.4). The time delay takes into account the time it takes a WBC to mature in the bone marrow before being released into the circulation.

Figure 13.6 plots the dynamics of x predicted by (13.4) as the gain in the feedback, governed by the parameter μ, increases. As μ increases, we first see a variety of periodic oscillations (these arise by a series of bifurcations). When $\mu \geq 9.70$, we obtain a chaotic solution. The Mackey–Glass equation was the first demonstration that chaotic dynamics could be generated by a plausible physiological control mechanism. In other words, irregularly varying fluctuations in circulating blood cell

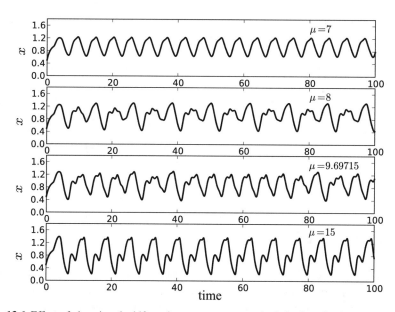

Fig. 13.6 Effect of changing the bifurcation parameter μ on the behavior of (13.4). Parameters: $k_1 = 1$, $k_2 = 2$, $K = 1$, and $\tau = 2$. The initial function is $x(0) = 0.5$ and $x(s) = 0.5$, $s \in [-\tau, 0)$.

numbers, such as observed in patients with CML, may not simply represent experimental error but can represent the dynamics of a disease process. It should also be noted that when μ is increased beyond the chaotic region, the fluctuations in x again become more periodic, indicating that a reverse bifurcation can occur.

Subsequent work has shown that the Mackey–Glass equation does not provide a valid description of WBC production. However, with improved models and better experimental observations has come the realization that treatment is possible (the interested reader is referred to [232]). The main present-day use of the Mackey–Glass equation is to provide a convenient way to produce chaotic signals for use as benchmarks in testing various time series analysis methods. The useful property is that in the chaotic regime, the complexity of the chaotic attractor increases as τ increases [217].

13.3 Chaos in the Laboratory

An unsettled question is whether the complex dynamics predicted by mathematical models of biological systems can be generated by living biological systems. In other words, it is possible that noisy dynamics actually reflect the effects of random perturbations. Thus considerable effort has been devoted to the design of experimental paradigms to test whether chaotic solutions can arise in real dynamical systems. One such paradigm, namely the periodic forcing of a biological oscillator, has already been discussed (see Figure 12.13). However, the question whether chaotic dynamics could arise when a control parameter is changed is not answered by these paradigms. The worrisome issue concerns the effect on random fluctuations in the parameters. By inspecting Figure 13.2, we see that the parameter range in which a given type of dynamics is observed becomes narrower and narrower as the complexity of the dynamics increases. At some point, inherently noisy parameter fluctuations would be expected to destroy the complex patterns. Could this same problem make it impossible to observe "purely chaotic" dynamics? Here we discuss two experimental paradigms that attempted to characterize the dynamics as parameters vary along a route to chaos. We have chosen to discuss these examples because in both cases, the investigators were concerned about the effects of random perturbations on the dynamics that they observed.

13.3.1 Flour Beetle Cannibalism

Perhaps the most elegant attempt to demonstrate chaos in a biological preparation involved the lowly flour beetle, *Tribolium castaneum* [143, 148, 169]. A dominant feature of the life cycle of flour beetles is cannibalism. The larval and pupal stages each last about two weeks and are followed by an adult stage. The larvae and adults both consume the eggs, and the adults also eat the pupae. In the description of age-structured populations, biologists use the term *recruitment* to indicate when a growing organism becomes a member of an identifiable population. Thus for flour beetles, we have recruitment into the larval, pupal, and adult age groups.

In the absence of cannibalism, the dynamics of this population would be described by the model

$$x_L(n+1) = bx_A(n),$$
$$x_P(n+1) = (1 - \mu_L)x_L(n),$$
$$x_A(n+1) = (1 - \mu_P)x_P(n) + (1 - \mu_A)x_A(n),$$
(13.5)

where x_L is the larval population, x_P the pupal population, and x_A the adult population; b is the larval recruitment rate per adult per unit time, and μ_L, μ_P, μ_A are the death rates of the three respective stages of the life cycle. The integer n denotes the time step in units of two weeks.

The full model, including the effects of cannibalism, becomes

$$x_L(n+1) = bx_A(n)\exp[-(c_{ea}x_A(n) + c_{el}x_L(n))],$$
$$x_P(n+1) = (1 - \mu_L)x_L(n),$$
$$x_A(n+1) = (1 - \mu_P)x_P(n)\exp[-c_{pa}x_A(n)] + (1 - \mu_A)x_A(n),$$
(13.6)

where c_{ea} and c_{el} are the adult and larval egg cannibalism rates and c_{pa} is the rate of adult cannibalism on the pupae. The possibility that an egg is not eaten in the presence of adults or larvae through the larval stage are given, respectively, by $\exp[-c_{ea}x_A(n)]$ and $\exp[-c_{el}x_L(n)]$. The survival of a pupa through the pupal stage in the presence of adults is $\exp[-c_{pa}x_A(n)]$. The terms of the form $e^{\alpha x}$ arise from the assumption that the cannibalistic acts occur randomly [74].

There are three parameters that can be manipulated experimentally. The adult mortality rate μ_A can be controlled by removing excess mature adults manually from the population at the time of census. The cannibalism rate c_{pa} can be changed by regulating the food supply: more food, less cannibalism. However, it is difficult to quantify this effect in terms of a numerical value for c_{pa}.

A much better approach to regulating c_{pa} is to control the recruitment rate $\exp[-c_{pa}x_A(n)]$ from pupae to adults by removing or adding young adults at the time of census. This is the approach that these investigators took. They kept the food supply constant by growing the flour beetle populations in half-pint milk cartons containing 20 g of food. Every two weeks (census time), the number of adults, pupae, and larvae were counted. The investigators manually removed or added adult flour beetles to maintain the desired μ_A and x_{pa}. The removed adults were added to a separate stock culture that was allowed to grow under natural conditions. This stock culture was used to supply flour beetles as needed to the experimental population. The importance of this procedure is that it minimizes the effects of genetic changes occurring as the flour beetle populations evolve.

The mathematical analysis of (13.5) and (13.6) is well beyond the scope of this introductory textbook (see [74] for a nice explanation). Here we focus on comparisons between the predictions of the model and experimental observations. Four types of solutions can be generated by (13.6) as a function of μ_A and c_{pa}: a stable fixed point, periodic oscillations, quasiperiodic oscillations, and chaotic oscillations. In order to account for the effects of random perturbations from environmental influences and other causes on flour beetle populations, the investigators modified (13.6) to

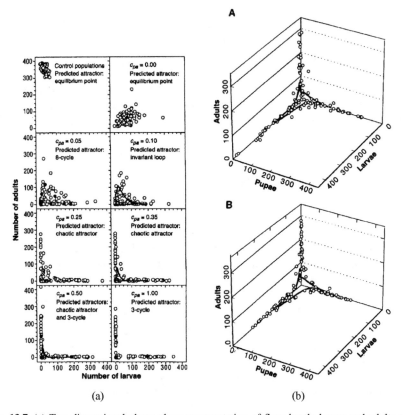

(a)   (b)

Fig. 13.7 (a) Two-dimensional phase-plane representation of flour beetle larvae and adults as a function of the cannibalism rate c_{pa}. Data points are represented by $\circ$, and predictions of (13.6) are represented by the solid symbols (line and $\bullet$). Figure reprinted from [143] with permission. (b) Comparison of (13.7) (upper panel) to experimental observation (lower panel) when c_{pa} is tuned to the chaotic regime. Figure reprinted from [169] with permission.

$$x_L(n+1) = bx_A(n)\exp[-(c_{ea}x_A(n)+c_{el}x_L(n)+\xi_1)],$$
$$x_P(n+1) = (1-\mu_L)x_L(n)\exp[\xi_2],$$
$$x_A(n+1) = [(1-\mu_P)x_P(n)\exp[-c_{pa}x_A(n)]+(1-\mu_A)x_A(n)]\exp[\xi_3],$$

(13.7)

where ξ_1, ξ_2, ξ_3 are random noise variables.

Figure 13.7a compares the observed and predicted changes in x_A and x_L for different choices of c_{pa} when $\mu_A = 0.96$. The investigators categorized the observed behaviors based on the dynamics expected for the noise-free population model given by (13.6). In these phase diagrams, a fixed point (referred to as an equilibrium point by the investigators) appears as a bullet (•). A periodic oscillation is represented by a number of bullets equal to the period of the oscillation (e.g., eight bullets for the 8-cycle). Both quasiperiodic and chaotic solutions yield an invariant loop, represented as a solid line (see also Figure 12.13). Although the qualitative changes in the experimental observations (○) as c_{pa} is changed agree with those predicted by (13.6), it is difficult to judge the extent of agreement.

However, the goal of this investigation was to determine whether chaotic fluctuations could arise in this flour beetle population. As we mentioned previously, the dynamical behaviors of a chaotic attractor are sensitive to the initial conditions. A parameter that can be used to characterize this initial condition sensitivity is the *Lyapunov exponent* λ_L. The precise definition of λ_L is quite technical; however, we can understand the basic idea by considering two solutions of a dynamical system each associated with a different initial condition. The slope of a plot of the logarithm of the separation distance between these two solutions as a function of time provides a measure of λ_L. If the dynamical system is stable, we expect that $\lambda_L < 0$, but if it is chaotic, then $\lambda_L > 0$. The special case of quasiperiodicity is characterized by $\lambda_L = 0$.

The measured values of λ_L for the different values of c_{pa} shown in Figure 13.7(a) were exactly as expected. For the stable solutions, namely $c_{pa} = 0, 0.05, 1.0$, one had $\lambda_L < 0$, and for the quasiperiodic solution, $\lambda_L = 0.1$. Positive λ_L were observed for all situations in which a chaotic solution was predicted by (13.6), namely $c_{pa} = 0.25, 0.35, 0.5$. Figure 13.7b shows the three-dimensional phase-plane diagram for the chaotic attractor that occurs when $c_{pa} = 0.35$. The upper panel shows the chaotic attractor predicted by the stochastic flour beetle model given by (13.7), and the bottom panel shows the experimental observations. Taken together, these observations provide strong evidence that chaotic dynamics arise in this flour beetle population.

13.3.2 Clamped PLR with "Mixed Feedback"

As we discussed, chaotic dynamics can arise in mathematical models of dynamical systems that include time-delayed mixed-feedback control mechanisms. The clamped pupil light reflex (PLR) we introduced in Chapter 9 can be used to test the hypothesis that time-delayed mixed-feedback control can lead to chaotic dynamics in the real world [498, 585]. The equation that describes the dynamics of the PLR clamped with piecewise constant mixed feedback (PCMF) is

$$\frac{dA}{dt} = \begin{cases} A_{OFF} - k_d A & \text{if } A(t-\tau) < A_{ref}^L, \\ A_{ON} - k_c A & \text{if } A_{ref}^L \le A(t-\tau) \le A_{ref}^U, \\ A_{OFF} - k_d A & \text{if } A(t-\tau) > A_{ref}^U, \end{cases} \tag{13.8}$$

where A is the pupil area, τ is the time delay, k_c, k_d are the rate constants for pupil-lary constriction and dilation, and $A_{ref}^U > A_{ref}^L$ are two area thresholds. When A_{ref}^U is larger than the largest A, PCMF is identical to piecewise constant negatuve feedback (PCNF). In other words, the feedback is negative, since the effect of turning the reti-nal illuminance ON is to decrease A. The difference between PCMF and PCNF is that when $A > A_{ref}^U$, the retinal illuminance is turned OFF. Since turning the light OFF causes A to increase, we have positive feedback.

Mathematical studies demonstrate that (13.8) can generate periodic oscillations and chaos [17, 18]. The only parameters that can be manipulated experimentally are the two area thresholds. Figure 13.8 compares the observed changes in A to those predicted by (13.8) when A_{ref}^L and A_{ref}^U are varied. In Figure 13.8a, A_{ref}^U has been set to a value higher than the largest A. Under these conditions, it is possible to measure $k_c, k_d, A_{ON}, A_{OFF}$ by varying A_{ref}^L [584]. These experimentally determined parameter values are used in the model simulations.

Complex oscillations are observed when A_{ref}^U is adjusted to be smaller than the maximal A. The more complicated oscillations in Figure 13.8b, c are in excellent agreement with the solutions of (13.8). Our focus is on the nature of the very com-plex behaviors shown in Figure 13.8d.

As in the case of flour beetles, it is necessary to account for the effects of noise on the dynamics of the clamped PLR. Noise in the PLR is multiplicative [795] and hence manifests itself through fluctuations in the parameters (see also Sec-

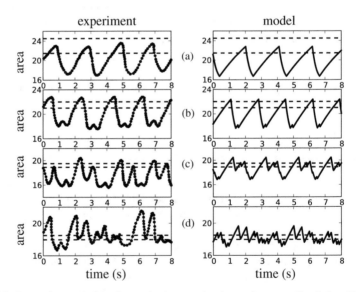

Fig. 13.8 Comparison of the observed changes in A to those predicted by (13.8) for different choices of (A_{ref}^L, A_{ref}^U). (a) $(21.5\ \text{mm}^2, 24.5\ \text{mm}^2)$, (b) $(21\ \text{mm}^2, 22\ \text{mm}^2)$, (c) $(18.9\ \text{mm}^2, 19.5\ \text{mm}^2)$, (d) $(17.95\ \text{mm}^2, 18.5\ \text{mm}^2)$ (represented in figure by horizontal dashed lines). The parameter values are $\alpha_c = 3.88\ \text{s}^{-1}$, $\alpha_d = 0.265\ \text{s}^{-1}$, $A_{ON} = 15.5\ \text{mm}^2$, $A_{OFF} = 34.2\ \text{mm}^2$, and $\tau = 0.411\ \text{s}$. Figure reproduced from [585] with permission.

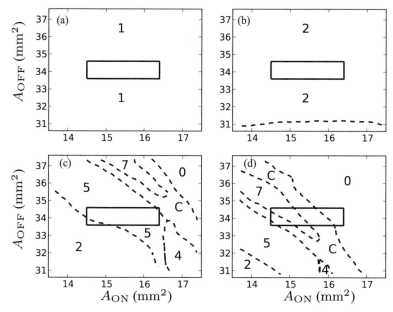

Fig. 13.9 The $(A_{ON} - A_{OFF})$ parameter space for (13.8). Panels (a), (b), (c), (d) correspond, respectively, to (a), (b), (c), (d) in Figure 13.8. The rectangle encloses the observed variations in A_{OFF} and A_{ON}. The integers refer to the number of maxima per period of the periodic solutions of (13.8). Complex periodic solutions and unstable mixing solutions occur in the region labeled C in close proximity. Figure reproduced from [585] with permission.

tion 15.3.2). It is observed experimentally that the fluctuations occur mainly in the parameters A_{OFF} and A_{ON}. Figure 13.9 shows the $A_{OFF} - A_{ON}$ parameter space for (13.8) for the choices of A_{ref}^L and A_{ref}^U shown in Figure 13.8. In this figure, the periodic solutions of (13.8) are characterized by the number of maxima per period. Thus the solution shown in Figure 13.8b is classified as a 2 and that in Figure 13.8c as a 5. The symbol C denotes a region of parameter space in which complex periodic solutions exist in close proximity to unstable mixing solutions. The rectangle encloses the range of the variations in A_{ON} and A_{OFF} measured cycle by cycle under PCNF conditions (Figure 13.8a). For the simpler periodic oscillations, the variations in A_{ON} and A_{OFF} are not large enough to move the dynamics outside a region characterized by a single type of periodic oscillation. Thus experimentally, we observe a solution whose backbone is the solution determined by (13.8).

However, for the complex solution in Figure 13.8d, the variations in A_{ON} and A_{OFF} are large enough to cause switches between qualitatively different solutions of (13.8. Thus the solution in Figure 13.8d is not likely to be a chaotic solution; it more likely represents a combination of several solutions plus transients.

Taken together, these observations indicate that it is possible to observe bifurcations in a human physiological control mechanism as parameters are varied. However, the effects of multiplicative noise are large enough that switches between qualitatively different solutions occur as a function of time. These experiments do not eliminate the possibility that mixed feedback can produce chaotic dynamics. This demonstration would require the construction of a continuous mixed feedback clamping function. We leave this experiment to a future generation. The investigators who did these studies (André Longtin and John Milton) subsequently devoted their careers to studying of the effects of noise on neurodynamical systems.

13.4 Microchaos

There are three sufficient conditions necessary to establish the existence of chaotic dynamics in a mathematical model [877]: 1) sensitivity to initial conditions, 2) the existence of closed invariant sets, and 3) mixing. Using these conditions, we will show that the presence of sensory dead zones in time-delayed feedback control mechanisms feedback can produce a novel type of chaotic dynamics. Since the amplitude of the chaotic fluctuations is typically very small, it is referred to as *microchaos* [152, 153, 200, 311, 592].

13.4.1 The quail map

To illustrate the effects of a threshold on dynamics consider the piecewise linear map (Figure 13.10)

$$x_{t+1} = \begin{cases} \alpha x_t & \text{if } x_t < \Pi, \\ \beta x_t & \text{if } x_t \geq \Pi, \end{cases} \tag{13.9}$$

where Π is the threshold, $\alpha > 1$ and $0 < \beta < 1$ [46]. Mathematically inclined readers will note that this map represents a homeomorphism on the circle. We refer to this map as the *quail map* since maps of this type were initially proposed to describe the growth of bobwhite quail populations [570]. In this context, the theshold x_t equals the number of available hiding spots. When $x_t < 1$, the quail are able to hide from predators and as a consequence the quail population grows. When $x_t > 1$, the quail are vulnerable to predators and hence the quail population declines.

In terms of our criteria for chaotic dynamics, the quail map satisfies two of the sufficient conditions. First, when $x_t < 1$ the fixed point $x_t = 0$ is unstable and hence we have a sensitivity to initial conditions and $\lambda_L > 0$. Second for all initial conditions $x_t > 0$, x_t is eventually bounded on interval $[\beta, \alpha]$. To show this, take $x_t = 1$. This means that there is the existence of a closed invariant set.

There are no stable fixed-point solutions of (13.9). Periodic solutions satisfy the condition

$$\alpha^j \beta^k = 1 \tag{13.10}$$

or

$$\frac{j}{k} = -\frac{\log \alpha}{\log \beta}, \tag{13.11}$$

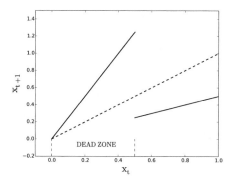

Fig. 13.10 Example of (13.9) when $\alpha = 0.5$, $\beta = 2.5$, and $\Pi = 0.5$. Figure reproduced from [589] with permission.

where j, k are positive integers. When (13.10) is satisfied, all points in $[\beta, \alpha]$ are periodic and marginally ("neutrally") stable. Even when (13.10) is not satisfied, the numbers of iterates between successive maxima differ by no more than 1. The patterns of the iterations can be characterized symbolically by using the continued fraction expansion of $-\frac{\log \alpha}{\log \beta}$ [46].

The quail map illustrates an important property of dynamical systems which possess a dead zone. On the small scale of the order of x_t the dead zone generates complex dynamics such as oscillations. However, on the large scale (i.e., $x_t \gg 1$) the system is stable.

The quail map does not exhibit mixing. The concept of "mixing" can be intuitively understood with the analogy of stirring cream into a cup of coffee. As we stir the cream-coffee mixture, the cream eventually becomes uniformly distributed throughout the coffee. In other words the cream has become mixed into the coffee. In mathematical terms, mixing means that all of the averages of the system's properties quickly decorrelate. This is not true for the quail map since the properties of the oscillatory solutions implies that the ensemble averages do not decorrelate and are independent of the initial conditions. Thus the quail map is ergodic.

13.4.2 Microchaotic dynamics

Microchaotic dynamics arise when a time-delayed digital processor attempts to stabilize an unstable fixed-point of a mechanical system. The effects of digital control introduce quantization into both the time domain (sampling) and the force ("round off"). Digitization in time is a linear effect that increases the dimension of the state space due to the inherent time delay. On the other hand, quantization in space makes the problem strongly nonlinear. Microchaos results from the interplay between a time delay and spatial quantization effects ("dead zone") which arise from analog-digital conversion [152, 153, 200, 220, 311, 374, 803, 839, 840].

To illustrate, let us examine the effects of a digital controller on the behavior of the Hayes equation we introduced in Section 10.5.2 [592, 803]

$$\dot{x}(t) = kx(t) - Gx(t-\tau) \tag{13.12}$$

where $k > 0$ and G is the feedback gain. When $G = 0$, the fixed point $x = 0$ is unstable. The stability boundaries when $G \neq 0$ in the (G, τ) plane are given by (see solid line in Figure 13.11)

1. the line $G = 1$
2. $G = \sqrt{\omega^2 + 1} \quad \tau = \frac{1}{\omega}\arctan(\omega)$.

with $\omega \in [0, \infty)$.

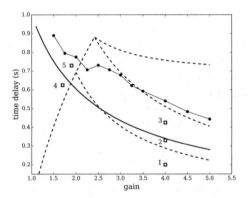

Fig. 13.11 Regions of stable behavior for the Hayes equation (solid line), the quantized Hayes equation (solid line with •) and the switching feedback model for balance control introduced in Problem 7 (dashed line). Numerical solutions for parameter points 1-5 are shown in Figure 13.12. The difference between the quantized Hayes equation and the switching feedback model is that in the latter there is only one quantized step.

In biological applications quantization in the force around zero occurs due to the effects of a dead zone. Thus the feedback forces are computed using integer multiplies of the quantization step, h. Hence we obtain the quantized Hayes equation

$$\dot{x}(t) = x(t) - Gh\mathrm{Int}\left(\frac{x(t-\tau)}{h}\right), \tag{13.13}$$

where $\mathrm{Int}()$ integer function rounds down a real number to the nearest integer. In contrast to (13.12), the stable solutions of (13.13) are limit cycle oscillations which can be determined by piecing together segments of exponential functions as

$$x(t) = \begin{cases} \cdots \\ 2Gh + (x_0 - 2Gh)e^{t-t_0} & \text{if} \quad 2h \le x(t-\tau) < 3h, \\ Gh + (x_0 - Gh)e^{t-t_0} & \text{if} \quad h \le x(t-\tau) < 2h, \\ x_0 e^{t-t_0} & \text{if} \quad -h < x(t-\tau) < h, \\ -Gh + (x_0 + Gh)e^{t-t_0} & \text{if} \quad -2h < x(t-\tau) \le -h, \\ -2Gh + (x_0 + 2Gh)e^{t-t_0} & \text{if} \quad -3h < x(t-\tau) \le -2h, \\ \cdots \end{cases}$$

where $x_0 = x(t_0)$ and t_0 refers to the initial time instant of each segments. Thus when $x(t)$ crosses a threshold at time $t = t_T$, an integer change occurs in the feedback τ later at instant $t = t_T + \tau$. The amplitude of the oscillations scale with h [803]. If $h \to 0$ then (13.13) gives (13.12).

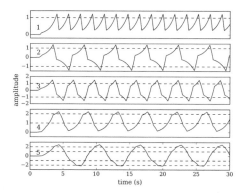

Fig. 13.12 Numerical solutions for the quantized Hayes equation for the parameter points 1-5 in Figure 13.11. The horizontal dashed lines indicate the values of the quantized step.

Figure 13.11 shows that "coarse grained" quantized feedback, i.e., $h = 1$, can stabilize an unstable control system. In the case of the Hayes equation, this stabilization is manifested by the existence of stable solution for values of G and τ which would correspond to unstable dynamics when the feedback is continuous. Abrupt changes in motor force can also arise in the setting of quantization of voluntary movements such as observed in visually-directed arm reaching movements in infants [90] and patients with brain injury [407].

The introduction of time discretization into the feedback requires the use of a time step Δt such that $\tau = R\Delta t$ here R is a positive integer. Hence (13.13) becomes the quantized-sampled Hayes equation

$$\dot{x}(t) = x(t) - Gh\mathrm{Int}\left(\frac{x(t_{j-R})}{h}\right), \tag{13.14}$$

where $t \in [t_j, t_{j+1})$ and $t_j = j\Delta t$, $j = 0, 1, 2\ldots$ gives the time instants where the state variables are sampled. If $\Delta t \to 0$ such that $R\Delta t \to \tau$ then (13.14) gives (13.13).

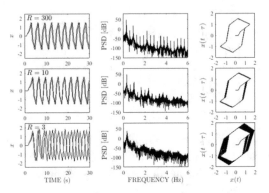

Fig. 13.13 Sensitivity of the solutions of (13.14) to initial conditions for different values of R as shown by (a) the time series, (b) the power spectral density, and (c) phase-plane projections. Parameters were $G = 4.0$, $\tau = 0.438$ and the initial conditions were $\Phi(s) = 0.5$ (black line) and 0.6 (gray line) for $s \in [-\tau, 0]$.

The sampling effect introduces a periodic parametric excitation into the time delay. This can be seen by writing $x(t_{j-R})$, $t \in [t_j, t_{j+1})$ as $x(t - \rho(t))$ where

$$\rho(t) = R\Delta t + t - \Delta t \operatorname{Int}\left(\frac{t}{\Delta t}\right), \tag{13.15}$$

is a Δt-periodic function. Although this observation suggests that (13.14) is equivalent to a time-periodic DDE (i.e., a DDE with time-periodic point delay) with principal period $\Delta t = \tau/R$, it is easier to regard (13.14) as a system of ODEs with a piecewise constant forcing on the right-hand side. The piecewise constant forcing arises when a zero-order hold is applied over each time interval Δt. The term "zero-order hold" means that the force is constant throughout the time interval Δt.

Figure 13.13 shows that the solutions of (13.14) are microchaotic. In particular, as R decreases the solutions become increasingly sensitive to choices of the initial condition, there is a decrease in the subharmonics in the power spectral density, and the enclosure of the orbits in the phase plane becomes larger. Although the amplitudes of the oscillations in (a) appear to be constant, they are actually fluctuating microchaotically [592]. These behaviors are also seen when (13.14) is implemented into an electronic circuit [592].

Example 13.1 An unexplained observation is that the fluctuations in measures related to the vertical displacement angle of quietly standing with eyes closed [181, 510, 894] and the movements of the trunk of a sitting infant [165] exhibited a $\lambda_L \leq 0.2$. Numerical simulations using (10.47) generate microchaotic dynamics characterized with a $\lambda_L \leq 0.2$ for $\tau = 0.1$s, $\Delta t \approx 0.005 - 0.03$s, and $R \approx 2 - 5$, where $\tau = R\Delta t$ [592]. These observations demonstrate that chaotic dynamics can be generated by a time-delayed intermittent control strategy in which there is a frequency-dependent force encoding. ◊

13.4.3 Microchaotic map

The microchaotic map

$$x(t_{j+1}) = ax(t_j) - bInt(x(t_j)) \qquad (13.16)$$

where

$$a = \exp(k\Delta t) > 1, \quad b = \frac{G}{k}(1 - \exp(k\Delta t))$$

generates chaotic and transient chaotic dynamics depending on the values of a and b [311, 803]. This equation can be derived in many ways, for example, by combining (13.13) and (13.14) an solving over the interval $[t_j, t_{j+1})$. Two cases have been extensively studied. When $0 < a - b < 1$ the map generates microchaos [311]. We see that when x_j is large, x_{j+1} always decreases. In this sense the system is stable on the "large scale." However, the fixed point 0 is unstable and hence the system is unstable on the "small scale." The term "microchaos" originated because the chaotic dynamics are confined to a small region near 0 (Figure 13.14).

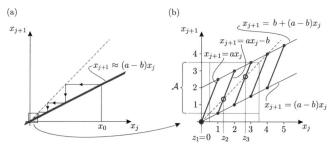

Fig. 13.14 The microchaotic map (13.16) when $a = 2.5$ and $b = 2$. For these parameter choices there are three unstable fixed points ($z_1 = 0$, $z_2 = 4/3$ and $z_3 = 8/3$). When $x_j \gg 1$, $x(j+1) \approx (a-b)x(j)$ (large scale). When $Int(x(j)) = 0$, $x(j+1) = ax(j)$ (small scale). Figure reproduced from [589] with permission.

The other case that has received attention is $a - b < -1$. Despite the fact that the fixed point is both locally and globally unstable, it is possible that a stable microchaotic solution exists [803]! This case anticipated the observation in Figure 13.11 that feedback quantization could enhance the region of stability for the Hayes equation which we have already discussed. In addition, transient microchaotic solutions are also possible. Here the term "transient microchaos" refers to metastable solutions that transiently survive close to 0 before diverging toward infinity [152, 459].

13.5 Multistability

Switch-like transitions occur frequently in biology [741]. For example, the neural networks that control movement [185, 486], autonomic [63], and behavioral [737] states. Moreover it has been suggested that switches in network dynamics may trigger pathological states such as migraine [157] and epilepsy [491, 587, 597, 700, 704]. Therapeutic interventions are continually being developed to exploit switch-like changes in dynamics. These strategies range from inducing immunity through vaccination with sub-pathologic viruses [751] to using electrical stimuli to defibrillate the heart [421], suppress parkinsonian tremors [313, 732] and abort epileptic seizures [158, 493, 574, 576, 661].

Up to this point, we have identified switch-like transitions with bistability in nonlinear ordinary differential equations. However, a remarkable property of time-delayed dynamical systems is their propensity to exhibit multistability, i.e., more than two coexistent attractors. Over the last 25 years a great of insight into the origin of multistability in the nervous system has been obtained by studying the dynamics of small neuronal microcircuits composed of 2-3 neurons (for a review see [597]). These neural microcircuits are currently referred to as *motifs* and are thought to be the basic building blocks of brain dynamics [10, 406, 567, 788].

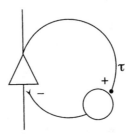

Fig. 13.15 Schematic representation of a recurrent inhibitory loop. An excitatory neuron, E, excites an inhibitory interneuron, I, which in turn inhibits E after a time delay, τ. The time delay accounts for the time between when E produces an action potential and an inhibitory postsynaptic potential is delivered by I to E.

Here we focus on the dynamics of a single recurrent inhibitory loop (RI) (Figure 13.15). Neuro-anatomists have long recognized that all excitatory neurons in the central nervous system are associated with at least one inhibitory feedback loop [109, 762, 875]. Consequently the dynamics of these inhibitory loops have attracted considerable attention [42, 101, 118, 216, 236–238, 515, 516, 521, 581, 597, 603, 686, 760, 880]. Mathematical and computer simulations demonstrate that under certain conditions the delayed recurrent inhibitory loops can be multistable. We briefly illustrate the observation with three types of models: 1) integrate-and-fire models, 2) Hodgkin–Huxley networks, and 3) 2-neuron coupled networks with delay.

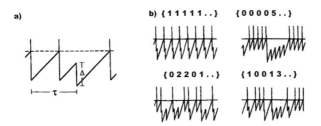

Fig. 13.16 a) The time course of the membrane potential v for the integrate-and-fire neuron E in a time-delayed recurrent inhibitory loop. The dashed line indicates the threshold. b) Four coexistence periodic attractors that occur when $\tau = 4.1$ and $\Delta = 0.8$. The different patterns are described by the number of inhibitory pulses between two successive neuron spikes. Thus for the lower left pattern in b) going from left to right, 0 inhibitory pulses between the first two spikes, 2 inhibitory pulses between the next two spikes, then 2 inhibitory spikes, then 0, then 1. After this the pattern repeats. Thus the shorthand label for this spike pattern is {02201}. Figure reproduced from [597] with permission.

13.5.1 Multistability: Integrate-and-fire models

The simplest model for RI that illustrates the interplay between τ and multistability is the integrate-and-fire model whose dynamics are shown in Figure 13.16a. The membrane potential, V, of E increases linearly at a rate, R, until it reaches the firing threshold, Π [237]. When $V = \Pi$, E spikes and V is reset to the resting membrane potential, V_0. The period is $T = \Pi/R$. The spike generated by E excites I, which in turn after a time delay, τ, delivers an inhibitory post-synaptic potential which lowers the membrane potential of E by an amount δ. It is assumed that the effect of δ is independent of the phase it was delivered. Consequently, the effect of the IPSP when $R > 0$ is to decrease V by an amount δ. For simplicity we take $V_0 = 0$ and define the following dimensionless variables: $\tau^* = \tau/T, t^* = t/T, v^* = V/\Pi, \Delta = \delta/\Pi$, so that the dimensionless firing threshold, period and voltage growth rate are, respectively, $\Pi^* = 1, T^* = 1, R^* = \Pi^*/T^* = 1$. Dropping the asterisks, we see that the dynamics of the recurrent loop depend only on two parameters, namely $\tau > 0$ and $\Delta \geq 0$. When $\tau < 1$, E spikes periodically with period $1+\Delta$. This is because decreasing the membrane voltage by an amount Δ is equivalent to increasing the interspike interval by $1+\Delta$.

The essential condition for multistability in this dimensionless model is that $\tau > 1$. Complex behaviors become possible since the inhibitory pulses are not necessarily the result of the immediately preceding excitatory pulse. When $\tau = 4.1$ and $\Delta = 0.89$ there are 4 different periodic spiking patterns possible (Figure 13.16b). All of these solutions are periodic with the same period and that there are exactly five excitatory spikes and hence five inhibitory pulses per period.

Example 13.2 Despite the simplicity of this model, it does make an intriguing suggestion concerning the bimodal incidence of all types of epilepsy [328] and the changes in brain myelination that occur with development [597]. Assume that the incidence of epilepsy is associated when the multistable nature of the brain. Brain maturation is associated with increased myelination of neuronal axons which increases their conduction velocities, thereby decreasing τ, and reducing the number of coexistent attractors. This observation could explain why epilepsy is particularly common in children and why seizures tend to decrease in frequency, and even disappear altogether, as the child gets older. On the other hand, in the elderly axonal conduction velocities increase due to the death of oligodendrocytes, namely the cells responsible for myelinating axons in the central nervous system. Although when an oligodendrocyte dies, other oligodendrocytes remyelinate the axon, the new myelin sheaths are thinner, the internodal distances are shorter and hence the conduction velocities are slower [682]. This would explain why the incidence of seizures increases in the elderly. Of course the reader should realize that this explanation might be an oversimplification! ◇

13.5.2 Multistability: Hodgkin–Huxley models with delayed recurrent loops

An extension of the delay recurrent loop model is to replace the integrate-and-fire neuron by a conductance-based model of the neuron described by the Hodgkin–Huxley (HH) neuron [236, 515, 516]. The resultant model takes the form

$$\begin{cases} C\dot{x}(t) = -g_{Na}m^3h(x(t) - E_{Na}) - g_K n^4(x(t) - E_k) \\ \qquad\qquad -g_L(x(t) - E_L) - F(x(t - \tau)) + I_s(t), \\ \dot{m}(t) = \alpha_m(x)(1 - m) - \beta_m(x)m, \\ \dot{n}(t) = \alpha_n(x)(1 - n) - \beta_n(x)n, \\ \dot{h}(t) = \alpha_h(x)(1 - h) - \beta_h(x)h, \end{cases} \qquad (13.17)$$

where $F(x)$ is the signal function which describes the effect of the inhibitory neuron I on the membrane potential of the excitatory neuron E. The initial function ϕ in the interval $[-\tau, 0]$ was assumed to have the form of neural spike trains. Namely, it is given by a sum of square pulse functions.

With sufficiently large I_s that makes the neuron fire successively, several coexisting periodic attractors exist (Figure 13.17). Solutions starting from domains of attraction of these periodic solutions exhibit exotic transient behaviors but eventually become periodic. These attractors take the form of spike trains with different patterns of interspike intervals. The panels on the right side of the figure show a blow up of the solutions in a given period (not delay τ) to more clearly illustrate the patterns of solutions. The number of coexistent attractors increase as the ratio of τ to the intrinsic spiking period increases. Indeed, when this ratio is ~ 9, there exists $> 10^6$ existent attractors. Multistability can arise in models of delayed recurrent loops which take into account the phase resetting properties of each neuron in the loop [238, 827]. The advantage of this approach is that the phase resetting curve can be measured experimentally and thus all parameters in the model are known.

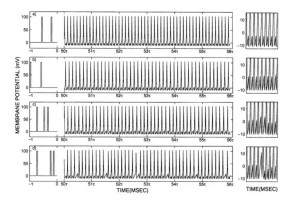

Fig. 13.17 Four coexisting attracting periodic solutions generated by the excitatory neuron E for the Hodgkin–Huxley model (HH) given by (13.17). The initial conditions are shown on the left-hand side and have the form of neural spike trains. These were constructed using square-wave pulses. The panels on the right-hand side are "blow ups" of the solutions to make it easier to see the differences between them. Figure reproduced from [597] with permission.

13.5.3 Multistability: 2-neuron coupled networks with delay

The next step is to examine models which describe the dynamics of 2-neuron coupled networks. In particular, we explore the dynamics exhibited by a motif described by coupled neuron networks of the form

$$\dot{x}_1 = -k_1 x_1(t) + k_{11} g_{11}(x_1(t - \tau_{11})) + k_{21} g_{21}(x_2(t - \tau_{21})) + I_1,$$
$$\dot{x}_2 = -k_2 x_2(t) + k_{22} g_{22}(x_2(t - \tau_{22})) + k_{12} g_{12}(x_1(t - \tau_{12})) + I_2, \quad (13.18)$$

where $x_j(t)$ $(j = 1, 2)$ are the spiking rates of the neurons at time t, k_{ij} represent the strength of the connections, and I_j $(j = 1, 2)$ are the external inputs to the neurons. The function $g(x)$ is sigmoidal and can be written in many ways, most commonly taken as $\tanh(cx)$, $x^n/(c + x^n)$, or $1/(1 + e^{-cx})$. For a recurrent inhibitory loop $k_{11}, k_{12} < 0$ and $k_{21}, k_{22} > 0$. The special case when $k_{12} = k_{21}$ corresponds to a 2-neuron Hopfield network [349].

The fixed points of (13.18), $\bar{x}_1, \bar{x}_2$, can be solved by setting all of the time delays equal to zero, $\dot{x}_1 = \dot{x}_2 = 0$, and solving

$$0 = -k_1 \bar{x}_1 + k_{11} g_{11}(\bar{x}_1) + k_{21} g_{21}(\bar{x}_2) + I_1, \quad (13.19a)$$
$$0 = -k_2 \bar{x}_2 + k_{22} g_{22}(\bar{x}_2) + k_{12} g_{12}(\bar{x}_1) + I_2. \quad (13.19b)$$

For recurrent inhibition there is typically only one fixed point, $(\bar{x}_1, \bar{x}_2)$. The stability of the fixed point can be determined by linearizing (13.18) about the fixed point to obtain

$$\dot{u}_1 = -k_1 u_1(t) + a_{11} u_1(t - \tau_{11}) + a_{21} u_2(t - \tau_{21}), \quad (13.20)$$
$$\dot{u}_2 = -k_2 u_2(t) + a_{22} u_2(t - \tau_{22}) + a_{12} u_1(t - \tau_{12}),$$

where $u_i(t) = x_i(t) - \bar{x}_i$ and $a_{ij} = k_{ij}g'_{ij}(\bar{x}_i)$. The analysis is complicated by the fact that a_{11} depends explicitly on g_{11} but also depends implicitly on the other k_{ij} through the value of the fixed point. Similarly, for the other a_{ij}. Assuming that $u(t) \sim e^{\lambda t}$, we obtain the characteristic equation

$$(\lambda + k_1 - a_{11}e^{-\lambda\tau_{11}})(\lambda + k_2 - a_{22}e^{-\lambda\tau_{22}}) - a_{12}a_{21}e^{-\lambda(\tau_{12}+\tau_{21})} = 0. \quad (13.21)$$

We see that the delay associated with the connections between neurons only appears in the combination $\tau_{12} + \tau_{21}$ and hence it is the total delay that will be important in determining the dynamics of the RI loop [100].

Since there is only a single fixed point, multistability must be the consequence of bifurcations. As we have seen a bifurcation arises when a change in the stability of the fixed point happens as the result of a parameter change. The bifurcations arise when at least one root of the characteristic equation has zero real part and all of the rest have negative real parts. The mathematical analysis of (13.18) is difficult. Some results related to the occurence of multistability are

1. (13.21) can have one zero root ($\lambda = 0$) when $(k_1 - a_{11})(k_2 - a_{22}) - a_{12}a_{21} = 0$ and $a_{22} > k_2$. The last conditions means that there must be strong enough self-coupling on the inhibitory neuron. This bifurcation is associated with the creation or destruction of a fixed point and hence can be important in the generation of multistability of fixed points.
2. (13.21) can have a pair of pure imaginary roots. Thus it is possible that there occurs a Andronov–Hopf bifurcation leading to the creation of a periodic solution [42].
3. (13.21) can have multiple roots with zero real part, particularly when multiple delays are present [42, 101, 760]. In such cases, multistability and even more complex dynamics are possible [291, 455]. For example, there can be i) multi-stability between a slowly varying periodic solution and one or more fixed points (double zero root) [216], ii) multistability between a periodic solution and one or more fixed points (zero root plus a pure imaginary pair), and iii) bistability between periodic orbits with an unstable torus (or reverse) (two pairs of pure imaginary eigenvalues without resonance [100, 760].

13.6 Delay-induced transient oscillations

We ask whether, in the presence of noisy perturbations, the dynamics of delayed multistable networks are dominated by the stable fixed points located within basins of attraction or by the unstable separatrices that separate the basins of attraction. The following example demonstrates that the dynamics of such networks may, in fact, be dominated by transient effects related to the separatrices that separate stable fixed points.

A simple neural circuit that exhibits bistability takes the form of two mutually inhibitory neurons (Figure 13.18):

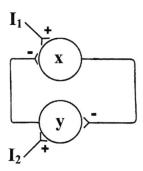

Fig. 13.18 Schematic representation of a neural network with mutually inhibitory neurons. The minus sign indicates an inhibitory connections, while the plus sign indicates an excitatory connection.

$$T_1 \frac{dx}{dt} = -x - S_2(y(t - \tau_2)) + I_1(t), \quad T_2 \frac{dy}{dt} = -y - S_1(x(t - \tau_1)) + I_2(t), \quad (13.22)$$

where x, y are the firing rates of the neurons, T_1, T_2 are the neural time constants, I_1, I_2 represent external inputs, τ_1, τ_2 are the conduction time delays between the two neurons, and

$$S_j(u) = \frac{k_j u^{n_j}}{K_j^{n_j} + u^{n_j}}, \quad j = 1, 2,$$

describe sigmoidal functions representing inhibitory influences (increasing and non-negative for $u \geq 0$). For simplicity, we assume that $\tau_1 = \tau_2$. This equation has arisen in discussions of the dynamics of neural networks [288, 665, 666], decision-making models [577] and certain types of epilepsy [587, 704].

For appropriate choices of the parameters, there can be three fixed points, two of which are stable. In the context of a decision-making model [577], each stable fixed point corresponds to a decision. The role of the unstable fixed point on this process depends on the value of τ. When $\tau = 0$, the unstable fixed point is associated with an unstable limit cycle [575, 666]. Since this limit cycle is unstable, it is not observed. However, when $\tau > 0$, there can be a transient stabilization of this unstable limit cycle [575, 666].

Figure 13.19 shows the effect of the unstable fixed point on the dynamics of (13.22) as a function of τ. In these computer simulations, the initial function has been chosen to place the dynamical system near, but not at, the unstable fixed point. Transient limit-cycle oscillations arise whose duration depends on τ. These delay-induced transient oscillations (DITOs) can last so long that it would be almost impossible to distinguish them from those generated by a limit-cycle attractor [288, 577, 665, 666, 704].

Bistable states naturally arise in biological systems as one attractor is replaced by another [423, 587, 704]. Thus one would anticipate that DITOs arise frequently in biological systems. Indeed, it has been suggested that DITOs might explain the onset of seizures that occur in a form of epilepsy referred to as nocturnal frontal lobe epilepsy (NFLE) [587, 704]. NFLE is a familial epilepsy that is related to a

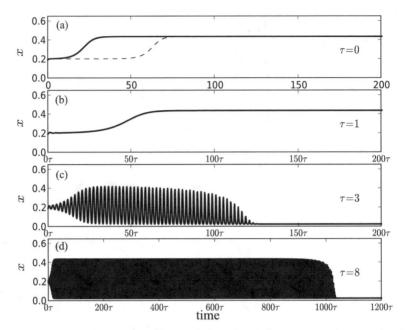

Fig. 13.19 Behavior of (13.22) for different choices of τ. Parameters were chosen so that there exist three fixed points (see Exercise 4). When $\tau = 0$, the initial condition was either $(0.19, 0.19)$ (solid line) or $(0.19999, 0.19999)$ (dotted line). When $\tau \neq 0$, the initial function was $(X(s), Y(s)) = (0.19, 0.19)$ for $s \in [-\tau, 0]$. Figure reproduced from [575] with permission.

molecular defect in central nicotinic acetylcholine receptors. The curious feature of this epileptic syndrome is that seizures occur only when the subject is sleeping, most commonly during the transition from Stage I to Stage II sleep. These observations raise the possibility that the clinically observed seizures in NFLE may be a manifestation of a DITO.

13.7 What Have We Learned?

1. What are four types of problems whose dynamics are described by a one-dimensional map?

> Situations that are amenable to descriptions in terms of one-dimensional maps: (1) the growth of a population characterized by a single discrete breeding time; (2) the use of a Poincaré section to describe the dynamics of a limit cycle (see Section 12.1); (3) the description of the dynamics of numerical algorithms used to integrate differential equations (see Section 4.4.2); and (4) the *singular perturbation limit* of a delay differential equation (Exercise 2).

2. What is the condition for stability of a continuous-time dynamical system and of a discrete-time dynamical system?

> The condition for stability of a fixed point in a continuous-time dynamical system is that the real parts of all of the eigenvalues λ be negative. For a discrete-time dynamical system, the condition is $|\lambda| < 1$ for all of the eigenvalues.

3. What is a dynamical disease?

> A dynamical disease is a disease that arises because a parameter has been changed from a value for which the system exhibits healthy dynamics to a value for which the system exhibits unhealthy dynamics.

4. What are the conditions that establish the existence of chaotic dynamics in a mathematical model?

> There are three sufficient conditions necessary to establish the existence of chaotic dynamics in a mathematical model [877]: 1) sensitivity to initial conditions, 2) the existence of closed invariant sets, and 3) mixing.

5. What are the properties of a chaotic attractor?

> A chaotic attractor is a time-dependent dynamical behavior that is bounded and aperiodic and that exhibits a sensitive dependence on changes in initial conditions. The dynamics are characterized by a positive Lyapunov exponent.

6. What are examples of biological experimental paradigms that have produced chaotic dynamics?

> Two examples of biological experimental paradigms that exhibit chaotic dynamics are flour beetle populations and sinusoidally stimulated excitable systems such as invertebrate neurons and the green alga *Nitella*.

7. What is the effect of digital control on the properties of a continuous dynamical system?

> The effects of digital control introduce quantization into both the time domain (sampling) and the force ('round off'). Digitization in time is a linear effect that increases the dimension of the state space due to the inherent time delay. On the other hand, quantization in space makes the problem strongly nonlinear.

8. What is microchaos?

> The term "microchaos" describes chaotic dynamics which are confined to a small region near 0. Thus the effects of microchaos on the observed dynamics are very small. Typically, micro-chaos results from the interplay between a time delay and spatial quantization effects ('dead zone') which arise from analog-digital conversion.

9. With respect to microchaotic dynamics what do the terms "small scale" and "large scale" mean?

> Microchaos most often arises where there is a threshold below which feedback control is not in effect. Thus when the controlled variables is below threshold (unstable on a "small scale") it increases in magnitude. However, once the controlled variable exceeds

the threshold (stable on a "large scale"), feedback control is exerted and the variable decreases in magnitude.

10. What is "transient microchaos?"

> The term "transient microchaos" refers to a metastable microchaotic solution that transiently survives close to 0 before diverging towards infinity.

11. What is a neuronal (neural) motif?

> Small neuronal microcircuits composed of 2-3 neurons which are thought to be the basic building blocks of larger ensembles of neurons.

12. What is multistability?

> The coexistence of 2 or more stable states.

13. In what paradigms of a 2-neuron delayed recurrent loop has it been possible to demonstrate the existence of multistabilty?

> Multistability has been demonstrated in four types of models for a delayed recurrent loop: 1) integrate-and-fire models, 2) Hodgkin-Huxley networks, 3) 2-neuron coupled networks with delay and 4) an *Aplysia* motoneuron dynamically clamped with inhibitory feedback [238].

14. What is a delay-induced transient oscillation?

> A delay-induced transient oscillation (DITO) is a transient oscillation that arises in a multistable time-delayed dynamical system that is tuned close to the separatrix. The importance of a DITO is that for sufficiently long delays, it can be indistinguishable from a limit cycle oscillation.

13.8 Exercises for Practice and Insight

1. Write a computer program to generate a cobweb diagram such as shown in Figure 13.4.
2. An important use of the cobweb diagram arises in the analysis of the $n : m$ phase-locking patterns observed when a limit-cycle oscillator is perturbed periodically. In particular, the cobweb diagram is used to determine the integer values of n and m that correspond to a solution with a given winding number ρ. Figure 13.20 shows that for the standard circle map described by (12.10), we have a 2 : 3 phase-locking pattern when $b = 2/3$ and $k_1 = 0.0025$.

   a. What happens to the cobweb diagram as k_1 increases (recall that $k_1 < 0.16$)?
   b. Refer to Figure 12.12 and choose b and k_1 that have the same winding number. Does the cobweb diagram for these choices of the parameters resemble that in Figure 13.20?
   c. In what way are the cobweb diagrams for all of these choices of b and k_1 similar?

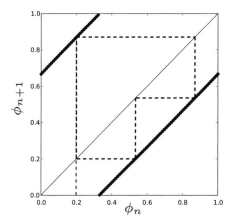

Fig. 13.20 Cobweb diagram for (12.10) for a choice of b corresponding to $\rho = 2/3$ with $k_1 = 0.0025$. This solution corresponds to a $2 : 3$ solution: there are two contacts with the left branch of the circle map for every three contacts of both branches to complete the cycle.

3. Use the computer program you wrote to produce a cobweb diagram to investigate the dynamics of the microchaotic map given by (13.16) for the following parameter choices:

   a. $a - b > 1$
   b. $0 < a - b < 1$
   c. $-1 < a - b < 0$
   d. $a - b < -1$

   See [803] to check your results.

4. The dynamics of a differential equation and its discrete-time approximation are very different. The continuous-time version of (13.2) is the Verhulst, or logistic, equation

$$\frac{dx}{dt} = kx(1 - x).$$

   a. What is the fixed point of the Verhulst equation, and what is its stability?
   b. Using the definition of a derivative (Tool 2), show that the Verhulst equation can be reduced to an equation of the form of (13.2).
   c. Is it possible for the Verhulst equation to generate an oscillation?
   d. The delayed Verhulst equation has the form [362]

$$\frac{dx}{dt} = rx(t) \left[1 - \frac{x(t - \tau)}{K} \right].$$

   i. What is the fixed point and determine its stability?
   ii. Can this equation generate an oscillation when the fixed point is unstable (write a computer program to see whether oscillations occur)?

5. It is interesting to explore the behavior of the Mackey–Glass equation described by (13.4).

    a. What are the fixed points?

    b. What is the condition for instability?

    c. Write a computer program to integrate (13.4). Compute the power spectrum for values of μ that yield a periodic solution and a chaotic one. How does the power spectrum change? Hint: This task is easiest to accomplish when the time step for the numerical integration is $1/2^n$.

6. Show that a Andronov–Hopf bifurcation cannot occur in (13.18) if all of the delays are equal to zero.

7. Write a computer program to examine the dynamics of (13.22). Choose the following parameters [575]: $k_x = 0.4$, $k_y = 0.4$, $K_y = 0.2 = K_x$, $n_x = n_y = 2$, $I_x = 0.5$, $I_y = 0.5$. Assumed that the conduction delay τ for $x \to y$ is the same as for $y \to x$ and choose the initial functions $\Phi_x(t)$ and $\Phi_y(t)$, where $t \in [-\tau, 0]$, to be constant.

    a. Verify that (13.22) has either one or three fixed points and evaluate their stability.

    b. Demonstrate the existence of bistability in (13.22).

    c. For the above parameter choices, the unstable point among the three coexistent fixed points occurs for $x^* = y^* = 0.2$. Chose the initial functions to be close, but not equal, to the unstable fixed point, say $\Phi_x(t) = 0.19 = \Phi_y(t)$. What happens as τ increases? The oscillation you see is called a *delay-induced transient oscillation*, or DITO [665, 666, 704].

    d. Do DITOs arise in the time-delayed versions of the two-repressor and three-repressor models discussed, respectively, in Exercises 6 and 7 in Chapter 9? Write a computer program to verify your prediction.

Chapter 14
Random Perturbations

Up to this point, our "mathematics in the laboratory" approach to the investigation of biological systems has been very simplistic. We have assumed that very complex biological systems, ranging from ecosystems to the human brain, can be adequately described by measuring changes in just a few variables. Moreover, we have ignored the fact that all living dynamical systems are continuously subjected to large numbers of influences. For example, it has been estimated that a typical neuron in the human brain receives $\approx 10^4$ synaptic connections from other neurons [806].

A first step toward a more thoroughgoing investigation of complex biological systems is to make use of the systems–surroundings concept that was introduced in Section 1.4. A key decision is to fix the line that demarcates the system from its surroundings: this line must be chosen such that the surroundings can influence the system, but not vice versa. A prudent choice for the system reduces the number of variables to be considered. If the number of independent perturbations[1] acting on the system from the surroundings is very large, then we can model the influence of the surroundings on the system as a kind of "collective noise." This system–surroundings approximation leads to considerations of *stochastic differential equations* of the form

$$\frac{dx}{dt} = f(x, \xi), \tag{14.1}$$

where ξ accounts for effects of the collective noise.

Considerations of the nature of biological noise and its effects on the control of biological systems have long been topics of interest for biologists (for discussions related to the nervous system, see [215, 218, 578, 609]). Over the last few decades, the perspective of biologists on noise has dramatically shifted from a focus on its detrimental effects on biological systems to considerations of its beneficial effects [567, 578, 609]. In part, the impetus has been provided by advances in technology. The accuracy at which measurements can be made is often better than the magnitude of the observed fluctuations in the variable. Thus it is no longer possible to blame

---

[1] The term *independent* means that a change in one variable does not lead to a change in other variables. We will give a more formal definition in Section 14.2.

© Springer Nature Switzerland AG 2021
J. Milton and T. Ohira, *Mathematics as a Laboratory Tool*,
https://doi.org/10.1007/978-3-030-69579-8_14

measurement noise as the source of stochasticity. Moreover, it is now possible to study the dynamics of a single cell and even of single molecules. In addition, issues related to statistical distributions with "broad shoulders" and power-law behaviors arise frequently in discussions of biological dynamics. Thus techniques for the analysis and description of stochastic dynamical systems have become essential tools for laboratory researchers.

Although the study of stochastic dynamical systems is difficult, we believe that a major obstacle to understanding such systems is the large gap that often exists between scientists and engineers who work at the benchtop and mathematicians who explore the theoretical foundations of the subject. This has produced a multiplicity of terminologies and few attempts to discuss stochastic processes in practical terms (for notable exceptions, see [49, 161, 518]). Our goals in this and the next two chapters is to bring this information together "under one roof."

Fortunately, we will be able to show that some of the techniques that we applied to deterministic dynamical systems, in particular Fourier analysis, can be used to study stochastic dynamical systems. Three themes will run through our presentation: What is the best way to characterize noise? What are the effects of noise on dynamical systems? Can noise have beneficial effects on the dynamics of biological systems, and if so, what is the nature of those benefits?

The questions we will answer in this chapter are:

1. What is the Bayes' theorem?
2. How the relation between the joint and conditional probabilities are used in the Bayes' theorem?
3. How is the Bayes' theorem used to infer a cause?
4. What are the priori and posterior probabilities?
5. What are the Bayesian updates?
6. What is the Monty Hall problem?
7. Why is the concept of probability density function (pdf) important?
8. What is the advantage of characterizing stochastic processes in the frequency domain?
9. What is the autocorrelation function, and why is it useful?
10. What is the cross-correlation function, and why is it useful?
11. Why can methods for determining statistical descriptions of time series behave poorly?
12. How do we determine the power spectrum $W(f)$ for a stochastic process?
13. What is the rate-limiting step in computer programs used to simulate large-scale data sets: computational speed or memory access?
14. What does it mean to prewhiten a time series, and why is it necessary to prewhiten time series in the analysis of the cross-correlation function?

14.1 Noise and Dynamics

The deterministic component of dynamical systems provides the backbone, or scaffolding, on which the effects of noise produce the dynamics observed at the benchtop. There are three reasons to consider the interplay between determinism and randomness. First, the analysis of noisy dynamics can uncover the nature of underlying control mechanisms, because we can consider noise to be a naturally occurring perturbation to system dynamics. At the very least, it behooves deterministic modelers to examine to what extent the observed dynamics can be mimicked by their models when noise is added. Second, it is possible that novel mechanisms arise from the interplay between stochastic and deterministic processes that are impossible in the absence of noise. Finally, certain biological problems are fundamentally stochastic in nature. Here we briefly introduce three illustrative examples.

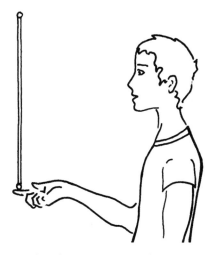

Fig. 14.1 Schematic representation of a person balancing a stick at the fingertip.

14.1.1 Stick Balancing at the Fingertip

In Chapter 10, we introduced the stabilization of a pendulum in its inverted position as a benchmark for control engineers: the shorter the pendulum that can be stabilized, the more robust the controller. The fact that the inverted upright position can be stabilized using a time-delayed feedback controller led one of the authors of this book to study to pole balancing at the fingertip (Figure 14.1). Curiously measurements of the vertical displacement angle using high-speed motion-capture cameras suggested that the fluctuations have an intermittent character: occasional large deviations are interspersed with many much smaller fluctuations. One of the authors of this book found that he was unable to explain the significance of this observation.

Then one day, a young Venezuelan physicist named Juan Luis Cabrera[2] walked into his office, and within minutes of having the problem explained to him, he had the answer. He explained that what we were observing was the dynamics of a system that was exhibiting *on–off intermittency* [94, 96].

As explained in Section 15.3.3, this intermittency is a result of the stochastic forcing of a parameter back and forth across a stability boundary. This interpretation led, in turn, to the hypothesis that the nervous system uses an intermittent control strategy, referred to as "drift and act," to control balance [95, 579, 595, 596]. In other words, for small displacements from the vertical, the system is allowed to *drift* without corrective movements being made actively. Under these conditions, control is maintained by the interplay between noise and delay [94, 580]. It is then necessary to *act* by means of corrective movements once the vertical displacement angle becomes larger than a certain threshold.

Since that time, several different hypotheses have been proposed for intermittent motor control (for reviews, see [23, 370, 579]). The importance of the intermittent control hypothesis is that it draws attention to the possibility that biological organisms likely favor energy-efficient control strategies over those that are energetically expensive. However, for the purpose of our present discussion, we emphasize that all of this research may never have been undertaken unless someone, Juan Luis in this case, had recognized the nature of the stochastic process that was exhibited by pole balancing at the fingertip.

14.1.2 Noise and Thresholds

A simple experiment was sufficient to demonstrate that the addition of noise to a dynamical system could have beneficial effects. Consider the perception of a pure tone whose amplitude is below the detection threshold of auditory neurons. This means that we should not be able to hear such a sound. However, the addition of noise of sufficient intensity to the pure tone signal enables a person to hear the sound. This phenomenon is an example of *stochastic resonance* and is discussed in Section 15.2.3.

14.1.3 Stochastic Gene Expression

It has become possible to investigate the properties of biochemical reactions that occur in a single cell. Since the volume of a cell is small, the number of molecules that participate in a given reaction becomes a consideration [301]. For example, in bacteria, there are typically only one or two copies of a gene, and transcription factors number only in the tens [400, 709]. Stochastic effects, or *stochasticity*, become a concern when the numbers of reactants are of such small sizes [195, 301, 711].

---

[2] Juan Luis Cabrera (PhD 1997), Venezuelan condensed matter biophysicist.

First, when numbers are small, the probability of a collision between reactants is also small, and hence reaction rates are low. Second, when collisions do occur, they alter the internal energies of the molecules and hence their propensity to react. As a consequence, the order of events becomes important: have there been previous collisions, and if so, how many? It has been possible to show that the stochastic effects on transcription dominate the stochastic effects on translation (see Section 15.1.4). A startling observation is the ability of certain bacteria to use a single molecular event to stochastically switch their phenotype [125], possibly to achieve an evolutionary advantage [484].[3]

14.2 Stochastic Processes Toolbox

Here we briefly introduce a number of basic concepts related to random variables and stochastic processes. Our presentation is in the form of a toolbox. Readers who wish a more detailed and rigorous presentation are referred to a number of excellent textbooks that we consulted in putting together our toolbox [27, 49, 161, 668, 670].

14.2.1 Random Variables and Their Properties

Suppose we perform an experiment and then repeat it many times. Let us identify the outcome of each experiment as a point s (possibly n-dimensional) in a sample space S. This sample space represents the aggregate of all possible outcomes (events) of the experiment.

Example 14.1 When we flip a coin, the possible outcomes are heads and tails. The sample space is $S = \{H, T\}$, consisting of the two points H and T.[4] ◇

Often, it is convenient to express the outcome of an experiment in real numbers. A *random variable* $x(s)$ is a real-valued function defined on a suitable collection of subsets of a sample space S [161].[5] Thus, a random variable is a function that maps the collection of events S to the set of real numbers.

Example 14.2 The sample space S for two coin flips is

$$S = \{HH, HT, TH, TT\},$$

---

[3] The ability of a bacterium to change its phenotype to achieve evolutionary advantage is reminiscent of a similar ability that has been observed in *Homo sapiens*, which has been known to change its hair color.

[4] $S = \{H, T\}$ is standard mathematical notation for the set S consisting of the elements H and T.

[5] Technically, the probability is defined on the sigma field or algebra of the sample space (for an introduction, see [670]).

where H indicates a head, and T a tail. The random variable "number of heads" associates the number 0 with the set $s_0 = \{TT\}$, the number 1 with the set $s_1 = \{HT, TH\}$, and the number 2 with the set $s_2 = \{HH\}$. Of course, we could choose other random variables, such as "number of tails," "two heads," and "number of heads minuts number of tails."   ◇

Let us consider an event A that was obtained as the outcome of a given experiment. The *probability* $P(A)$ of the event A is the probability $P(S_A)$ that the sample point s corresponding to the outcome of the experiment falls in the subset of sample points S_A corresponding to the event A:

$$P(A) = P(s \in S_A) = P(S_A),$$

where the notation $s \in S_A$ means that a point s is an element of the point set S_A. It is necessary to specify a priori the probability for each elementary event s in order to compute $P(A)$. In other words, if we are considering the example of coin flipping, then in order to estimate the probabilities, we need to specify first whether, for example, the coin flips are unbiased or biased. The following two examples illustrate how a priori probabilities assigned to each elementary event change the probability of a particular event.

Example 14.3 Let us consider again two coin flips, where

$$S = \{HH, HT, TH, TT\}.$$

We want to know the probability of the event A that the number of heads after two coin flips is 1. Thus $S_A = \{HT, TH\}$. If the coin is fair, the probability of getting one head and one tail from two flips is $1/2$. Thus,

$$P(A) = P(s \in S_A) = P(S_A) = \frac{1}{2}.$$

◇

Example 14.4 Now suppose that the coin is biased and that the probability of a head and tail are, respectively, $3/4$ and $1/4$. In that case, we have $P(A) = 3/8$.   ◇

A random variable may be discrete or continuous. A *finite discrete random variable* x takes on only a finite number of values. Figure 14.2a shows an example of a discrete random variable that takes on only five possible values, $\{x_1, x_2, x_3, x_4, x_5\}$. The probability associated with each x_j is $P(x_j)$. The function of X whose value is the probability $P(x \leq X)$ that the random variable x is less than or equal to X is called the *probability distribution function*

$$P(x \leq X) = \sum_{x_j \leq X} P(x_j). \tag{14.2}$$

From Figure 14.2b, we see that $P(x \leq X)$ is a staircase function that is nondecreasing and is bounded between 0 and 1. It is perhaps easier to think of $P(x \leq X)$ as the

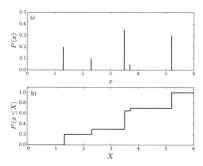

Fig. 14.2 (a) Probabilities $P(x_i)$ for a stochastic system consisting of only five discrete random outcomes. (b) The probability distribution function $P(x \leq X)$ for this system.

cumulative probability distribution function as X goes from left to right; however, this terminology is considered redundant.

In general, any random variable may be thought of as having both a discrete and a continuous part. This is not surprising, since even a continuous function can be approximated as a staircase function with very closely spaced steps. However, in experimental situations, we most often encounter probability distributions that are not only continuous but also differentiable with a continuous derivative except, perhaps, at a discrete set of points. In this case, more powerful methods of analysis are available, since we can define a *probability density function $p(x)$* as

$$p(x) = \frac{dP(x \leq X)}{dX},$$ (14.3)

where $p(x)$ satisfies

$$\int_{-\infty}^{+\infty} p(s)\,ds = 1.$$ (14.4)

Equation (14.4) expresses the fact that the probability that some event out of all possible events will occur is 1. That is, the probability of getting *some* result is 100 %. In other words, when you perform an experiment, you always obtain a result.

Example 14.5 To illustrate the difference between $P(x \leq X)$ and $p(x)$, consider the example of randomly choosing a point on the interval $[a, b]$ (Figure 14.3). Suppose that the choice of every point is equally probable. Thus $P(x \leq X)$ is given by (Figure 14.3a)

$$P(x \leq X) = \begin{cases} 0, & X < a, \\ \dfrac{X - a}{b - a}, & a \leq X \leq b, \\ 1, & X > b, \end{cases}$$ (14.5)

and $p(x)$ (Figure 14.3b) is

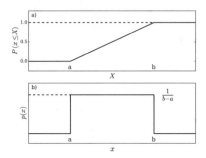

Fig. 14.3 (a) Cumulative probability distribution and (b) probability density for the uniform probability distribution.

$$p(x) = \begin{cases} (b-a)^{-1}, & a \leq x \leq b, \\ 0, & \text{otherwise}. \end{cases} \tag{14.6}$$

Together, these functions describe the uniform probability distribution.   ◇

It can become quite confusing if the term "distribution" is used without it being specified whether $P(x \leq X)$ or $p(x)$ is meant. In this text, we use the notation pdf to refer to the probability density function $p(x)$. The importance of the pdf is that it can be used to describe the average behavior of random variables. For a useful handbook that summarizes the properties of a large number of pdfs, see [213].

The probability that a continuous random variable has a value falling in the interval $(a < x \leq b)$ is given by the integral of the pdf over that interval:

$$P(a < x \leq b) = \int_a^b p(s)\,ds. \tag{14.7}$$

Note that when $a = b$, this integral will be zero. This means that the probability that a continuous random variable takes on a specific value is zero. That is why we defined the probability on the interval $(a < x \leq b)$ rather than $(a \leq x \leq b)$. In other words, in a random dynamical system, we cannot predict specific outcomes but only an "average" outcome following the performance of a very large number of experiments. Herein is the Achilles' heel[6] for the study of stochastic dynamical systems. We must be prepared to perform either a large number of experiments or a large number of numerical simulations in order to identify a system's behavior "on average."

---

[6] Achilles (fl. ca. 1250 B.C.E.), Greek hero of the Trojan War. According to legend, his only physical vulnerability was his heel, and he is said to have been killed in the Trojan War by Paris, who shot him in the heel with an arrow.

14.2.2 Stochastic Processes

We can generate other stochastic variables from a single stochastic variable x by taking some function of x, namely

$$v = f(x),$$

where v is a new stochastic variable. We can also extend this function to depend on two variables such that

$$v(t) = f(x,t), \tag{14.8}$$

where t is a variable, but not a random variable. This function $v(t)$ is called a *random function* or *stochastic process*, with t viewed as the time variable.

With each realization of x, we can obtain a realization of $v(t)$; each realization is called a *sample path*. With these definitions, we can view each stochastic time series obtained from an experiment or a computer simulation as a sample path of some stochastic process. It is often the case that from a collection of such stochastic time series, we would like to extract the nature of a stochastic process that can generate such collections.

Equation (14.8) indicates that the definition of a stochastic process involves a large number of random variables. This large number of variables arises because each new measurement of a variable must be regarded as a different random variable. Obviously, we must extend the concept of the pdf introduced in the previous section to take this into account. In order to do this properly, we need the concepts of joint and conditional probability and of statistical independence.

Joint Probability

We begin with the key concept of *joint probability*. Consider two events A and B with $P(A)$ and $P(B)$ giving the respective probabilities that event A occurs and that event B occurs. The joint probability, denoted by $P(A;B)$, describes the probability that both events take place.

Example 14.6 Consider a classroom with 25 boys and 20 girls. The number of boys who are taller than five feet is 10, while the number of girls taller than five feet is 5. Now we randomly select one student from this class and consider the following events: A, that the student is a boy, and B, that the student's height is greater than 5 feet. Then since there are 45 pupils altogether, of whom 10 are boys more than five feet tall, we have $P(A;B) = 10/45$. ◇

Now let us further consider $\overline{A}$ and $\overline{B}$, which indicate the complements of A and B, namely the respective events that A and B do not occur. The probability of the complementary event is the probability that the original event does not occur. Therefore, $P(\overline{A}) = 1 - P(A)$ and $P(\overline{B}) = 1 - P(B)$ (since the probability that an event either does or does not occur is 1). The fundamental relation for the joint probability, also known as the law of total probability, is given as follows

$$P(A) = P(A;B) + P(A;\overline{B}), \quad P(B) = P(A;B) + P(\overline{A};B). \tag{14.9}$$

Example 14.7 For Example 14.6 above, show that $P(A) = 25/45$, $P(B) = 15/45$, $P(A;\overline{B}) = 15/45$, and $P(\overline{A};B) = 5/45$. Thus we see that (14.9) is correct. $\diamond$

We can extend this concept of joint probability to random variables.

The Joint Probability Density Function

Recall that the joint probability $P(x_1 = a_1; x_2 = a_2)$ of discrete random variables x_1 and x_2 is the probability that the two variables will assume the particular values a_1 and a_2. A simple example is the joint probability for the appearance of two numbers when we throw a pair of dice.

We denote the *joint probability (density) function*, or joint pdf, by $p(x_1; x_2)$. The joint probability density has the following properties, analogous to (14.9).

$$p(x_1) = \int_{-\infty}^{+\infty} p(x_1; x_2)\, dx_2, \quad p(x_2) = \int_{-\infty}^{+\infty} p(x_1; x_2)\, dx_1. \tag{14.10}$$

Thus, we can obtain a pdf of a single variable from a joint pdf of two variables. In general, we can deduce lower-order pdfs from higher-order ones, but not vice versa.

Conditional Probability

The next concept we introduce is the *conditional probability*. Conditional probability is not the same as joint probability. Let us again consider the events A and B. Conditional probability, denoted by $P(A \mid B)$, describes how the occurrence of event A depends on that of B. The *conditional probability of A given B*, which is the probability that A will occur given that B has already occurred, is defined as

$$P(A \mid B) = \frac{P(A;B)}{P(B)}. \tag{14.11}$$

Similarly, we can define

$$P(B \mid A) = \frac{P(A;B)}{P(A)}. \tag{14.12}$$

In general, $P(A \mid B) \neq P(B \mid A)$.

Example 14.8 Consider again the classroom of Example 14.6 with 25 boys and 20 girls. The number of boys who are more than five feet tall is 10, and the number of such girls is 5. Again we randomly pick one student from this class and again consider the following events: A: The pupil is a boy, and B: the pupil is more than five feet tall. Then,

$$P(A \mid B) = \frac{P(A;B)}{P(B)} = \frac{10/45}{15/45} = \frac{2}{3}.$$

This makes sense, because two-thirds of the pupils who are more than five feet tall are boys, so given that the randomly chosen student was over five feet tall, the probability that it was a boy is $2/3$. However, we have

$$P(B \mid A) = \frac{P(A;B)}{P(A)} = \frac{10/45}{25/45} = \frac{2}{5}.$$

So $P(A \mid B)$ and $P(B \mid A)$ are not equal in this example. ◇

We can also extend this concept to random variables. A conditional probability (density) function describes the situation in which the occurrence of x_1 depends also on the occurrence of a certain realization of x_2. The conditional probability is given by

$$p(x_1 \mid x_2) = \frac{p(x_1;x_2)}{p(x_2)}. \tag{14.13}$$

Similarly, we can define

$$p(x_2 \mid x_1) = \frac{p(x_1;x_2)}{p(x_1)}. \tag{14.14}$$

The above two conditional pdfs are not in general the same, and the notion of conditional pdf is not the same as that of joint pdf.

Statistical Independence

The third concept is that of *statistical independence* between two random variables. Two events are called statistically independent if

$$P(A;B) = P(A)P(B). \tag{14.15}$$

Intuitively, independence implies that the statistical behavior of A has nothing to do with that of B. The assumption of independence leads to a number of simpler statistical properties such as

$$P(A \mid B) = P(A) \quad \text{and} \quad P(B \mid A) = P(B). \tag{14.16}$$

Thus independence means that the realization of one event does not affect the other's probability. We can again map this concept to random variables. Two random variables x_1 and x_2 are said to be statistically independent if their joint pdf has the following properties:

$$p(x_1;x_2) = p(x_1)p(x_2) \tag{14.17}$$

for all realizations of its values, and

$$p(x_1 \mid x_2) = p(x_1) \quad \text{and} \quad p(x_2 \mid x_1) = p(x_2). \tag{14.18}$$

With these concepts of joint and conditional pdfs, we can define the *pdf for a stochastic process* as follows: The *n*-point joint pdf of the stochastic process $x(t)$, denoted by

$$p(x_1, t_1; x_2, t_2; x_3, t_3; \ldots; x_n, t_n),$$

gives the joint probability density that

$$x(t_1) = x_1, \quad x(t_2) = x_2, \quad x(t_3) = x_3, \quad \ldots, \quad x(t_n) = x_n,$$

where $t_1, t_2, \ldots, t_n$ are time points.

14.2.3 Statistical Averages

In general, the properties of stochastic processes are time-dependent, i.e., the current value of $x(t)$ depends on the time that has elapsed since the process began. However, experimentalists often perform measurements on dynamical systems under constant laboratory conditions that have run so long that a steady, or stationary, state has been attained (Chapter 3). The mathematical description of stationary states for a stochastic process requires that the pdf does not change with time. For example, for a stochastic process $x(t)$,

$$\begin{aligned}
p(x_1, t_1; x_2, t_2; x_3, t_3; \ldots, x_n, t_n) \\
= p(x_1, t_1 + t'; x_2, t_2 + t'; x_3, t_3 + t'; \ldots, x_n, t_n + t'),
\end{aligned} \tag{14.19}$$

for all combinations of x_i, t_i, and positive time shift t'. In particular, the single-point pdf $p(x, t) = p(x)$ is independent of time, and the two-point joint pdf

$$p(x_1, t_1; x_2, t_2) = p(x_1, x_2; t_2 - t_1) \tag{14.20}$$

depends only on the time difference $t_2 - t_1$, not on the actual values of t_1 and t_2. Stationarity is a mathematical idealization. Analysis of a stationary process is frequently much simpler than for a similar process that is time-dependent, or nonstationary. In particular, when a stochastic process is nonstationary, the moments (see below) themselves become functions of time.

The pdf can be used to calculate the average, or expected, values of certain quantities of interest such as the mean. Statisticians refer to these averages as *moments*. The *n*th moment of the random variable x is the statistical average of the *n*th power of x, i.e.,

$$E[x^n] := \langle x^n \rangle = \int_{-\infty}^{+\infty} x^n p(x) \, dx, \tag{14.21}$$

where the notation $E[\cdot]$ denotes the expected value, or expectation. For example, the first moment is the mean

$$E[x] := \langle x \rangle = \int_{-\infty}^{\infty} x p(x) \, dx, \tag{14.22}$$

and the second moment is

$$E[x^2] := \langle x^2 \rangle = \int_{-\infty}^{\infty} x^2 p(x)\, dx. \tag{14.23}$$

With these two moments, we can obtain the *variance*

$$E\left[(x - E[x])^2\right] := \langle (x - \langle x \rangle)^2 \rangle = \int_{-\infty}^{+\infty} (x - \langle x \rangle)^2 p(x)\, dx. \tag{14.24}$$

We can also show that

$$\langle (x - \langle x \rangle)^2 \rangle = \langle x^2 \rangle - \langle x \rangle^2. \tag{14.25}$$

In time series analysis, it is standard procedure to remove the mean before performing some other statistical analysis. The advantage of this procedure is that the second moment with the mean removed equals the variance. We use the symbol σ^2 to denote the variance calculated when the mean is removed.

Example 14.9 To illustrate the use of (14.21), let us use the pdf to calculate the mean and the variance for the uniform probability density function described by (14.6). The mean is

$$\langle x \rangle = \int_a^b \frac{x\, dx}{b - a} = \frac{b + a}{2}, \tag{14.26}$$

and the variance is

$$\langle (x - \langle x \rangle)^2 \rangle = \frac{(b - a)^2}{12}. \tag{14.27}$$

$\diamondsuit$

Another type of statistical average that is of considerable importance is the characteristic function $\varphi(f)$, which is the statistical average of $e^{-j2\pi f x}$:

$$\varphi(f) = \int_{-\infty}^{\infty} e^{-j2\pi f x} p(x)\, dx, \tag{14.28}$$

where f is real. Since $p(x)$ is nonnegative and $e^{-j2\pi f x}$ has unit magnitude, it follows that

$$\left| \int_{-\infty}^{\infty} e^{-j2\pi f x} p(x)\, dx \right| \leq \int_{-\infty}^{\infty} p(x)\, dx = 1,$$

and hence the characteristic function always exists. The characteristic function $\varphi(f)$ and the pdf form a Fourier transform pair [669],

$$p(x) = \int_{-\infty}^{\infty} e^{j2\pi f x} \varphi(f)\, df. \tag{14.29}$$

Thus $\varphi(f)$ plays a key role in the description of stochastic processes, especially in the description of random walks (Chapter 16).

The characteristic function $\varphi(f)$ is the Fourier transform of the pdf, and it can be used to calculate the moments. If we evaluate $\varphi(f)$ at $f = 0$, we obtain

$$\varphi(0) = \int_{-\infty}^{\infty} p(x)e^{-j2\pi fx}\,dx\Big|_{f=0} = \int_{-\infty}^{\infty} p(x)\,dx = 1.$$

In other words, the area under the curve of a function from $-\infty$ to $+\infty$ is equal to the value of its Fourier transform at the origin. The value of $\varphi(0)$ is often referred to as the *central ordinate of the transform*.

What can we say about the first derivative $\varphi'(0)$ of $\varphi(0)$? Using a result that we obtained in Section 8.4.1, we have

$$\varphi'(f) = \int_{-\infty}^{\infty} -j2\pi x p(x)e^{-2j\pi xf}\,dx. \tag{14.30}$$

Taking $f = 0$, we see that the first moment is

$$\int_{-\infty}^{\infty} xp(x)\,dx = -\frac{\varphi'(0)}{j2\pi}. \tag{14.31}$$

Similarly, we can show that the second moment is

$$\int_{-\infty}^{\infty} x^2 p(x)\,dx = -\frac{\varphi''(0)}{4\pi^2}, \tag{14.32}$$

and the nth moment is

$$\int_{-\infty}^{\infty} x^n p(x)\,dx = \frac{\varphi^{(n)}(0)}{(-2\pi j)^n}. \tag{14.33}$$

Example 14.10 Evaluate

$$\int_{-\infty}^{\infty} \frac{\sin \pi x}{\pi x}\,dx = \int_{-\infty}^{\infty} \mathrm{sinc}(x)\,dx.$$

Solution. From a table of Fourier integrals, we see that the Fourier transform of $\mathrm{sinc}(x)$ is the rectangular function $\Pi(f)$. Hence we have

$$\int_{-\infty}^{\infty} \mathrm{sinc}(x)\,dx = 1.$$

Example 14.11 Suppose that $p(x)$ is the Gaussian[7] (normal) distribution, i.e.,

$$g(x) = \frac{1}{\sqrt{2\pi}} e^{-x^2/2}.$$

It should be noted that the factor $1/\sqrt{2\pi}$ ensures that $p(x)$ is normalized, and the factor $1/2$ in the exponent defines the width of the distribution as the distance between the inflection points, which is 1 in this case. What are the mean and variance?

---

[7] The Gaussian distribution is just one of a multitude of mathematical concepts (including the gauss, the cgs unit of magnetic field strength) named after Johann Carl Friedrich Gauss (1777–1855), German mathematician.

Solution. The Fourier transform of $g(x)$ is

$$\varphi(f) = e^{-(2\pi f)^2/2}.$$

(It is useful to keep in mind that the Gaussian distribution is the only statistical distribution for which the pdf and its Fourier transform are both Gaussian.) To estimate the first moment, we calculate

$$\varphi'(f) = -4\pi^2 f e^{-(2\pi f)^2/2},$$

and hence

$$\int_{-\infty}^{\infty} x p(x)\, dx = \frac{\varphi'(0)}{-2j\pi} = 0.$$

In order to determine the second moment, we calculate

$$\varphi''(f) = 4\pi^2 e^{-(2\pi f)^2/2}(4\pi^2 f^2 - 1)$$

to obtain

$$\int_{-\infty}^{\infty} x^2 p(x)\, dx = \frac{\varphi''(0)}{-4\pi^2} = 1.$$

Since the mean is zero, the second moment equals the variance. ◇

14.3 Bayesian Inferences

In our black box analogy, we are interested in how the input is changed by the box to produce an associated output. Here, we focus on investigating the nature or characteristics of the output given the input. Often, however, we would like to turn our black box around and ask that given the output, what is the nature of the input? This situation naturally arises when the goal is to infer a cause given something has happened. Examples include analysis of accidents, disaster, malfunctions, unexpected success, phylogeny, and so on (see, for example, [192, 356, 899]) Also, when there are possible multiple causes, we would like to know the order of importance of factors involved (see, for example, [159, 281, 872]).

Normally, there are uncertainties involved in making inferences about causes and hence concepts from probability theory and statistics become necessary. The methods for such inferences are thus called *Statistical Inferences*. Bayes' theorem provides a representative framework for making statistical inferences.

The basic notions of Bayes' theorem are straightforward. The starting point of this theorem is connected to the definitions of the conditional probability. Let us state these again. We have two events A and B on which we would like to consider their probabilistic nature. The joint probability $P(A;B)$ is the probability that event A and B both occurs. The conditional probability of the event A given (conditioned by) the occurrence of the event B is defined as

$$P(A|B) = \frac{P(A;B)}{P(B)} \quad (P(B) > 0). \tag{14.34}$$

In the same manner, we can obtain the conditional probability of the event B conditioned by the occurrence of the event A as

$$P(B|A) = \frac{P(B;A)}{P(A)} \quad (P(A) > 0). \tag{14.35}$$

Noting that the joint probability is symmetric by definition, i.e., $P(A;B) = P(B;A)$, we can obtain the following from the above definitions of the conditional probabilities.

$$P(A|B) = \frac{P(A;B)}{P(B)} = \frac{P(B;A)}{P(B)} = \frac{P(B|A)P(A)}{P(B)} \quad (P(B) > 0). \tag{14.36}$$

This equation is the main ingredient of Bayes' theorem.

Let us consider the meaning of (14.36) in the light of inferences. If we set the event B as the result and A as a possible cause, the conditional probability $P(A|B)$ can be interpreted as the probability that, given the occurrence of the event (result) B, the event A is its cause. In other words, it represents the probability that the event A is responsible for the observed event B. Hence, when we want to infer a connection to an observation, this is the main probability we want to compute.

The right-hand side of (14.36) shows how this probability is connected to the related probabilities. In particular, we note that it is connected to the conditional probability $P(B|A)$, which shows the likeliness of the event B follows from the event A, as well as $P(A)$ and $P(B)$. These probabilities can often be obtained from experimental measurements or other observations as we will explain later.

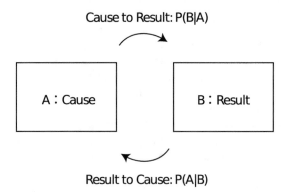

Fig. 14.4 Concepts of conditional probabilities for the Bayesian Inference. See text for Discussion

We illustrate the meaning of Bayes' theorem with an example (Figure 14.4).

Example 14.12 Let us consider a situation where we want to track a producer of a defective sample found in a pile of a product (Figure 14.5). All the goods in the pile are either made in the U.S. or abroad. We would like to infer the probability this defective sample is produced in the U.S. We label the related events as follows

A : it is produced in the U.S

$\overline{A}$: it is not produced in the U. S., i.e., it is produced abroad

B : a sample is defective

We also assume we have a knowledge on production and defective rate from the analysis of past data as follows.

$P(A)$: the probability of the product produced in the U.S

$P(\overline{A})$: the probability of the product produced abroad, i.e., $P(\overline{A}) = 1 - P(A)$

$P(B|A)$: the probability that the product is defective given it is produced in the U.S

$P(B|\overline{A})$: the probability that the product is defective given it is produced abroad

We can estimate these probabilities from ratios or fraction of the produce produced in the US or abroad divided by total amount of product produced: 60% of the produced is produced in the U.S. and 40% is produced abroad. From past data, we know that the 5% of the product produced in the U.S. is defective and 10% of that produced abroad is defective. Thus we have

$$P(A) = 0.6$$
$$P(\overline{A}) = 0.4$$
$$P(B|A) = 0.05$$
$$P(B|\overline{A}) = 0.10$$

From these numbers, we can first calculate

- $P(B)$: the total defective rate (probability) of the product as

$$P(B) = P(B;A) + P(B;\overline{A}) = P(B|A)P(A) + P(B|\overline{A})P(\overline{A})$$
$$= (0.05)(0.6) + (0.10)(0.4) = 0.07. \tag{14.37}$$

This means that the total (produced in the U.S. and abroad combined) defective rate is 7%.

Now we can use (14.36) to calculate $P(A|B)$ which is the probability that a defective sample found in a pile is produced in the U.S (Figure 14.6).

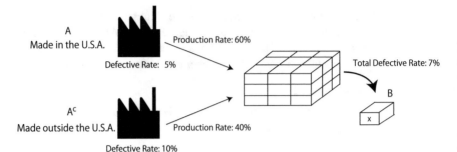

Fig. 14.5 Schematic View of the Defective Rate

$$P(A|B) = \frac{P(B|A)P(A)}{P(B)} = \frac{(0.05)(0.6)}{0.07} \approx 0.43, \qquad (14.38)$$

showing that it is approximately 43 percentile. One can see that this result is reasonable, given the ratio of production is higher but the defective rate is lower in the U.S.   ◊

In the above calculation, we used Bayes' theorem in its simplest form by combining (14.36) and (14.37).

$$P(A|B) = \frac{P(B|A)P(A)}{P(B|A)P(A) + P(B|\overline{A})P(\overline{A})} \quad (P(B) > 0). \qquad (14.39)$$

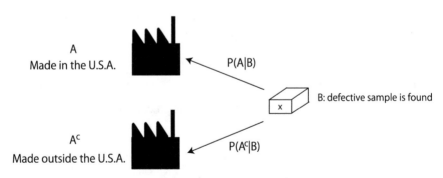

Fig. 14.6 Schematic View of the Bayesian Inference of the origin of the Defective Sample

Let's consider another example.

Example 14.13 At the airport, there are security gates to check our inflight luggage. Let us consider the case that we leave a bottle of water in a bag. It is likely that the detective machines can detect it and gives out a warning. Suppose that a bag is detected as suspicious. What is the probability that it actually contained a bottle of water, given the following?

- The detective machine has a 95 % accuracy.
- Nowadays, people are careful and forget to leave a bottle in their bag only 1 % of the time.

As in the previous example, the key to use the Bayes' theorem is first to label events and then to identify associated probabilities. Here, we label as follows.

- A: A bottle of water is in the bag
- B: detector is activated to indicate suspicion

We want compute is the conditional probability, $P(A|B)$, that a bottle of water is actually in the bag when a detector is activated. Using the above information, we calculate the necessary probabilities as

$$P(A) = 0.01$$
$$P(\overline{A}) = 0.99$$
$$P(B|A) = 0.95$$
$$P(B|\overline{A}) = 0.05$$

Now, we apply the Bayes' theorem.

$$P(A|B) = \frac{P(B|A)P(A)}{P(B|A)P(A) + P(B|\overline{A})P(\overline{A})}$$
$$= \frac{(0.95)(0.01)}{(0.95)(0.01) + (0.05)(0.99)} \approx 0.161.$$

Thus, the probability of actually finding a bottle in the luggage is roughly 0.16. ◇

Many people may find this result strange and ask, "why can it be only a 16% chance give that the detector is accurate 95 % of the time?" To answer to this question we note that before the bag is scanned, the probability of a bottle being in the luggage is only 0.01. By using the detector, the probability of a bottle being in the luggage improved by more than 15 times!

Detecting a rare event is difficult and similar situations occur when you are checked for a rare disease using a highly accurate medical test. Well, this is one of the interesting aspects of probability: when one actually compute numbers, you can sometimes get rather puzzling numbers (at first sight) for probabilities.

14.3.1 Extensions to general Bayes' theorem

We can extend our simple form of the Bayes' theorem (14.39) further to accommodate inferences with multiple possible causes. For example, in the previous example, we only consider two cases of made in the U.S.A.(A) or abroad ($\overline{A}$). If we further decompose abroad into foreign countries, we need to extend our simple form of the theorem.

To illustrate let us assume the following: There are n possible causes and the events A_i are mutually exclusive, i.e., $A_i \cap A_j = \emptyset$ $(\forall i \neq j)$, and

$$P(\cup_{i=1}^{n} A_i) = \sum_{i=1}^{n} P(A_i) = 1, \quad (P(A_i) > 0, \forall i) \tag{14.40}$$

That is there is only one real cause A_i and all of the possible causes are listed in $A_1, A_2, \ldots, A_n$. Note that, in the context of the previous example, a defective sample can only be produced either in the U.S.A. or abroad, and there are no other possibilities. Therefore, the above condition is satisfied with $A_1 = A$ and $A_2 = \overline{A}$.

With the above considerations, the Bayes' theorem can now be stated as follows.

$$P(A_i|B) = \frac{P(B|A_i)P(A_i)}{\sum_{i=1}^{n} P(B|A_i)P(A_i)} \quad (P(B) > 0). \tag{14.41}$$

We observe again that the conditional probability to estimate each of the causes given the result B is again connected to the conditional probability $P(B|A_i)$ of B for each possible cause of A_i and the probability $P(A_i)$ of A_i. With this extention, we can now infer more than two causes.

Example 14.14 I had a bad flu and my physician suggested that I take three different medications A_1, A_2, A_3. After taking them, I began to have a stomach ache. I checked the Internet to see what the side effects of these medications were and found out that these pills can cause stomach ache as A_1: 5%, A_2: 10%, A_3: 15%. By assuming that a single pill is responsible, we can use (14.41) to estimate the most probable cause of my stomach ache.

As before, let us label the events as follows.

- A_i: the pill A_i is responsible for the ache ($i = 1, 2, 3$)
- B: stomach ache happens

From the information provided, related conditional probabilities are obtained as

$$P(B|A_1) = 0.05$$
$$P(B|A_2) = 0.10$$
$$P(B|A_3) = 0.15$$

In order to estimate $P(A_i|B)$ by the Bayes' theorem, what we need are the probabilities $P(A_i)$. Without any further information, we need to estimate these to begin with. In reality, it may be related to the amount of each pills taken at the same time, or other factors. It is often the case that we assume to take these as equal: We assume that the fact we do not know which pill is responsible means that they are equally likely. This is called the "principle of insufficient reason." If we employ this principle (assumption), we have

$$P(A_1) = P(A_2) = P(A_3) = 1/3. \tag{14.42}$$

Now we are in position to use the Bayes' theorem (14.41). After simple computations (the reader should confirm these), we have

$$P(A_1|B) = 1/6 \approx 0.17$$
$$P(A_2|B) = 1/3 \approx 0.33$$
$$P(A_3|B) = 1/2 = 0.5$$

We note that in this case the ratio between the conditional probabilities are the same, namely

$$P(A_1|B) : P(A_2|B) : P(A_3|B) = P(B|A_1) : P(B|A_2) : P(B|A_3) = 1 : 2 : 3. \tag{14.43}$$

This is a consequence of the assumption (14.42). ◇

At this point, let us summarize the concrete steps (recipes) in order to utilize the Bayes' theorem.

- From the information given, label the events, both the observed result and possible causes clearly.
- Check if these events satisfy the assumption such as mutual exclusiveness of the causes and (14.40) needed to apply the theorem.
- List the related probabilities appearing at the right-hand side of the theorem (14.41). If needed, infer with reasons to estimate the related probabilities.
- Use the theorem to compute the desired conditional probabilities.

14.3.2 Priori and Posteriori probabilities

We now introduce a bit of terminologies relating to applications of the Bayes' theorem, which are rather confusing at first sight. In the theorem (14.41), the probabilities related to possible causes A_i are connected before ($P(A_i)$) and after ($P(A_i|B)$). In a way, we can interpret this that we updated the probabilistic knowledge on A_i through the observation of B. An analogy can be found in detective stories, where the level of suspicions on multiple suspects are updated as we gain evidences. With the above view, the probabilities $P(A_i)$ (before the event B) are called "Priori" probabilities, and $P(A_i|B)$ (after the event B) "Posteriori" probabilities. These terminologies and concepts sometimes cause problems for students. However, if you could remember the detective analogy, it is not so difficult to use these probabilities. The Bayes' theorem connects, or updates, the priori and posteriori probabilities through an event (evidence, observation) in a natural and familiar way.

14.3.3 Bayesian Updates

Now let us go back to the analogy with detective stories. Normally, the story goes in a way such that evidences come in one after another so that the doubtfulness of possible suspects changes repeatedly (Figure 14.7). With the Bayes' theorem, we can do the same. This updating procedure is referred to as Bayesian updates. The idea is that we can chain or iterate (14.41) in such a way that posteriori probabilities obtained in the previous inference are used as priori probabilities in the next step. We illustrate the Bayesian updates by an example in the line of detective stories.

Example 14.15 A crime of a pick pocket is reported. From the report, there are three possible suspects 1, 2, 3 are identified who are nearby at the time of the incident.

We label that A_i as the event that suspect i is the criminal.

At this point, all of them are equally likely to be responsible by applying the principle of insufficient reason. So the priori probabilities are considered as follows

$$P(A_1) = P(A_2) = P(A_3) = 1/3. \tag{14.44}$$

Now information B came in about the distance between the location of the incident and the location of each suspect at half-hour later. From this, the conditional probabilities of three suspect can be at their locations after the crime are estimated as

$$P(B|A_1) = 0.10$$
$$P(B|A_2) = 0.15$$
$$P(B|A_3) = 0.25$$

Now we can use the Bayes' theorem to compute the conditional probabilities for three suspects given this evidence B.

$$P(A_1|B) = 0.2$$
$$P(A_2|B) = 0.3 \tag{14.45}$$
$$P(A_3|B) = 0.5$$

These are the posteriori probabilities given evidence B. We note that by the information B, the suspicion levels are now updated so that they are not equal.

Now three suspects are further investigated, so that the amount of cash they have is now obtained. This is the new information C. Based on the difference from their estimated normal carrying cash amount, we can estimate the conditional probabilities as

$$P(C|A_1) = 0.05$$
$$P(C|A_2) = 0.70 \tag{14.46}$$
$$P(C|A_3) = 0.10$$

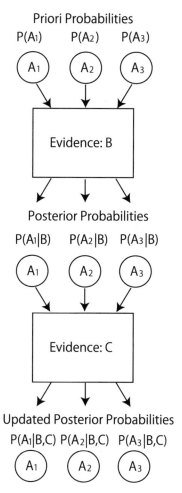

Fig. 14.7 Flow of the Bayesian Updates

Here is the crucial point of the Bayesian update. What do we set for $P(A_i)$ in applying the theorem? The answer is to use the posteriori probabilities based on B. This is natural as by the evidence of B, the suspicion levels are no longer equal but given by (14.45).

Thus, we update as

$$P(A_1) = P(A_1|B) = 0.2$$
$$P(A_2) = P(A_2|B) = 0.3 \qquad (14.47)$$
$$P(A_3) = P(A_3|B) = 0.5$$

We can now use (14.46) and (14.47) to compute the following updated probabilities (the reader should confirm these estimates themselves).

$$P(A_1|B;C) = 0.037$$
$$P(A_2|B;C) = 0.778 \qquad (14.48)$$
$$P(A_3|B;C) = 0.185$$

At this point, the probability the two evidences B and C has led to the inference that the suspect 2 is more likely to be the one responsible.

We can calculate the above probabilities (14.48) in slightly different way. For this, we consider that the two evidences B and C are obtained simultaneously or considered together. Then, we infer that

$$P(B;C|A_1) = P(B|A_1)P(C|A_1) = (0.10)(0.05) = 0.005$$
$$P(B;C|A_2) = P(B|A_2)P(C|A_2) = (0.15)(0.70) = 0.105$$
$$P(B;C|A_3) = P(B|A_3)P(C|A_3) = (0.25)(0.10) = 0.025$$

With these along with

$$P(A_1) = P(A_2) = P(A_3) = 1/3. \qquad (14.49)$$

we can use the Bayes' theorem to obtain (14.48), which in turn justifies our notation of $P(A_i|B;C)$.   ◇

We can continue on repeating the Bayesian updates, as more information or evidences comes in. Starting with an initial priori probability, we can chain through the Bayes theorem to obtain a final posteriori probability taking account of all related evidences. Again, although the terminologies and mathematics may appear confusing at first sight, we are actually proceeding like a detective and are pursuing a natural inference process.

14.3.4 Monty Hall Problem

We now consider a famous problem called the Monty Hall Problem (Figure 14.8). This problem demonstrates vividly the subtleties involved in the concept of conditional probabilities. The problem originated from a TV show whose host was Monty Hall[8]. However, it has caused a debate among professional mathematicians to the extent a book devoted to this problem has been published [729].

We analyze this problem from a perspective of the Bayes' theorem. Let us start with the statement of the problem.

---

[8] Monty Hall (1921-2017). Canadian-American game show host of *Let's Make a Deal*.

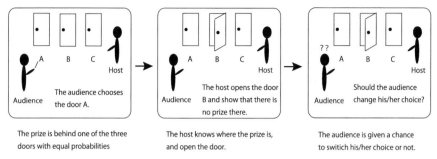

Fig. 14.8 Outline of the Monty Hall Problem. See text for discussion.

1. There are three closed doors A, B, C in the show. The game is to guess the correct door behind which there is a prize. The prize was placed behind one of the three doors with the equal probability of $1/3$.
2. The host knows where the prize is and asks someone from the audience to guess the door. Without loss of the generality, let us label the door chosen as A.
3. The host then opens one of the doors from B, C, which does not have a prize behind (If there is a choice, i.e., the prize is behind A, one of the two doors is chosen with equal probability of $1/2$.). Again, without loss of generality, let us label the door opened as B. Now the audience knows that the prize is either behind A or C.
4. The host then gives a chance for the audience to switch their choice from door A to C. The audience can switch to C or keep their original choice of A.

With the above set up, the question is the probabilities for the audience to win the prize with his choices. There are three possibilities.

1. It does not matter whether he switches or not.
2. It is better for the audience to keep his/her choice of A.
3. It is better for the audience to switch to C.

Let us consider this problem using the Bayes' theorem. As always, we start from labeling events.

$$A : \text{the prize is behind door } A$$
$$B : \text{the prize is behind door } B$$
$$C : \text{the prize is behind door } C$$
$$B_o : \text{the door } B \text{ opened}.$$

What do we want to compute to answer the above question? Initially all the doors have equal probability to have a prize behind. So, initially, we have $P(A) = P(B) = P(C) = 1/3$.

Then, the event B_o of door B opening takes place, showing that there is no prize at the door B. Thus, the conditional probability $P(B|B_o)$ is 0. Hence, we would like to compute and compare the other conditional probabilities $P(A|B_o)$ and $P(C|B_o)$ for the location of the prize given the event B_o.

We can apply the Bayes' theorem as follows.

$$P(A|B_o) = \frac{P(B_o|A)P(A)}{P(B_o)}$$

$$= \frac{P(B_o|A)P(A)}{P(B_o|A)P(A) + P(B_o|B)P(B) + P(B_o|C)P(C)}, \qquad (14.50)$$

$$P(C|B_o) = \frac{P(B_o|C)P(C)}{P(B_o)}$$

$$= \frac{P(B_o|C)P(C)}{P(B_o|A)P(A) + P(B_o|B)P(B) + P(B_o|C)P(C)}. \qquad (14.51)$$

The location of the prize is deduced from the proceedings of the game by the host. Thus the conditional probabilites of the curtain B opened are as follows:

$$P(B_o|A)P(A) = 1/2, \quad P(B_o|B)P(B) = 0, \quad P(B_o|C)P(C) = 1. \qquad (14.52)$$

Together with $P(A) = P(B) = P(C) = 1/3$, the values for the desired conditional probabilities are calculated by (14.50) and (14.51).

$$P(A|B_o) = 1/3, \quad P(C|B_o) = 2/3. \qquad (14.53)$$

Thus, it is better for the audience to switch to C rather than to keep his original choice of A.

Some people have argued that this is strange conclusion because the prize was initially placed behind the doors with the equal probability of $1/3$. It is true at the beginning of the game, but the fact that the door B is opened by the host updated the probabilities. For those who are still not convinced, please write a simple computer program and simulate the game. You will find, as in the past computer simulations of this problem, the chance to win the prize is about twice as big by changing the choice of the door as calculated above.

14.3.5 Applications to neuroscience

Bayesian statistics provides a powerful way to show how new information can be combined with prior beliefs to improve estimates pertaining to current behavior. A problem is to identify the components of Bayes' theorems, namely the priori, likelihood and posterior probabilities, with what can be identified and measured in a laboratory setting. Here we give two examples.

Example 14.16 One of the earliest applications of Bayesian inference to neuroscience has been to understand the role of adaptation in motor learning [439, 440].

Here the term "adaptation" is used to describe the short term learning process which underlies the gradual improvement in motor performance in response to changes in sensory information. In this context, the priori probability describes previous knowledge about the physical state of the object that is being observed. The likelihood probability describes the new sensory information that is used to update the priori probability. Finally, the posterior probability describes the updated estimate of the physical state. ◇

Example 14.17 How does the nervous system integrate information from different sensory modalities to interpret the external world? This integration appears to follow the principles of Bayesian estimation [837]. A simple model of audio-visual integration makes it possible to identify the Bayesian variables [838]. This model assumes that the two areas of the nervous system responsible for auditory and visual localization are topologically organized. Each of these regions receives modality specific information from the external environment as well as cross-modal inputs. The widths of the receptor fields correspond to the likelihood probability. The position in each receptive field is the priori spatial probability. Finally, the posterior probability is the probability of co-occurrence of the audio-visual stimuli and is encoded in the cross-modal synapses. ◇

14.4 Laboratory Evaluations

Traditionally, the characterization of fluctuations, referred to herein as *noise*, involved the measurement of the three main attributes of noise, namely its intensity, pdf, and autocorrelation function. Depending on the context, it may also be necessary to consider whether noise exhibits power-law behavior. Before beginning our discussion, we emphasize that *experimentalists typically subtract the mean value from a time series before analyzing its statistical properties.*

14.4.1 Intensity

The intensity of a stochastic process is equal to its variance. However, biomedical engineers most often use the root mean square value, or quadratic mean,

$$x_{\text{rms}} = \sqrt{\frac{x_1^2 + x_2^2 + \cdots + x_n^2}{n}}, \tag{14.54}$$

since this is the variable that is used to calculate the power of a signal. When the mean is zero, $\langle x \rangle = 0$, then x_{rms} is the same as the standard deviation, the square root of the variance; however, when $\langle x \rangle \neq 0$, then x_{rms} does not equal the standard deviation. Moreover, the sum of the rms values for two independent signals is

$$(x+y)_{\text{rms}} = \sqrt{x^2_{\text{rms}} + y^2_{\text{rms}}}.$$

14.4.2 Estimation of the Probability Density Function

The standard procedure to determine a pdf from experimental measurements is first
to divide, or *bin*, the observed range of the variable into n equal disjoint intervals.
Suppose a single variable x is restricted to take only values between 0 and 1 (Figure 14.9), say $x \in [0,1)$. Then the ith bin is [462]

$$\left[\frac{(i-1)}{n}, \frac{i}{n} \right), \quad i = 1, \dots, n,$$

where we have excluded the endpoint.

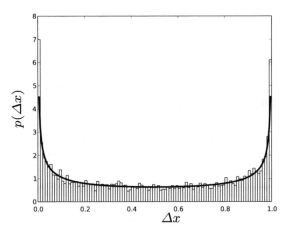

Fig. 14.9 Probability density function for the quadratic map (13.2) estimated using the bin method.
The solid line gives the predicted pdf, which is $\left[\pi \sqrt{x(1-x)} \right]^{-1}$ [462, 836].

Now we estimate the pdf by dividing the number of events x_i in each bin by N,
the total number of data points (Figure 14.9):

$$f_i = \frac{x_i}{N}, \tag{14.55}$$

which is the ratio of points in the ith bin, so that

$$\sum_{i=1}^{n} f_i = 1.$$

However, there are two problems with this direct approach to estimating the pdf.
First, since the choice of bin size is arbitrary, it is possible that information can be

lost. Second, the number of points required to fill the bins with enough statistical significance grows exponentially as the number of different variables in the pdf increases.

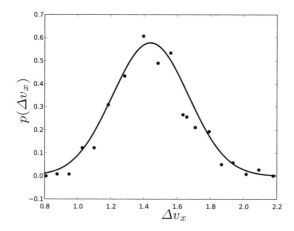

Fig. 14.10 Probability density function (pdf) of the difference in the velocity in the *x*-direction from the mean for a crowd of pedestrians crossing a street in Sydney, Australia. The solid line represents the pdf calculated for the Maxwell–Boltzmann distribution, and the pedestrian data (•) are taken from [338].

Example 14.18 The movements of large groups of animals, including humans, can often be understood on the basis of very simple interaction rules [33, 140, 338, 401, 402, 613]. One of the earliest demonstrations involved determining the pdf for the walking speeds *v* of pedestrians in a crowd. Intuitively, one might expect that since pedestrians have individual preferences and destinations, crowd behavior would be very complex. However, as shown in Figure 14.10, the distribution of walking speeds in a crowd of people can be approximated by the Maxwell–Boltzmann[9] distribution

$$p(v) = \sqrt{\frac{2}{\pi}} \left(\frac{v^2 e^{-v^2/2a^2}}{a^3} \right),$$

where $a = \sqrt{kT/m}$, k is Boltzmann's constant, and T is temperature. Thus the behavior of the crowd is predictable. As crowd density increases, the motion of individuals becomes affected by repulsive interactions with other pedestrians [336, 613]. This leads to a variety of self-organizing phenomena that are important to understand in designing living spaces and in crowd control. ◇

Example 14.19 Up to now, we have considered that the pdf characterizes a stochastic process. However, certain chaotic deterministic dynamical systems can also be

---

[9] James Clerk Maxwell (1831–1879), Scottish mathematical physicist; Ludwig Eduard Boltzmann (1844–1906), Austrian physicist and philosopher.

characterized by a pdf. Figure 14.9 shows that when $\mu = 4$, the quadratic map (13.2) can generate a distribution function that is indistinguishable from the arcsine pdf generated by a stochastic process [836]. Indeed, Lasota[10] and Mackey have proven that for every pdf generated by a stochastic process, there is an infinite number of chaotic deterministic processes that can generate the same pdf [462].   ◇

The development of methods to estimate the pdf is an active area of research. One possibility is to use approaches based on the use of artificial neural networks [259, 485] or kernel estimation [149]. Presently, a very popular method is based on the use of the characteristic function $\varphi(f)$ [435, 908]. It is often easier to calculate the characteristic function and then take the Fourier transform than to obtain the probability density using (14.29) (see Chapter 16). Methods to estimate the empirical characteristic functions are being developed [435, 908], and several software packages are already available (see, for example, MATLAB's STABLE software package and the R gmm package).

14.4.3 Estimation of Moments

How well can moments be estimated from finite data sets? A typical approach to measuring the mean or variance is to perform multiple trials of the same experiment. For each trial, we compute the value of the moment ε_i, and after N trials, we calculate the expected value $E[\varepsilon]$ of ε as

$$E[\varepsilon] = \frac{1}{N} \sum_{i=1}^{N} \varepsilon_i .$$

This is the sample mean of ε. The hope is that $E[\varepsilon] = \hat{\varepsilon}$, where $\hat{\varepsilon}$ is the true value determined, for example, by computing the moments from the pdf. If this result is obtained, then the estimate $E[\varepsilon]$ is said to be *unbiased*. For example, the calculation of an average to estimate the mean provides an unbiased estimate of the first moment. On the other hand, if $E[\varepsilon] \neq \hat{\varepsilon}$, then

$$b[\varepsilon] = E[\varepsilon] - \hat{\varepsilon} ,$$

where $b[\varepsilon]$ is the *bias*. When $b[\varepsilon] \neq 0$, the estimate is said to be *biased*.

The total estimation error is given by the *mean square error*, namely the expected value of the squared difference from $\hat{\varepsilon}$, i.e., $E[(\varepsilon_i - \hat{\varepsilon})^2]$. An estimate can be biased because the mean square error contains two contributions: a nonsystematic random component and a systematic biased component. As an illustration, consider a target shooter whose shots cluster to the right of the bull's-eye. There is a clear bias ("the gun pulls to the right"); however, if the spacing between the holes in the target is small, the random component will be small as well.

---

[10] Andrzej Aleksander Lasota (1932–2006), Polish mathematician.

Statisticians have devoted considerable effort to developing unbiased methods to estimate the various moments. For example, the unbiased estimate for the variance of a sample of values of x_i when the mean is zero is

$$E[(x - E[x])^2]_{\text{unbiased}} = \frac{1}{N-1} \sum_{i=1}^{N} x_i^2,$$

whereas the biased estimate is

$$E[(x - E[x])^2]_{\text{biased}} = \frac{1}{N} \sum_{i=1}^{N} x_i^2.$$

Although it seems reasonable to expect that unbiased estimates would always be better, this expectation is not always borne out (see, for example, Section 14.5.1). This is because it may not always be possible to minimize both contributions to the mean square error simultaneously. In other words, tradeoffs become necessary. The recent emphasis on the use of maximum likelihood estimators is motivated by the need to develop methods that minimize the mean square error (see [692] for more details).

14.4.4 Signal-averaging

The response of a black box to an input is often obscured by the presence of uncorrelated, noisy fluctuations. The ratio of the power in a signal, P_{sig}, to that in the noise, P_{noise}, is called the signal-to-noise ratio (SNR)

$$
\begin{aligned}
\text{SNR} &= \frac{P_{\text{sig}}}{P_{\text{noise}}} \qquad\qquad (14.56) \\
&= \frac{E[x_{\text{sig}}^2]}{E[x_{\text{noise}}^2]} \\
&= \frac{E[x_{\text{sig}}^2]}{\sigma^2}
\end{aligned}
$$

where we have assumed that the noise is white with variance σ^2.

Signal-averaging techniques are used to reduce the contributions of noise in the recording while enhancing the contributions of the desired signal. These techniques apply repeated, identical stimuli to an "open-loop" dynamical system with two goals in mind: 1) measuring the time delay, or latency, between when the stimulus was applied, and 2) to characterize the response of the system to the applied stimulus. Well-studied examples arise in the use of evoked potentials in clinical neurophysiology [888].

The power in the noise for n sampled signals, z can be readily estimated for white noise as

$$E\left[\frac{1}{n}E\left[\sum_{i=1}^{n}x_i^2\right]\right] = \frac{1}{n^2}E\left[\sum_{i=1}^{n}x_i^2\right]$$

$$= \frac{1}{n^2}\sum_{i=1}^{n}E_i[x^2]$$

$$= \frac{1}{n^2}n\sigma^2 = \frac{1}{n}\sigma^2. \qquad (14.57)$$

Thus we see that as n increases, $E[x_{\text{noise}}^2]$ decreases.

Suppose that response to the stimulus at a frequency that is higher that its Nyquist frequency. This is referred to as *oversampling*. We know that a bandwidth-limited signal can be completely recovered under these conditions and hence

$$\frac{E[x^2]}{\frac{1}{n}\sigma^2} = n\frac{E[x^2]}{\sigma^2} = n\frac{P_{\text{sig}}}{P_{\text{noise}}} = n\text{SNR}. \qquad (14.58)$$

Thus we see that the SNR increases as n increases.

Example 14.20 In Example 10.1, we showed that the time delay for the feedback control for expert pole balancing can be measured from the responses to a visual blank. However, during the visual blank out the subject moves their hand in a way that is not correlated with the movements of the pole. These movements obscure an estimate of when the nervous system responds to correct the position of the pole after the blank out. By averaging many trials, it is possible to eliminate these effects and obtain a good estimate of the time delay (see Figure 10.2). ◊

A number of techniques are used to improve this signal-averaging procedure. First, the signal is typically time-locked to the stimulus. The simplest way to accomplish this is to use the onset of the stimulus to trigger data collection system. Second, investigators often average the odd and even trials separately. The difference between the odd and even time average provides an estimate of the noise power and the sum is the signal power using all trials. Finally, for the signal-averaging to produce the desired result the noise must be uncorrelated. A common problem in laboratory and clinical settings is the presence of electrical noise (50 or 60 Hz depending on where the laboratory is located). When electrical signal contamination cannot be sufficiently reduced, investigators try a number of "tricks" including randomizing the stimulus intensity and/or using non-integer stimulus rates in order to reduce its effects.

14.4.5 Broad-Shouldered Probability Density Functions

A curious property of pdfs measured for some biological variables is that values more than three standard deviations (SD) from the mean occur much more frequently than would be expected for a Gaussian pdf [95, 137, 287, 784, 820]. In other words, the observed pdf has "broad shoulders" (see, for example, Figure 16.11 in Section 16.7). Indeed, as better measuring devices and larger data sets become available, it is becoming recognized that such behavior may in fact be ubiquitous in nature [477].

A reader who has completed a typical undergraduate course in statistics will likely be disturbed by these comments. The central limit theorem states that the limit of normalized sums of independent identically distributed terms with finite variance must be Gaussian. The solution to this paradox is provided by relaxing the requirement that the variance be finite. The generalized central limit theorem states that the only possible nontrivial limits of normalized sums of independent identically distributed terms are Lévy-stable.[11] All Lévy-stable, or α-stable, distributions have the property that if x and y are random variables from this family, then the random variable formed by $x + y$ also has a distribution from this family.[12] The Gaussian distribution ($\alpha = 2$) is the special case of an alpha-stable distribution for which both the mean and variance exist: only the mean exists for $1 < \alpha < 2$, and neither the mean nor the variance exists when $\alpha \leq 1$. Figure 14.11 shows examples of Lévy-stable distributions for different values of the Lévy exponent α.

The possibility that random variables in biology are characterized by Lévy-type pdfs poses both experimental and modeling challenges. Despite the increased likelihood of events larger than three standard deviations from the mean, such events are still quite rare. Consequently, very large data sets must be collected in order to ensure that such rare events are detected, e.g., $\geq 10^5$ data points [873]. In collecting time series, these rare events can be missed unless proper attention is given to the cutoff frequencies for low-pass and high-pass filters. Moreover, it is crucial that the sampling frequency be sufficiently high. Indeed, it can be shown that as the sampling frequency decreases, the distribution of the fluctuations necessarily approaches the Gaussian distribution [537].

A particularly important issue is that of *truncation*. The tails of the Lévy distributions for $\alpha \neq 2$ shown in Figure 14.11 extend to $\pm\infty$. In other words, from a mathematical point of view, values of $x - x_0$ of arbitrary size are possible. In the real world, however, it may not be physically possible for $x - x_0$ to exceed a certain value, e.g., objects can change speed only within a certain limit. The effect of such physical limitations is to truncate the Lévy distribution: compare Figure 16.11c with Figure 14.11. Surprisingly, there has been little research on the proper way to handle truncation in the context of a Lévy flight (i.e., a random walk in which the step length is chosen from a Lévy distribution). For notable exceptions, see [299, 300, 438]. A simple approach is simply to set all $x - x_0$ greater than a preset threshold equal to zero (this is what was done in making Figure 16.11d). It has been suggested that

---

[11] Paul Pierre Lévy (1886–1971), French mathematician who introduced many concepts in probability theory used in this book, including alpha-stable distributions and Lévy flights.

[12] Keep in mind that this statement is true when the mean values have been removed before the distributions are added.

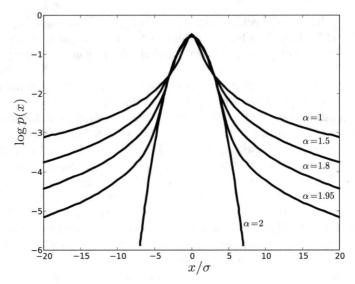

Fig. 14.11 Examples of probability density functions for Lévy-stable distributions as a function of the Lévy exponent α. The standard deviation σ is determined after the mean has been subtracted.

when faced with the control of a Lévy flight, biology tries to alter the truncation [95, 137].

A practical difficulty in working with Lévy-stable distributions is that closed-form expressions for the pdf exist only for three special choices of the parameters (Figure 14.11): the Gaussian distribution corresponds to the Lévy distribution with $\alpha = 2$ and $\gamma = \sigma^2/2$; the Cauchy[13] distribution corresponds to the case $\alpha = 1$; and the Lévy distribution (an unfortunately confusing choice of terms) corresponds to the case $\alpha = 0.5$. In contrast, a closed form for the Lévy characteristic function $\varphi_L(f)$ exists for all α. Thus most investigations of the properties of Lévy flights make use of $\varphi_L(f)$. This characteristic function can take various forms [920]. The form most useful for fitting experimental data to Lévy distributions when $0.5 \leq \alpha \leq 2$ is [435]

$$\varphi_L(f) = \exp\left[2\pi j\zeta f - |2\pi\gamma f|^\alpha + j\beta g(f,\alpha,\gamma)\right],$$

where

$$g(f,\alpha,\gamma) = \begin{cases} 2\pi\gamma f \tan\dfrac{\pi\alpha}{2}\left(|2\pi f|^{\alpha-1} - 1\right) & \text{if } \alpha \neq 1, \\ -4\gamma f \log|2\pi\gamma f| & \text{if } \alpha = 1, \end{cases}$$

---

[13] Augustin–Louis Cauchy (1789–1857), French mathematician.

where β is the symmetry index, and γ scales the spread of the distribution.[14]

14.4.6 Survival functions

An important statistic is the times-to-failure [464]. For example, patients with cancer want to know their expected life expectancy. However, survival estimates are not limited to medical applications. Applications range from determining the lifetime of radioactive isotopes, the time that an appliance or motor vehicle is expected to last before it fails, to the survival of metastable states [96, 97].

Survival times, t_{esc}, are usually considered to be random variables and hence can be described by a probability distribution. We define the *cumulative distribution function*, $\Phi(t)$, as the probability than a state fails before time t,

$$\Phi(t) = \mathrm{Prob}(t_{esc} < t), \tag{14.59}$$

and the *probability density function*, $p(t)$, as the probability that a failure occurs in a small time interval, where

$$\phi(t) = \frac{d\Phi(t)}{dt}. \tag{14.60}$$

The probability, $p(t)$, that an event lasts longer than a time t is

$$p(t) = 1 - \Phi(t) = \mathrm{Prob}(t_{esc} > t).$$

The *survivorship function*, $s(t)$ provides an estimate of $p(t)$ and is determined experimentally by plotting the fraction of individual events that survive longer than a time t as a function of t. Since $s(t) \approx 1 - \Phi(t)$, we obtain estimates of both $\Phi(t)$ and $\phi(t)$. This makes it possible to determine the *hazard function*, $h(t)$, where

$$h(t) = \frac{\phi(t)}{p(t)}.$$

The hazard function gives the probability of failure per unit time. This procedure is referred to as Kaplan–Meier analysis [183].

Often $p(t)$ has the form

$$p(t) = \exp[-(kt)^\gamma]$$

where k is a constant with units of reciprocal time. When $\gamma = 1$, we have $s(t) \approx p(t) = \exp(-kt)$, $\phi(t) = k\exp(-kt)$ and $h(t) = k$. Hence the risk of failure remains constant. When $\gamma > 1$ the probability of failure is an increasing function of time. In other words, the longer the process lasts the greater the risk of failure. When $\gamma < 1$

---

[14] For MATLAB programmers, a very useful software package has been developed by M. Veillette. It is freely downloadable from the Internet site http://math.bu.edu/people/mveillet/html/alphastablepub.html. Similar tools for Python have been developed within the package PyLevy.

the probability of failure is a decreasing function of time. In other words, the longer a process survives the greater the probability that it will continue to survive.

Example 14.21 The measurement of survival times has played an important role in unraveling details about how the nervous system becomes expert at pole balancing at the fingertip. The survival function, $s(t)$, is estimated by having a subject perform a large number of trials (typically 25-150) and calculating the fraction of trials for which the pole remains balanced as a function of time [96, 97]. Figure 14.12a shows $s(t)$ measured for a novice pole balancer during their first 9 days of practice. The increase in skill is reflected by a progressive shift to the right in the pole balancing survival curve, i.e., a plot of the fraction of poles still balanced at time t. In other words, the mean pole balancing time, $t_{\frac{1}{2}}$, determined on a single day becomes progressively longer with days of practice. When time is rescaled by $t/t_{\frac{1}{2}}$ it can be seen that the survival curves collapse onto a single survival curve (Figure 14.12b). This observation suggests that $t_{\frac{1}{2}}$ is a relevant time scale for the development of pole balancing skill. Curiously no significant change in the survival curve on a given day is seen if a second practice session is done on the same day [97]. This observation supports suggestions that sleep-dependent consolidation and refinement of motor control is part of the skill acquisition process [302, 453]. $\Diamond$

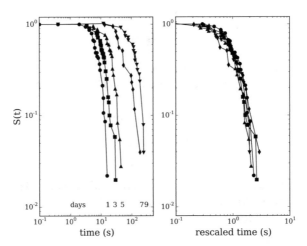

Fig. 14.12 (a) Survival functions for stick balancing for a novice pole balancer during the first 9 days of practice. (b) Data in (a) was rescaled by replacing t by $t/t_{\frac{1}{2}}$, where $t_{\frac{1}{2}}$ is the mean pole balancing survival time for a given day.

14.5 Correlation Functions

In the context of time series analysis, the term *correlation* refers to the extent that a change in one variable at a given time is associated with a change in either its behavior in the past (autocorrelation function, denoted by $C_{xx}(\Delta)$) or by changes in another variable (cross-correlation function, denoted by $C_{xy}(\Delta)$). The autocorrelation function describes the extent to which the value of a variable measured at one time can be predicted from knowledge of its value measured at some earlier time. For a deterministic system, namely a system whose time evolution is completely described by a differential equation, it is obvious that the values at various times are intimately connected. Indeed, if the initial conditions are defined for the differential equation, then we can completely predict $x(t)$ for all subsequent times. However, if the values of $x(t)$ contain a random, or stochastic, component, then the value of $x(t_0)$ predicts less well the value of $x(t)$, where $t > t_0$. In particular, we can anticipate that as stochastic influences dominate, one would be completely unable to predict future values of $x(t)$ based on a knowledge of $x(t_0)$. Thus measurements of $C_{xx}(\Delta)$ are used to estimate the time-dependence of the changes in the self-influences, where the lag is $\Delta = t - t_0$.

In contrast, the cross-correlation function describes the extent to which the value of one variable $x(t)$ at time t can be predicted from knowledge of the value of a different variable $y(t')$ measured at some other time t'. Thus measurements of $C_{xy}(\Delta)$ arise in estimating the extent to which changes in one variable are related to changes in another.[15]

A comment on notation will make reading the following easier. Since our custom is to subtract the mean from the time series before analysis, $C_{xx}(\Delta)$ is the same as the autocovariance function. For consistency, we will use the term "correlation" rather than covariance. Finally, we use K to refer to the autocorrelation or cross-correlation divided by the autocorrelation or cross-correlation when $\Delta = 0$.

We begin by defining $C_{xx}(\Delta)$. The autocorrelation integral is related to the convolution integral $z(\Delta)$ that we studied in Section 6.4:

$$z(\Delta) = \int_{-\infty}^{\infty} x(u)h(\Delta - u)\,du.$$

The convolution of $x(t)$ with itself is

$$y(\Delta) = \int_{-\infty}^{\infty} x(u)x(\Delta - u)\,du.$$

In contrast, the autocorrelation integral is

$$C_{xx}(\Delta) = \int_{-\infty}^{\infty} x(u)x(u - \Delta)\,du = \int_{-\infty}^{\infty} x(u)x(u + \Delta)\,du. \tag{14.61}$$

---

[15] The reader should keep in mind that correlation does not imply causality.

The key difference between $z(\Delta)$ and $C_{xx}(\Delta)$ is that $C_{xx}(\Delta)$ incorporates $x(u - \Delta)$ rather than $x(\Delta - u)$. Thus in terms of the graphical interpretation of the convolution integral we used in Section 6.4, namely folding $\rightarrow$ displacement $\rightarrow$ multiplication $\rightarrow$ integration, there is no folding step. The consequence of a function simply displaced with respect to itself by an amount Δ without folding is that the integral of the product will be the same whether Δ is positive or negative. Thus, provided that $x(t)$ is a real function, $C_{xx}(\Delta)$ will be an even function. The major exception occurs when $x(t)$ itself is an even function. In that case, convolution and autocorrelation are equivalent, because an even function is identical to its mirror image.

Figure 14.13 shows examples of the types of autocorrelation commonly encountered in the laboratory (for a larger list of examples, see [49]). The following example shows how $C_{xx}(\Delta)$ can be calculated using (14.61).

Example 14.22 Consider the sine wave process

$$\{x_k(t)\} = \{A\sin[2\pi f_0 t + \phi(k)]\}, \tag{14.62}$$

where A, f_0 are constants. Let $\phi(k)$ be a random variable drawn from a uniform density on the interval $[0, 2\pi]$, where $p(\phi) = (2\pi)^{-1}$. Determine $C_{xx}(\Delta)$.

Solution. For every fixed value of Δ, we have

$$x_k(t) = A\sin[2\pi f_0 t + \phi(k)],$$
$$x_k(t + \Delta) = A\sin[2\pi f_0(t + \Delta) + \phi(k)].$$

Define the following variables:

$$\alpha = 2\pi f_0 t, \quad \beta = 2\pi f_0(t + \Delta),$$
$$\alpha - \beta = -2\pi f_0 \Delta, \quad \alpha + \beta = 4\pi f_0 t + 2\pi f_0 \Delta.$$

Thus we have

$$C_{xx}(\Delta) = A^2 \int_0^{2\pi} \sin(\alpha + \phi)\sin(\beta + \phi)p(\phi)d\phi$$
$$= \frac{A^2}{2\pi}\left[\frac{2\phi}{4}\cos 2\pi f_0 \Delta - \sin(\alpha + \beta + 2\phi)\right]\Big|_0^{2\pi} = \frac{A^2}{2}\cos 2\pi f_0 \Delta.$$

$\diamond$

Figure 14.13 shows examples of the types of autocorrelation commonly encountered in the laboratory (for a larger list of examples, see [49]). For experimentalists, the falloff of the autocorrelation versus Δ is most interesting. Thus it is useful to divide $C_{xx}(\Delta)$ by the variance $C_{xx}(0)$ to obtain $K_{xx}(\Delta)$.

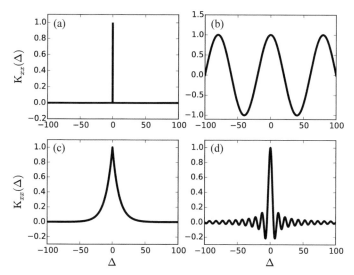

Fig. 14.13 Examples of $K_{xx}(\Delta)$ often observed in experimental time series: (a) white noise ($K_{xx}(\Delta) = k\delta(\Delta)$), (b) sine wave ($K_{xx}(\Delta) = 0.5A^2\cos 2\pi f_0\Delta$), (c) exponential ($K_{xx}(\Delta) = \exp(-k|\Delta|)$), (d) low-pass-filtered white noise ($K_{xx}(\Delta) = kB\sin 2\pi B\Delta/2\pi B\Delta$), where f_0, k, A, B are constants.

14.5.1 Estimating Autocorrelation

Most typically, $C_{xx}(\Delta)$ is computed using the fast Fourier transform algorithm (see Section 8.4.2). However, there are several caveats to be kept in mind in measuring $C_{xx}(\Delta)$ from experimental data. First, the time series are sampled at fixed time intervals $t_1 - t_2$, and hence we have $\Delta = m(t_1 - t_2)$, where m is an integer. Second, collected time series necessarily are of finite length. Thus the estimated autocorrelation function is best referred to as the sample autocorrelation. Does the sample correlation $c_{xx}(\Delta)$ provide a good estimate of $C_{xx}(\Delta)$?

Finally, since $x(t)$ is not typically a periodic function, it is necessary to worry about end effects. As in our calculations of the convolution integral in Section 6.4, we can overcome this problem through the use of zero padding.

In order to determine correlations up to a total lag of R, it is necessary to append an R-length buffer of zeros to the end of the data set [692]. On the other hand, the fast Fourier transform is most efficient for data sets of length 2^n, where n is a positive integer. This observation suggests that the length of the zero padding should be equal to the length of the original time series. Although both approaches are theoretically correct, the first approach will typically be the one that computes $C_{xx}(\Delta)$ in the shortest amount of time for very large data sets. This is true because we must consider the CPU time required to compute the fast Fourier transform plus the time required to access and write to memory, especially when the size of the data set approaches or exceeds the available RAM. For present-day computers, the fast step is CPU computation, and the slow steps involve memory storage and access.

Figure 14.14 compares the sample autocorrelation function from which the mean is not removed (Figure 14.14a) to the function from which the mean is removed (Figure 14.14b). The graph of $K_{xx}(\Delta)$ in Figure 14.14b has the characteristic shape observed for white noise. Since significant correlation exists only for $\Delta = 0$, white noise is often referred to as delta-correlated noise.

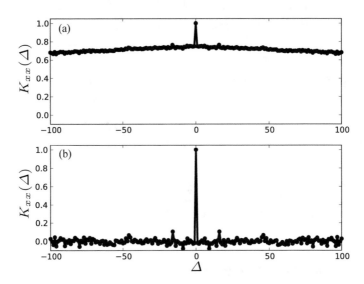

Fig. 14.14 The autocorrelation function $K_{xx}(\Delta)$ determined for uniformly distributed white noise with (a) nonzero mean and (b) the mean removed. The downward falloff of $K_{xx}(\Delta)$ in (a) is a numerical artifact that arises because of the finite length of the time series.

Correlation Time

The second measure that describes the extent to which the measured noise in a noisy time series differs from white noise is the correlation time t_{corr}, defined as

$$t_{\text{corr}} = \frac{1}{C_{xx}(0)} \int_0^\infty C_{xx}(\Delta) \, d\Delta. \tag{14.63}$$

The rate at which $C_{xx}(\Delta)$ approaches 0 as Δ approaches ∞ is a measure of the memory for the stochastic process. The following example gives some insight into t_{corr}.

Example 14.23 Suppose we have exponentially correlated noise with

$$C_{xx}(\Delta) = e^{-k\Delta}.$$

What is t_{corr}?

Solution.

$$t_{\text{corr}} = \int_0^\infty e^{-k\Delta}\,d\Delta = \frac{1}{k}.$$

Thus we see that the correlation time is the time it takes $C_{xx}(\Delta)$ to fall to the fraction e^{-1} of its initial value (see also Section 2.2.2). $\diamondsuit$

Bias

The sample autocorrelation $c_{xx}(\Delta)$, calculated using `np.correlate()` in Python, calculates the raw correlation

$$c_{xx}^{\text{raw}}(\Delta) = \sum_{n=0}^{N-|\Delta|-1} x(n)x(n+\Delta). \tag{14.64}$$

In the statistics literature, $c_{xx}(\Delta)$ is typically normalized in one of two ways. First, the biased estimate of $c_{xx}(\Delta)$ is

$$c_{xx}^{\text{biased}}(\Delta) = \frac{1}{N} \sum_{n=0}^{N-|\Delta|-1} x(n)x(n+\Delta),$$

and the unbiased estimate is

$$c_{xx}^{\text{unbiased}}(\Delta) = \frac{1}{N-|\Delta|} \sum_{n=0}^{N-|\Delta|-1} x(n)x(n+\Delta).$$

The biased estimate $c_{xx}^{\text{biased}}(\Delta)$ has advantages over the unbiased estimate. In particular, using the biased estimate avoids random large variations which are often observed at the end point of the correlation sequence when the unbiased estimate is calculated. As Δ approaches N, $(N-\Delta)^{-1}$ approaches ∞, and the number of nonzero terms used in computing the sum decreases. Thus there will be random and potentially very large variations at the endpoints of the correlation interval. On the other hand, the only point of $c_{xx}^{\text{biased}}(\Delta)$ that is biased is $c_{xx}^{\text{biased}}(0)$. Thus this estimate becomes more desirable.

Significance

For a white-noise signal, it has been shown that for a sufficiently long time series, the fluctuations in $C_{xx}(\Delta)$ when $\Delta \neq 0$ are approximately normally (Gaussian) distributed with mean zero and variance N^{-1} [19]. Therefore, the 95% confidence interval is $\pm 1.96/\sqrt{N}$ (see also [388] and Section 16.1). If $C_{xx}(\Delta)$ lies outside this confidence interval for a given Δ, then the null hypothesis that the original signal is white is rejected.

An alternative strategy to test for significant correlations is to use a "bootstrap" method to estimate the confidence interval. The first step is to randomly shuffle the time series data (in `numpy`, the function is `np.random.shuffle()`) to remove any correlation. The envelope of $C_{xx}(\Delta)$ calculated for, say, 100 different shuffles of the same time series gives the values of $C_{xx}(\Delta)$ that would be obtained "just by chance." If $C_{xx}(\Delta)$ lies outside this empirically determined confidence interval, then it is likely that a significant autocorrelation exists, in this example at the $\approx 99\%$ level.

14.6 Power Spectrum of Stochastic Signals

In Section 8.4.2, we saw that the estimation of the sample power spectrum $w(f)$ for a deterministic time series is facilitated by the use of a very powerful algorithm, the fast Fourier transform. However, this procedure is not a valid approach to estimating $w(f)$ for a stochastic time series [388]. In order to understand this issue, it is necessary to note that $w(f)$ refers to the power spectrum determined from a finite sample of the time series. As the length of this time series becomes longer and longer, we expect that $w(f)$ approaches $W(f)$, where $W(f)$ is the power spectrum determined for a time series of infinite length. Is this expectation also true for a stochastic time series? In fact, it is not (see Figure 14.15)!

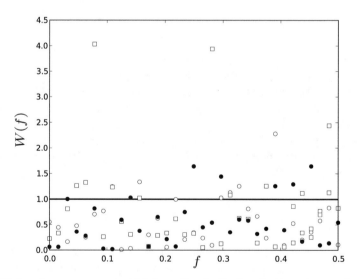

Fig. 14.15 Sample power spectrum $w(f)$ for uniformly distributed white noise as a function of length of the time series: $\bullet = 128$ points, $\circ = 512$ points, and $\square = 1,024$ points. The horizontal solid line gives $W(f)$.

The problem is that $w(f)$ measured for a finite-length stochastic process does not converge in any statistical sense to a limiting value as the sample length becomes longer and longer [388] (Figure 14.15). This observation should not be particularly surprising. Fourier analysis is based on the assumption of fixed amplitudes, frequencies, and phases. However, time series of stochastic processes are characterized by random changes of amplitudes, frequencies, and phases. Fortunately, these issues can be overcome by making use of the autocorrelation function $C_{xx}(\Delta)$ together with the following theorem.

The *Wiener–Khinchin theorem* states that the sample power spectral density function $w(f)$ and the sample autocorrelation function $C_{xx}(\Delta)$ form a Fourier transform pair, namely,

$$w(f) = \int_{-\infty}^{\infty} C_{xx}(\Delta) e^{-j2\pi f\Delta} \, d\Delta, \qquad (14.65)$$

and taking the inverse Fourier transform yields

$$C_{xx}(\Delta) = \int_{-\infty}^{\infty} w(f) e^{j2\pi f\Delta} \, df. \qquad (14.66)$$

Since $C_{xx}(\Delta)$ is an even function, it follows that (16.32) and (16.31) can be rewritten in a more convenient form as

$$\hat{w}(f) = 4 \int_{0}^{\infty} C_{xx}(\Delta)(\cos 2\pi f\Delta) \, d\Delta \qquad (14.67)$$

and

$$C_{xx}(\Delta) = \int_{0}^{\infty} \hat{w}(f)(\cos 2\pi f\Delta) \, df, \qquad (14.68)$$

where $\hat{w}$ is the one-sided power spectrum, i.e., the power spectrum defined for non-negative frequencies.[16] The *autospectral density function*, denoted by $S_{xx}(f)$, is frequently used to emphasize that the power spectrum is defined using the same measure of power used to define the autocorrelation function. Thus $S_{xx}(f)$ corresponds to the periodogram introduced in Section 8.3.3.

The Fourier transform pairs given by (16.32) and (16.31) or (14.67) and (14.68) are valid whether $x(t)$ represents a deterministic time series or a realization of a stochastic process [388]. The sample autocorrelation function for a finte time series, $c_{xx}(\Delta)$, provides a good estimate of $C_{xx}(\Delta)$ for an infinitely long time series. In particular, it can be shown that the mean square error for $c_{xx}(\Delta)$ is of order $1/T$, and hence its distribution tends to be more clustered about $C_{xx}(\Delta)$ as t approaches ∞ [388]. This observation is a restatement of the ergodic hypothesis, i.e., that the ensemble average $C_{xx}(\Delta)$ can be estimated from the time average $c_{xx}(\Delta)$, which

---

[16] It should be noted that mathematicians and investigators working in communication theory often use the two-sided power spectrum [49]. In this case, we have, for example,

$$\hat{w}(f) = \int_{-\infty}^{\infty} c_{xx}(\Delta)\cos(2\pi f\Delta) \, d\Delta = 2\int_{0}^{\infty} c_{xx}(\Delta)\cos(2\pi f\Delta) \, d\Delta.$$

However, the use of the one-sided power spectrum is favored in the stochastic signal literature.

will be true provided that $C_{xx}(\Delta)$ tends to zero at a sufficiently high rate. Even though this ergodic property applies to $c_{xx}(\Delta)$, it does not apply to $w(f)$.

Example 14.24 Calculate the autospectral density function of the sine wave process described by (14.62).

Solution. From the definition of $S_{xx}(f)$, we have

$$S_{xx}(f) = \frac{A^2}{2} \int_{-\infty}^{\infty} e^{-j2\pi f \Delta} \cos(2\pi f_0 \Delta) \, d\Delta \, ,$$

where $c_{xx}(\Delta)$ was calculated in the previous example. Using Euler's relation, we can write

$$S_{xx}(f) = \frac{A^2}{4} \int_{-\infty}^{\infty} \left(e^{-j2\pi (f+f_0)\Delta} + e^{-j2\pi (f-f_0)\Delta} \right) d\Delta$$

$$= \frac{A^2}{4} \left[\delta(f+f_0) + \delta(f-f_0) \right] .$$

This is the same result that we obtained in Section 8.4.2.   ◇

In general, signals collected in the laboratory contain both deterministic and stochastic contributions. Little work has been done on determining the best strategy for estimating $W(f)$ in these situations. Since most biologists use $w(f)$ to estimate the frequency of the periodic components in the signal, it is reasonable simply to use the periodogram or the power spectral density. However, if it is necessary to have a good estimate of power, it may be reasonable first to fit the autocorrelation to an appropriate functional form (see Figure 14.13 and [49]) and then calculate $W(f)$ by hand using the Wiener–Khinchin theorem. This, in fact, is what is often done in statistical physics [257, 413].

14.7 Examples of Noise

Although the subject of fluctuations, or noise, seems rather esoteric, it sometimes turns out to provide deep insights into mechanisms [518]. For example, the observation that membrane noise recorded from excitable cells has the characteristics of "shot noise" provided the first evidence of the existence of membrane pores, now referred to as ion channels [850]. Recent attention has focused on determining the explanation for the widespread occurrence of "$1/f$ noise" in physical and biological systems [172, 323, 620]. Thus a working knowledge of the terminologies that describe noise is useful. The reader should realize that noise is a subject of interest in many different fields, in particular engineering and physics. This leads to the use of terms whose meaning at times seems obscure to biologists.

White noise refers to the situation in which the autocorrelation function is a delta function: it is equal to 1 when $\Delta = 0$ and equals 0 for all $\Delta \neq 0$ (see Figure 14.14). Delta-correlated noise can be further classified on the basis of its pdf. Typically, the phrase "white noise" refers to noise that is uniformly distributed. We used white noise of this type in Section 8.4.1 to illustrate the effects of a low-pass filter. An advantage in using white noise is that the Fourier transform is simply a constant. This type of noise is physically unrealistic, since its power would be infinite. However, over small frequency ranges, it may be possible to approximate noise as white. Integrating white noise produces *brown noise*, which simulates Brownian motion, the subject of Chapter 16. On the other hand, if we differentiate white noise, we obtain *violet noise*.

A second type of delta-correlated noise that frequently arises is *Gaussian-distributed white noise*. The central limit theorem gives the conditions under which the distribution obtained from a sufficiently large number of statistically independent random variables, each identically distributed with finite mean and variance, will approximate a Gaussian (normal) distribution. Thus in our system–surroundings formulation, it is common to represent the surroundings as Gaussian-distributed white noise. An advantage of the use of Gaussian-distributed noise is that its Fourier transform is also Gaussian.

Exponentially correlated noise, whose correlation function decays exponentially as a function of Δ, is often observed. It has been shown theoretically that exponentially correlated noise characterizes Brownian motion when inertial effects are taken into account. This situation is referred to as an *Ornstein–Uhlenbeck*[17] *process* and is discussed further in Section 15.1.

Colored noise is noise whose correlation function decays nonexponentially.[18] Particular attention is given to power-law noise, whose power spectrum has the form $1/f^\beta$, where β is a constant. The $1/f^\beta$ terminology arises from the observation that the power spectral density per unit bandwidth is proportional to $1/f^\beta$. This means that a log–log plot of power versus frequency will be linear with slope $-\beta$. The color terminology refers to a loose analogy between the spectrum of frequencies present in a sound and the equivalent spectrum of light-wave frequencies. Thus the sound wave pattern of white noise ($\beta = 0$) translated into light waves would be perceived by a viewer as white light, and so on. However, this analogy is not always valid, since brown noise is named for its association with Brownian motion, not for its color.

By far the most attention has been devoted to *pink*, or $1/f$, *noise* ($\beta = 1$) [172, 323, 620, 847]. The power density falls off by 10 dB per decade, or 3 dB per octave. Since pink noise is linear in logarithmic space, there is equal power in frequency bands that are proportionately wide. Thus pink noise has the same power in the frequency band 40–60 Hz that it has in the frequency band 4,000–6,000 Hz.

Example 14.25 An octave is a frequency band whose highest frequency is exactly twice its lowest frequency. Most people perceive the sound of pink noise as being even, or flat, because it has equal power per octave, suggesting that humans hear in

---

[17] Leonard Salomon Ornstein (1880–1941), Dutch physicist; George Eugene Uhlenbeck (1900–1988), Dutch–American theoretical physicist.

[18] It seems more reasonable to define colored noise as noise that is not delta-correlated. However, in current usage of the term, colored noise does not include exponentially correlated noise.

a proportional space. For this reason, pink noise is widely used as a reference signal by the audio industry, for example in the design of stereo systems.   ◇

There is another property of pink noise that distinguishes it from other types of colored noise. Since there is an infinite number of logarithmic bands at both the low-frequency (DC) and high-frequency ends of its power spectrum, every finite energy spectrum must have less energy than pink noise at both ends. Pink noise is the only power-law spectral density that has this property. When $\beta > 1$, power spectra are finite if integrated to the high-frequency end; when $\beta < 1$, they become finite when integrated to the DC, or low-frequency, end.

14.8 Cross-Correlation Function

The cross-correlation function $C_{xy}(\Delta)$ is defined in exactly the same way as $C_{xx}(\Delta)$, except that we consider two functions $x(t)$ and $y(t)$,

$$C_{xy}(\Delta) = \int_{-\infty}^{\infty} x(t)y(t-\Delta)\,dt = \int_{-\infty}^{\infty} x(t+\Delta)y(t)\,dt.$$

For a real process, $C_{xy}(\Delta)$ satisfies the property

$$C_{xy}(\Delta) = C_{yx}(-\Delta),$$

but in general,[19]

$$C_{xy}(\Delta) \neq C_{xy}(-\Delta).$$

It follows that when x and y are correlated, $C_{xy}(\Delta)$ is not typically symmetric about $\Delta = 0$. A frequent cause of this asymmetry is the presence of a time delay. The cross-correlation function is useful because it provides a way to use white noise to estimate the impulse function and the time delay. Moreover, together with the autocorrelation functions, we can define a coherence function to determine to what extent a biological system behaves linearly (see also Section 9.2).

When a statistical test for nonzero cross-correlation $C_{xy}(\Delta)$ is required, it is necessary to remove all causal information from both $x(t)$ and $y(t)$ [388]. In other words, $C_{xx}(\Delta)$ and $C_{yy}(\Delta)$ must both be delta-correlated. This process of removing causal information from $x(t)$ and $y(t)$ is called *prewhitening*. There are two commonly used strategies to prewhiten a time series [388, 862]. The simplest is to use a difference filter $d(t) = x(t+1) - x(t)$ [388]; in computer programs, this function is usually called diff(). It may be necessary to difference the time series more than once to obtain a $C_{xx}(\Delta)$ that is delta-correlated.

The second approach used to prewhiten a time series is first to fit the original time series to an appropriate linear autoregressive (AR) model [388, 692, 862]. These models have the general form

---

[19] The major exception occurs when the two functions are not correlated, and hence $C_{xy}(\Delta) = 0$ for all Δ.

$$\hat{X}_t = \alpha_0 + \alpha_1 X_{t-1} + \alpha_2 X_{t-2} + \cdots + \alpha_n X_{t-n},$$

where α_i are constants and n is the largest significant lag in $C_{xx}(\Delta)$. The residuals $X_t - \hat{X}_t$, where X_t is the raw data, form the prewhitened time series. This form of prewhitening is frequently employed in the context of data compression. For example, cell phones do not directly transmit your digitized voice over the air. Instead, they first whiten your voice signal and transmit the whitened time series along with the filter coefficients obtained from the appropriate AR model. When sample sizes are small, care must be taken to ensure that these procedures do not introduce a bias (see, for example, [814]). The development and use of AR models is beyond the scope of this textbook.

14.8.1 Impulse Response

The cross-covariance or cross-correlation function for a linear system with a white-noise input is [49, 688]

$$y(t) = \int_0^\infty I(s)x(t-s)\,ds + \xi(t) = \int_{-\infty}^t I(t-v)x(v)\,dv + \xi(t), \qquad (14.69)$$

where $x(t), y(t)$ are an input and output and $\xi(t)$ is a noise term not correlated with $x(t)$. Then we can obtain $C_{xy}(\Delta)$ by multiplying both sides by $x(t-u)$ and taking an average:

$$\langle x(t-\Delta)y(t)\rangle = \int_0^\infty I(s)\langle x(t-\Delta)x(t-s)\rangle\,ds + \langle x(t-\Delta)\xi(t)\rangle, \qquad (14.70)$$

or

$$C_{xy}(\Delta) = \int_0^\infty I(s)C_{xx}(s-\Delta)\,ds.$$

The last term vanishes due to our assumption that there is no correlation between $x(t)$ and $\xi(t)$. Since the mean has been subtracted from the two signals, we can interpret $C_{xx}(\Delta)$ as either the autocorrelation or autocovariance function. Note that the above equations are written in terms of the cross-covariance. In particular, if the input $x(t)$ is white noise with variance σ^2, then

$$C_{xx}(s-\Delta) = \sigma^2\delta(s-\Delta).$$

Finally, we obtain

$$I(\Delta) = \frac{C_{xy}(\Delta)}{\sigma^2}.$$

This means that we can obtain the impulse function of the system in terms of the cross-covariance of the white-noise input and the associated output. This estimation method of the impulse function is called cross-correlation analysis. Real-time series are not typically delta-correlated. Thus the time series must be prewhitened before this technique can be applied.

14.8.2 Measuring Time Delays

The time delay τ can be estimated using either time-domain or frequency-domain methods. In the time domain, the delay is measured as the time interval between the onset of a stimulus and the onset of the response (see Figure 10.3). In the frequency domain, τ is typically estimated by measuring the cross-covariance function $C_{xy}(\Delta)$ between the input to the system $x(t)$ and the output $y(t)$ [284]. The use of $C_{xy}(\Delta)$ to estimate τ requires that the involved feedback loop have been opened and that the system introduce only a pure delay to its response.

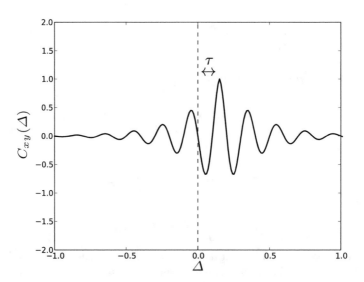

Fig. 14.16 Example of estimating the delay τ using the cross-covariance function $C_{xy}(\Delta)$.

Consider

$$y(t) = kx(t-\tau) + \xi(t),$$

where k is a constant and $\xi(t)$ is delta-correlated measurement noise. In this case, we can write

$$C_{xy}(\Delta) = \int_{-\infty}^{\infty} x(t)y(t+\Delta)\,dt = \int_{-\infty}^{\infty} x(t)[kx(t-\tau)+\xi(t)]\,dt$$

$$= k\int_{-\infty}^{\infty} x(t)x(t+\Delta-\tau)\,dt = kC_{xx}(\Delta-\tau). \tag{14.71}$$

Thus we see that $C_{xy}(\Delta)$ is equal to $kC_{xx}(\Delta)$ shifted by an amount τ, that is, $kC_{xx}(\Delta - \tau)$. The maximum value of $C_{xy}(\Delta)$ occurs when $\Delta = \tau$ and

$$C_{xy}(\Delta)_{max} = C_{xy}(\tau) = kC_{xx}(0) = k\sigma_x^2 .$$

The demonstration that $C_{xy}(\Delta)$ and $C_{xx}(\Delta)$ are identical in all respects expect for a shift by an amount τ validates the estimate of the delay (Figure 14.16).

14.8.3 Coherence

In Section 9.2, we introduced the use of coherence for the identification of a linear dynamical system. The squared coherence function $\gamma^2(f)$ is

$$\gamma^2(f) = \frac{|W_{xy}(f)|^2}{W_{xx}(f)W_{yy}(f)}, \tag{14.72}$$

where $W_{xx}(f), W_{yy}(f), W_{xy}(f)$ are the respective power spectrum functions. It should be noted that filtering the time series can affect both the amplitude and phase spectra of the coherence [388].

Experimentalists typically use functions such as cohere() to estimate $\gamma^2(f)$ quickly. The function cohere() estimates $W(f)$ from the periodogram. Although this function can be accessed from either pylab or matplotlib.mlab, we recommend the latter, since the result is consistent with that obtained with MATLAB. In all cases, the algorithms do not prewhiten the time series before estimating $W_{xy}(f)$. Thus, if the issue of linearity is in doubt, prudent investigators will write their own computer programs to estimate $\gamma^2(f)$ by examining the effects of various prewhitening strategies.

14.9 What Have We Learned?

1. **What is the Bayes' theorem?**

   The Bayes' theorem is a method to statistically infer the probabilities of causes A_i of some observed event B. Namely, it gives us a way to compute the conditional probabilities $P(A_i|B)$.

2. **How is the relation between the joint and conditional probabilities used in the Bayes' theorem?**

   The basic structure of the theorem is the definition of conditional probability from the joint probability, i.e., $P(A|B)P(B) = P(A;B) = P(B;A) = P(B|A)P(A)$

3. **How is the Bayes' theorem used to infer a cause?**

When there are multiple mutually exclusive causes A_i for an event B, the theorem gives a recipe to calculate the probability for each A_i to be a cause of B in the form of conditional probabilities $P(A_i|B)$

4. **What are the priori and posterior probabilities?**

A priori probability of the event A is the probability before the observation of the related fact or event B. A posterior probability of the event A is the updated conditional probability given the event B. They are connected by the Bayes' theorem.

5. **What is the method of Bayesian updates?**

It is a method to update the desired conditional probabilities with a repeated application of the Bayes' theorem as new related facts or events emerges. The key idea is to chain the theorem so that the posterior probability in the previous step becomes the priori probability in the next step.

6. **What is the Monty Hall problem?**

It originated from a TV show with Mr. Monty Hall as a host. A chosen member of the audience is asked to pick a door behind which there is a prize. After the choice is made out of the three doors, the host gives a clue by opening a door where there is no prize. The person is then given a chance to switch or keep their original choice of the door. A careful application of the Bayes' theorem shows that the person has a better chance of finding a prize if they change their original door selection.

7. **Why is the concept of probability density function (pdf) important?**

If the pdf is known, then it is possible to determine the moments, which, in turn, relate to descriptors of the data, such as the mean and variance.

8. **What is the advantage of characterizing stochastic processes in the frequency domain?**

Characterizing stochastic processes in the frequency domain makes it possible to use a number of powerful techniques of time series analysis, many of which are based on the fast Fourier transform. From a theoretical perspective, it can be much easier to estimate the moments from the Fourier transform of the pdf than by solving the appropriate expectation integrals.

9. **What is the autocorrelation function, and why it is useful?**

The autocorrelation function provides a measure of the extent to which a change in one variable at a given time is influenced by a change in the same variable at some other time in the past. It is one of the measures that is used to characterize noise; for example, delta correlation is a characteristic of white noise. In principle, the autocorrelation function can be used to estimate the power spectrum (see below).

10. **What is the cross-correlation function, and why is it useful?**

The cross-correlation function provides a measure of the extent to which a change in one variable at a given time is influenced by a changes in another variable at some other time in the past. The cross-correlation provides a method to determine the impulse function and to estimate the time delay. (The reader is cautioned to keep in mind that correlation does not imply causality.)

11. Why can methods for determining statistical descriptions of time series behave poorly?

> The use of techniques based on the fast Fourier transform assumes that the time series is continuous, periodic, and infinitely long. In contrast, time series datasets collected in the laboratory are discrete (for example, the digital output of an A/D board), of finite length, and usually aperiodic.

12. How do we determine the power spectrum $W(f)$ for a stochastic process?

> Theoretically, the best way to determine the power spectrum $W(f)$ is as the Fourier transform of the autocorrelation function. This is a result of the Wiener–Khinchin theorem, and the power spectrum determined in this manner is referred to as the autospectral density function $S(f)$. However, when time series are digital and of finite length, this procedure does not give a good estimate of $S(f)$. Thus other techniques are used, such as multitapering [694].

13. What is the rate-limiting step in computer programs used to simulate large-scale datasets: computational speed or memory access?

> In a modern computer, the slower steps in a computation are those that involve retrieving and storing data. The faster steps are those that involve computation. All other things being equal, a program that minimizes the number of times that memory is accessed will likely run significantly faster than one that does not.

14. What does it mean to prewhiten a time series, and why is it necessary to prewhiten time series in the analysis of the cross-correlation function?

> Before calculating the cross-correlation function between two times series $x(t)$ and $y(t)$, it is necessary to remove all causal information from each of them. If this is not done, then it is possible to obtain a falsely positive cross-correlation.

14.10 Problems for Practice and Insight

1. Suppose there are 100 students in the freshman class of the college, which offers 5 different majors. The number of (male, female) students in each major (A_1 to A_5) are given by the following.

$A_1 : (10,6), A_2 : (8,5), A_3 : (6,8), A_4 : (6,4), A_5 : (5,5)$

We also set that F is the event that student is a female, if you choose one student at random from this class. We can easily compute the probability as $P(F) = 28/100$.

a. Compute the following joint probabilities:
   $P(F;A_1), \quad P(F;A_2), \quad P(F;A_3), \quad P(F;A_4), \quad P(F;A_5)$

b. Compute the following conditional probabilities:
   $P(F|A_1), \quad P(F|A_2), \quad P(F|A_3), \quad P(F|A_4), \quad P(F|A_5)$

    c. Compute the following probabilities:

$$P(A_1), \quad P(A_2), \quad P(A_3), \quad P(A_4), \quad P(A_5)$$

    d. Confirm that in this case the following equation is satisfied.

$$P(F) = P(F;A_1) + P(F;A_2) + \cdots + P(F;A_5) = \sum_{i=1}^{5} P(F;A_i) = \sum_{i=1}^{5} P(F|A_i)P(A_i)$$

2. Compute the mean and the variance of the Gaussian distribution, whose p.d.f. is given by

$$p(x) = \frac{1}{\sqrt{2\pi\sigma^2}} e^{-\frac{(x-m)^2}{2\sigma^2}}$$

with m and σ are constants.

For this, you need to compute the integrals:

For the mean;

$$E[X] = \int_{-\infty}^{+\infty} x p(x) dx \tag{14.73}$$

For the variance;

$$V[X] = \int_{-\infty}^{+\infty} (x - E[x])^2 p(x) dx$$

You can use the following formula of the gaussian integral for your calculation.

$$\int_{-\infty}^{+\infty} e^{-\alpha x^2} dx = \sqrt{\frac{\pi}{\alpha}}$$

By evaluating the above integrals, convince yourself the followings:

$$E[X] = m, \quad V[X] = \sigma^2$$

3. In the example, we used the Bayes' theorem to compute the probability that a bottle of water is in the bag, when detector is activated. Now, compute the probability that a bottle of the water is not in the bag when the alarm did not set off with the same conditions. You may further assume $P(\bar{B}|A) = 0.05$ and $P(\bar{B}|\bar{A}) = 0.95$. Interpret the value comparing with the result in the text.
4. In the text, we consider a Bayesian update with evidences B and C come in that order. Consider the case that the order is reversed., i.e., evidence C come in before B. Is this result consistent with reality always?
5. Write a computer program to simulate Monty Hall Problem. Verify numerically that the switching the choice of the door gives twice better chance of finding the prize as derived in the text.

6. The stochastic variable ξ has the finite expectation $E(\xi) = m$ and finite variance $V(\xi) = \sigma^2$. Prove the following inequality, which holds for every positive real number α:

$$P(|\xi - m| \geq \alpha) \leq \frac{\sigma^2}{\alpha^2}.$$

This inequality shows that it is rather unlikely that a stochastic variable takes a value that is far from its mean.

7. For the uniform pdf on the interval $[a,b]$, compute the second moment $\langle x^2 \rangle$ and the third moment $\langle x^3 \rangle$. Confirm that when $a = -b$, the mean is zero and the variance is equal to the second moment.

8. A random variable whose pdf is the Cauchy distribution

$$p(x) = \frac{1}{\pi(1+x^2)}$$

is a representative example that does not possess a finite mean and variance. Convince yourself of this fact.

9. Compute the power spectrum $W(f)$ for uniformly distributed white noise generated using the python program from Section 8.4.2 as a function of the length of the time series:

a. 10^2 points.
b. 10^3 points.
c. 10^4 points.
d. 10^6 points.

Does the power spectrum appear to be converging to a stationary form?

10. A random variable is said to be normal when it obeys the Gaussian distribution. The pdf of a normal stochastic variable with mean m and variance σ^2 is given by

$$p(x) = \frac{1}{\sqrt{2\pi\sigma^2}} e^{-(x-m)^2/2\sigma^2}.$$

Show that its characteristic function is given by

$$\varphi(t) = e^{-(2\pi f)^2 \sigma^2/2 - 2j\pi fm}.$$

11. Consider n independent random variables X_i with mean 0 and variance σ^2. Now consider the new random variable

$$Z_n = \frac{1}{\sqrt{n}} \sum_{k=1}^{n} X_k.$$

a. Show that the characteristic function of Z_n is given by

$$\varphi_n(f) = E\left[e^{-j(2\pi f)Z_n}\right] = \left\{E\left[\exp\left(\frac{-j(2\pi f)X_k}{\sqrt{n}}\right)\right]\right\}^n.$$

b. Consider

$$g(s) = \begin{cases} \dfrac{e^{-js} - 1 + js}{(-js)^2/2}, & s \neq 0, \\ 1, & s = 0. \end{cases}$$

Use the fact that $g(s)$ is continuous to show that

$$E\left[\exp\left(\frac{-j(2\pi f)X_k}{\sqrt{n}}\right)\right] = 1 - \frac{(2\pi f)^2}{2n} E\left[X_k^2 g\left(\frac{(2\pi f)X_k}{\sqrt{n}}\right)\right].$$

c. By taking the limit as $n \to \infty$, show that the characteristic function of Z_n is given by

$$\lim_{n \to \infty} \varphi_n(f) = e^{-(2\pi f)^2 \sigma^2/2}.$$

d. Convince yourself that the above is the characteristic function of the normally distributed stochastic variable with mean 0 and variance σ^2. What does this tell you?

12. Noise can create the appearance that an oscillation is present in a dynamical system even though an underlying limit cycle oscillation does not exist. This situation most often arises when a stable spiral point is influenced by noisy perturbations. To illustrate this point, consider the second-order linear dynamical system

$$\frac{dx}{dt} = y, \quad \frac{dy}{dt} = k_1 x + k_2 y + \sigma^2 \xi,$$

where ξ represents white-noise input with mean zero and variance σ^2 (see Chapter 14).

a. Choose k_1, k_2 such that the fixed point at $\sigma^2 = 0$ is a stable spiral point. What is the value of $\omega = 2\pi j f$ for your choice of k_1, k_2?

b. Write a Python program to integrate this equation. Note that random-noise generators can be imported using the command

```
import numpy.random.rnd.
```

The command `rnd.random()` generates uniformly distributed white noise on the interval $[0, 1]$. You will have to subtract 0.5 to make the mean equal to zero.

c. Choose $\sigma^2 \neq 0$. You should see an oscillatory solution with a varying amplitude. Estimate the period. Is it the same as predicted above?

d. Why does the solution of the differential equation with noise look smoother than the signal from the noisy input?

e. Determine the power spectrum as a function of σ^2.

13. A simple model of patient survival proposed by M. Mackey and J. Milton [524] provide insights into the meaning of $S(t), h(t)$. Write a computer program to compute $S(t)$ and $h(t)$ for a population of 1000 subjects whose evolution is governed by

$$x_{t+1} = 4x_t(1 - x_t)$$

where D is an interval of width $b - a$ centered at $x = 0.5$ and $\mu(D)$ describes how this interval changes as a function of time. Each individual in this population corresponds to a random choice of the initial value of x, x_0 on the interval $[0, 1]$ such that $x \notin \mu(D)$. An individual is said to die whenever $x_t \in \mu(D)$ for some t. In this way we can compute 1000 survival times and then use them to determine $S(t)$ and $h(t)$. This model predicts that $S(t) = \exp[-(\lambda)^\gamma]$ and that $S(t)$ and $h(t)$ depend on the properties of $\mu(D)$. Take $\lambda = 0.1$ and use the computer program to demonstrate the following.

a. Show that if $\mu(D) = b - a$, then $\gamma = 1$. In other words, $S(t)$ decreases exponentially and $h(t)$ is a constant. How does the e^{-1} time depend on $b - a$?

b. Assume that $\mu(D) = (b - a)(\varepsilon t + 1)^\alpha$ with $\alpha > 0$. Hence $\mu(D)$ increases with time. Take $\varepsilon = 0.55$ and $\alpha = 0.5$ and show that $\gamma > 1$.

c. Assume that $\mu(D) = (b - a)(\varepsilon t + 1)^\alpha$ with $-1 < \alpha < 0$. Hence $\mu(D)$ decreases with time. Take $\varepsilon = 0.85$ and $\alpha = -0.5$. Show that $\gamma < 1$.

d. Choices of the parameters that describe $S(t)$ and $h(t)$ for various cancers before 1990 are given in [524]. Search for survival statistics for these cancers in the Internet since 1990 and then fit the statistics to this model. Has anything changed in the survival statistics of these cancers since 1990? Did survival improve because of changes in γ and/or $\mu(D)$?

e. What happens to the survival if the dynamics of $\mu(D)$ are described using a different equation?

Chapter 15
Noisy Dynamical Systems

While examining the form of these particles immersed in water, I observed many of them very evidently in motion. These motions were such to satisfy me that they arose neither from currents in the field, nor from its gradual evaporation, but belonged to the particle itself.

—Robert Brown

The study of fluctuations in dynamical systems had rather humble beginnings. With the advent of the light microscope in the late 1600s, many investigators noticed that small particles suspended in liquid were continually moving, even though no macroscopic movements of the fluid could be detected [675]. Initially, it was thought that these movements indicated that the particles were alive. However, in 1827, Robert Brown[1] [84] proved that this hypothesis was wrong: qualitatively similar movements were seen whether the particles were derived from animate (e.g., pollen grains) or inanimate (e.g., dust collected from the Sphinx) sources.[2]

In 1905, Albert Einstein[3] brought this subject matter to the attention of physicists by suggesting that the movements observed by Brown arose directly from the incessant random pushes, or perturbations, to the particle made by molecules in the surrounding fluid [190]. Since the movements are the result of collisions with other molecules, they are ballistic in nature. The term "ballistic" means that there are rapid changes in both direction and speed. Einstein estimated that for a Brownian particle 10^{-6} m in diameter, the collisions occur every 10^{-7}s, and hence the particle moves only 10^{-10}m before the next collision occurs [702]. The ballistic nature of Brownian motions has been confirmed experimentally in both gases [481] and liquids [353].

---

[1] Robert Brown (1773–1858), Scottish botanist.

[2] Robert Brown was not the first person to observe these movements. In his original paper, Brown cites ten investigators who had earlier observed these movements; subsequently, a scientific historian added the name J. Ingen-Housz to this list [843]. However, it is clear that Robert Brown was the first to investigate the origins of this phenomenon in a systematic manner, and consequently, he deserves to have his name attached to these particle motions [675].

[3] Albert Einstein (1879–1955), German-born American physicist.

© Springer Nature Switzerland AG 2021
J. Milton and T. Ohira, *Mathematics as a Laboratory Tool*,
https://doi.org/10.1007/978-3-030-69579-8_15

His observations provided the foundations for the study of random walks that we address in the following chapter.

However, the movements of a Brownian particle are also influenced by inertial effects and the viscosity of the liquid environment. In 1908, Paul Langevin[4] formulated the first stochastic differential equation to describe Brownian motion [461]. The importance of the Langevin equation is that it emphasizes that the dynamical behaviors observed by an experimentalist can be shaped by the interplay between deterministic processes and noise.

The study of stochastic differential equations, collectively referred to as Langevin equations, lies at the cutting edge of present-day biological research. This subject area is challenging, yet a knowledge of the properties of Langevin equations even at the descriptive level is sufficient to enable scientists to design useful experiments. We have attempted to strike a balance between useful benchtop insights and mathematical rigor by organizing this chapter as follows. First, we analyze the simplest version of the Langevin equation for the case of Gaussian-distributed white noise. In this case, we can readily obtain solutions using concepts and tools that we developed in earlier chapters. The main results we wish to emphasize for benchtop researchers are illustrated with applications. There is often a tendency for investigators to regard noise as a nuisance or as something that should be avoided. Thus, we discuss examples that highlight the beneficial effects of noise. A recurring theme throughout this chapter is the importance of understanding the effects of noise when dynamical systems are tuned near or at stability boundaries, or more colorfully, systems tuned *at the edge of stability*.

The questions we will answer in this chapter are:

1. What is additive noise?
2. What is parametric noise?
3. What is the Langevin equation?
4. What are the autocorrelation function $C_{xx}(\Delta)$ and the power spectrum $W(f)$ determined from the Langevin equation?
5. What is the net mean square displacement as a function of time of a particle whose movements are described by the Langevin equation?
6. How was the discrepancy between the predicted and observed velocity of the movements of a particle resolved?
7. What is an Ornstein–Uhlenbeck process?
8. What dynamical behavior observed for parametric noise has not yet been observed for additive noise, and what is its characteristic?
9. What is the effect of a time delay on the power spectrum and autocorrelation function for a first-order linear time-delayed dynamical system with additive noise?

---

[4] Paul Langevin (1872–1946), French physicist.

15.1 The Langevin Equation: Additive Noise

The Langevin equation was initially developed to describe the movement of particles subject to inertial effects and viscosity within a liquid environment. Consider a particle confined to move in the x-direction. Using Newton's law of motion, we have

$$M\frac{dv}{dt} = \mathscr{F}(t), \tag{15.1}$$

where M is the mass of the particle, v is its velocity, and $\mathscr{F}$ is the total force acting on the particle at any instant due to the effects of the environment. If we choose

$$\mathscr{F}(t) := -\frac{v}{b} + F(t),$$

then we obtain the *Langevin equation*

$$\frac{dv}{dt} + \frac{v}{t_{\text{relax}}} = \xi(t), \tag{15.2}$$

where $\xi(t) := F(t)/M$ describes the random accelerations (noise) on the particle arising from environmental perturbations. The macroscopic relaxation time t_{relax} of the particle is the $1/e$ time we introduced in Chapter 2 with Tool 3. We assume that the magnitude of the perturbation $F(t)$ at each instant in time is chosen randomly from a prescribed probability density function (pdf). Consequently, the perturbations (collisions) are delta-correlated and hence are white noise.

Since the magnitudes of these collisions are independent of the state variable v, the effects of this noise are state-independent, or additive. This assumption means that each random perturbation occurs rapidly compared to t_{relax} and that the perturbations have short memory (i.e., the correlation time is short). Mathematically, these assumptions are summarized as

$$\langle \xi(t) \rangle = 0, \quad \langle \xi(t)\xi(t') \rangle = \sigma_n^2 \delta(t - t'), \tag{15.3}$$

where σ_n^2 is the variance of the noise $\xi(t)$, and we have assumed that higher-order terms (cumulants) are zero. The variance term σ_n^2 arises because the fluctuating force from $\xi(t)$ has significant magnitude at every instant in time. The above properties of $\xi(t)$ lead to Gaussian-distributed white noise, i.e., the pdf of ξ at each t is given as follows:

$$p(\xi) = \frac{1}{\sqrt{2\pi\sigma_n^2}} \exp\left[-\frac{\xi^2}{2\sigma_n^2}\right]. \tag{15.4}$$

The Langevin equation thus defined is also called an *Ornstein–Uhlenbeck process* in the mathematical literature.

It is important to keep in mind that the $\xi(t)$ term in (15.2) is nowhere differentiable; however, its integral exists. This means that solutions of (15.2) must be obtained with care. We can understand many of properties of the Langevin equation by using the integrating factor tool (Tool 6) [518].

$$\frac{d}{dt}\left(ve^{t/t_{\text{relax}}}\right) = \xi(t)e^{t/t_{\text{relax}}},$$

where $e^{t/t_{\text{relax}}}$ is the integrating factor. The solution obtained by integration is

$$v(t) = V_0 e^{-t/t_{\text{relax}}} + e^{-t/t_{\text{relax}}}\int_0^t e^{s/t_{\text{relax}}}\xi(s)\,ds,\qquad(15.5)$$

where V_0 is the value of v at $t = 0$, and $\xi(t)$ is a random perturbation whose pdf is $p(\xi)$.

What do we mean by a solution of (15.5)? Each simulation of (15.5), referred to as a realization or sample path, will be different, since at every time step, we need to choose $\xi(t)$ randomly from $p(\xi)$. Thus, it would be a hopeless task to try to decide whether (15.5) agreed with an experimental measurement, since the probability that two random realizations are identical is essentially zero. On the other hand, we can use (15.5) to predict the average statistical quantities, such as the mean and variance, and then use these moments to decide whether the experimental observations are in agreement with (15.5). Thus, when we talk about obtaining the solution of a stochastic differential equation, we mean obtaining the time-dependent behavior of the moments.

Once we specify $p(\xi)$, the determination of the moments requires that we evaluate the stochastic integral on the right-hand side of (15.5). There are two versions of stochastic calculus (Itō's[5] and Stratonovich's[6]) and hence two different ways to evaluate the integral in (15.5). However, when σ_n^2 is constant, these formulations are equivalent, since they lead to the same Fokker[7]–Planck equation [722]. Thus, we make the simplifying assumption that we can estimate the moments by performing a large number of realizations [518]. This assumption means that we can replace $\xi(t)$ in the stochastic integral by its appropriate moments and then treat the integration in the same way as we would for a deterministic variable.[8]

In this manner, we can readily obtain the following properties of (15.2). It should be noted that these properties are valid if $\xi(t)$ describes delta-correlated noise with mean zero:

1. The mean velocity, or first moment, is

$$\langle v\rangle = \int_{-\infty}^{\infty}\left[V_0 e^{-t/t_{\text{relax}}} + e^{-t/t_{\text{relax}}}\int_0^t e^{s/t_{\text{relax}}}\xi(s)\,ds\right]p(\xi)\,d\xi\qquad(15.6)$$

$$= V_0 e^{-t/t_{\text{relax}}}\int_{-\infty}^{\infty}p(\xi)\,d\xi + e^{-t/t_{\text{relax}}}\int_0^t e^{s/t_{\text{relax}}}\langle\xi\rangle\,ds = V_0 e^{-t/t_{\text{relax}}},$$

---

[5] Kiyoshi Itō (1915–2008), Japanese mathematician.

[6] Ruslan Leont'evich Stratonovich (1930–1997), Russian physicist and engineer.

[7] Adriaan Daniël Fokker (1887–1972), Dutch physicist and musician.

[8] When the noise is multiplicative and/or the equation is nonlinear, the Ito and Stratonovich formulations are not typically equivalent.

where we have used (15.3) and the fact that

$$\int_{-\infty}^{\infty} p(\xi)d\xi = 1.$$

2. The mean square, or second moment, $\langle v^2 \rangle$ is

$$\langle v^2 \rangle = V_0^2 e^{-2t/t_{\text{relax}}} + V_0 e^{-2t/t_{\text{relax}}} \int_0^t \int_0^t e^{(s_1+s_2)/t_{\text{relax}}} \langle \xi(s_1)\xi(s_2)\rangle \, ds_1 \, ds_2$$

$$= V_0^2 e^{-2t/t_{\text{relax}}} + \frac{\sigma_n^2 t_{\text{relax}}}{2}\left(1 - e^{-2t/t_{\text{relax}}}\right). \tag{15.7}$$

3. The variance σ_v^2 for the velocity can be determined by first noting that

$$\langle(v(t) - \langle v\rangle)^2\rangle = \langle v^2\rangle - \langle v\rangle^2,$$

so that we obtain

$$\sigma_v^2 = \frac{\sigma_n^2 t_{\text{relax}}}{2}(1 - e^{-2t/t_{\text{relax}}}). \tag{15.8}$$

It should be noted that σ_v^2 increases as t_{relax} becomes longer.

4. The pdf $p(v,t)$ for the Langevin equation has the form of a Gaussian distribution (see Exercises 2 and 3). A Gaussian distribution is completely determined by two parameters: the mean and the variance. Since we have already obtained these two statistical variables, we can write down the pdf $p(v,t)$ as follows:

$$p(v,t) = \frac{1}{\sqrt{2\pi\sigma_n^2(t)}} \exp\left[-\frac{(v - \langle v(t)\rangle)^2}{2\sigma_n^2(t)}\right]$$

$$= \sqrt{\frac{1}{\pi\sigma_n^2 t_{\text{relax}}(1 - e^{-2t/t_{\text{relax}}})}} \exp\left[-\frac{(v - V_0 e^{-t/t_{\text{relax}}})^2}{\sigma_n^2 t_{\text{relax}}(1 - e^{-2t/t_{\text{relax}}})}\right].$$

5. The autocorrelation function $C_{xx}(\Delta)$ is

$$C_{xx}(\Delta) := \langle v(t)v(t+\Delta)\rangle$$

$$= V_0^2 e^{-(2t+u)/t_{\text{relax}}}$$

$$+ e^{-(2t+\Delta)/t_{\text{relax}}} \int_0^t \int_\Delta^{t+\Delta} e^{(v+w)/t_{\text{relax}}} \langle \xi(v)\xi(w)\rangle \, dv \, dw$$

$$= V_0^2 e^{-(2t+\Delta)/t_{\text{relax}}} + \frac{\sigma_n^2 t_{\text{relax}}}{2} e^{-\Delta/t_{\text{relax}}}\left[1 - e^{-2t/t_{\text{relax}}}\right].$$

If we consider a sufficiently long-time average (stationary state) in which effects related to the initial conditions have vanished, we obtain

$$C_{xx}(\Delta) = \frac{\sigma_n^2 t_{\text{relax}}}{2} e^{-\Delta/t_{\text{relax}}}. \tag{15.9}$$

6. The one-sided power spectrum $W(f)$ for a particle's frequency of movement in a liquid environment can be determined using the Wiener–Khinchin theorem,

$$W(f) = 4\sigma_n^2 t_{\text{relax}} \int_0^\infty \frac{e^{-s/t_{\text{relax}}}}{2} \cos 2\pi f s \, ds = \frac{2\sigma_n^2 t_{\text{relax}}^2}{1 + (2\pi f t_{\text{relax}})^2} . \qquad (15.10)$$

We remind the reader that $W(f)$ depends only on the autocorrelation function. Since (15.2) describes a low-pass filtering of white noise, (15.10) can also be obtained using techniques based on the Fourier transform (see (8.37) in Section 8.4.1).

7. There is a relationship between $\langle v^2 \rangle$ and $W(f)$, namely [518]

$$\langle v^2 \rangle = \int_0^\infty W(f) \, df . \qquad (15.11)$$

This relationship can be obtained using the Fourier series (Tool 10). To illustrate we represent a stochastic time series as

$$v(t) = \sum_n a_n \cos 2\pi n f_0 t + \sum_n b_n \sin 2\pi n f_0 t ,$$

where we have subtracted the mean. If $v(t)$ is stochastically fluctuating but statistically stationary, then the fluctuations in a_n and b_n will also be statistically stationary about the mean zero. In other words, $\langle v(t) \rangle = 0$, and the mean square is

$$\langle v^2(t) \rangle = \sum_n \frac{\langle a_n^2 \rangle}{2} + \sum_n \frac{\langle b_n^2 \rangle}{2} .$$

Since the two sums are necessarily nonnegative, we see that $\langle v^2(t) \rangle \neq 0$. However, since the phase of any particular frequency component varies stochastically with time, we have $\langle a_n^2 \rangle = \langle b_n^2 \rangle$. If we define $f = n f_0$, then

$$w(n f_0) f_0 = \langle a_n^2 \rangle = \langle b_n^2 \rangle ,$$

and hence

$$\langle v^2(t) \rangle = \sum_n w(n f_0) f_0 .$$

Finally, if we let f_0 approach 0 (i.e., $1/f_0$ approaches ∞), we obtain (15.11).

15.1.1 The Retina As a Recording Device

Few scientists realize that Einstein's prediction of the velocity of Brownian movements was approximately three orders of magnitude faster than what is measured by the human visual system [517, 518, 680]. The resolution of this discrepancy between measurement and prediction was the first demonstration of the importance of the interplay between noise and time delays for shaping the responses of the nervous system. The core of the issue concerns the effect of the response time of a measuring instrument, in this case the human eye, on the detection of a particle that exhibits Brownian motion.

The frequency response of a measuring device provides a measure of its limitations (see Section 8.4.1). The frequency response $\hat{W}(f)$ for the Langevin equation is determined from its response to a delta-function input

$$\langle v^2 \rangle = 2\sigma_n^2 t_{relax}^2 \int_0^\infty \frac{df}{1+(2\pi f t_{relax})^2} = \frac{\sigma_n^2 t_{relax}}{2}. \tag{15.12}$$

Strictly speaking, this is the response of a measuring device that has a very fast response time, i.e., a device whose $1/e$ time, t_m, is very small. However, the human retina has a finite response time of about 0.1 s. Thus the frequency response becomes

$$\langle v^2 \rangle = 2\sigma_n^2 t_{relax}^2 \int_0^{1/t_m} \frac{df}{1+(2\pi f t_{relax})^2} = \sigma_n^2 t_{relax}^2 \left(\frac{\arctan(2\pi t_{relax}/t_m)}{2\pi t_{relax}} \right)$$
$$\approx \frac{\sigma_n^2 t_{relax}}{2} \left(\frac{t_{relax}}{t_m} \right), \tag{15.13}$$

where we have assumed that $t_m \ll t_{relax}$.

In order to appreciate the consequences of these observations, let us assume that the motions of a Brownian particle resemble those of an ensemble of molecules that obey classical mechanics. Thus from statistical-mechanical considerations in one dimension, we have

$$\langle v^2 \rangle = \frac{k_B T}{M}, \tag{15.14}$$

where T is the temperature (in kelvins) and k_B is Boltzmann's constant ($1.38 \times 10^{-16}\,\mathrm{g \cdot cm^2 \cdot s^{-2}}$). If we assume that $M \approx 10^{-12}$ g, we obtain an average particle speed of $v \approx 0.2\,\mathrm{cm \cdot s^{-1}}$. This estimate ($\approx 2 \times 10^{-1}\,\mathrm{cm \cdot s^{-1}}$) is approximately 1,000 times higher than what is observed experimentally ($\approx 4 \times 10^{-4}\,\mathrm{cm \cdot s^{-1}}$) [680].

Equating (15.12) and (15.14), we find that

$$\frac{\sigma_n^2 t_{relax}}{2} = \frac{k_B T}{M}. \tag{15.15}$$

This relation can be recognized as a version of the *fluctuation–dissipation theorem*. In contrast, from (15.13), we have

$$\frac{\sigma_n^2 t_{relax}}{2} \left(\frac{t_{relax}}{t_m} \right) = \frac{k_B T}{M}. \tag{15.16}$$

Thus, the speed of motion of a Brownian particle estimated using the human eye would be reduced by a factor of $\sqrt{t_{relax}/t_m}$.

To complete our considerations, we need to estimate t_{relax}. When viscous forces dominate inertial forces, that is, at low Reynolds[9] number, the drag force on a parti-

---

[9] Osborne Reynolds (1842–1912), Anglo-Irish physicist.

cle is given by Stokes's law[10] [675, 858]

$$F_{drag} = 6\pi\eta rV_{settling}, \tag{15.17}$$

where η is the viscosity (in units of poise, i.e., $1\,P = 1\,g \cdot cm^{-1} \cdot s^{-1}$), r is the radius in centimeters, and $V_{settling}$ is the settling velocity $(cm \cdot s^{-1})$. Thus, we can estimate $t_{settling}$, the $1/e$ time for settling, from the equation

$$M\frac{dV_{settling}}{dt} = -F_{drag} = -6\pi\eta rV_{settling}$$

to obtain

$$t_{relax} \approx t_{settling} = \frac{M}{6\pi\eta r}.$$

If the particle has diameter $1\,\mu$, then $r = 0.5 \times 10^{-4}$ cm, and if the viscosity of the fluid is $\approx 10^{-2}$ P, we have

$$\left(\frac{t_{relax}}{t_m}\right)^{1/2} \approx 10^{-3}.$$

Thus, the discrepancy between the visually measured and the theoretically estimated speed for Brownian particles can be completely accounted for by taking into account the response time of the human visual system.

15.1.2 Time-Delayed Langevin Equation

The time-delayed Langevin equation

$$\frac{dv}{dt} + kv(t - \tau) = \xi_w(t) \tag{15.18}$$

arises in considerations of the effects of the interplay between white noise ξ_w and delay on feedback control [298, 651, 654]. The Fourier transform of a delay is

$$G(g(t - \tau))(f) = e^{-j2\pi f\tau}G(g(t))(f).$$

Taking the transform of (15.18), we obtain

$$2\pi jfV(f) + ke^{-j2\pi f\tau}V(f) = \beta, \tag{15.19}$$

for which β is a constant, and

$$V(f) = \frac{\beta}{2\pi jf + ke^{-j2\pi f\tau}}. \tag{15.20}$$

Thus, the power spectrum is

---

[10] George Gabriel Stokes (1819–1903), Irish mathematician, physicist, politician, and theologian.

$$W(f) = \frac{\beta^2}{(2\pi f)^2 - 4\pi k f \sin(2\pi f \tau) + k^2}. \tag{15.21}$$

Figure 15.1 plots the power spectrum given by (15.21). Overall, the power spectrum decreases as frequency increases with a clear "periodic ripple." However, in the low-frequency range, $W(f)$ peaks at $(2\tau)^{-1}$.

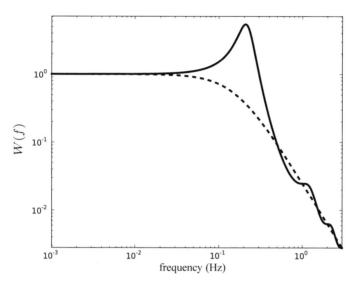

Fig. 15.1 Power spectrum function $W(f)$ calculated using (15.21) for a first-order delay differential equation given by (15.18). Parameter values are $\alpha = \beta = \tau = 1$. Dashed line shows $W(f)$ when $\tau = 0$.

The amplification predicted in Figure 15.1 was first observed experimentally in the retina of the horseshoe crab *Limulus* [712, 713]. In the *Limulus* retina, interactions between the light-sensitive excitable cells, referred to as *ommatidia*, are inhibitory and time-delayed. The time delay is ≈ 0.1 s. The observed amplification formed the basis of the "center ON, surround OFF" concept of neural receptive fields and won H.K. Hartline[11] and his colleagues the Nobel Prize in physiology or medicine in 1967. The amplification in the frequency-dependent suppression of seizures in hippocampal slice preparations at 1–2 Hz [390] may similarly be related to the time-delayed recurrent inhibitory interactions that occur in this preparation [358].

The autocorrelation function $C_{xx}(\Delta)$ can also be determined using the Wiener–Khinchin theorem [298, 651, 654]:

$$C_{xx}(\Delta) = 4 \int_{-\infty}^{\infty} \frac{\beta^2 \cos(2\pi f \Delta)}{(2\pi f)^2 - 4\pi k f \sin(2\pi f \tau) + k^2} df. \tag{15.22}$$

---

[11] Haldan Keffer Hartline (1903–1983), American physiologist.

This integral cannot be solved analytically. Using a different approach, Küchler and Mensch determined that for $\Delta < \tau$, the stationary correlation function is [449]

$$C_{xx}(\Delta) = C_{xx}(0)\cos(k\Delta) - \frac{1}{2\Delta}\sin(k\Delta), \qquad (15.23)$$

where

$$C_{xx}(0) = \frac{1 + \sin(k\tau)}{2k\cos(k\tau)}. \qquad (15.24)$$

(Note that we set $\beta = 1$.) A similar expression is obtained by numerical integration of (15.22) [298, 654].

15.1.3 Skill Acquisition

In individual sporting events such as golf and archery, skill levels can be estimated from the variance between successive repetitions of the same task [214, 594]: the lower the variance, the higher the skill level. For example, the variance in the contact point between the surface of the putter head and the golf ball for a professional golfer is of order one-third the diameter of a dimple on the golf ball [676]. Skill increases with practice, because the brain continues to refine its internal maps.

Recently, attention has turned to the use of the potential-like diagrams we introduced in Chapter 4 to motivate a dynamical system formulation for skill acquisition and development [423, 594]. To illustrate this approach, let us identify a component

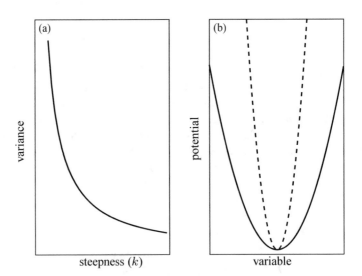

Fig. 15.2 (a) Variance of the fluctuations predicted by (15.2) as a function of the steepness $k = 1/t_{\text{relax}}$ of the sides of the basin of attraction. (b) The potential (equal to kx^2) for high steepness (solid line) and low steepness (dashed line).

of skill with a fixed-point attractor. Thus, we have the picture of a valley $U(x)$ described by a parabolic potential $U(x) = kx^2$. The skill level at a given time corresponds to the value of k. Of course, each time a skill is repeated, there will be variations, and these we attribute to the effects of additive noise. From this perspective, we can interpret skill acquisition using what we have learned from the Langevin equation. The decrease in the variance as skill increases is equivalent to increasing k, which increases the steepness of the walls of the valley (Figure 15.2). As k increases, the variance given by (15.8) decreases.

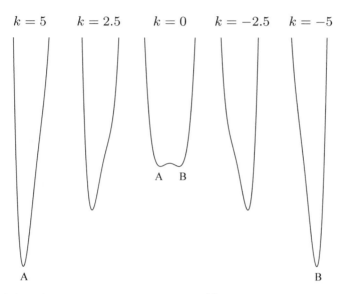

$$k = 5 \qquad k = 2.5 \qquad k = 0 \qquad k = -2.5 \qquad k = -5$$

A   B

A   B

Fig. 15.3 Schematic representation of the changes in $V(x)$ described by (15.25) as a parameter k changes slowly to cause one attractor (A) to disappear as another (B) appears.

However, skill does not necessarily increase linearly with practice. Scott Kelso[12] and colleagues suggested that the development of skill can be thought of as a process in which new attractors replace old ones [423]. We can represent the process of a new fixed-point attractor replacing an old one schematically, as shown in Figure 15.3. Here we have chosen

$$V(x) = kx - \frac{x^2}{2} + \frac{x^4}{4} \tag{15.25}$$

and varied the constant k. As k goes from 5 to -5, the attractor A gradually disappears and is replaced by a new attractor B. As this process evolves, the slope of the sides of one valley decreases as the slope of the walls of the newly forming attractor increases. Thus, before switching to the new attractor, one would expect the variance to increase and the skill level consequently to decrease transiently.

[12] L.A. Scott Kelso (b. 1947), Irish-American neuroscientist.

This happens because the walls of the basin of attraction are less steep just before the attractor disappears. This phenomenon of getting worse before getting better has been observed in a few subjects as they develop skill in stick balancing. It is also well documented in the acquisition of new skills by both children and adults [249, 769, 889]. The take-home message is that one should not be disheartened if setbacks occur in learning new skills. Indeed, momentary confusion may actually be the first sign of gaining increased comprehension. The authors hope that this observation will motivate those readers who may be struggling with the material in this chapter to persevere.

15.1.4 Stochastic Gene Expression

Stochastic effects due to small numbers within a given cell (referred to as intrinsic stochasticity) must be distinguished from stochastic effects that reflect variation between different cells (referred to as extrinsic stochasticity). The relative importance of intrinsic and extrinsic stochasticity was investigated in a bacterium using the green fluorescent protein (GFP) technique [195, 711]. In this procedure, GFP is incorporated into a bacterial chromosome downstream of a promoter that is activated by the system of interest, in this case, activity in the *lac* operon. Thus, it is possible to detect when the *lac* operon is activated by the presence of GFP fluorescence. How can this technique be used to estimate the intensity[13] of the intrinsic noise σ_{int}^2? The answer is to use two identical strains of bacteria and two variants of GFP that produce two distinct fluorescence spectra. Each strain gets a different GFP allele. By placing the GFP alleles in identical locations on the genome downstream from the *lac* operon, gene activity can be monitored in each strain from differences in the fluorescence spectra. The intensity of the total noise σ_{tot}^2 was estimated from the total fluorescence from the two GFP alleles as a function of time. The difference in the fluorescence from the two GFP alleles was used to estimate σ_{int}^2. In other words, σ_{int}^2 is a measure of the extent to which the activities of two identical copies of the gene (each monitored by one of the GFP variants) in the same intracellular environment fail to correlate. If the fluorescence measured from the two GFP probes for a large number of cells correlated exactly as a function of time, then we would conclude that $\sigma_{int}^2 = 0$. The greater the lack of correlation, the greater the value of σ_{int}^2. Finally, the intensity σ_{ext}^2 of the external noise was calculated from the expression [816]

$$\sigma_{ext}^2 = \sigma_{tot}^2 - \sigma_{int}^2.$$

The intensity of gene expression noise depends on the transcription rate, regulatory dynamics, and genetic factors [195]. The investigators varied the transcription rate by adding differing amounts of β-D-thiogalactopyranoside (IPTG) to the growing

---

[13] The noise intensity can be measured in a variety of ways, including its variance, root-mean-square value, its standard deviation, and the coefficient of variation, which is the standard deviation divided by the mean.

cultures: IPTG binds to and inactivates the *lac* repressor. The striking observation is that the dependence of σ^2_{int} on transcription rate is different from that of σ^2_{ext}. Whereas σ^2_{int} decreases monotonically with increasing transcription rate as predicted [816], σ^2_{ext} attains a maximum at intermediate transcription rates.

At first glance, it is surprising that there is no mention of the effects of the rate of translation on σ^2_{tot}. Swain[14] and coworkers [816] used an analysis based on the Langevin equation to demonstrate that the rate of transcription is a more likely source of σ^2_{int} than the rate of translation. Their starting point was the Goodwin model that we introduced in Section 9.3.1. If we ignore the feedback term, the stochastic version of the Goodwin equation becomes

$$\frac{dM}{dt} = v_0 - d_0 M + \xi_1(t), \quad \frac{dN}{dt} = v_1 M - d_1 N + \xi_2(t), \quad (15.26)$$

where $\xi_1(t), \xi_2(t)$ are two delta-correlated additive noise inputs that satisfy

$$\langle \xi_1(t_1)\xi_1(t_2) \rangle = 2d_0 M_{\text{ss}} \delta(t_1 - t_2),$$
$$\langle \xi_2(t_1)\xi_2(t_2) \rangle = 2d_1 N_{\text{ss}} \delta(t_1 - t_2),$$
$$\langle \xi_1(t_1)\xi_2(t_2) \rangle = 0,$$

and M_{ss} and N_{ss} are the steady-state values of M, N satisfying

$$M_{\text{ss}} = \frac{v_0}{d_0}, \quad N_{\text{ss}} = \frac{v_1}{d_1} M_{\text{ss}}. \quad (15.27)$$

The first step in Swain et al.'s analysis was to reduce the second-order Langevin equation described by (15.26) to a first-order Langevin equation. In other words, they reduced a more complex equation to a simpler equation whose behavior is better understood. It is observed that many mRNA fluctuations occur during one protein fluctuation. Thus, the mean of the mRNA fluctuations approaches M_{ss} much more quickly than the mean of the protein fluctuations approaches N_{ss}. Thus, we can assume the steady-state approximation

$$\frac{dM}{dt} \approx 0,$$

which implies that

$$M = M_{\text{ss}} + \frac{\xi_1}{d_0},$$

and hence (15.26) reduces to

$$\frac{dN}{dt} = v_1 M_{\text{ss}} - d_1 N + \Psi, \quad (15.28)$$

where

---

[14] Peter Swain (b. 1970), British-Canadian systems biologist.

$$\Psi = \frac{v_1}{d_0}\xi_1 + \xi_2, \quad \langle\Psi\rangle = 0, \quad \langle\Psi(t_1)\Psi(t_2)\rangle = 2d_1 N_{ss}\left[1 + \frac{v_1}{d_0}\right]\delta(t_1 - t_2).$$

Swain and colleagues were able to demonstrate that [816]

$$\sigma_M^2 = \frac{1}{\langle M\rangle}$$

and

$$\sigma_N^2 = \frac{1}{\langle N\rangle} + \frac{d_1}{d_0}\frac{1}{\langle M\rangle}. \tag{15.29}$$

Typical numbers for constitutive (unregulated) expression in *E. coli* are $d_1 \approx 0.0003\,\mathrm{s}^{-1}$, $d_0 \approx 0.006\,\mathrm{s}^{-1}$, $\langle N\rangle = 10^3$, and $\langle M\rangle = 5$. Thus, we see that

$$\sigma_N^2 = 0.001 + 0.01,$$

and hence the mRNA term determines the overall magnitude of the noise. Experimentally, it is also possible to estimate v_1. The translation rate of ribosomes is $\approx 40\,\mathrm{nt}\cdot\mathrm{s}^{-1}$ [77], where nt is the length of the gene measured as the number of nucleotides. This means that for a 1,000-nt protein,[15] we have

$$v_1 = \frac{40}{1000}\,\mathrm{s}^{-1} \approx 0.04\,\mathrm{s}^{-1}.$$

We can eliminate d_1 and $\langle N\rangle$ from (15.29) using the steady-state values of N and M given by (15.27) to obtain

$$\sigma_N^2 = \frac{d_1}{v_1 M_{ss}} + \frac{d_1}{d_0}\frac{1}{M_{ss}}.$$

Since the rate of translation occurs only in the first term on the right-hand side, the hypothesis that the effects of transcription on σ_N^2 dominate those of translation will be correct if

$$\frac{d_1}{v_1 M_{ss}} \ll \frac{d_1}{d_0}\frac{1}{M_{ss}},$$

or

$$v_1 \gg d_0.$$

This is what is observed.

---

[15] A 1,000-nt protein corresponds to a protein sequence containing ≈ 333 amino acids, since each amino acid is encoded in the gene by three consecutive nucleotides.

15.1.5 Numerical Integration of Langevin Equations

Care must be taken in using numerical methods to simulate differential equations of the form of Langevin equation. By simple discretization, the equation

$$\frac{dv}{dt} = -kv + \sigma_n^2 \xi,$$

where $k = 1/t_{relax}$ can be approximated by

$$v(t + \Delta t) - v(t) = -kv(t)\Delta t + \sigma_n^2 \xi' \sqrt{\Delta t}, \tag{15.30}$$

where for ξ', we take a value from the Gaussian distribution with mean 0 and variance 1 at each step of incrementation. The notable point is the appearance of $\sqrt{\Delta t}$. When the noise term represents Brownian motion, it was shown by Weiner that the noise is distributed as Gaussian white noise with 0 mean and variance Δt. Thus

$$\xi(x) \approx \frac{1}{\sqrt{\Delta t}} e^{x^2/2\Delta t}$$

or

$$\xi(x)\sqrt{\Delta t} \approx e^{x^2/2\Delta t}.$$

This follows from the fact that the variance of simple stochastic processes increases linearly with time (see also Section 16.1).

It is less clear whether the $\sqrt{\Delta}$ factor is required when simulations use other types of noise such as uniformly distributed white noise or band-limited white noise. A detailed discussion of this point is beyond the scope of this book. The interested reader can consult books on stochastic differential equations under the topic of Itō calculus or Itō integration [432].

15.2 Noise and Thresholds

A dynamical system tuned near a threshold is particularly sensitive to the effects of random perturbations (noise) [94, 252, 253, 334, 609, 909]. The threshold could correspond, for example, to a stability boundary or the spiking threshold of a neuron, or it could arise as a result of quantization by an A/D board (Figure 15.4a).

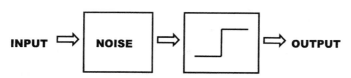

Fig. 15.4 A black-box representation that describes the effects of additive noise on one quantization step of an A/D board.

15.2.1 Linearization

A black-box representation of a single digitization step of an A/D board is shown in Figure 15.4. We can describe the function of the black box by the Heaviside function (see Section 6.1)

$$H(x) = \begin{cases} 1 & \text{if } x \geq 0.5, \\ 0 & \text{otherwise}. \end{cases}$$

What is the output when we apply a constant input of 0.25? Obviously, the output will be 0. Now suppose we add noise $\xi(t)$ to the constant input that is uniformly distributed on the interval $[-0.5, 0.5]$. Now for each trial, or realization, the output will be either 0 or 1. In particular, the output will be 1 if the magnitude of the added noise is ≥ 0.25, since in that case, the sum of the noise and the input to the A/D board will be ≥ 0.5. The probability that the added noise has magnitude greater than or equal to X is given by the complementary cumulative distribution function $\Phi(x \geq X)$, where

$$\Phi(x \geq X) = 1 - P(x \leq X),$$

and $P(x \leq X)$ is given by (14.5). The average output for a large number of realizations is equal to the number of trials for which the output is 1 divided by the total number of trials. Thus, for $\Phi(x \geq 0.25)$, the averaged output will be 0.25. Figure 15.5 summarizes the mean output of the black box shown in Figure 15.4 as a function of a constant input and the variance σ_n^2 of the added uniformly distributed white noise. As we see, the effect of noise is to linearize the behavior of $H(x)$ [252, 909].

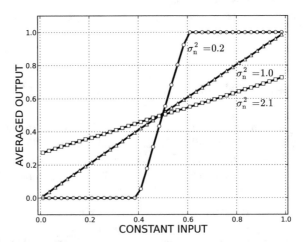

Fig. 15.5 Plot of the average output of the black box shown in Figure 15.4 for inputs of the form $K + \sigma_n^2 \xi(t)$, where K is a constant and the threshold is 0.5. The noise $\xi(t)$ is uniformly distributed delta-correlated ("white") noise on the interval $[-0.5, 0.5]$. The average output was computed from 10,000 realizations.

An application of these concepts arises in the context of the relationship between a neuron's input and its output. To a first approximation, a neuron is a threshold device. Surprisingly, living neurons exhibit a wide dynamic range: input mean spike rates can be continuously modulated over a nearly 1,000-fold range, i.e., from nearly 0 spikes per second to almost 1,000 spikes per second. How is it possible for neurons to exhibit such a large dynamic range?

All neural network modelers are familiar with sigmoidal approximations of the input–output relationship of neurons, for example as $\tanh(x)$ (Figure 15.6a). However, few realize that the validity of this approximation represents an effect of noise. Figure 15.6b (solid line) shows the calculated input–output relationship for a leaky integrate-and-fire (IF) neuron [222, 567]. An IF neuron does not, by definition, spike unless the integrated membrane potential exceeds a spiking threshold. Thus, an IF neuron cannot modulate input spiking rates that are lower than a certain critical rate. However, the addition of noise (dashed lines) extends the dynamic range of the neuron [909] and in particular, converts the input–output of the relationship of the neuron to a form that closely resembles $\tanh(x)$. In other words, $\tanh(x)$ describes the input–output relationship for a noisy IF neuron.

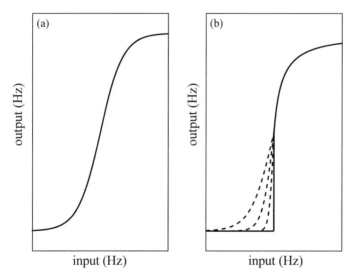

Fig. 15.6 (a) Representation of the typical $\tanh(x)$ form of the input–output relationship of a neuron used in neural network calculations. (b) Solid line shows the input–output relationship of a model neuron (integrate-and-fire, Hodgkin–Huxley) in the absence of noise [222, 567]. The dashed lines show the effects of adding noise of increasing intensity to the neural input [909]. As can be seen, the combination of noise plus the calculated input–output relationship approaches the sigmoidal shape shown in (a).

15.2.2 Dithering

Two errors enter during amplitude quantization by an A/D board: (1) signal quantization leads to unavoidable distortion, i.e., the presence of spurious signals in a frequency band different from the original one, and (2) loss of signal detail that is small compared to the quantization step (the dynamic range of the digital signal is finite). Define the quantization error η as

$$\eta = X - Y ,$$

where X is the analog input to the A/D board and Y is the digital representation of the analog signal input from the A/D board to the computer. If we had a linear response of the A/D board, i.e., $X = Y$, then $\eta = 0$. The minimum loss of statistical data from the input X occurs when the quantization error can be made independent of X. A simple way to accomplish this is to add noise to the signal before quantization. This technique is referred to as *dithering* [252].

A great deal of work has been devoted to determining the best choices of noise for dithering. The main conclusions are these: The addition of a proper dither signal can cause the independence and whitening of the quantization error. This results in both a reduction of signal distortion and an improvement of the system dynamic range. The best choice for the dither signal is white noise that is uniformly distributed over an interval whose amplitude is equal to the quantization step. An alternative choice is to use "blue noise," that is, noise for which $W(f) \approx f$ for a finite range of f. Retinal cells in the primate eye are arranged in a blue-noiselike pattern, suggesting that this organization may have evolved to favor good visual resolution [901].

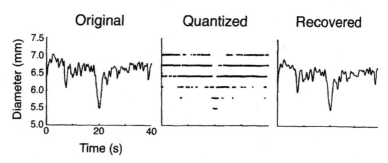

Fig. 15.7 Application of dithering for measuring pupil size in a noisy pupillometer with low resolution. Figure from [568] with permission.

The essential requirement for dithering is that the time series be oversampled in the presence of noise. If the frequency content of the noise is much higher than that for the primary signal, then over short intervals, the average population of the two quantized levels is a measure of how close the true signal is to the threshold that separates two adjacent bins. Thus, it follows that the true signal can be recovered by low-pass filtering (equivalent to averaging) the quantized data. This procedure is illustrated for a noisy pupillometer in Figure 15.7 [360]. An analog input signal proportional to pupil diameter (left-hand graph) is discretized in the presence of noise at 25 Hz (middle graph). It should be noted that the major frequency components of the open-loop irregular fluctuations in pupil size are 0.3 Hz [501, 795]. The quantized signal is then low-pass filtered to yield a time series that closely resembles the original time series (compare left- and right-hand graphs). The difference is that the low-pass filtering removes the high-frequency components of the original signal.

15.2.3 Stochastic Resonance

Although the linearizing effects of noise on a threshold are useful in some contexts, this phenomenon can be a nuisance in situations in which a threshold crossing is used to trigger another event. This is because in the presence of noise, there will be multiple threshold crossings when the signal is close to the threshold. Electrical engineers solved this problem by developing a special type of voltage comparator, called a Schmitt[16] trigger, that is less likely to produce multiple-trigger events. A Schmitt trigger is a bistable device that has two thresholds that in turn depend on the output state. Fauve and Heslot [219] studied the behavior of a Schmitt trigger that simultaneously received a weak periodic voltage and noise. Surprisingly, they observed that the information in the input could be amplified and optimized in the output by adjusting the intensity of the noise. This phenomenon is called *stochastic resonance*. It can be readily appreciated that dithering is the special case of stochastic resonance in which the two voltage threshold levels in the Schmitt trigger coincide [253].

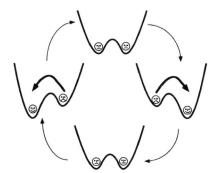

Fig. 15.8 An amusing schematic representation of the stochastic resonance mechanism described by (15.31). In the words of the authors [253], "A suitable dose of noise (i.e., when the period of the driving approximately equals twice the noise-induced escape time) will make the "sad face" happy by allowing synchronized hopping to the attractor with the lower energy (strictly speaking, this holds only for the statistical average)." Figure reproduced from [253] with permission.

---

[16] Otto Herbert Schmitt (1913–1998), American inventor, engineer, and biophysicist.

In neuroscience, the significance of stochastic resonance is that it emphasizes the importance of noise for the detection of weak periodic signals by sensory neural receptors [26, 253, 609]. Consider the properties of a general bistable system driven by a sinusoidal function with added noise. A simple example of such a dynamical system is

$$\frac{dx}{dt} = -\frac{dU(x)}{x} + m\sin 2\pi ft + \sigma_n^2 \xi(t), \qquad (15.31)$$

where $U(x)$ is a double-well potential and ξ describes Gaussian distributed white noise with intensity (variance) σ^2. The effect of the sinusoidal modulation is to rock the potential surface to and fro (Figure 15.8). In the absence of the sinusoidal forcing, switches between the two basins of attraction are possible through the effects of noise. For small-intensity noise and a sufficiently high barrier between the two basins of attraction, such switches occur rarely. However, for $m \neq 0$, the switching times between the two potential wells become correlated with the periodic forcing, and hence the transition probabilities become periodically modulated. It is easy to see that switches become more probable once the intensity of the noise plus the amplitude of the weak periodic signals exceeds the height of the potential barrier that separates the two wells.

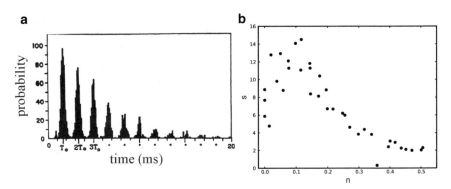

Fig. 15.9 (a) Interspike histogram measured from a single auditory nerve fiber of a squirrel monkey with sinusoidal 80-dB sound-pressure-level stimulus of period $T_0 = 1.66$ ms applied to the ear. Figure reproduced from [727] with permission. (b) Signal-to-noise ratio (SNR) of a hydrodynamically sensitive hair mechanoreceptor in the tail of a crayfish in the presence of periodic water motion to which noise of increasing intensity has been added. Figure reproduced from [176] with permission.

This stochastic resonance mechanism predicts two phenomena that have been observed experimentally [497]. First, it explains two curious properties of multimodal interspike histograms that have been recorded for a variety of neurons including retinal ganglion neurons in the cat and in auditory fibers of monkeys [727] (Figure 15.9a): (1) the modes occur at an integer multiple of a fundamental period; and (2) the mode amplitude decreases approximately exponentially. Second, the signal-to-noise ratio (SNR) for a stochastically resonant dynamical system exhibits a nonmonotone dependence on noise intensity [176]. As the intensity of the noise increases, the output of the sensory neuron increases (Figure 15.9b). This is because switches between the basins of attraction are possible only if noise is present. The SNR reaches a maximum at roughly the point where the noise intensity is on average sufficient to cause a transition at the point in time when the potential barrier between the basins of attraction is most shallow. Thereafter, further increases in noise intensity deteriorate the SNR (as is always the case in dynamical systems that do not exhibit stochastic resonance).

The conclusion that noise is essential for enabling the nervous system to detect weak sensory inputs has generated a large and ever growing number of practical applications. For example, the addition of noise improves the performance of cochlear implants [608], balance control in subjects with peripheral neuropathies and may help prevent falls in the elderly [489]. For example, the addition of noise improves the performance of cochlear implants [608] and balance control in subjects with peripheral neuropathies [698]. The interested reader is directed to the special issue of the journal *Chaos* dedicated to the life and work of the stochastic resonance pioneer Frank Moss[17] [26].

15.3 Parametric Noise

Up to this point, we have assumed that the effects of noise are additive and hence that the stochastic differential equation takes the form

$$\frac{dx}{dt} = f(x, x(t - \tau), \mu) + \sigma^2 \xi, \tag{15.32}$$

where μ is a parameter and $\xi(t)$ denotes noise with variance σ^2. Additive noise is also referred to as state-independent noise to emphasize that its effects do not depend on the magnitude of the state variable x. In other words, x does not appear in the term $\sigma^2 \xi$. However, it is often the case that the effects of noise depend on the state of the dynamical system [95, 317, 393, 501, 795]. In these situations, the stochastic differential equation takes the form

$$\frac{dx}{dt} = f\left(x, x(t - \tau), \mu, \sigma^2 \xi(t)\right), \tag{15.33}$$

and the effects of noise are state dependent, since they depend in some way on the state variable x. The effects of state-dependent and state-independent noise on a dynamical system are completely different from the effects of *measurement noise*, familiar to all laboratory investigators. Measurement noise refers to the uncertainty in measurements introduced by the experimenter and/or apparatus while a process is being monitored. Thus, measurement noise is added to the solution of the differential equation. In contrast, state-dependent and state-independent noise enter into the solution of the stochastic differential equation.

---

[17] Frank Edward Moss (1934–2011), American physicist, pioneered applications of stochastic resonance to biology.

Parametric, or *multiplicative*, noise may actually be more common in biology than additive noise [95, 317, 393, 501, 795]. The simplest case arises when noise enters through a parameter. For example, the membrane noise of a neuron reflects fluctuations in conductance. In the Hodgkin–Huxley equations for a neuron, the current flow across the membrane is equal to the product of the conductance and the driving potential. Consequently, the effects of membrane noise are state dependent and hence parametric. Parametric noise also underlies the spontaneous fluctuations observed in pupil size [795]. Recently, the role of parametric noise in motor [317, 393] and balance [94, 95] control has been emphasized. In mathematical models of feedback control, multiplicative noise is used to capture the effect of parametric uncertainty on closed-loop performance [58, 370, 405].

Many of the effects observed in dynamical systems subject to additive noise have analogous effects in linear dynamical systems subject to parametric noise [501]. For example, although we have discussed stochastic resonance in the context of additive noise, the same phenomenon occurs in dynamical systems subject to the effects of parametric noise (for a review, see [253]). There is one scenario in which the effects of additive and parametric noise differ. In ordinary differential equations, parametric, but not additive, noise can postpone the onset of instability (see also Section 5.2.1). Curiously, in time-delayed dynamical systems, both types of noise can cause this postponement [501].

The effects of additive and parametric noise become of great interest for dynamical systems tuned at or near the edge of stability. Near the edge of stability, it is possible that parametric noise can stochastically (or chaotically) force the parameter back and forth across a stability boundary [94, 334, 687]. As a consequence, such stochastic dynamical systems exhibit two behaviors: *intermittency* and *power-law behaviors*, which are not seen in the corresponding dynamical system subject to additive noise.[18]

The importance of the distinction between additive and parametric noise can be readily appreciated by considering the effects of noise on the potential $U(x)$. For additive noise, the potential does not change, but it changes in the presence of parametric noise. If there were no change in stability, that is, all minima remained minima, then differential equations with additive and parametric noise would share many features [270]. However, with parametric noise, there is the possibility that the potential changes dramatically, e.g., minima might become maxima or disappear altogether as parameters cross stability boundaries.

Under these conditions, the properties of differential equations with parametric noise will differ from those with additive noise. Since the mantra of dynamical systems theory is that complex systems tend to self-organize near stability boundaries [291], we see that the effects of parametric noise become of great interest. Indeed, it is possible that unstable dynamical systems can be transiently stabilized by parametric noise on these boundaries [94, 580].

---

[18] Intermittency is the term used by physicists to describe the burstiness of biological signals discussed in Section 17.5.2. Technically, the type of intermittency we describe here is referred to as *on–off intermittency* in order to distinguish it from other forms of intermittency [689].

15.3.1 On–Off Intermittency: Quadratic Map

We can appreciate the role of parametric noise for generating power-law behaviors using computer simulations of the noisy quadratic map

$$x_{t+1} = r(\xi_t)x_t(1 - x_t),\tag{15.34}$$

where

$$r(\xi_t) = ky_t$$

and k is a constant and y_t is a random variable with a uniform distribution on the interval $[0, 1]$. Here we briefly summarize the mathematical analysis of this equation (for more details, the interested reader is referred to [334]).

When $k < e$, where e is the base of the natural logarithm, the fixed point $x = 0$ is stable. A transcritical bifurcation occurs at $k = e$. Thus, we can think of $r(\xi_t)$ as randomly forcing the system back and forth across this stability boundary. Figure 15.10a shows the time series generated by (15.34) when $k = 2.8$. Numerical experiments indicate that the longer the laminar phases, the smaller the magnitude of the fluctuations. This observation suggests that the dynamics are almost completely determined by the linear part of the map. The nonlinear terms serve only to bound or reinject the dynamics back toward small values of x. In other words, the nonlinearities are essential for sustaining the intermittent behavior but not for initiating the bursts.

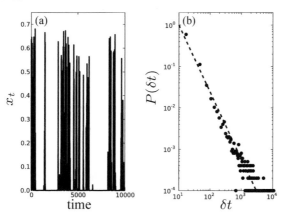

Fig. 15.10 (a) Bursting dynamics of (15.34) when $k = 2.75$. (b) Log–log plot $P(\delta t)$ versus δt of the probability of the time between successive crossings of a threshold, demonstrating the $-3/2$ characteristic on–off intermittency (dashed line).

The importance of choosing $k \geq e$ to generate the dynamics in Figure 15.10 can be readily understood [334]. Consider the linear map

$$x_{t+1} = r(\xi_t)x_t.\tag{15.35}$$

As long as X remains small, the long-time behavior of this map is given by

$$x_n = \prod_{j=0}^{n-1} r_j x_0 \, ,$$

where the r_j are the values chosen from $r(\xi_t)$ for a single realization, and x_0 is the value of x_t at $t = 0$. These observations mean that the long-time behavior is determined by the asymptotic behavior of the random product

$$\Psi_n = \prod_{j=0}^{n-1} r_j \, . \tag{15.36}$$

Taking logarithms of both sides and making use of the law of large numbers, we obtain[19]

$$\log \Psi_n = \sum_{j=0}^{n-1} \log r_j \approx n \langle \log r \rangle \, .$$

Since we have assumed that the density of the distribution for the parametric noise is uniform, we can readily calculate the statistical average as

$$\langle \log r \rangle = \int_0^k p_u(r) \, \log r \, dr = \frac{1}{k} \int_0^k \log r \, dr = \log k - 1 \, .$$

Thus, we see that the asymptotic limit of (15.35) is

$$x_n \approx e^{n \langle \log r \rangle} X_0 \approx e^{n(\log k - 1)} x_0 = \left(\frac{k}{e} \right)^n x_0 \, .$$

When $x_n \approx X_0$, we have $k = e$. When $\langle \log r \rangle > 0$, the fixed point $x = 0$ is, on average, exponentially unstable. This instability is the source of the intermittent bursts. However, we must keep in mind that this result is only local. This instability does not preclude the occurrence of long orbit segments in the neighborhood of $x = 0$, i.e., the laminar phases. Thus, the critical value of k for the onset of intermittency is

$$e = 2.71828 \ldots \, .$$

In order to demonstrate the power-law behavior of (15.34), it is necessary to define a *laminar phase* (see Section 17.5.2). Briefly, we choose an arbitrary threshold (horizontal dashed line in Figure 15.10a) and then measure the times between successive threshold crossings. Here we have chosen to define the laminar phase to be the length δt between two successive crossings of the threshold in the upward direction. Figure 15.10b shows a log–log plot $P(\delta t)$ versus δt of the probability that a laminar phase has length δt. There is clearly a region of linearity whose slope is $-3/2$. The existence of a $-3/2$ power-law behavior in (15.34) when $k \geq e$ has been predicted theoretically [334].

---

[19] The law of large numbers states that the arithmetic mean of a very large sum of independent observations of a random variable $x(s)$ is equal to the mean of $x(s)$ [223, 668].

15.3.2 Langevin Equation with Parametric Noise

The form of the Langevin equation that can generate power-law behaviors includes both additive and multiplicative noise,

$$\frac{dv}{dt} = \eta(t)v(t) + \xi(t), \tag{15.37}$$

where $v(t)$ is a dynamic variable, $\eta(t)$ is parametric white noise with mean $\overline{\eta}$ and intensity σ_η^2, and $\xi(t)$ is additive white noise with mean zero and intensity σ_ξ^2. The mathematical analysis of (15.37) is well beyond the scope of our introductory discussion. However, it is useful to be aware of a number of observations that have been made concerning the behavior of (15.37) [170, 623, 738, 784, 819, 849]. It has been shown that the stationary probability density $p(v)$ is

$$p(v) \approx \left(\sigma_\xi + \sigma_\eta v^2\right)^{-1/2 + \overline{\eta}/2\sigma_\eta}, \tag{15.38}$$

and that the tails for large v exhibit power laws of the form

$$p(v) \approx |v|^{-\beta - 1}, \tag{15.39}$$

where $\beta = -\overline{\eta}/\sigma_\eta$. This observation has a number of consequences: First, the power-law exponent depends only on the statistical properties of the parametric noise, in particular, its mean and intensity. Second, the parametric noise is not typically delta-correlated. Third, power laws can be observed for a completely linear dynamical system. Fourth, the power-law behaviors cannot be maintained for an arbitrarily long time unless there is additive noise. These predictions have been verified in a simple analog electronic circuit [739].

15.3.3 Pole Balancing at the Fingertip

An experimental paradigm in which parametric noise is thought to play a role is pole balancing at the fingertip [94, 393] (Figure 15.11a). The equation for stabilizing an inverted pendulum with time-delayed feedback is

$$\frac{d^2\theta}{dt^2} = -a\frac{d\theta}{dt} + q\sin\theta - R(\theta(t - \tau)), \tag{15.40}$$

where θ is the vertical displacement angle, a, q are constants, R is the gain, and τ is the neural delay. An essential requirement for stability of the fixed point for (15.40) is that τ be shorter than a critical delay that is proportional to the square root of the length of the stick [801]. However, contrary to expectation, even long sticks at the fingertip can fall.

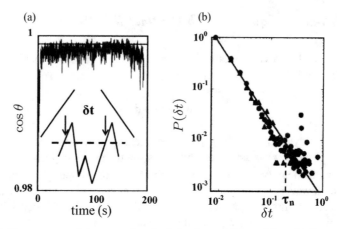

Fig. 15.11 Measurement of the power law for balancing a 0.62-m stick at the fingertip. (a) The laminar phases δt are estimated from the temporal series of $\cos\theta$ by placing a threshold (solid horizontal line) and then measuring the time between successive crossings of this threshold in the upward direction. (b) The power-law exponent is estimated from the slope of the solid line of a log–log plot of $P(\delta t)$ versus δt. Figure reproduced from [578] with permission.

Figure 15.11a shows the log–log plot $P(\Delta t)$ versus Δt of the probability that the time interval between successive corrective movements is linear with slope $-3/2$. In particular, it can be seen that more than 98 % of the Δt are shorter than τ. In view of the presence of a $-3/2$ power law, it has been suggested that these faster corrective movements represent on–off intermittency [94]. In particular, numerical simulations of (15.40) reproduced the power law when

$$R(t) = R_0 + \xi(t),$$

where $\xi(t)$ is Gaussian-distributed white noise with intensity (variance) σ_n^2. Parametric noise of this form may arise, by analogy with an inverted pendulum [5, 66], from noisy perturbations in the vertical movements of the fingertip. Although these observations argue that power laws observed in stick balancing are related primarily to parametric noise, this is by no means a certainty. As we mentioned in Section 15.3, when time delays are present, the effects of additive and parametric noise are often similar. Thus, more work must be done before these power-law behaviors are understood.

15.4 What Have We Learned?

1. What is additive noise?

    Additive noise is noise whose effects are independent of the state of the system. Examples are given in (15.32) and (15.2).

2. **What is parametric noise?**

> Parametric noise is noise whose effects are dependent on the state of the system. An alternative term is multiplicative noise. Examples are given in (15.33), (15.34), and the stochastic version of (15.40).

3. **What is the Langevin equation?**

> The Langevin equation is a first-order differential equation with additive white noise that is studied to understand the properties of the movements of particles in a liquid environment.

4. **What are the autocorrelation function $C_{xx}(\Delta)$ and the power spectrum $W(f)$ determined from the Langevin equation?**

> The autocorrelation function for the Langevin equation is given by (15.9), and the power spectrum by (15.10).

5. **What is the net mean square displacement as a function of time of a particle whose movements are described by the Langevin equation?**

> It is given by (15.7).

6. **How was the discrepancy between the predicted and observed velocities of the movements of a particle resolved?**

> It was necessary to account for the time delay in the human retina.

7. **What is an Ornstein–Uhlenbeck process?**

> It is another name for a stochastic process described by the Langevin equation.

8. **What dynamical behavior observed for parametric noise has not yet been observed for additive noise and what is its characteristic?**

> On–off intermittency has not yet been observed for additive noise, and it is characterized by the presence of a $-3/2$ power law.

9. **What is the effect of a time delay on the power spectrum and autocorrelation function for a first-order linear time-delayed dynamical system with additive noise?**

> The effect of the delay is to add a "periodic ripple" to both the power spectrum and autocorrelation function.

15.5 Problems for Practice and Insight

1. Consider the Langevin equation

$$\frac{d}{dt}v(t) = -\mu v(t) + \xi(t),$$

where μ is a constant and $\xi(t)$ is a Gaussian (normal) white noise as discussed in the text with

$$\langle \xi(t) \rangle = 0, \quad \langle \xi(t_1)\xi(t_2) \rangle = \sigma_n^2 \delta(t_1 - t_2),$$

and $\langle x \rangle$ is the mean value.

a. Show by direct integration that

$$\langle v^2(t) \rangle = v_0^2 e^{-2\mu t} + \frac{1}{2\mu}\sigma_n^2 \left(1 - e^{-2\mu t}\right).$$

b. Show that the variance is given by

$$\sigma_v^2 = \frac{1}{2\mu}\sigma_n^2 \left(1 - e^{-2\mu t}\right),$$

and that in the limit $t \to \infty$,

$$\sigma_{v_{\text{settling}}}^2 = \frac{1}{2\mu}\sigma_n^2.$$

Does this result make sense physically?

2. The Fokker–Planck equation, which corresponds to the Langevin equation in the previous problem, is given by

$$\frac{\partial}{\partial t}P(v,t) = \frac{\partial}{\partial v}(\mu v P(v,t)) + \frac{\sigma_n^2}{2}\frac{\partial^2}{\partial v^2}P(v,t).$$

This is a dynamical equation of the pdf for the stochastic process described in the Langevin equation. Let us explore this correspondence.

a. We would like to solve the Fokker–Planck equation using the Fourier transform with respect to v:

$$R(k,t) = \int_{-\infty}^{\infty} e^{-jkv}P(v,t)\,dv.$$

By taking the Fourier transform, show that we can obtain the following:

$$\frac{\partial}{\partial t}R(k,t) = -\mu k\frac{\partial}{\partial k}R(k,t) - \frac{\sigma_n^2}{2}k^2 R(k,t).$$

b. Let us set the initial condition of the pdf to

$$P(v,0) = \delta(v - v_0).$$

Show that this condition translates to the following with the Fourier transform:

$$R(k,0) = e^{-jkv_0}.$$

c. One can solve the above transformed equations using the method of characteristics for partial differential equations. Then, using the inverse Fourier transform, one can obtain the desired pdf as a solution of the Fokker–Planck equation. (You are welcome to try it.) The result is given by

$$P(v,t) = \sqrt{\frac{\mu}{\pi \sigma_n^2 \left(1 - e^{-2\mu t}\right)}} \exp\left(-\frac{\mu \left(v - e^{-\mu t} v_0\right)^2}{\sigma_n^2 \left(1 - e^{-2\mu t}\right)}\right).$$

Verify by direct substitution that this satisfies the Fokker–Planck equation.

d. Compute from the above pdf the mean and variance when $v_0 = 0$. Verify that this computation agrees with the results we obtained using the Langevin equation (Section 15.1).

Chapter 16
Random Walks

We began this book with the notion that the mechanisms that underlie biological processes are deterministic. This hypothesis underscores the use of models based on ordinary and delay differential equations. Then, in the last two chapters, we admitted the possibility that there is likely to be a stochastic (random) element as well. In particular, the stochastic differential equations discussed in the previous chapter assert that biological dynamics reflect the interplay between deterministic and stochastic processes. Here we consider the final possibility, that some biological processes are dominated by random processes.

Random walks provide a model of how random events on a microscopic scale are manifested macroscopically in the form of transport processes such as diffusion. However, the application of random walks is not limited to problems related to molecular diffusion. Investigators have applied this approach to the study of phenomena ranging from the movements of cells [287, 563, 564, 753] to foraging patterns of bacteria, birds, mammals (for a review, see [855]), and fish [357] to the movements of the fingertip during stick balancing [95, 137] and even the movements of a reader's eyes—such as yours—in scanning this page in search of something interesting [83]. So pervasive is the concept of random walks that variants arise in descriptions of the behavior of Wall Street [530]; predictions of rare events such as earthquakes, severe weather, and epileptic seizures [785]; in analyses of the genome [677]; and even in quantum mechanics [424].

The study of random walks provides an accessible introduction to the investigation of stochastic dynamical systems. It is the simplest case of randomness, since the motions are due solely to the effects of random perturbations, and the dynamics depend only on the position of the particle [549]. What can we predict about the movements of a random walker? Intuitively, the answer must be in the form of probabilities, such as the probability that the walker is located at a certain distance from the origin at a certain time. Thus, we again anticipate that the probability density function (pdf) must play a central role in shaping answers to these questions. Complexities arise because for a random walk and for stochastic processes in general, the pdf changes in time, and hence so do the averaged quantities calculated from it.

© Springer Nature Switzerland AG 2021
J. Milton and T. Ohira, *Mathematics as a Laboratory Tool*,
https://doi.org/10.1007/978-3-030-69579-8_16

Historically, scientists were interested in diffusive processes that involve large numbers of molecules. Consequently, problems related to diffusion were considered within the framework of two partial differential equations referred to as Fick's[1] laws of diffusion (for reviews, see [51, 774]). However, the present-day focus is on the movements of very small numbers of molecules. In these cases, Fick's laws of diffusion do not provide an appropriate description, and models must be in the form of random walks. For example, for a given gene, there may be only five to ten molecules of messenger RNA present in the cytoplasm of a eukaryotic cell [415]. With recent advances in technology, it has become possible to track the movements of single molecules at the surface of macromolecules such as DNA [168, 859], within plasma membranes [463], within single cells and cellular networks [85, 352] and even to measure the power spectral density of a single diffusing molecule [443, 444]. Thus, scientists can explore a wide range of cellular phenomena including the assembly of proteins into membranes [454, 735], cellular signaling pathways [352], and actin-based intracellular cargo transporters [629], and they are able to characterize nanoscale environments within developing embryos [85, 625]. Moreover, with the advent of global positioning devices and high-speed motion-capture technologies, it has been possible to investigate phenomena ranging from the search patterns of organisms to the control of bipedal locomotion and balance using analogies based on random walks.

This chapter is designed to provide a descriptive overview of present-day applications of random walks to stochastic processes in biology. Broadly speaking, there have been four approaches. The first emphasizes the time-dependence of the mean square displacement $\langle X^2 \rangle$ of the walker. The phrase "simple random walk" refers to situations in which $\langle X^2 \rangle$ increases linearly with time. When $\langle X^2 \rangle$ does not increase linearly with time, we use the term "anomalous diffusion." Presently emphasized examples of anomalous diffusion include correlated random walks, delayed random walks, and Lévy flights.

The second approach is motivated by the qualitative nature of the trajectories of random walks in two and three dimensions. Counterintuitively, random walks tend to be confined to small regions of space for relatively long periods of time, then move rapidly to a distant region of space, where again they are temporarily confined, and so on. This qualitative nature of random-walk trajectories is suggestive of a search pattern, and hence random-walk models are frequently used to study the movements of cells and organisms. The theory of anomalous diffusion is very much a work in progress. However, the theoretical limitations do not prevent the study of such phenomena at the benchtop.

The third approach has been to determine how the relative positions of a group of random walkers changes as a function of time. Suppose that the random walkers are initially confined to a small region of space and that we construct a pdf, $P(x,t_0)$, to describe the distribution of the positions of the walkers at time t_0. As time progresses, each of the random walkers wanders about independently of the motion of the other walkers. The consequence is that $P(x,t)$ spreads out; a process referred to

---

[1] Derived in 1855 by Adolf Eugen Fick (1829–1901), German physician and physiologist.

as diffusive spreading, or more simply as diffusion. The Fokker–Planck equation is a partial differential equation that describes the time evolution of $P(x,t)$. Although the study of such equations is beyond our scope, surprisingly a great deal of insight can be obtained using the characteristic function we introduced in Section 14.2.3. This follows from the fact that the pdf and the characteristic function are a Fourier transform pair.

Finally, a limitation of simple random walks is that they neglect the possibility that the walkers might resist the effects of perturbations. The notion of stability leads to the Ehrenfest model for a random walk and then to the investigation of time-delayed random walks.

The questions we will answer in this chapter are:

1. For a simple random walk how does the variance change with time?
2. What is an anomalous random walk?
3. What is a correlated random walk? What are examples of biological processes that can be modeled as a correlated random walk?
4. What is the Ehrenfest random walk and what is its relationship to the simple random walk?
5. What is a delayed random walk? What are examples of biological processes that can be modeled as a delayed random walk?
6. What is a Lévy flight?
7. What is the characteristic feature of random walks in two and three dimensions? What are examples for which this property of random walks is important?
8. What is the relationship between the pdf and the characteristic function?
9. What is the relationship between the autocorrelation function and the power spectrum for a random walk?
10. How does the power spectrum for a random walk depend upon frequency?
11. Why does the convolution integral appear in problems related to a random walk?
12. What is the Fokker–Planck equation?
13. What is the master equation?
14. What is a generating function?
15. What is the relationship between a characteristic function and the moments of a probability distribution?
16. What is the relationship between a generating function and the moments of a probability distribution?

16.1 A Simple Random Walk

A simple random walk describes the path of a walker constrained to move along a line by taking identical discrete steps at identical discrete time intervals (Figure 16.1). Let $X(t)$ be the position of the walker at time t, and assume that at time zero, the walker starts at the origin, i.e., the initial condition is $X(0) = 0$. The simple random walker is assumed to take a step of size ℓ to the right (i.e., $X(t+1) = X(t) + \ell$) with probability p, and to the left (i.e., $X(t+1) = X(t) - \ell$) with probability $q = 1 - p$. For now, we shall assume that the probability of stepping to the right is the same as that of stepping to the left, i.e., $p = q = 1/2$.

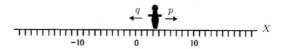

Fig. 16.1 Schematic representation of a simple random walk.

The total displacement X from the origin after n steps is

$$X = \sum_{i=1}^{n} \ell_i, \tag{16.1}$$

where $\ell_i = \pm\ell$. Since $p = q$, we have

$$\langle X \rangle = \sum_{i=1}^{n} \langle \ell_i \rangle = 0, \tag{16.2}$$

where the notation $\langle \cdot \rangle$ signifies the ensemble average. The ensemble average is the average obtained from realizations of a large number of random walks.

Example 16.1 A drunkard exits a bar and begins the (straight-line) journey home. At each street corner, the drunkard flips a coin to determine which way to walk: rightward if the result is heads, leftward if it is tails. After very many coin flips, where on average is the drunkard most likely to be?

Solution. We suspect that many readers will be tempted to refer to (16.2) and conclude that the drunkard will most likely be at the bar. However, this is not the correct interpretation of (16.2). All that (16.2) implies is that after a large number of steps, we can expect that for a given displacement of a random walker to the right of the origin, we are equally likely to find a corresponding displacement to the left. It is in this sense that the average of the displacements from the origin is 0. In order to determine how far from the bar the drunkard is on average, we need to do further analysis (see below and Example 16.5).   ◊

The displacement of a random walker from the origin after n steps can be derived from the variance X^2, i.e.,

$$X^2 = (\ell_1 + \ell_2 + \cdots + \ell_n)(\ell_1 + \ell_2 + \cdots + \ell_n) = \sum_{i=1}^{n} \ell_i^2 + \sum_{i \neq j}^{n} \ell_i \ell_j.$$

If we now average over many realizations of the random walk, we obtain

$$\langle X^2 \rangle = \sum_{i=1}^{n} \langle \ell_i^2 \rangle + \sum_{i \neq j}^{n} \langle \ell_i \ell_j \rangle . \tag{16.3}$$

The first term on the right-hand side is $n\ell^2$. In order to evaluate the second term, we must keep in mind that the direction of a step that the walker takes does not depend on the direction of the previous step. We can evaluate the second term by looking at its four components:

$$(+\ell, +\ell) = +\ell^2 , \quad (+\ell, -\ell) = -\ell^2 , \quad (-\ell, +\ell) = -\ell^2 , \quad (-\ell, -\ell) = +\ell^2 .$$

Thus, we see that the sum of these contributions is zero, and hence

$$\langle X^2 \rangle = n\ell^2 . \tag{16.4}$$

In terms of the standard deviation, which provides an estimate of the magnitude of the displacement, we have

$$\sqrt{\langle X^2 \rangle} = \sqrt{n}\ell . \tag{16.5}$$

16.1.1 Random walked diffusion

Random walks provide a simple model for diffusion. If we assume that the walker makes v steps per second, so that the number of steps is given by $n = vt$, then

$$\langle X^2 \rangle = v\ell^2 t . \tag{16.6}$$

In other words, a plot of $\langle X^2 \rangle$ versus t is linear. If we define $2D := v\ell^2$, where D is the diffusion coefficient, we have

$$\langle X^2 \rangle = 2Dt$$

for a one-dimensional random walk.[2] There are two ways that $\langle X^2 \rangle$ can be estimated as a function of time.

First, we can calculate $\langle X^2 \rangle$ using many separate realizations of a random walk. In Figure 16.2a, the random walk is simulated using a computer program. Each time a simulation is performed, we obtain a single realization, or sample path, of the random walk. At each time t, we can compute $\langle X^2(t) \rangle$ as the variance of all of the realizations at that point in time, denoted by $\langle X^2(t) \rangle$, namely

$$\langle X^2(t) \rangle = \frac{1}{N} \sum_{i=1}^{N} x_i^2(t) .$$

By repeating this procedure as a function of time, we obtain the result shown in Figure 16.2b (solid line). The slope of $\langle X^2(t) \rangle$ as a function of time is $2D$.

---

[2] The reason for $2D$ rather than D is explained in [51]. Briefly, including the factor 2 simplifies the calculations.

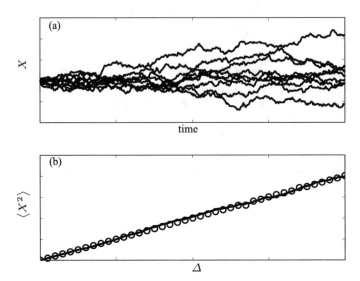

Fig. 16.2 (a) A plot of nine consecutive realizations of a random walk. (b) The mean square displacement $\langle X^2 \rangle$ determined using two methods: the solid line is $\langle X^2(t) \rangle$ calculated from 1,000 consecutive realizations, and $\circ$ is $\langle X^2(\Delta) \rangle$ calculated from a single 5,000-step random walk. In all cases, the initial position X_0 of the walker at $t = 0$ is 0. Note that for $\langle X^2(t) \rangle$, $\Delta = t - t_0 = t$.

We can also estimate $\langle X^2 \rangle$ by using a single sufficiently long realization of the random walk [443, 444]. In this case, we have

$$\langle X^2(\Delta) \rangle = \frac{1}{N-n} \sum_{i=1}^{N-n} [x(t_i + \Delta) - x(t_i)]^2 , \tag{16.7}$$

where N is the total number of points in the time series, $N - n$ is the number of points used to estimate $\langle X^2(\Delta) \rangle$, and $x(t_i)$ is the position at time t_i. Equation (16.7) implies that $\langle X^2(\Delta) \rangle$ is calculated for all possible pairs of times separated by $\Delta = n\delta t$, where δt is the time step.

A remarkable property of a simple random walk is that

$$\langle X^2(t) \rangle = \langle X^2(\Delta) \rangle ; \tag{16.8}$$

see Figure 16.2b. This means that the statistical properties are the same whether the moments are calculated using a large number of different realizations of the random walk, e.g., $\langle X^2(t) \rangle$, or by following a single realization for a very long time, e.g., $\langle X^2(\Delta) \rangle$ (compare the solid line and $\circ$ in Figure 16.2b). A random walk exhibiting this property is said to be *ergodic*.

16.1.2 Sampling

Random walks arise in the context of a statistical technique known as *sampling*, namely the use of a subset of individuals selected from a large population to estimate characteristics of the whole population. Sampling techniques are necessary in situations in which the population is so large that it is impractical to measure every single individual to obtain the "true" value of any statistical moment. In order to make the discussion more concrete, suppose that we want to measure the mean weight among the individual members of a large population. What is the expected difference, or error, between the mean weight of a subset of the population selected at random and the "true" average weight of the entire population, which could be determined only by weighing every member of the population?

A random-walk model can provide a first-order approximation of the dependence of sampling error on the size of the sampled subset. In this case, a "step" in the "random walk" is the unknown distance between the true mean weight of the population and the actual weight of the sampled individual. Intuitively, we expect sampling error to decrease as the sample size increases, and in particular, the error will be zero when the sample set is the entire population. A definition of sampling error that satisfies this expectation is the fractional error

$$\varepsilon := \frac{|\sqrt{\langle X^2 \rangle}|}{N\ell},$$

where $N\ell$ is the length of the random walk. Using (16.5), we obtain

$$\varepsilon = \frac{|\sqrt{N}|\ell}{N\ell} = \frac{1}{\sqrt{N}}. \tag{16.9}$$

In other words, even though $\langle X^2 \rangle$ increases linearly with N, the fractional error ε decreases.

We can apply these concepts to our example of estimating the average weight in a population by identifying the length of the random walk with the number of individuals weighed. Suppose it is sufficient to estimate the average weight in a population to within 1% of its true value, i.e., $\varepsilon = 0.01$. Equation (16.9) implies that we would need to weigh 10,000 individuals. An important property of this sample-size estimate is that it depends only on ε and not on the size of the population. Indeed, even though the population of Canada and the United State of America differ by a factor of 10, both countries use the same sample size to estimate unemployment statistics [518].

It is important to keep in mind that the predicted sample size is an estimate and as such, does not guarantee that the average weight of an individual determined from the sample is within 1% of the true value. Rather, the result implies that we can be reasonably confident (in a sense that can be made statistically precise) that the average weight of an individual in the sampled population is within 1% of its true value.

16.1.3 Walking Molecules as Cellular Probes

The underlying assumption in single-molecule tracking experiments is that the random walk is ergodic. Thus, in two dimensions, the mean square displacement is

$$\langle R^2(\Delta) \rangle = \frac{1}{N-n} \sum_{i=1}^{N-n} \left[(x(t_i+\Delta) - x(t_i))^2 + (y(t_i+\Delta) - y(t_i))^2 \right], \quad (16.10)$$

where $R^2 = x^2 + y^2$. Since the fluctuations in the x and y coordinates occur independently, we have

$$\langle R^2(\Delta) \rangle = 4D\Delta.$$

In a similar manner, we obtain $\langle R^2(\Delta) \rangle = 6D\Delta$ for random walks that occur in three dimensions. In the experimental literature, the mean square displacement is also referred to as the *two-point correlation function* [142, 209, 596].

In one dimension, the mean square displacement can be related to the autocorrelation function $C_{xx}(\Delta)$, which we introduced in Section 14.5, by the expression

$$\langle R^2(\Delta) \rangle = -2C_{xx}(\Delta) + 2\langle X^2 \rangle.$$

In Chapter 14, we pointed out that $C_{xx}(\Delta)$ is a commonly used descriptor of noise. However, in the case of a random walk, $\langle R^2(\Delta) \rangle$ provides a more reliable characterization than $C_{xx}(\Delta)$ [677, 740]. This is because $C_{xx}(\Delta)$ is more sensitive than $\langle R^2(\Delta) \rangle$ to the inherent limitations in experimental time series, namely measurement error, the limited precision of measuring instruments, local trends of unknown cause, and the finite length of the time series [677, 790]. A variety of modifications for estimating $\langle R^2(\Delta) \rangle$ have been developed to minimize these limitations of time series, including detrended fluctuation analysis (DFA) [326, 327, 678] and real-time square displacement (RTSD) [85, 625]. In the examples that follow, we will add a subscript to $\langle R^2(\Delta) \rangle$ to indicate how it was evaluated. Thus, $\langle R^2(\Delta) \rangle_{\text{RTSD}}$ means that $\langle R^2(\Delta) \rangle$ was estimated using the RTSD method, and so on.

A powerful technique for measuring the movements of molecules in single cells involves the use of gold and silver nanoparticles [85, 352, 625]. These particles exhibit unique optical properties, including localized surface plasmon resonance that is sensitive to the size and shape of these particles and their environments and strong Rayleigh scattering, which facilitates identifying the location of these nanoparticles in real time using dark field microscopy and spectroscopy. Since silver and gold are noble metals, their nanoparticles are histocompatible and can be used to probe diffusion characteristics of different parts of living cellular networks as well as the motions of molecules to which they are attached in living cells.

Figure 16.3 shows examples of the random walks exhibited by silver nanoparticles placed in different parts of a developing zebrafish embryo [85]. Zebrafish embryos were chosen for this study because they are transparent and develop outside their mothers. Thus, the movements of nanoparticles as organs develop can be readily followed using light-microscopic techniques. The striking observation is

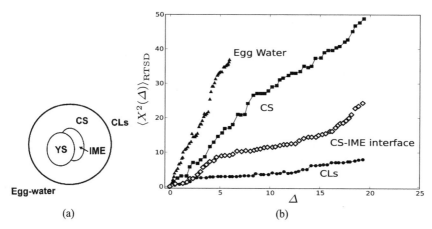

Fig. 16.3 (a) Schematic representation of the cleavage stage of the zebrafish embryo. (b) Random walks exhibited by gold nanoparticles placed in different regions of the cleavage stage of the zebrafish embryo. Figure (b) was prepared from data published in [85] with permission.

that the nature of the random walk clearly depends on the location of the silver nanoparticle within the embryo.

During the cleavage stage of embryological development, the zebrafish embryo contains four distinct regions (Figure 16.3a): the yolk sac (YS), the inner mass of the embryo (IME), the chorionic layers (CLs), and the chorionic space (CS). When gold nanoparticles are placed in the CS, the egg–water solution that surrounds the embryo, $\langle R^2(\Delta)\rangle_{\mathrm{RTSD}}$ increases linearly with Δ. This means that the movements of the nanoparticle in these locations are those of a simple random walker, albeit with different diffusion coefficients. In contrast, complex nonlinear random walks are seen for silver nanoparticles placed at the exterior surface of the CLs and at the CS–IME interface.

Mathematical modeling studies provide insight into the possible mechanisms that can produce the nonlinear random walks shown in Figure 16.3 [454, 735]. Currently, the random walks on the exterior surface of CLs are thought to reflect restricted diffusion [85]. In other words, the particles behave like simple random walkers on short time scales, but they cannot escape this region over longer time scales. The structure of CLs is quite complex; it contains many membrane layers as well as pores through which nanoparticles can pass into the CS. Thus, the type of restricted random walk suggests that molecules become temporarily trapped within the CLs domain before passing into the CS (for similar scenarios, see below). Finally, the movements monitored at the CS–IME interface appear to be intermediate between an unrestricted and a restricted random walk.

16.2 Random walk: Probability distribution function

Another way to study a random walk is to calculate the probability, $P(r,n)$ that after n steps the walker attains a position $X = r$. This probability distribution function satisfies

$$\sum_{r=-\infty}^{\infty} P(r,n) = 1.$$

The importance of $P(r,n)$ is that averaged quantities of interest, or expectations, such as the mean and variance can be readily determined from it. For a discrete variable, X, the moments can be determined by using the generating function, $Q(s,n)$, i.e.

$$Q(s,n) = \sum_{r=-\infty}^{+\infty} s^r P(r,n) \tag{16.11}$$

We will discuss the generating function in more detail in Section 16.9. Here we point out that the averaged quantities of interest to experimentalists can be calculated by differentiating $Q(s,n)$ with respect to s.

Example 16.2 For example $Q(s,n)$ for a simple random walker is (see Exercise 6)

$$Q(s,n) = \left[ps + \frac{q}{s} \right]^n \tag{16.12}$$

The mean is obtained by differentiating $Q(s,n)$ with respect to s

$$\frac{\partial}{\partial s} Q(s,n) \mid_{s=1} \equiv \langle X(n) \rangle = n(p-q) \tag{16.13}$$

The second differentiation leads to

$$\frac{\partial^2}{\partial s^2} Q(s,n) \mid_{s=1} = \langle X(n)^2 \rangle - \langle X(n) \rangle \tag{16.14}$$

The variance can be calculated from these two equations as $\sigma^2(n) = 4npq$. The variance, i.e., the AC component, gives the intensity of the varying component of the random process. The positive square root of the variance is the standard deviation which is typically referred to as the root-mean-square (rms) value of the AC component of the random process. When $\langle X(n) \rangle = 0$, the variance equals the mean square displacement.   ◇

It is easiest to estimate $P(r,n)$ by calculating the Fourier transform of the characteristic function [73, 161]. The discrete characteristic function is

$$\varphi(\theta,n) = \sum_{r=-\infty}^{+\infty} P(r,n) e^{-j\theta r} \tag{16.15}$$

where θ is the continuous "frequency" parameter. Since $\varphi(\theta,n)$ is defined in terms of a Fourier series whose coefficients are the $P(r,n)$, the probability distribution after n steps can be represented in integral form as

$$P(r,n) = \frac{1}{2\pi} \int_{-\pi}^{\pi} \varphi(\theta,n) e^{j\theta r} d\theta. \tag{16.16}$$

In order to use $\varphi(\theta,n)$ to calculate $P(r,n)$, we first write down an equation that describes the dynamics of the changes in $P(r,n)$ as a function of the number of steps, i.e.

$$P(r,0) = \delta_{r,0} \tag{16.17}$$
$$P(r,n) = pP(r-1,n-1) + qP(r+1,n-1).$$

where $\delta_{r,0}$ is the Kronecker delta function defined by

$$\delta_{r,0} = 1, \quad (r = 0)$$
$$\delta_{r,0} = 0, \quad (r \neq 0).$$

Second, we multiply both sides of (16.17) by $e^{-j\theta r}$, and sum over r to obtain

$$\varphi(\theta,0) = 1$$
$$\varphi(\theta,n) = (pe^{-j\theta} + qe^{+j\theta}) R(\theta, n-1).$$

The solution of these equations is

$$\varphi(\theta,n) = (pe^{-j\theta} + qe^{+j\theta})^n \tag{16.18}$$

Taking the inverse Fourier transform of (16.18) we eventually obtain

$$P(r,n) = \binom{n}{m} p^m q^{n-m}, \quad (r+n = 2m) \tag{16.19}$$
$$= 0, \quad (r+n \neq 2m),$$

where m is a nonnegative integer and[3] where

$$\binom{n}{m} = \binom{n}{n-m} \equiv \frac{n!}{m!(n-m)!}.$$

For the special case that $p = q = 0.5$ this expression for $P(r,n)$ simplifies to

$$P(r,n) = \binom{n}{\frac{r+n}{2}} \left(\frac{1}{2}\right)^n, \quad (r+n = 2m) \tag{16.20}$$
$$= 0, \quad (r+n \neq 2m),$$

and we obtain

---

[3] The notation $(\cdots)$ is also used to denote a matrix; however, in this book we do not introduce matrix notation.

$$\langle X(n)\rangle = 0 \tag{16.21}$$
$$\sigma^2(n) = n. \tag{16.22}$$

Figure 16.4 shows how $P(r,n)$ changes as a function of time when $p = q$.

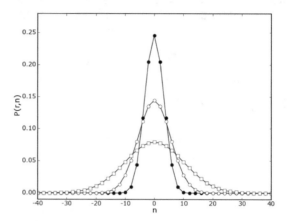

Fig. 16.4 The probability density function, $P(r,n)$, as a function of time for a simple random walk that starts at the origin with $p = 0.5$: (•) $n = 10$, (○) $n = 30$, (□) $n = 100$. We have plotted $P(r,n)$ for even n only: for these values of n, $P(r,n) = 0$ when n is odd.

We can also consider a random walk with drift. When $p > q$ the walker drifts towards the right and when $p < q$ the walker drifts towards the left. The evolution of $P(r,n)$ as a function of time when $p = 0.6$ is shown in Figure 16.5.

16.3 Random walk: Power spectrum

The stationary discrete autocorrelation function, $C(\Delta)$, provides a measure of how much average influence random variables separated Δ steps apart have on each other. The autocorrelation function is useful since it provides a reliable method to estimate the power spectral density using the Weiner-Khinchin theorem. However, the formal derivation of $C(\Delta)$ is difficult because it involves notions concerning the joint probability distribution which we have only touched upon. Here we discuss a simpler method for estimating the power spectrum of a random walk which circumvents issues related to the joint probability distribution.

Suppose we measure the direction that the simple random walker moves each step. Designate a step to the right as R, a step to the left as L, and a "flip" as an abrupt change in the direction that the walker moves. Then the time series for a simple random walker takes the form

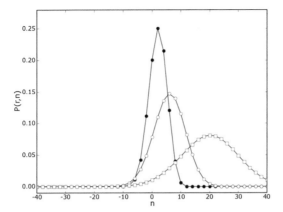

Fig. 16.5 The probability density function, $P(r,n)$, as a function of time for a simple random walk that starts at the origin with $p = 0.6$:($\bullet$) $n = 10$, ($\circ$) $n = 30$, ($\square$) $n = 100$. We have plotted $P(r,n)$ for even r only: for these values of n, $P(r,n) = 0$ when r is odd.

$$R \cdots \overbrace{RL}^{flip} \cdots \underbrace{LR}_{flip} \cdots \overbrace{RL}^{flip} \cdots \qquad (16.23)$$

Assume that the step interval, δn, is so small that the probability that two flips occur within the same step is approximately zero. The autocorrelation function, $C(\Delta)$, where $|\Delta| \geq |\delta n|$, for this process will be

$$C(\Delta) \equiv \langle X(n)X(n+|\Delta|)\rangle = A^2(p_0(\Delta) - p_1(\Delta) + p_2(\Delta) - p_3(\Delta) + \cdots) \quad (16.24)$$

where $p_\kappa(\Delta)$ is the probability that in a time interval Δ that exactly κ flips occur and A is the length of each step.

In order to calculate the p_κ, we proceed as follows: The probability that a flip occurs in δn is $\lambda \delta n$, where λ is some suitably defined parameter. Hence, the probability that no flip occurs is $1 - \lambda \delta n$. If $n > 0$, then the state involving precisely κ flips in the interval $(n, n + \delta n)$ arises from either $\kappa - 1$ events in the interval $(0, n)$ with one flip in time δn, or from κ events in the interval $(0, n)$ and no new flips in δn. Thus

$$p_\kappa(n + \delta n) = p_{\kappa-1}(n)\lambda \delta n + p_\kappa(n)(1 - \lambda \delta n) \qquad (16.25)$$

and hence we have

$$\lim_{\delta n \to 0} \frac{p_\kappa(n + \delta n) - p_\kappa(n)}{\delta n} \equiv \frac{dp_\kappa(n)}{dn} = \lambda \left[p_{\kappa-1}(n) - p_\kappa(n) \right] \qquad (16.26)$$

for $\kappa > 0$. When $\kappa = 0$, we have

$$\frac{dp_0(n)}{dn} = -\lambda p_0(n) \tag{16.27}$$

and at $n = 0$

$$p_0(0) = 1. \tag{16.28}$$

Equations (16.26)–(16.28) describe an iterative procedure to determine p_κ. In particular we have

$$p_\kappa(\Delta) = \frac{(\lambda |\Delta|)^\kappa e^{-\lambda |\Delta|}}{\kappa!}. \tag{16.29}$$

The autocorrelation function $C(\Delta)$ is obtained by combining (16.24) and (16.29) as

$$C(\Delta) = A^2 e^{-2\lambda |\Delta|}. \tag{16.30}$$

We can use (16.30) together with the Wiener–Khinchin theorem to calculate the power spectrum, $W(f)$. However, our interest is to determine $W(f)$ for a continuous random walk. To accomplish this task, we assume that the length of the step taken by the random walker is small enough so that we can replace δn by dt. Thus

$$C(\Delta) = \int_{-\infty}^{\infty} W(f) e^{j2\pi f \Delta} df \tag{16.31}$$

and hence

$$W(f) = \int_{-\infty}^{\infty} C(\Delta) e^{-j2\pi f \Delta} d\Delta. \tag{16.32}$$

Completing the integration we obtain

$$W(f) = \frac{2A^2}{\lambda} \left[\frac{1}{1 + (\pi f / \lambda)^2} \right] \tag{16.33}$$

Figure 16.6 shows $C(\Delta)$ and $W(f)$ for this random process, where f is Δ^{-1}. We can see that the noise spectrum for this random process is essentially "flat" for $f \ll \lambda$ and thereafter decays rapidly to zero with a power law $1/f^2$ when $f \gg \lambda/\pi$.

It is interesting to compare $W(f)$ to the power spectrum calculated for Brownian motion, W_{brown}. Brownian motion is the motion that is described by the Langevin equation as $t_{\text{relax}} \to \infty$. Thus, from (15.10), we have

$$W_{\text{brown}}(f) = \lim_{t_{\text{relax}} \to \infty} \frac{2\sigma_n^2 t_{\text{relax}}^2}{1 + (2\pi f t_{\text{relax}})^2} = \frac{S_0}{f^2} \tag{16.34}$$

where $S_0 = \sigma_n^2 / 2\pi^2$. Thus, for both the discrete random walk and Brownian motion, the power spectrum is proportional to f^{-2} provided that $f \gg \lambda/\pi$. It is often the case in experimental applications that $W(f)$ is white for low frequencies and then falls off as f^{-2} (see, for example, [893]).

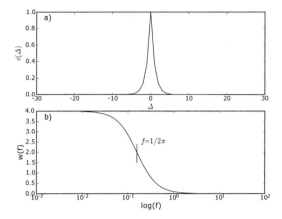

Fig. 16.6 (a) Auto–correlation function, $C(\Delta)$ and (b) power spectrum, $w(f)$, for the random process described by (16.23). Parameters: $A = 1$, $\lambda = 0.5$.

16.4 Anomalous Random Walks

An anomalous random walk is a random walk for which $\langle X^2 \rangle$ does not increase linearly with time. Sometimes, the term "superdiffusion" is used to describe situations in which $\langle X^2 \rangle$ increases more rapidly than linear as a function of time. Anomalous diffusion is observed very frequently in biology, and the identification of the mechanisms that cause the nonlinearity is often a major focus of research. Detailed discussions of anomalous diffusion, such as those provided by Sornette[4] [784], are well beyond the scope of this introductory text. Thus, we focus on experimental evidence that is consistent with one type of anomalous diffusion, namely correlated random walks, which has received considerable experimental attention.

16.5 Correlated Random Walks

A correlated random walk has the property

$$\langle R^2(\Delta) \rangle \approx \Delta^\kappa, \tag{16.35}$$

where $\kappa \neq 1$. Interest in correlated random walks traces its origins to the pioneering work of Mandelbrot and Van Ness[5] [535] on fractional Brownian motions.[6] When

---

[4] Didier Sornette (b. 1957), French physicist and economist.

[5] John W. Van Ness (PhD 1964), American mathematician and artist.

[6] Benoît Mandelbrot (1924–2010), Polish-born French and American mathematician noted for developing the field of fractal geometry.

$\kappa \neq 1$, then $\langle R^2 \rangle$ does not increase linearly with time. A property of correlated random walks is that long-range correlations exist potentially over all length scales. Here we discuss three examples of correlated random walks that can be readily explored by the reader. It is useful to keep in mind while reading the literature on correlated random walks that the interpretation of κ depends on the method used to estimate $\langle R^2(\Delta) \rangle$.

16.5.1 Random Walking While Standing Still

An application of a correlated random walk arises in the context of the control of human balance during quiet standing with eyes closed. As we discussed in Section 10.7, the elderly are at increased risk for falling. Measurements of the fluctuation in the center of pressure during quiet standing help identify those elders who have increased risk of falling while moving. One approach to analyzing postural sway (Figure 16.7b) uses the two-point correlation function [126, 142, 209, 596, 652]

$$\langle R^2(\Delta) \rangle \approx \Delta^{2H} ,$$

where H is a scaling exponent. A simple random walk corresponds to $H = 0.5$. Consequently, in this case, a log–log plot of $\langle R^2(\Delta) \rangle$ versus Δ is linear. When $H > 0.5$, there is *persistence*: when the walker moves in a particular direction, it tends to continue moving in that direction. When $H < 0.5$, we have *antipersistence*, which implies that increasing trends in the past produce decreasing trends in the future.

Figure 16.7b shows that a log–log plot of $\langle R^2(\Delta) \rangle$ versus Δ is often nonlinear [596]. This observation has been interpreted as evidence for the presence of multiple scaling regions each associated with a different value of H. As can be seen in Figure 16.7b, three different patterns have been identified [596]. The Type I pattern has two distinct nonoscillatory regions: one region with $H > 0.5$, the other with $H \approx 0$. The Type II pattern has three distinct nonoscillatory regions; and the Type III pattern has two distinct regions, one of which is oscillatory. Of these patterns, the Type II pattern has attracted the most interest [126, 142, 209, 596]. If we identify persistence with open-loop feedback control ($H > 0.5$) and antipersistence with closed-loop feedback control ($H < 0.5$), then we can see the beginnings of a "drift and act" type of controller for balance control [596].

A number of explanations have been proposed to explain these different patterns [126, 142, 548, 596]. For example, the model proposed Eurich and Milton [209] (see Exercise 7) suggests that these patterns may be related to the interplay between noise and bistability. In the absence of noise, there is a range of choices of C and τ for which there is bistability in the form of two coexistent oscillators. In this bistable regime, the effect of additive noise is to cause switches between the two basins of attraction. Noise-induced switching can account for the scaling regions shown in Figure 16.8b. In particular, when Δ is less than the period of the limit cycle oscillator, switches occur in only one direction, such as limit cycle 1 to limit

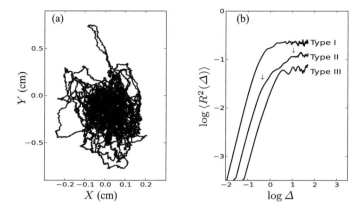

Fig. 16.7 (a) Fluctuations in the center of pressure as a subject stands quietly on a force platform with eyes closed for 2 min: X corresponds to displacements in the mediolateral direction and Y to displacements in the anterior–posterior direction. Data were collected with the assistance of Janelle Gyorffy (Pomona College) and Meredith King (Claremont McKenna College). (b) The mean square displacement $\langle R^2(\Delta) \rangle$ for the fluctuations in the center of pressure measured as a subject stands quietly with eyes closed on a force platform for 2 min. For the Type II pattern, the down arrows ($\downarrow$) subdivide $\langle R^2(\Delta) \rangle$ into three regions, each of which can be approximated with a linear slope. Data shown for three subjects. The sampling frequency was 200 Hz, and the data were then resampled at 100 Hz. Data collected and analyzed with the assistance of Jennifer Townsend (Scripps College).

cycle 2. On these time scales, the effect of increasing Δ is to increase $\langle R^2(\Delta) \rangle$. When Δ becomes larger than the order of the limit cycle period, transitions of the form $1 \rightarrow 2 \rightarrow 1$ become possible and cause a decrease in the rate of increase of $\langle R^2(\Delta) \rangle$. Finally, once Δ becomes sufficiently long, it becomes equally probable that the walker is in either basin of attraction after a large number of switches, and hence $\langle R^2(\Delta) \rangle$ equals the mean displacement.

The above observations suggest that random perturbations (noise) play a role in shaping the fluctuations in human postural sway during quiet standing. Three additional observations support this interpretation [596]: the same person can exhibit different patterns of postural sway on different days; applying a 250-Hz vibration whose amplitude is below the perceptual threshold to the Achilles tendon can change the postural sway pattern; and it is possible to improve human balance, measured as the reduction in the variability of the center of pressure fluctuations, using noisy [698] or periodic [573, 595] stimuli.

16.5.2 Gait Stride Variability: DFA

In Section 12.1.1, we discussed gait stride variability as an example in which ideas based on the Poincaré section could be translated into an investigation of human gait dynamics. The fluctuations in gait stride variability can also be modeled as a

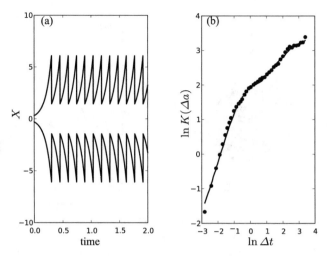

Fig. 16.8 (a) Two coexistent limit cycle attractors occur in (10.55) when τ, C are chosen appropriately. (b) Comparison of $\langle R^2(\Delta) \rangle$ observed experimentally (•) to that generated by (10.55) (solid line). The parameters are $k = 0.60\,\text{s}^{-1}$, $\Pi = 5.95\,\text{mm}$, $C = 19.67\,\text{mm} \cdot \text{s}^{-1}$, $\tau = 0.233\,\text{ms}$, and $\sigma = 1.12\,\text{mm} \cdot \text{s}^{-1}$.

correlated random walk [326, 327]. A practical problem with gait stride time series is that they are noisy and contain short-range nonstationarities. Thus, (16.10) does not provide a useful measure of the mean square displacement. However, a more reliable estimate can be obtained using a procedure called *detrended fluctuation analysis* (DFA). It has been observed that [117, 327]

$$\sqrt{\langle X^2(\Delta) \rangle_{\text{DFA}}} \approx \Delta^H, \tag{16.36}$$

where the notation $\langle X^2(\Delta) \rangle_{\text{DFA}}$ indicates that $\langle X^2(\Delta) \rangle$ is determined using the DFA method.

From the point of view of (16.36), a simple random walk corresponds to $H = 0.5$, and fractional Brownian motions characterized by long-range correlations to $0.5 < H \leq 1$. Figure 16.9 shows the DFA analysis for self-paced walking of a healthy subject on a level surface. For healthy active individuals, H is in the upper part of the range $0.5 < H \leq 1.0$. Hausdorff[7] interpreted this finding as implying that a supraspinal locomotor clock likely played an important role in generating the correlated random walk [322, 327]. Another possibility is that, a central clock, presumably activated by the metronome, can suppress the correlated random walk. Finally, in older people with a history of falling and in those with diseases of the basal ganglia (e.g., Parkinson's and Huntington's diseases), H also approaches 0.5 [324]. A practical problem with DFA analysis is the sensitivity of H to low-frequency trends in the data. It has been suggested that the use of frequency-weighted power spectra

---

[7] Jeffrey M. Hausdorff (PhD 1995), American biomechanist.

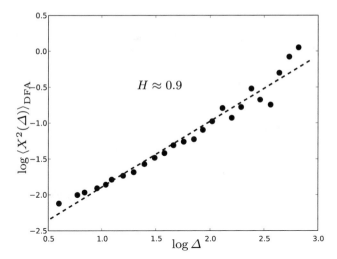

Fig. 16.9 Detrended fluctuation analysis (DFA) for self-paced walking of a healthy 20-year-old male for two laps of a 400-m running track. Slope of the line relating the log of the average fluctuation $\langle X^2(\Delta) \rangle_{DFA}$ for box size Δ to $\log \Delta$ was evaluated by a linear regression fit. Data collected with the assistance of Adam Coleman and David Nichols (Claremont McKenna College).

may offer a more reliable method to determine H for physiological time series, such as the heart rate [878].

16.5.3 Walking on the DNA Sequence

Applications of random walk analysis are not restricted to situations in which physical displacements occur. Indeed, it is possible to use the analogy of a random walker to analyze stochastic processes in which no physical movements occur. One example is fluctuations related to the stock market [530]. A more biologically interesting application arises in the analysis of the genetic code. It is possible to convert a sequence of DNA bases into a random walk as follows [677, 678]: the walker takes one step to the right if there is a pyrimidine (thymine (T) or cytosine (C)) at position i, and one step to the left if there is a purine (adenine (A) or guanine (G)). The DFA analysis we just described for the analysis of gait stride variability can be applied to this random walk. Using this approach, it was found that long-range correlations ($0.5 < H \leq 1.0$) exist in some DNA sequences but not all [677, 678].

16.6 Portraits in Diffusion

By focusing on considerations of $\langle R^2(\Delta) \rangle$ alone, investigators can easily overlook many of the interesting and counterintuitive properties of random walks [51].

Figure 16.10 shows four realizations of a random walk in two dimensions generated using a computer program. In this program, the length of the steps was drawn from a uniform distribution on the interval $[-1, 1]$. We strongly encourage the reader to try different distributions for the step length (see Exercise 4) and moreover, to animate the random walks as they evolve in time.

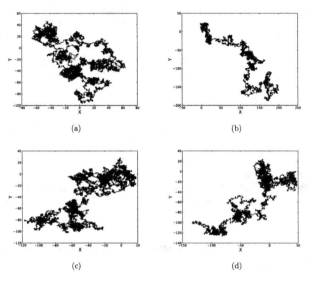

(a)   (b)

(c)   (d)

Fig. 16.10 Four realizations of a two-dimensional random walk in which the lengths of the steps were drawn from a uniform probability distribution on the interval $[-1, 1]$. In all cases, the initial position was $x = y = 0$.

We suspect that the qualitative nature of the random-walk trajectories shown in Figure 16.10 are very different from what the reader might have anticipated. The darker the regions, the longer time that the random walker spends exploring that region. Viewed from this perspective, we see that the walker is confined to certain regions of space longer than other regions. This behavior reflects the stochastic property of a random walk. In order to move from one region of space to another located many steps away requires that a number of consecutive steps be all in the same direction. The probability that this happens is quite small, and hence moves between distant regions of space occur infrequently. It is tempting to interpret this trajectory pattern in terms of a search pattern, namely, that the random walker carefully searches a small region of space before quickly moving to another region. Motivated by these interpretations, investigators have asked two types of questions. First, what is the impact of this nature of a random walk in enabling a walker to find its target? Second, what are optimal search patterns for finding a target?

16.6.1 Finding a Mate

Female silk moths attract males by releasing sex-attractant molecules, called pheromones, into the air [248]. These molecules are carried and dispersed by air currents until they are detected by the antennae of the male moth. It has been estimated that a male moth can detect a female moth from distances up to several kilometers. This observation suggests that the male's antennae are sensitive enough to detect a single molecule of pheromone. Let us explore the association of the properties of the random walk shown in Figure 16.10 with this remarkable courtship signaling process.

The moth's pheromone receptors are located in tapered sensory hairs called the sensilla trichodea (for a review, see [248]). The length of each of these hairs is about 50 times its diameter. The cuticles of the sensilla are penetrated by about 2,600 pores; molecules of pheromone enter the pores either by diffusion on the surface of the sensilla or by floating in the air before eventually arriving at the sensory receptor. Surprisingly, this structure is so well adapted that essentially every molecule of pheromone that passes through it is captured and gives rise to a nerve impulse [65].

Berg and Purcell[8] drew attention to role of the properties of random walks in making the insect sensilla so efficient [53]. They showed that the probability that a particle released some distance from a surface will reach the surface depends primarily on the surface's linear dimension, not its surface area (for a very readable discussion of this point, see [51]). We can model each of the pheromone receptors as a sphere. In three dimensions, the probability p_a that a particle released at distance b from a sphere of radius a contacts the surface of the sphere before drifting away to a distance $c > b$ is

$$p_a = \frac{a}{b}.$$

This estimate is valid even though the surface of the sphere contains no receptors. In the world of silk moths, sensory receptors are scattered on the surface of the sphere, and the distance between different receptors is likely large compared to the cross-sectional area of the receptors. Thus, the question is not simply whether the pheromone comes into contact with the surface of the sensilla, but how many times it contacts the surface before drifting away for good. Obviously, the more times the pheromone contacts the surface, the greater the probability that the pheromone contacts a receptor and hence is detected by the silk moth's nervous system.

Recall our initial question: how does the pheromone contact the receptors on the sensilla? The answer requires the assumption that the surface of the sphere is reflecting; each time a molecule hits the surface of the sphere, it is reflected away. If a molecule is released at distance $r = b$ from the surface of the sphere, how many times is it reflected from the surface? The probability p that the particle visits the sphere at least once before wandering away, never to contact it again, is simply p_a. The probability that the pheromone visits the sphere exactly one time, returns to $r = b$, and then wanders off never to return is

---

[8] Howard Curtis Berg (b. 1934), American biophysicist; Edward Mills Purcell (1912–1997), American physicist.

$$p_a(1 - p_a).$$

The probability that the pheromone molecule makes the round trip $r = b$ to $r = a$ to $r = b$ twice before wandering away is

$$p_a^2(1 - p_a).$$

Continuing in this way, we see that the probability that the molecule makes the round trip n times before wandering away is

$$p_a^n(1 - p_a).$$

Therefore, the mean number of times a pheromone molecule makes the round trip before wandering off is (see Exercise 5)

$$\langle n \rangle = \sum_{n=0}^{\infty} n p_a^n (1 - p_a) = \frac{a}{b - a}. \tag{16.37}$$

Thus, if the pheromone molecule comes sufficiently close to the sensilla (i.e., when $b - a > 0$ becomes small), the number of contacts that the pheromone molecule makes with the surface of the sensilla increases dramatically. By implication, once a molecule of pheromone comes close to the sensilla, the likelihood that it will bind to a sensory receptor becomes very high. Moreover, this process works much better if there are many small receptors than a few large ones [51, 53]. This corresponds exactly to the way in which the male's pheromone sensory organ is constructed [248].

16.6.2 Facilitated Diffusion on the Double Helix

In Section 9.3.1, we discussed simple models for the control of gene expression that involved the binding of a repressor molecule to a special site on the DNA molecule, a site referred to as the *operator site*. The operator site is much smaller than the DNA molecule (it has been estimated that the length of a single DNA molecule is ≈ 3 m). What is the mechanism by which a repressor molecule locates the promoter site on the DNA molecule? Surprisingly, the answer, in part, is closely related to (16.37).

The Smoluchowski[9] equation gives the limiting association rate constant k_{assoc} for a bimolecular reaction limited by diffusion [549] as

$$k_{\text{assoc}} = 4\pi Da, \tag{16.38}$$

where a (length) is the size of the target and D (length^2/time) is the diffusion coefficient. The units for k_{assoc} are volume/(moles $\cdot$ time). Surprisingly, experimental

---

[9] Marian Smoluchowski (1872–1917), Polish physicist.

measurements indicate that k_{assoc} is about 100 times larger than the diffusion limit predicted by (16.38) [719].

The answer to this apparent violation is *facilitated diffusion*. The essential idea is that while looking for its binding site the repressor molecule involves a "1D/3D mechanism" as the repressor molecules binds and unbinds to the DNA molecule [55, 600, 719] When the repressor molecule is not bound to the DNA its diffuses three dimensionally in solution until it "bumps into the DNA molecule". When the repressor molecule comes into contact with the DNA molecule it can be attracted nonspecifically to regions of the DNA other than the operator site. This attraction is strong enough to prevent the repressor molecule from immediately drifting away from the DNA molecule, but is not so strong as to prevent the repressor molecule from moving along the DNA fiber [823]. In the model the repressor molecule moves along the DNA molecule as a 1D random walker. The effect of including nonspecific binding regions that flank the operator site is to increase the target size a [310] and thus increase the number of contacts by decreasing $b - a$ in (16.37). The investigation of this mechanism has been facilitated by the development of single-molecule fluorescence imaging techniques that make it possible to measure the movements of the repressor molecule along the DNA helix [859]. It has been estimated that the time that the repressor molecule slides along the DNA is on the order of milliseconds. During this time it travels along ≈ 50 base pairs (a distance of $\approx 20\,nm$) interspersed with longer hops greater than 200 nm [168].

16.7 Optimal Search Patterns: Lévy Flights

Questions concerning the optimality of a random-walk search pattern are typically discussed within the context of a *Lévy flight* [95, 287, 357, 855, 856]. A Lévy flight is a random walk for which the probability $P(\ell)$ of taking a step of length ℓ obeys a power law. These observations raise the question whether organisms use Lévy-type search patterns in order to locate their targets.

A recent example that illustrates the possible relationship between a Lévy flight and an optimal search pattern arises in the context of the movements of microglia in the brain [287]. Microglia are migratory cells in the nervous system that play a role in the brain's defense against disease. In addition, activated microglia influence neuronal activity, and in turn, neuronal activity influences microglial mobility [287]. The behavior of the microglia has been compared to that of a vigilant police force [643, 707]. In the absence of trouble, most microglia remain stationary ("on post"), carefully sensing their environment with their processes; however, $\approx 3\%$ move on patrol. When appropriately activated by, for example, an infection or by the spreading depression that accompanies migraine, the number of moving microglia increases dramatically. Presumably, this increase in moving microglial cells allows them to find sites where they are needed and congregate there.

Figure 16.11a, b shows the movements of microglia in a cultured rat hippocampal slice in the absence of an activating stimulus. The movements are characterized

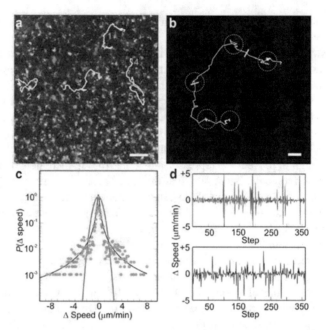

Fig. 16.11 (a) Control motion picture frame with white lines showing paths of microglia (1)–(5) that moved $\geq 50\,\mu m$. (b) Cell 1 from frame (a) with path enlarged to show movement beginning at upper right and ending at lower left; the yellow bars mark 2 and 4 h. Orange dashed lines encircle regions with multiple small steps compared to occasional long steps, suggestive of a Lévy flight pattern. (c) Control microglial cell ΔV showed a Lévy power-law distribution with exponent $\alpha = 0.9$ and scale factor $\gamma = 0.3$ (black line). The red line shows a Gaussian distribution ($\alpha = 2$; variance $= 0.5$). (d) Realization of a Lévy flight constructed using the distribution (c) determined from the full population of moving microglia (top) qualitatively reproduced the same intermittent behavior of an exemplary long-distance moving microglial cell (bottom). The calibration (white) bars in the lower right-hand corner of (a) and (b) correspond to, respectively, 50 and 10 μm. Figure reproduced from [287] with permission.

by long intervals during which the cells move very slowly interspersed with brief intervals during which they move more quickly. Since the microglia in this experiment are moving within a thin slice of hippocampus, we can model the movements as a two-dimensional random walk as follows [95]. The distance d_j traveled by the jth microglium in the (x, y)-plane is

$$d_j(t) = \sqrt{(x_j(t + \Delta t) - x_j(t))^2 + (y_j(t + \Delta t) - y_j(t))^2},$$

where $\Delta t = 1$ min for microglial movement. The speed $V_j(t_j)$ is

$$V_j(t_j) = \frac{d_j(t)}{\Delta t},$$

and the change in speed ΔV is

$$\Delta V = V_j(t_i + \Delta t) - V_j(t_i).$$

The quantity ΔV gives the scalar acceleration, since the direction of the movement is not considered. By analogy with the numerical calculation of a two-dimensional random walk, ΔV provides an estimate of the step length and direction drawn from the underlying pdf.

Figure 16.11c compares the distribution $P(\Delta V)$ of ΔV obtained by pooling the movements of a large number of migratory microglia to that predicted by a Gaussian pdf (red line). Clearly, the microglia can make ΔV that are larger than predicted from a Gaussian pdf. In other words, the pdf for $P(\Delta V)$ for microglial movements has "broad shoulders." The relationship between Lévy flights and distributions with broad shoulders can be seen as follows: A Lévy flight is a random walk for which the ΔV are chosen from a power-law distribution

$$P(\Delta V > s, \Delta t) \approx s^{-\alpha}, \tag{16.39}$$

where α is the Lévy index. After many steps, the probability distribution for a Lévy flight converges to the Lévy-stable distribution

$$L_\alpha(\Delta V, \Delta t) = \frac{1}{\pi} \int_{-\infty}^{\infty} \exp\left(-\gamma \Delta t q^\alpha\right) \cos(q \Delta V) dq, \tag{16.40}$$

where γ is a scaling factor. This observation suggests that the distribution $P(\Delta V)$ has "broad shoulders," which is what is observed. The choice of α that most closely fits the $P(\Delta V)$ for microglial cells is $\alpha \approx 1$. It is thought that Lévy flights with $\alpha \approx 1$ are optimal for a random search pattern [855, 856].

As better measuring devices and larger datasets have become available, scientists have recognized that Lévy distributions are prevalent in nature [477]. However, it is insufficient simply to demonstrate that experimental data fit a Lévy distribution to conclude that the biological process under consideration exhibits a Lévy flight. Since all datasets are of finite length, it is possible that other statistical distributions may in fact fit the data better. In our opinion, these observations simply lead to confusion: compare, for example, [187, 855]. The conclusion that a Lévy flight exists becomes more convincing when there is an appropriate model that predicts that a Lévy flight should be present. Examples in which both a model and experimental observation support the existence of a Lévy flight include stick balancing at the fingertip [95, 137], fish foraging [357], and cell movement [287, 716, 717].

16.8 Fokker–Planck equation

The Langevin approach to diffusion discussed in Chapter 15 assumes that the effects of noise are described by a continuous random variable $\xi(t)$. In other words, the movements of the Brownian particle are the result of a continuous bombardment by solvent molecules. The time evolution of $P(x,t)$ can, at least in principle, be obtained by solving the partial differential equation

$$\frac{\partial}{\partial t}P(x,t) = \frac{1}{t_r}\frac{\partial}{\partial x}(xP(x,t)) + D\frac{\partial^2}{\partial x^2}P(x,t),$$

where D is the diffusion coefficient. This partial differential equation is called the *Fokker–Planck equation* [722]. Obtaining the solution to this equation is not necessary for the discussion that follows. It is sufficient that the reader recognize that the goal of the Fokker–Planck equation is to predict the time evolution of $P(x,t)$.

In a very lucid manner, van Kampen[10] [413] showed that a diffusion process can always be approximated by a discrete, or jump, process in which the bombardments, namely the changes in x, occur at discrete time intervals. A process involving discrete jumps is the mental picture that underlies the derivation of the simple random walk [190]. In the case of a random walk with continuous time, the time evolution of $P(x,t)$ is described by a *master equation*.

16.8.1 Master Equation

The philosophy behind the derivation of a master equation is similar in spirit to the application of the law of mass action (Tool 5) and the development of production–destruction models (Section 9.4) for deterministic dynamical systems. The essential difference is that the variables are replaced by the probability of their occurrences. Thus, the derivation of a master equation involves identifying all the possible pathways that could have formed or destroyed a given state.

There are two advantages to the use of master equations for studying the evolution of $P(x,t)$. First, there is no need to distinguish between the effects of additive and parametric noise. Second, it avoids the necessity of evoking fluctuation dissipation arguments to describe the interplay between the deterministic motions and the random fluctuations. Under appropriate conditions, a Fokker–Planck equation can be formulated that is equivalent to that of the master equation [257].

The master equation can be used to describe stochastic processes that are continuous in time but for which the position x takes on, as for the simple random walk, only discrete states, labeled by integers $N = \ldots, -3, -2, -1, 0, 1, 2, 3, \ldots$. The master equation describes the rate of change of $P(n,t)$ as a function of time in terms of a balance of probability flow

$$\frac{\partial}{\partial t}P(n,t) = \sum_m k_{nm}P(m,t) - \sum_m k_{mn}P(n,t), \tag{16.41}$$

where k_{nm} is the rate of probability flow from state m to n, and k_{mn} is the rate of probability flow from state n to m. Thus, the first term can be interpreted as the summation of all the inflow of probability into state n from other states, and the

---

[10] Nicolaas Godfried van Kampen (1921–2013), Dutch theoretical physicist.

second term is all the outflow of probability from state n to other states. Naturally, the difference between these probability flows is the amount of change in $P(n,t)$ on the left-hand side of the equation.[11]

The most commonly used computer algorithm to simulate the solution of a master equation is the Gillespie[12] algorithm [269]. A number of implementations are available on the Internet. However, in order to understand how the master equation can be extended into cases of anomalous diffusion, it is useful to see how mathematicians solve these equations. A tool for solving master equations such as (16.41) is the *generating function*.

16.9 Tool 12: Generating Functions

A generating function is a mathematical expression that represents the function $F(z,t)$ as the product of a power series expansion z^n and another function, in this case $P(n,t)$ (for a nicely written introduction to generating functions and random walks, see [733]). The series expansion of e^x that we discussed in Section 7.2.1 can be thought of as an example of a generating function

$$e^x = 1 + \frac{x}{1!} + \frac{x^2}{2!} + \frac{x^3}{3!} + \cdots + \frac{x^n}{n!} + \cdots = \sum_{n=0}^{\infty} x^n \frac{1}{n!}.$$

For discrete stochastic processes, the generating function takes the form

$$F(z,t) = \sum_{n=-\infty}^{\infty} z^n P(n,t), \qquad (16.42)$$

where $F(0,t) = 1$. These generating functions are useful because we can readily calculate the moments by differentiating them. For example

$$\frac{\partial}{\partial z}F(z,t)|_{z=1} = \langle n(t) \rangle,$$

$$\frac{\partial^2}{\partial z^2}F(z,t)|_{z=1} = \langle n(n-1)(t) \rangle = \langle n^2(t) \rangle - \langle n(t) \rangle.$$

There is a connection between the generating function defined by (16.42) and the characteristic function $\varphi(t)$ introduced in Section 14.2.3:

$$\varphi(t) = \int_{-\infty}^{\infty} e^{-jtx} p(x)\,dx.$$

---

[11] We can consider the master equation for a continuous state variable. In that case, the sum is replaced by an integral. If we further expand this continuous-time and continuous-state master equation with small changes of state in short time intervals, it leads to the Fokker–Planck equation.

[12] Daniel Thomas Gillespie (b. 1938), American physicist.

If we define $z = -jt$ and make use of the series expansion of e^{zx}, we obtain

$$\varphi(t) = \int_{-\infty}^{\infty} p(x)\,dx + z\int_{-\infty}^{\infty} xp(x)\,dx + \frac{z^2}{2}\int_{-\infty}^{\infty} x^2 p(x)\,dx + \cdots$$
$$= \sum_{k=0}^{\infty} \frac{z^k}{k!} p_k,$$

where p_k is the kth moment of $p(x)$. Hence

$$\varphi(t) = \sum_{k=0}^{\infty} \frac{(-jt)^k}{k!} p_k.$$

The following examples illustrate the use of generating functions to solve master equations.

Example 16.3 (Protein synthesis) To illustrate how a master equation is derived, we consider a simple model for protein synthesis

$$\xrightarrow{\ c\ } \text{protein}(n) \xrightarrow{\ d(n)\ } .$$ (16.43)

We define the state of the system at time t as the number n of molecules of protein that are present. It should be noted that the creation rate c does not depend on the number of protein molecules, since proteins are synthesized from other types of molecules. However, the death rate $d(n)$ does depend on the number of protein molecules. Models of this type are called *birth and death processes*, since the protein can exist in only two states: it is either synthesized (born) or destroyed (dies). Our goal is to determine the probability that there are n molecules of protein at a given time and then determine how this probability changes as a function of time.

   In a birth and death model, the reaction rates are interpreted in terms of the probability per unit time of the relevant reaction. Let $P(n,t)$ be the probability that a process is in state n at time t. Next, we assume that changes between states are allowed only between neighboring states, i.e., $n \to n \pm 1$, and that they occur in a short time δt. This is the assumption of a jump, or one-step process. Thus, we can write

$$P(n, t + \delta t) = P(\text{no reaction from } n)(t) + P(\text{birth from } n - 1)(t)$$
$$+ P(\text{death from } n + 1)(t) - P(\text{death from } n)(t)$$
$$- P(\text{birth from } n)(t).$$

This equation can be interpreted in the context of a type of production–destruction equation. There are three ways in which a state consisting of n protein molecules can be created at time $t + \delta t$:

1. $P(\text{no reaction from } n)(t)$ indicates that at time t, there were n protein molecules, and no birth or death occurred in the time interval $t + \delta t$.

2. $P(\text{birth from } n-1)(t)$ indicates that at time t, there were $n-1$ protein molecules, and one protein was created in the time interval $t + \delta t$.
3. $P(\text{death from } n+1)(t)$ indicates that at time t, there were $n+1$ protein molecules, and one molecule was destroyed in the time interval $t + \delta t$.

There are two ways in which the state of n protein molecules can be destroyed:

1. $P(\text{death from } n)(t)$ indicates that at time t, there were n protein molecules, and one molecule was destroyed to yield a state with $n-1$ protein molecules at time $t + \delta t$.
2. $P(\text{birth from } n)(t)$ indicates that at time t, there were n protein molecules, and one molecule was created to yield a state with $n+1$ protein molecules at time $t + \delta t$.

In terms of the master equation formulation, we obtain

$$\frac{\partial}{\partial t}P(n,t) = \{cP(n-1,t)+d(n+1)P(n+1,t)\} - \{cP(n,t)+d(n)P(n,t)\}, \quad (16.44)$$

where $d(n)$ is the death rate when there are n proteins and $d(n+1)$ is the death rate when there are $n+1$ proteins. Since $n \geq 0$, we also have

$$\frac{\partial}{\partial t}P(0,t) = dP(1,t) - cP(0,t). \quad (16.45)$$

By multiplying the master equation (16.44) by z^n and taking the sum over n, we can obtain an equation in terms of $F(z,t)$

$$\frac{\partial}{\partial t}F(z,t) = \left(czF(z,t)+d\frac{\partial}{\partial z}F(z,t)\right) - \left(cF(z,t)+dz\frac{\partial}{\partial z}F(z,t)\right)$$

$$= (z-1)\left(cF(z,t)-d\frac{\partial}{\partial z}F(z,t)\right).$$

We shall not go into the problem of solving this equation in general, but we shall consider the case of a steady-state distribution, in which the left-hand side of the equation is 0, since there is no change over time. Then

$$\frac{\partial}{\partial z}F(z) = \frac{c}{d}F(z), \quad (16.46)$$

where we have dropped the t from the notation. Together with the property that $F(1) = 1$, we can solve this equation

$$F(z) = e^{c/d(z-1)}. \quad (16.47)$$

We can obtain a stationary probability from the above result. By expanding $F(z,t)$ over z, we obtain

$$F(z) = e^{-c/d}\sum_{n=0}^{\infty}\frac{(c/d)^n}{n!}z^n. \quad (16.48)$$

Then the stationary probability of having n proteins in the coefficient of z^n is given as

$$P(n) = e^{-c/d}\frac{(c/d)^n}{n!},\tag{16.49}$$

which is a *Poisson*[13] *distribution* with both mean and variance (c/d). (This can be verified by taking the derivative of the generating function as described above.) ◇

Example 16.4 (Simple random walk). Another example is a simple symmetric random walk in continuous time. In this case, the rate of transition can be set as $k_{nm} = C\delta_{n,m\pm1}$, where $\delta_{n,m}$ is the Kronecker delta function, defined in (8.3), and the master equation is

$$\frac{\partial}{\partial t}P(n,t) = C\{P(n-1,t)+P(n+1,t)\}-2CP(n,t).\tag{16.50}$$

Let us solve this equation using the generating function approach for $P(n,t)$ given by (16.42) with $C = 1/2$ and with the initial condition of starting from the origin. Multiplying the master equation by z^n and summing over n, we obtain

$$\frac{\partial}{\partial t}F(z,t) = \frac{1}{2}\left(z+\frac{1}{z}-2\right)F(z,t),\tag{16.51}$$

which can be solved as

$$F(z,t) = \Omega(z)\exp^{\frac{1}{2}(z+\frac{1}{z}-2)t},\tag{16.52}$$

where $\Omega(s)$ is a function of z determined by a boundary condition. Since the initial condition starts at the origin, we have $F(z,0) = 1$, and hence $\Omega(z) = 1$. Consequently,

$$F(z,t) = \exp^{\frac{1}{2}(z+\frac{1}{z}-2)t}.\tag{16.53}$$

If we expand over z, we obtain

$$F(z,t) = e^{-2t}\sum_{k,l=0}^{\infty}\frac{t^{k+l}}{k!l!}z^{k-l}.\tag{16.54}$$

Hence, setting $n = k - l$ for $-\infty < n < \infty$, we obtain

$$P(n,t) = e^{-2t}\sum_{l}\frac{t^{2l+n}}{(l+n)!l!}.\tag{16.55}$$

Note that the sum is over l, so that both l and $n+l$ are nonnegative. Though this result does not look appealing, it is of Gaussian form, as before. One can verify this by computing the mean, variance, and other moments directly from the moment-generating function [869].

---

[13] Siméon Denis Poisson (1781–1840), French mathematician and physicist.

As in this case, we can think of stochastic processes with different combinations of continuity and discreteness in time and space. There are subtle differences of which one should be aware. However, from the physical point of view, they should correspond to each other within appropriate limits, and qualitatively, the same behaviors are expected.

The master equation approach is useful in situations in which we have information on transition rates, which is typical in the case of chemical reactions. We do not have to worry about the details or types of noise as in the Langevin or other stochastic differential equations. On the other hand, the master equation is a dynamical equation of a pdf in probability space and hence is a bit more abstract than a physical dynamical equation. ◇

16.9.1 Approaches Using the Characteristic Function

The generating function given by (16.42) can be considered a special case of the characteristic function. Even more powerful methods for solving the master equation are available based on the characteristic function [733, 869]. In order to appreciate how these methods work, we need to reinterpret the meaning of the convolution integral. This reinterpretation is illustrated by the following example.

Example 16.5 Suppose a dieter is asked to restrict his caloric intake to 600 calories per meal and to design their meals from a menu by selecting one carbohydrate portion and one fat/protein portion (in addition to a large portion of vegetables, which we shall assume in this example to contribute zero calories). However, since there are so many choices, the dieter decides to choose the two portions randomly. What is the probability that the total meal composed in this manner meets the requirement of 600 calories?

Solution. Let $P_{carb}(X)$ describe the distribution of calories for the carbohydrate portions, and $P_{fat}(X)$ the distribution of calories for the fat/protein portions. The dieter first chooses the carbohydrate portion, which has caloric value X with probability $P_{carb}(X)$. This means that in order for the total caloric value to equal 600 calories, the fat portion must have caloric value $600 - X$. The frequency of occurrence of meals that have a total of 600 calories is the product of the occurrence of X in the carbohydrate choices and $600 - X$ in the fat choices integrated over all possible choices of X, namely

$$P(600) = \int_0^{600} P_{carb}(X) P_{fat}(600 - X) \, dX. \tag{16.56}$$

◇

Following along the lines of this example, we can describe a random walk using the equation

$$p(x, t + t') = \int_{-\infty}^{\infty} m(\ell, t; t') p(x - \ell, t) \, d\ell, \tag{16.57}$$

where $m(\ell,t;t')$ is the transition probability of taking a step length of ℓ at time t to $t+t'$. This equation states that in order to be at position x at time $t+t'$, the random walker was at position $x-\ell$ at time t and then made a step of the correct length ℓ and direction to reach the position $x(t+t')$. The integral is over all possible starting positions.

Equation (16.57) describes the evolution of $p(x,t)$ as a function of time. This equation does not predict how a single realization of a random walk evolves. Instead, it summarizes the results obtained from a very large number of realizations in a statistical way. Equation 16.57 draws attention to the use of the Fourier transform, and in particular, the characteristic function for the study of random walks [733, 869].

Let us consider the special case $t'=1$. For this choice of t', we have a random walk with discrete time steps, and the transition probability is independent of time, i.e., $m(\ell,t;t')=m(\ell)$. Since the right-hand side of (16.57) is a convolution integral, we can take the Fourier transform with respect to the position x to obtain

$$\hat{p}_{t+1}(f) = \varphi_e(f)\hat{p}_t(f),$$

where

$$\hat{p}_t(f) = \int_{-\infty}^{\infty} p(x,t)e^{-j2\pi fx}dx, \quad \varphi_e(f) = \int_{-\infty}^{\infty} m(x)e^{-j2\pi fx}dx.$$

We note that $\varphi_e(f)$ is a characteristic function for each elementary transition step probability $m(x)$. Now we can work backward in time to construct the ensemble of random walks at time $t>0$ that were initially at $x=0$

$$\hat{p}_t(f) = \varphi_e(f)\hat{p}_{t-1}(f) = \varphi_e^2(f)\hat{p}_{t-2}(f) = \varphi_e^3(f)\hat{p}_{t-3}(f) = \cdots$$
$$= \varphi_e^t(f)\hat{p}_0(f).$$

If we take the initial condition as $p(x,0) = \delta(x)$ (in other words, all walkers initially start from the origin), we have $\hat{p}_0(f) = 1$. Hence, we obtain

$$p(x,t) = \int_{-\infty}^{\infty} e^{j2\pi fx}\varphi(f)df, \tag{16.58}$$

where we have identified $\varphi(f) := \varphi_e^t(f)$.

Example 16.6 Let us consider our simple random walk with a unit discrete step in the above formulation. Let us take, for simplicity, the case of a symmetric random walk $p=q=1/2$ and have the walker start from the origin. Because the step allowed for the walker is ± 1, we can see that only two terms in the Fourier integral are nonzero, and the characteristic function for each step is

$$\varphi_e(f) = \left(\frac{1}{2}e^{j2\pi f} + \frac{1}{2}e^{-j2\pi f}\right) = \cos(2\pi f).$$

Hence, we have

$$p(x,t) = \int_{-1/2}^{1/2} e^{j2\pi fx} \cos^t(2\pi f) \, df. \tag{16.59}$$

Note the change in the limits of integration. This is due to the fact that we are now considering steps of discrete length [869].

This integration can be performed to give

$$p(x,t) = \sum_{s=0}^{t} \binom{t}{s} \left(\frac{1}{2}\right)^t \delta_{(t+x),2s}. \tag{16.60}$$

It should be noted that going from (16.59) to (16.60) requires quite a few algebraic manipulations, which include the following:

1. Integral representation of the Kronecker delta function δ_{x,x_0},

$$\delta_{x,x_0} = \int_{-1/2}^{1/2} e^{j2\pi f(x-x_0)} \, df. \tag{16.61}$$

2. The binomial theorem,

$$(a+b)^n = \sum_{k=0}^{n} \binom{n}{k} a^k b^{n-k},$$

where

$$\binom{n}{k} = \binom{n}{n-k} \equiv \frac{n!}{k!(n-k)!}.$$

Finally, (16.60) can be rewritten as

$$P(x,t) = \begin{cases} \binom{t}{(t+x)/2} \left(\frac{1}{2}\right)^t, & t+x = 2m, \\ 0, & t+x \neq 2m, \end{cases} \tag{16.62}$$

where m is a nonnegative integer. ◇

The usefulness of the characteristic function approach now becomes apparent. As we showed in Section 14.2.3, the moments of $p(x,t)$ can be determined by differentiating its Fourier transform and evaluating it at $f = 0$. Since $p(x,t)$ and $\varphi(f)$ are Fourier transform pairs, we can also obtain the moments from $\varphi(f)$. In fact, it is often much easier to work with $\varphi(f)$. In the case of this simple random walk, we have, after a large number n of steps

$$\langle x \rangle = \frac{-1}{2\pi j} \frac{d}{df} \varphi(f)\Big|_{f=0} = \frac{1}{2\pi j} \frac{d}{df} (\cos 2\pi f)^n\Big|_{f=0}$$

$$= \frac{n}{j} \sin(2\pi f) \cos^{n-1}(2\pi f)\Big|_{f=0} = 0$$

and

$$\langle x^2 \rangle = \frac{-1}{4\pi^2} \frac{d^2}{df^2} \varphi(f)\Big|_{f=0} = \frac{n\cos^{n-2}(2\pi f)}{2}[n\cos(4\pi f) - n + 2]\Big|_{f=0} = n.$$

Equation (16.58) is the exact representation for $p(x,t)$, and hence it is the most important relationship for the study of random walks. This equation can also be derived in other ways; these alternative approaches provide additional insight into the analysis of random walks. Below, we briefly discuss two alternative approaches to the derivation of (16.58).

The first approach relies on the representation of $p(x,t)$ in terms of a multiple integral. Suppose that a random walker starts at position x_1. After t steps, the walker will be at

$$x = x_1 + x_2 + x_3 + \cdots + x_t. \tag{16.63}$$

If the positions of the random walker can take on any value from $-\infty$ to $+\infty$ with the same pdf $p(x)$, then

$$p(x,t) = \int_{-\infty}^{\infty} \cdots \int_{-\infty}^{\infty} p(x_t)\cdots p(x_2)p(x_1)\delta\left(\sum_{s=1}^{t} x_s - x\right) dx_1 \cdots dx_t, \tag{16.64}$$

where the Dirac delta function is required to enforce the condition given by (16.63). We note that in writing the product of the probabilities in the integrand, we are again appealing to the independence of each step. Taking into consideration the integral representation of the Dirac delta function,

$$\delta(x - x_0) = \int_{-\infty}^{\infty} e^{j2\pi f(x-x_0)} df, \tag{16.65}$$

we have

$$p(x,t) = \int_{-\infty}^{\infty} e^{j2\pi fx} df \int_{-\infty}^{\infty} p(x_1)e^{-j2\pi fx_1} dx_1 \cdots \int_{-\infty}^{\infty} p(x_t)e^{-j2\pi fx_t} dx_t$$

$$= \int_{-\infty}^{\infty} e^{j2\pi fx}\varphi(f)df,$$

which is the same as (16.58).

The second approach makes use of the concept of moments that we used to compute $\langle x \rangle$ and $\langle x^2 \rangle$ for the simple random walk. The expression

$$\varphi(f) = \int_{-\infty}^{\infty} p(x,t)e^{-j2\pi fx} dx$$

can also be viewed as the average of the function $e^{-j2\pi fx}$, namely

$$\langle e^{-j2\pi fx} \rangle = \int_{-\infty}^{\infty} p(x,t)e^{-j2\pi fx} dx.$$

Hence

$$\varphi(f) = \left\langle e^{-j2\pi fx} \right\rangle .$$

If we decompose x in the above equation according to the independence of each step, we can appeal to the fact that the average of independent stochastic variables can be factored into products. Hence, by assuming that all the steps have the same pdf, we obtain

$$\varphi(f) = \left\langle e^{-j2\pi fx} \right\rangle = \left\langle e^{-j2\pi fx_1} \right\rangle \left\langle e^{-j2\pi fx_2} \right\rangle \cdots \left\langle e^{-j2\pi fx_t} \right\rangle = \varphi_e^t(f), \qquad (16.66)$$

which again agrees with our previous results.

The above discussion emphasizes the importance of the characteristic function for the analysis of random walks. In particular, we can extend these approaches even if $\varphi(f)$ is different for each step of the random walk. It is not necessary to assume that the individual steps of the random walk are identically distributed variables. The key assumption is the statistical independence of each of the walker's steps, as shown in the following example.

Example 16.7 In one dimension, the characteristic function for a Gaussian pdf with mean zero is [213]

$$\varphi(f) = \exp\left(\frac{-(2\pi\sigma f)^2}{2}\right),$$

where σ^2 is the variance. Suppose that at each step, the random walker obtains the step length from a Gaussian pdf with variance σ_i^2. What is the effect on $p(x,t)$?

Solution. From (16.66), we see that

$$\varphi(f) = e^{-(2\sigma_1\pi f)^2/2}e^{-(2\sigma_2\pi f)^2/2}\cdots e^{-(2\sigma_t\pi f)^2/2} = e^{-(2\sigma_T\pi f)^2/2},$$

where

$$\sigma_T^2 = \sigma_1^2 + \sigma_2^2 + \cdots + \sigma_t^2.$$

In other words, $p(x,t)$ is also a Gaussian pdf, but the variance of the random walk becomes the sum of the variances of each step, which is a consequence of the independence of steps.   ◇

Since the characteristic function for Lévy-stable distributions is known (Section 14.4.5), we can, at least in principle, apply the methods we have discussed here to the analysis of Lévy flights. However, as might be expected, the mathematical analysis is much more difficult (the interested reader is referred to [869]).

16.10 Stable random walks

The major limitation for applying simple random walk models to questions related to feedback is that they lack the notion of *stability*, i.e., the resistance of dynamical systems to the effects of perturbations. The urn model developed by Paul and

Tatyana Ehrenfest showed how a quadratic potential could be incorporated into a discrete random walk [188, 399, 416].

16.10.1 Ehrenfest urn model

The "2-urn model" was initially developed by Paul and Tatyana Ehrenfest[14] to obtain insights into the second law of thermodynamics. Consider two urns, A and B. There are a total of N balls and we label the N balls, $1, \ldots, N$ in order to keep track of them. The rule is that at each step we pick a ball and move it to the other urn. Assume at step 0 there are N balls in urn A and 0 balls in urn B. For the first step it is obvious that the picked ball moves $A \to B$, since there are no balls in urn B. For the second step we have that the probability that the ball goes $A \to B$ is $(N-1)/N$ and from $B \to A$ is $1/N$. After step 3 the probabilities are, respectively, $(N-2)/N$ and $2/N$. The probability that there are 3 balls in urn B after 3 steps is $(N-1)(N-2)/N^2$. And so on.

When N is very large it is easy to see that the probability of moves in the $A \to B$ direction are much, much higher than moves in the $B \to A$ direction. Thus, initially, almost all of the ball movements occur in the $A \to B$ direction. However, as the number of balls in urn A approaches $N/2$, the probability of moving a ball in either direction becomes about the same. The key insight was to realize that the dynamics of the urn model and a random walk resemble each other if we identify the excess of balls in urn A over $N/2$ with the displacement of the walker in a discrete random walk [399]. Indeed, it is possible to show that the Fokker–Planck equation obtained for the Ehrenfest model is exactly the same as that obtained for the Langevin equation [399].

After a moment's reflection, we can imagine that the greater the number of balls in a given urn over $N/2$, the more likely that the next step transitions a ball to the other urn. This situation resembles a system evolving in a quadratic potential that we discussed in Section 4.1. By analogy we can incorporate the influence of a quadratic–shaped potential on a random walker by assuming that the transition probability towards the origin increases linearly with distance from the origin (of course up to a point) [651, 654]. In particular, the transition probability for the walker to move toward the origin increases linearly at a rate of β as the distance increases from the origin up to the position $\pm a$ beyond which it is constant (since the transition probability is between 0 and 1). Equation (16.17) becomes

$$P(r,0) = \delta_{r,0} \tag{16.67}$$
$$P(r,n) = g(r-1)P(r-1,n-1)$$
$$\qquad\qquad + f(r+1)P(r+1,n-1)$$

---

[14] Paul Ehrenfest (1880–1933) was an Austrian-Dutch theoretical physicist and his daughter Tatyana (1905–1984) was a Dutch mathematician.

where a and d are positive parameters, $\beta = 2d/a$, and

$$f(x) = \begin{cases} \frac{1+2d}{2} & x > a \\ \frac{1+\beta x}{2} & -a \le x \le a \\ \frac{1-2d}{2} & x < -a \end{cases}$$

$$g(x) = \begin{cases} \frac{1-2d}{2} & x > a \\ \frac{1-\beta x}{2} & -a \le x \le a \\ \frac{1+2d}{2} & x < -a \end{cases}$$

where $f(x), g(x)$ are, respectively, the transition probabilities to take a step in the negative and positive directions at position x such that

$$f(x) + g(x) = 1.$$

The random walk is symmetric with respect to the origin provided that

$$f(-x) = g(x) \quad \text{for all } x.$$

We classify random walks by their tendency toward move towards the origin. The random walk is said to be *attractive* when

$$f(x) > g(x) \quad (x > 0).$$

and *repulsive* when

$$f(x) < g(x) \quad (x > 0).$$

We note that when

$$f(x) = g(x) = \frac{1}{2} \quad \text{for all } x,$$

the general random walk given by (16.68) reduces to the simple random walk discussed Section 16.1. Here we consider only the attractive case.

16.10.2 Auto–Correlation function: Ehrenfest random walk

Assume that with sufficiently large a, we can ignore the probability that the walker is outside of the range (-a, a). In this case, the probability distribution function $P(r,n)$ approximately satisfies the equation

$$P(r,n) = \frac{1}{2}(1 - \beta(r-1))P(r-1,n-1) \tag{16.68}$$
$$+ \frac{1}{2}(1 + \beta(r+1))P(r+1,n-1).$$

By symmetry, we have

$$P(r,n) = P(-r,n)$$
$$\langle X(t) \rangle = 0$$

The variance, obtained by multiplying (16.68) by r^2 and summing over all r, is

$$\sigma^2(n) = \langle X^2(n) \rangle = \frac{1}{2\beta}(1 - (1 - 2\beta)^n) \tag{16.69}$$

Thus, the variance in the stationary state is

$$\sigma_s^2 = \langle X^2 \rangle_s = \frac{1}{2\beta}. \tag{16.70}$$

The autocorrelation function for this random walk can be obtained by rewriting (16.68) in terms of joint probabilities to obtain

$$P(r,n;m,n-\Delta) = \frac{1}{2}(1 - \beta(r-1))P(r-1,n-1;m,n-\Delta) \tag{16.71}$$
$$+ \frac{1}{2}(1 + \beta(r+1))P(r+1,n-1;m,n-\Delta).$$

By defining

$$P_s(r;m,\Delta) \equiv \lim_{n \to \infty} P(r,n;m,n-\Delta)$$

the joint probability obtained in the stationary state, i.e., the long time limit, becomes

$$P_s(r;m,\Delta) = \frac{1}{2}(1 - \beta(r-1))P_s(r-1;m,\Delta-1) \tag{16.72}$$
$$+ \frac{1}{2}(1 + \beta(r+1))P_s(r+1;m,\Delta-1).$$

Multiplying by rm and summing over all r and m yields

$$C(\Delta) = (1 - \beta)C(\Delta - 1) \tag{16.73}$$

which we can rewrite as

$$C(\Delta) = (1 - \beta)^\Delta C(0) = \frac{1}{2\beta}(1 - \beta)^\Delta \tag{16.74}$$

where $C(0)$ is equal to the mean-square displacement, $\langle X^2 \rangle_s$. When $\beta << 1$, we can approximate,

$$C(\Delta) \approx \frac{1}{2\beta}e^{-\beta|\Delta|}. \tag{16.75}$$

Then, from the Wiener–Khinchine theorem, we have

$$W(f) \approx \frac{2}{\beta^2}\left[\frac{1}{1 + (2\pi f/\beta)^2}\right] \tag{16.76}$$

These expressions for $C(\Delta)$ and $W(f)$ are in the same form as those obtained for a random process, i.e., (16.30) and (16.33). Hence, $C(\Delta)$ and $W(f)$ are qualitatively the same as those shown in Figure 16.6.

16.10.3 Delayed Random Walks

A delayed random walk is defined as a random walk in which the transition probabilities depend on the position of the walker at some time τ in the past [651, 652, 654]. This means that the probability that a walker is at position x depends on a joint probability that takes into account where the walker was τ steps earlier. If we generalize (16.68) to include a time delay, we obtain

$$P(r,n+1) = \sum_m g(m)P(r-1,n;m,n-\tau)$$
$$+ \sum_m f(m)P(r+1,n;m,n-\tau), \qquad (16.77)$$

where the position of the walker at time n is $X(n)$, $P(r,n)$ is the joint probability for the walker to be at $X(n) = r$ and $P(r,n;m,n-\tau)$ is the joint probability such that $X(n) = r$ and $X(n-\tau) = m$ takes place. $f(x)$ and $g(x)$ are transition probabilities for the walker to take the step to the negative (-1) and positive $(+1)$ directions, respectively, and are the same as those used for the random walk on a quadratic potential described in the previous section. Examples in which DRWs have been used to gain insight into delayed stochastic dynamics include eye movements [557], Neolithic transitions [234], postural sway (see below), stock market fluctuations [285, 653], and estimating the pdf for stochastic delay differential equations [60].

Three properties delayed random walks are particularly important for the discussion that follows. First, by the symmetry with respect to the origin, we have that the average position of the walker is 0. In particular, for an attractive delayed random walk, the stationary state (i.e., when $n \to \infty$)

$$P(r,n+1;r+1,n) = P(r+1,n+1;r,n). \qquad (16.78)$$

We can show the above as follows: By the definition of the stationarity, we have

$$P(r,n+1;r+1,n) + P(r,n+1;r-1,n) = \qquad (16.79)$$
$$P(r+1,n+1;r,n) + P(r-1,n+1;r,n). \qquad (16.80)$$

For $r = 0$, we note that due to the symmetry, we have

$$P(0,n+1;1,n) = P(1,n+1;0,n). \qquad (16.81)$$

Using these two equations inductively leads us to the desired relation (16.78).

Second, the generating function

$$\langle \cos(\alpha X(n)) \rangle = \tag{16.82}$$
$$\cos(\alpha)\langle\cos(\alpha X(n))\rangle + \sin(\alpha)\langle\sin(\alpha X(n))\{f(X(n-\tau)) - g(X(n-\tau))\}\rangle$$

can be obtained by multiplying Eq. (16.77) for the stationary state by $\cos(\alpha r)$ and then summing over r and m.

Finally, we have the following invariant relationship with respect to the delay

$$\frac{1}{2} = \langle X(n)\{f(X(n-\tau)) - g(X(n-\tau))\}\rangle \tag{16.83}$$

When we choose f, g as before, this invariant relation in Eq. (16.83) becomes

$$\langle X(n+\tau)X(n) \rangle = C(\tau) = \frac{1}{2\beta}. \tag{16.84}$$

This invariance with respect to τ of the correlation function with τ steps apart is a simple characteristic of this quadratic potential delayed random walk model. This property is a key to obtaining the analytical expression for the correlation function. Below we discuss $C(\Delta)$ for the stationary state.

For the stationary state and $0 \le \Delta \le \tau$, the following is obtained from the definition (16.77):

$$P_s(r, n+\Delta; m, n) = \sum_{\ell} g(\ell) P_s(r-1, n+\Delta; m, n+1; \ell, n+\Delta-\tau)$$
$$+ \sum_{\ell} f(\ell) P_s(r+1, n+\Delta; m, n+1; \ell, n+\Delta-\tau)$$

$$\tag{16.85}$$

We can derive the following equation for the correlation function by multiplication of this equation by rm and summing over.

$$C(\Delta) = C(\Delta-1) - \beta C(\tau+1-\Delta), \qquad (0 \le \Delta \le \tau). \tag{16.86}$$

A similar argument can be given for $\tau < \Delta$,

$$C(\Delta) = C(\Delta-1) - \beta C(\Delta-1-\tau), \qquad (\tau < \Delta). \tag{16.87}$$

Equations (16.86) and (16.87) can be solved explicitly using (16.84). In particular, for $0 \le \Delta \le \tau$, we obtain

$$C(\Delta) = C(0)\frac{(z_+^\Delta - z_+^{\Delta-1}) - (z_-^\Delta - z_-^{\Delta-1})}{z_+ - z_-} - \frac{1}{2}\frac{(z_+^\Delta - z_-^\Delta)}{z_+ - z_-}$$
$$C(0) = \frac{1}{2\beta}\frac{(z_+ - z_-) + \beta(z_+^\tau - z_-^\tau)}{(z_+^\tau - z_+^{\tau-1}) - (z_-^\tau - z_-^{\tau-1})}$$

$$\tag{16.88}$$

where

$$z_\pm = (1 - \frac{\beta^2}{2}) \pm \frac{\beta}{2}\sqrt{\beta^2 - 4}.$$

For $\tau < \Delta$, it is possible to write $C(\Delta)$ in a multiple summation form, though the expression becomes rather complex. For example, with $\tau < \Delta \le 2\tau$,

$$C(\Delta) = \frac{1}{2\beta} - \beta \sum_{i=1}^{\Delta-\tau} C(i) \qquad (16.89)$$

where the $C(i)$ are given by (16.88).

Figure 16.12 compares $C(\Delta)$ and $W(f)$ for different values of the delay. As we increase τ, oscillatory behavior of the correlation function appears. The decay of the peak envelope is found numerically to be exponential. The decay rate of the envelope for the small u is approximately $1/(2C(0))$.

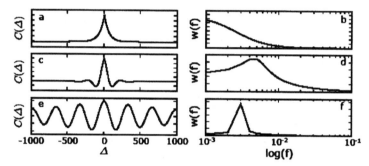

Fig. 16.12 Stationary auto–correlation function, $C(\Delta)$ (left–hand column) and power spectra, $W(f)$ (right–hand column) for an attractive delayed random walk on a quadratic potential for different time delays, τ. Stationary correlation function $C(\Delta)$ calculated using (16.89) and $W(f)$ calculated from this using the Weiner–Khinchin theorem. The parameters were $a = 50$, $d = 0.4$, and τ was 0 for (a) and (b), 40 for (c) and (d), and 80 for (e) and (f).

Finally, we can show that the expression for $C(\Delta)$ obtained from the delayed random walk (16.89) for $0 \le \Delta \le \tau$ are equivalent to those determined for the delayed Langevin equation in Chapter 15. In particular, for small β, we have

$$\frac{(z_+^\Delta - z_+^{\Delta-1}) - (z_-^\Delta - z_-^{\Delta-1})}{z_+ - z_-} \sim \cos(\beta\Delta)$$

$$\frac{\beta(z_+^\Delta - z_-^\Delta)}{z_+ - z_-} \sim \sin(\beta\Delta)$$

Thus

$$C(\Delta) \sim C(0)\cos(\beta\Delta) - \frac{1}{2\Delta}\sin(\beta\Delta) \qquad (16.90)$$

with

$$C(0) \sim \frac{1+\sin(\beta\tau)}{2\beta\cos(\beta\tau)} \tag{16.91}$$

16.11 Postural sway

The fluctuations in the center of pressure (COP) during quiet standing with eyes closed (see Section 16.5.1) can be modeled as a delayed random walk [652]. It was assumed that the probability $p_+(n)$ for the walker to take a step at time n to the right (positive direction) was given by

$$p_+(n) = \begin{cases} p & X(n-\tau) > 0 \\ 0.5 & X(n-\tau) = 0 \\ 1-p & X(n-\tau) < 0 \end{cases} \tag{16.92}$$

where $0 < p < 1$. The origin is attractive when $p < 0.5$. By symmetry with respect to the origin, we have $\langle X(n) \rangle = 0$. As shown in Figure 16.7, this simple model was remarkably capable of reproducing some of the features observed for the fluctuations in the center–of–pressure observed for certain human subjects.

There are a number of interesting properties of this random walk (Figure 16.13). First, for all choices of $\tau \geq 0$, $\sqrt{\langle X^2(n) \rangle}$ approaches a limiting value, Ψ. Second, the qualitative nature of the approach of $\sqrt{\langle X^2(n) \rangle}$ to Ψ depends on the value of τ. In particular, for short τ there is a non–oscillatory approach to Ψ, whereas for longer τ damped oscillations occur (Figure 16.14) whose period is approximately twice the delay. Numerical simulations of this random walk led to the approximation

$$\Psi(\tau) \sim (0.59 - 1.18p)\tau + \frac{1}{\sqrt{2}(1-2p)} \tag{16.93}$$

This approximation was used to fit the delayed random walk model to the experimentally measured fluctuations in postural sway shown in Figure 16.7.

In the context of a generalized delayed random walk described by (16.68), (16.92) corresponds to choosing $f(x)$ and $g(x)$ to be

$$f(x) = \frac{1}{2}[1 + \eta\theta(x)] \tag{16.94}$$

$$g(x) = \frac{1}{2}[1 - \eta\theta(x)] \tag{16.95}$$

where

$$\eta = 1 - 2p.$$

and θ is a step-function defined by

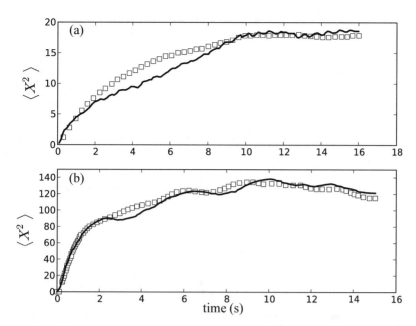

Fig. 16.13 Comparison of the two–point correlation function, $\langle X^2(s)\rangle = \langle (X(n) - X(n-s))^2\rangle$, for the fluctuations in the center–of–pressure observed for two healthy subjects (solid line) with that predicted using a delayed random walk model ($\circ$). In a) the parameters for the delayed random walk were $p = 0.35$ and $\tau = 1$ with an estimated unit step length of 1.2 mm and a unit time of 320 ms. In b) the parameters were $p = 0.40$ and $\tau = 10$ with an estimated step length and unit time step of, respectively, 1.4 mm and 40 ms. For more details, see [652].

$$\theta(x) = \begin{cases} 1 & \text{if } x > 0 \\ 0 & \text{if } x = 0 \\ -1 & \text{if } x < 0 \end{cases} \tag{16.96}$$

In other words, the delayed random walk occurs on a V–shaped potential (which, of course is simply a linear approximation to a quadratic potential). Below, we briefly describe the properties of this delayed random walk (for more details see [654]).

By using symmetry arguments, it can be shown that the stationary probability distributions $P_s(X)$ when $\tau = 0$ can be obtained by solving the system of equations with the long time limit.

$$\begin{aligned} P(0, n+1) &= 2(1-p)P(1,n) \\ P(1, n+1) &= \frac{1}{2}P(0,n) + (1-p)P(2,n) \\ P(r, n+1) &= pP(r-1,n) + (1-p)P(r+1,n) \quad (2 \le r) \end{aligned} \tag{16.97}$$

where $P(r,n)$ is the probability to be at position r at time n, using the trial function $P_s(r) = Z^r$, where

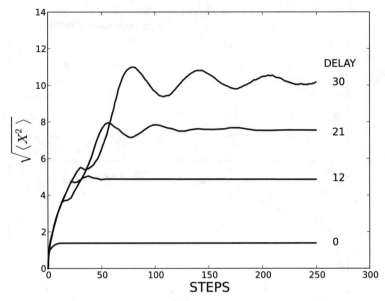

Fig. 16.14 Examples of dynamics of the root-mean-square position $C(0,t)^{1/2}$ for various choices of τ when $p = 0.25$.

$$P_s(r) = \lim_{n \to \infty} P(r,n)$$

In this way, we obtain

$$P_s(0) = 2C_0 p$$

$$P_s(r) = C_0 \left(\frac{p}{1-p} \right)^r \quad (1 \leq r)$$

where

$$C_0 = \frac{(1-2p)}{4p(1-p)}$$

Since we know the pdf, we can easily calculate the variance when $\tau = 0$, $\sigma^2(0)$, as

$$\sigma^2(0) = \frac{1}{2(1-2p)^2} \tag{16.98}$$

The stationary probability distributions when $\tau > 0$ can be obtained by solving the set of equations with the long time limit.

for $(0 \leq r < \tau + 2)$

$$P(r,n+1) = pP(r-1,n;r>0,n-\tau)+\frac{1}{2}P(r-1,n;r=0,n-\tau)$$
$$+(1-p)P(r-1,n;r<0,n-\tau)+pP(r+1,n;X<0,n-\tau)$$
$$+\frac{1}{2}P(r+1,n;r=0,n-\tau)+(1-p)P(r+1,n;r>0,n-\tau)$$

for $(\tau+2\leq r)$

$$P(r,n+1) = pP(r-1,n)+(1-p)P(r+1,n)$$

These equations are very tedious to solve and not very illuminating. Indeed, we have only been able to obtain the following results for $\tau=1$

$$\langle X^2\rangle = \frac{1}{2(1-2p)^2}\left(\frac{7-24p+32p^2-16p^3}{3-4p}\right)$$

and for $\tau=2$

$$\langle X^2\rangle = \frac{1}{2(1-2p)^2}\left(\frac{25-94p+96p^2+64p^3-160p^4+64p^5}{5+2p-24p^2+16p^3}\right)$$

16.11.1 Transient auto–correlation function

Up to this point we have assumed that the fluctuations in COP were realizations of a stationary stochastic dynamical system. However, this assumption is by no means clear. An advantage of a delayed random walk model is that it is possible to gain some insight into the nature of the auto–correlation function for the transient state, $C_t(\Delta)$. In particular, for the transient state we can calculate in the similar manner to (16.86)–(16.87), the set of coupled dynamical equations

$$C_t(0,n+1) = C_t(0,n)+1-2\beta C_t(\tau,n-\tau) \qquad (16.99)$$
$$C_t(\Delta,n+1) = C_t(\Delta-1,n+1)-\beta C_t(\tau-(\Delta-1),n+\Delta-\tau), \quad (1\leq\Delta\leq\tau)$$
$$C_t(\Delta,n+1) = C_t(\Delta-1,n+1)-\beta C_t((\Delta-1)-\tau,n+1), \quad (\Delta>\tau)$$

For the initial condition, we need to specify the correlation function for the interval of initial τ steps. When the random walker begins at the origin we have a simple symmetric random walk for $n\in(1,\tau)$. This translates to the initial condition for the correlation function as

$$C_t(0,n) = n \quad (0\leq\Delta\leq\tau), \qquad (16.100)$$

with

$$C_t(u,0) = 0 \quad \text{for all } n.$$

The solution can be iteratively generated using (16.100) and this initial condition. We have plotted some examples for the dynamics of the mean square displacement $C_t(0, n)$ in Figure 16.15. Again, the oscillatory behavior arises with increasing τ. Hence, the model discussed here shows the oscillatory behavior with increasing delay which appears in both its stationary and transient states.

We also note that from (16.100), we can infer the corresponding set of equations for the transient autocorrelation function of the delayed Langevin equation with a continuous time, $c_t(\Delta)$. They are given as follows:

$$\frac{\partial}{\partial t}c_t(0,t) = -2kc_t(\tau, t-\tau) + 1$$

$$\frac{\partial}{\partial \Delta}c_t(\Delta, t) = -kc_t(\tau - \Delta, t + \Delta - \tau) \quad (0 < \Delta \le \tau) \qquad (16.101)$$

$$\frac{\partial}{\partial \Delta}c_t(\Delta, t) = -kc_t(\Delta - \tau, t) \quad (\tau < \Delta)$$

Studies on these coupled partial differential equations with delay are yet to be done.

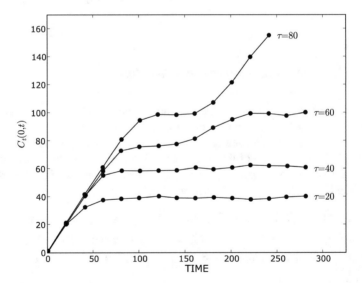

Fig. 16.15 Examples of dynamics of the transient variance $C_t(0,t)$ for different delays τ. Data averaged from 10,000 simulations($\bullet$) is compared to that determined analytically from (16.100) (solid line). The parameters were $a = 50, d = 0.45$. Figure reproduced from [651] with permission.

16.12 Delayed Fokker–Planck equation

Here we derive the Fokker–Planck equation for the delayed random walk in a quadratic potential. This method is a direct use of the procedure used by M. Kac to obtain the Fokker–Planck equation for the case that $\tau = 0$ [399] For f and g as defined in the previous section, (16.77) becomes

$$P(r,n+1) = \sum_m \frac{1}{2}(1 - \beta m)P(r-1,n;m,n-\tau_d)$$
$$+ \sum_m \frac{1}{2}(1 + \beta m)P(r+1,n;m,n-\tau_d), \qquad (16.102)$$

To make a connection, we treat that the random walk takes a step with a size of Δx at a time interval of Δt, both of which are very small compared to the scale of space and time we are interested in. We take $x = r\Delta x$ and $y = m\Delta x$, $t = n\Delta t$ and $\tau = \tau_d \Delta t$. With this stipulation, we can rewrite (16.102) as follows:

$$\frac{P(x,t+\Delta t) - P(x,t)}{\Delta t} =$$
$$\frac{1}{2}\left(\frac{P(x-\Delta x,t) + P(x+\Delta x,t) - 2P(x,t)}{(\Delta x)^2}\right)\left(\frac{(\Delta x)^2}{\Delta t}\right)$$
$$+ \frac{1}{2}\sum_{\frac{y}{\Delta x}} \frac{y}{\Delta x}(P(x,t;y,t-\tau) - P(x-\Delta x,t;y,t-\tau))\frac{\beta}{\Delta t}$$
$$+ \frac{1}{2}\sum_{\frac{y}{\Delta x}} \frac{y}{\Delta x}(P(x+\Delta x,t;y,t-\tau) - P(x,t;y,t-\tau))\frac{\beta}{\Delta t} \qquad (16.103)$$

In the limits

$$\Delta x \to 0, \quad \Delta t \to 0, \quad \beta \to 0$$
$$\frac{(\Delta x)^2}{2\Delta t} \to D, \quad \frac{\beta}{\Delta t} \to \gamma, \quad n\Delta t \to t, \quad \tau_d \Delta t \to \tau$$
$$n\Delta x \to x, \quad m\Delta x \to y, \qquad (16.104)$$

the difference equation (16.103) goes over to the following integro–partial differential equation

$$\frac{\partial}{\partial t}\int_{-\infty}^{\infty} P(x,t;y,t-\tau)dy =$$
$$\int_{-\infty}^{\infty} \gamma\frac{\partial}{\partial x}(yP(x,t;y,t-\tau))dy + D\int_{-\infty}^{\infty} \frac{\partial^2}{\partial x^2}P(x,t;y,t-\tau)dy. \qquad (16.105)$$

This is the same Fokker-Planck equation that is obtained from the delayed Langevin equation [246]. When $\tau = 0$, we have

$$P(x,t;y,t-\tau) \rightarrow P(x,t)\delta(x-y), \tag{16.106}$$

and the above Fokker–Planck equation reduces to the following familiar form.

$$\frac{\partial}{\partial t}P(x,t) = \gamma\frac{\partial}{\partial x}(xP(x,t)) + D\frac{\partial^2}{\partial x^2}P(x,t). \tag{16.107}$$

Thus, the correspondence between Ehrenfest's model and Langevin equation carries over to that between the delayed random walk model and the delayed Langevin equation.

We can rewrite (16.105) by making the definition

$$P_\tau(x,t) \equiv \int_{-\infty}^{\infty} P(x,t;y,t-\tau)dy$$

to obtain

$$\frac{\partial}{\partial t}P_\tau(x,t) = \int_{-\infty}^{\infty} \gamma\frac{\partial}{\partial x}[yP(x,t;y,t-\tau)]\,dy + D\frac{\partial^2}{\partial x^2}P_\tau(x,t) \tag{16.108}$$

In this form, we can more clearly see the effect of the delay on the drift of the random walker.

The Fokker–Planck equation we derived above is only "approximate" in the sense that if we want to solve it, we need to approximate the two-point joint probabilities [271]. As in the case of solving delay differential equations, we need to specify the joint probabilities for the entire interval of length τ. Thus, we are faced with the same difficulty as in the specification of initial interval of length τ, which is infinite dimensional; i.e., it is like a function over the interval which is changing stochastically, rather than a random variable at a single time point. If it is a discrete time-delayed random walk with delay steps of τ, it is a vector with $\tau + 1$ components doing a random walk. This is called "$(\tau + 1)$-th order Markovian". The persistent and anti-persistent random walks are examples of the second order Markov systems with $\tau = 1$. Deriving a corresponding full delayed Fokker–Planck equation is yet to be explored.

16.13 What Have We Learned?

1. For a simple random walk how the variance change with time?

   The variance grows linearly with time.

2. What is an anomalous random walk?

   If the variance does not grow linearly with time, it is referred to as an anomalous random walk. The Lévy flights discussed in this chapter are biological examples of anomalous random walks.

3. **What is a correlated random walk? What are examples of biological processes that can be modeled as a correlated random walk?**

   The definition of a correlated random walk is given by (16.35). Correlated random walks are observed in various situations. For a useful list of applications, see [117].

4. **What is the Ehrenfest random walk and what is its relationship to the simple random walk?**

   The Ehrenfest 2-urn problem describes the movements of balls between two urns, labeled A and B, in discrete time steps. If there are N total balls, the excess of $N/2$ balls in urn A can be interpreted as the displacement in a simple random walk. The Ehrenfest model can also interpreted as describing the movements of a particle in a quadratic potential. This makes it possible to introduce the concept of stability into a random walk.

5. **What is a delayed random walk? What are examples of biological processes that can be modeled as a delayed random walk?**

   A delayed random walk is a random walk in which the transition probabilities at time t depend on where the walker was at time $t - \tau$ in the past. Delayed random walks most commonly arise in the setting of feedback control, where they describe fluctuations about a fixed point.

6. **What is the characteristic feature of random walks in two and three dimensions? What are examples for which this property of random walks is important?**

   A random walker tends to remain in a confined region of space for a relatively long time and then over a short time interval move to another region of space, where the walker is again confined for a long time before moving off again. This pattern of behavior is seen frequently in biology, for example in the movements of cells, in search patterns, and in the movements of your eyes as you scanned this chapter trying to figure out the answers to our questions.

7. **What is a Lévy flight?**

   A Lévy flight is a random walk in which the probability $P(\ell)$ of taking a step of length ℓ obeys a power law, i.e., $P(\ell) \approx \ell^{-\beta}$.

8. **What is the relationship between the pdf and the characteristic function?**

   They are a Fourier transform pair.

9. **What is the relationship between the autocorrelation function and the power spectrum for a random walk?**

   They are a Fourier transform pair.

10. **How does the power spectrum for a random walk depend upon frequency?**

    The power, variance, decreases as f^{-2}.

11. Why does the convolution integral appear in problems related to a random walk?

   The probability for a random walker to be at position $x \pm \ell$ and the probability that the walker will take a step of length ℓ are independent. Thus we need to multiply them and sum over all possible lengths of ℓ to obtain the probability of being at x. This process leads to the convolution integral.

12. What is the Fokker-Planck equation?

   A partial differential equation which describes the evolution of the pdf, $P(x,t)$.

13. What is the master equation?

   The master equation is a set of first-order differential equations which describes the time evolution of a stochastic system which can occupy only a number of discrete states. For example, a two state system is one which a particle is either synthesized (born) or destroyed (death). An example arises in a model of protein synthesis described in Section 16.9.

14. What is a generating function?

   A generating function is a mathematical expression that represents the function $F(z,t)$ as the product of a power series expansion z^n and another function, $P(n,t)$. It is used to solve a master equation. It can be related to the characteristic function.

15. What is the relationship between a characteristic function and the moments of a probability distribution?

   The moments can be determined by differentiating the characteristic function.

16. What is the relationship between a generating function and the moments of a probability distribution?

   The moments can be determined by differentiating the generating function. This should not be surprising since the generating function can be related to the characteristic function which, in turn, is the Fourier transform of the pdf.

16.14 Problems for Practice and Insight

1. One can do no better as preparation for the study of random walks in biology than to spend a weekend reading [51].
2. Complete the steps of the derivation is passing from (16.59) to (16.60).
3. Repeat the calculations of Example 16.5 for a random walk when $p \neq q$. Why do we have to consider the cases $t + x = 2m$ and $t + x \neq 2m$ separately? Why does the last case give probability zero?
4. Suppose that molecule A is present in a colony of bacteria with concentration 1 nm. If the volume of a single bacterium is $\approx 10^{-6} \, cm^3$, how many molecules of A will the bacterium contain?

5. Show that (see [51] for hints)

$$\sum_{n=0}^{\infty} np^n(1-p) = \frac{p}{1-p}.$$

6. Show that (16.12) can be derived from (16.18) by setting $s = e^x$.
7. The number of ways you can choose k objects (where the order is irrelevant) from a collection of n objects is given by the following:

$$_nC_k = \binom{n}{k} := \frac{n!}{k!(n-k)!}.$$

a. Show that the number of arrangements of a identical red balls and b identical black balls in a row is given by

$$_{a+b}C_a = \binom{a+b}{a} = \binom{a+b}{b} = \frac{(a+b)!}{a!b!}. \tag{16.109}$$

Hint: Number the positions on the line that the balls will occupy from 1 to $a+b$ and think of those numbers as the items being selected.

b. In a box, we have n_1 red balls and n_2 black balls. Let $n = n_1 + n_2$. We now choose r balls at random from this box. Show that the probability that there are exactly k red balls among the r balls is

$$P_k = \frac{\binom{n_1}{k}\binom{n_2}{r-k}}{\binom{n}{r}} = \frac{\binom{n_1}{k}\binom{n-n_1}{r-k}}{\binom{n}{r}}.$$

c. Show also that P_k is equal to

$$P_k = \frac{\binom{r}{k}\binom{n-r}{n_1-k}}{\binom{n}{n_1}}.$$

d. You are required to estimate the number of fish in a large pond. One way is to catch n_1 fish, tag them with some method such as putting a red mark on their tails, and return them to the pond. We shall assume that the caught fish are well mixed among the other fish in the pond. Now catch r fish, and suppose that among them, k are marked. Using the above equations, you can write P_k as a function of n, which is unknown and what we want estimate. Plot P_k as a function of n when $n_1 = 1000$, $r = 1000$, and $k = 100$.
e. One way to estimate n is to use the value that makes the above plot of P_k maximum. What is it and does this make sense? (This procedure is one of the simplest forms of sampling, or estimation, theory.)

8. A simple symmetric random walk can be visualized by drawing a path. In this "path visualization," the point (n,x) indicates that at time step n, the random walker is at position x. We can ask various questions regarding the nature of these random-walk paths, the answers to which are quite surprising.

a. Calculate the total number of paths $S(n,x)$ that the random walker can take from the point $(0,0)$ to (n,x). In other words, compute the total number of ways a walker starting at the origin can reach position x at time step n. Clearly, $|x| \leq n$.

b. Using the above results, compute explicitly $S(10,0), S(20,0)$, i.e., compute the number of ways (paths) for the walker to return to the origin at the tenth and twentieth steps.

c. We can classify the above paths as follows:

   i. Positive (negative) path: a path that does not cross or touch the horizontal axis $(x = 0)$.

   ii. Nonnegative (nonpositive) path: a path that does not cross the horizontal axis.

   Now convince yourself that the following statement is true. This is called the *reflection principle*. Consider paths going from point $A = (n_0, x_0)$ to $B = (n,x)$, where we assume $0 \leq n_0 < n$, $x_0 > 0$, $x > 0$. Then the total number of paths that either touch or cross the horizontal axis is equal to the total number of paths going from $A' = (n_0, -x_0)$ to $B = (n,x)$. Note that the point A' is the reflected point of A with respect to the horizontal axis, whence the name of this principle.

d. Compute the number of positive paths from the origin $(0,0)$ to (n,x), $0 < x \leq n$. Here you can use the reflection principle.

e. Compute the total number of nonnegative paths of length $2n$ starting from the origin.

9. Team A and Team B are playing a series of baseball games. Each team has an equal chance of winning each game. After ten games, we note that team A has won six of the ten games. What is the probability that during this series of games, Team A was always ahead of Team B, that is, that after the first game, Team A had always won more games than team B?

10. If a random walker starts from the origin, what is the average number of time steps for the walker to reach a certain position x, where $x > 0$, for the first time? This is called the first passage time. You can use the results of Exercise 8 above.

11. The following questions address a simple symmetric walk in which the walker starts at the origin. Here, we consider the walker's return to the origin.

a. Compute the probability r_{2m} that the walker returns to the origin (not necessarily for the first time) at time step $2m$.

b. Compute the probability that the walker never returns to the origin during his first $2m$ steps.

c. Show that the probability f_{2m} that the walker first returns to the origin at time step $2m$ is

$$f_{2m} = r_{2(m-1)} - r_{2m}.$$

    d. Use the above results to show that the random walker will eventually return to the origin with statistical certainty. (This is true with a simple symmetric random walker in one and two dimensions. However, this probability will decrease in higher dimensions. It is about 0.34 in three dimensions.)

12. Let $Q(2m, 2k)$ be the probability that a random walker starting at the origin will have taken $2k$ steps above the horizontal axis during the first $2m$ time steps. (For $x = 0$, we count it as positive if the next step gives $x = +1$; otherwise, it is negative.) One can show that

$$Q(2m, 2k) = r_{2k} r_{2m-2k} .$$

(Prove this if you like.)

    a. Compute and graph $Q(2m, 2k)$ as a function of k for $m = 10, m = 20$.
    b. What does the plot of $Q(2m, 2k)$ tell you? Is this what you expected from a simple symmetric random walk? (This leads to what is known as the "arcsine law of random walks" by considering the limit as $m \to \infty$.)

Chapter 17
Thermodynamic Perspectives

All biological processes require energy. Models of biological processes must, therefore, ultimately be shaped by considerations of energy budgets: Is there enough energy available for the process to take place? How is the heat generated by the process dissipated? And so on. Indeed, since an organism is a complex processing plant in which multitudes of chemical and physical processes occur concurrently, it is not hard to conclude that evolution would likely have favored mechanisms that are energetically efficient over those that are not [260, 580, 913]. Biologists are well aware of the importance of thermodynamics (see, for example, [604, 605]); however, their typical introduction to that topic is through courses in chemistry and physics that emphasize the study of systems at, or very near, equilibrium (see Section 17.3.1). In such courses, connections between thermodynamics and dynamical systems are seldom addressed.

However, the only biological systems that are truly at equilibrium are those that have been dead for a very long time. Excitable systems, such as those that describe beating hearts and spiking neurons, operate far from equilibrium, and the challenge in biology is to understand the thermodynamics of such systems. Indeed, the differential equations that describe excitable systems are nonlinear, and the fixed points are often unstable. The study of dynamical systems that operate far from equilibrium is very much a work in progress. The overall goal of this chapter is to draw attention to the observation that dissipative systems operating far from equilibrium often generate intermittent types of dynamical behaviors that are best described by power laws. Power-law behaviors are being observed increasingly in large-scale biological systems [28, 243, 635, 663, 784], and these can be studied in the laboratory. In passing, we note that a variety of self-organized structures can also emerge, including traveling spiral waves [883]. The study of these phenomena requires the use of partial differential equations and hence is beyond the scope of this book.

This chapter is organized as follows: First, we briefly discuss some fundamental concepts from equilibrium thermodynamics. Although for many readers, this material reviews what they learned in their introductory chemistry courses, we use our presentation to distinguish carefully between the notions of equilibrium, irreversibil-

© Springer Nature Switzerland AG 2021
J. Milton and T. Ohira, *Mathematics as a Laboratory Tool*,
https://doi.org/10.1007/978-3-030-69579-8_17

ity, and nonequilibrium. The concept that emerges from this discussion is that the thermodynamic variable of most interest is the *entropy production*, denoted by Λ.

From this perspective, a system at equilibrium (for which the entropy production of the system Λ_{sys} and that of the surroundings Λ_{surr} are equal to zero) is clearly different from a system at steady state (in which $\Lambda_{sys} = \Lambda_{surr} > 0$) (see also the discussion in Chapter 3). Next, we discuss three examples that illustrate the interplay between dynamics and thermodynamics, namely, the temperature-dependence of enzyme reactions, the equilibrium fixed point, and the *principle of detailed balancing*. Then we outline two attempts to calculate Λ for systems operating near equilibrium. A historically seductive concept was that such systems were characterized by *minimum entropy production*. However, this is now known not to be true, and in fact, these systems are most often characterized by *maximum entropy production* [543]. The remainder of the chapter focuses on the generation of power-law behaviors in slowly driven dissipative dynamical systems. Particular emphasis is given to the characterization of power-law behaviors at the benchtop. The application of these techniques is illustrated by a study of seizure recurrence in patients with medically intractable epilepsy.

The questions we will answer in this chapter are:

1. What is a state function and how is it defined mathematically?
2. Why are work and heat not state functions?
3. What is an equation of state?
4. What are the similarities between a thermodynamic equilibrium and a stable fixed point in a dynamical system?
5. To what extent do dynamic and thermodynamic descriptions of an equilibrium state agree?
6. What are the differences between a thermodynamic equilibrium and a steady state?
7. What is the principle of detailed balancing and why is it important?
8. What are examples of biological systems that operate far from equilibrium?
9. What is the thermodynamic variable that is currently thought to best describe nonequilibria?
10. How does the master-equation approach for random walks arise in the description of nonequilibria?
11. What is self-organized criticality and what experimental measurements suggest that such phenomena may be present?
12. What is a power law? What do we mean by scale-invariance?
13. How do we distinguish between a power-law distribution and other types of statistical distributions?

17.1 Equilibrium

A thermodynamic equilibrium is a time-independent state occurring in a closed system. Not only do all variables have constant values, but those values do not depend on how the equilibrium state was attained. A first step toward developing an understanding of the thermodynamics of time-dependent states is to explore the mathematics of time-independent states [539, 818].

17.1.1 Mathematical Background

Thermodynamic variables can be classified as either intensive (independent of the mass of the system) or extensive (dependent on the mass of the system). Examples of intensive variables are pressure (P), temperature (T), density, and surface tension. Extensive variables include the number of moles (n), energy, and volume (V). The fundamental postulate of thermodynamics is the *state principle* [724]: for every real system at equilibrium, there exists a positive integer n such that if n intensive variables are fixed, then all other intensive variables are fixed. Thus, if X_i represents an intensive thermodynamic variable, then there exists for a defined state an *equation of state* of the form

$$X_{i+1} = f(X_1, X_2, \ldots, X_i).$$

Finally, if the value of one extensive variable is known, then the mass of the system is also known.

Example 17.1 Equations of state cannot be deduced directly from thermodynamic principles; they must be determined from experimental measurements. A number of equations of state are known (for a list, see [724]), including (expressed in intensive variables) the following:

1. The ideal gas law

$$PV_{\mathrm{m}} = RT, \tag{17.1}$$

where $V_{\mathrm{m}} = V/n$ is the molar volume and R is the gas constant.
2. The van der Waals[1] gas equation

$$\left(P + \frac{a}{V_{\mathrm{m}}^2}\right)(V_{\mathrm{m}} - nb) = RT, \tag{17.2}$$

where a, b are constants.
3. The equation of an ideal elastic substance

$$\mathscr{T} = cT \left(\frac{L}{L_0} - \frac{L^2}{L_0^2}\right),$$

---

[1] Johannes Diderik van der Waals (1837–1923), Dutch theoretical physicist.

where $\mathscr{T}$ is the tension, L is the length, L_0 is the length under zero tension, and c is a constant. $\diamond$

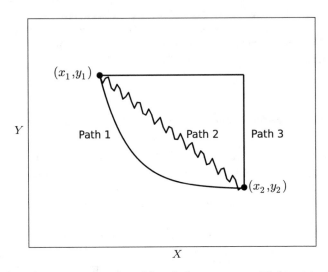

Fig. 17.1 Schematic representation of possible paths between two equilibrium states described by two state variables X and Y.

We can use an equation of state to describe the change from one equilibrium to another. Figure 17.1 illustrates the change between two hypothetical equilibrium states (X_1, Y_1) and (X_2, Y_2) whose equation of state is described by two intensive variables X and Y. It takes time to perform this change, and hence we can design an experiment, at least in principle, to perform it in many ways. Just as in the case of the phase-plane diagram we used in Section 4.3.3, time does not appear explicitly in this diagram. However, the time-dependent ways in which we perform the change between the two equilibria, the *paths*, are represented by the various lines connecting (X_1, Y_1) and (X_2, Y_2) in Figure 17.1. For example, one path might begin by changing X while holding Y constant and then adjusting Y while holding X constant (Path 3 in the figure). Of course, we could imagine a variety of paths of greater complexity (Paths 1 and 2 in the figure). The point is that if X and Y are state variables, then the change in equilibrium does not depend on the path taken between the equilibrium states. In other words

$$\Delta X = X_2 - X_1, \quad \Delta Y = Y_2 - Y_1.$$

Examples of state variables include V_{m}, P, T, internal energy (E), enthalpy (H), Gibbs free energy (G), Helmholtz[2] free energy (A), and entropy (S).

---

[2] Hermann Ludwig Ferdinand von Helmholtz (1821–1894), German physician, physicist, and inventor of the ophthalmoscope.

The path-independence of the change in state variable between two equilibrium states is established by solving an exact differential equation [539, 818]. Consider a function $f(x,y)$ of two variables x and y. The exact differential of $f(x,y)$ is

$$df = \frac{\partial f}{\partial x}dx + \frac{\partial f}{\partial y}dy, \tag{17.3}$$

where the symbol d denotes an infinitesimal change and the symbol ∂ denotes the partial derivative. The physical meaning of this statement is that the total sum of influences that determine df at any instant is the sum of every separate contribution [556]. This is nothing more than a restatement of the principle of superposition (Tool 1). Based on these observations, we can expect that classical thermodynamics has much in common with linear dynamical systems theory.

Our goal is to evaluate the integral

$$\int_{\text{path}} df$$

along a path that links two states a and b. This path is described parametrically by

$$x = x(t), \quad y = y(t).$$

Thus, as t increases from a to b, the corresponding point on the path goes from $(x(a),y(a))$ to $(x(b),y(b))$. If $f(x,y)$ depends only on the current values of x,y and not on the path taken, then we can write

$$\frac{\partial f}{\partial x} = M(x,y) \quad \text{and} \quad \frac{\partial f}{\partial y} = N(x,y)$$

if and only if

$$\frac{\partial M}{\partial y} = \frac{\partial N}{\partial x}. \tag{17.4}$$

In addition, if the path is closed, then we have

$$\oint df = 0, \tag{17.5}$$

where the symbol $\oint$ denotes an integral about a closed path. Together, equations (17.4) and (17.5) define a *state function*. Thus, at the foundation of the thermodynamics of equilibria lies the concept that state variables obey an *exact (Pfaffian)[3] differential equation* given by (17.3) together with (17.5).

Example 17.2 Evaluate the line integral

$$\int_{\text{path}} \left(x^2 - y^2\right) dx - 2xy\,dy,$$

where the path is described by

---

[3] Named for Johann Friedrich Pfaff (1765–1825), German mathematician.

$$x(t) = t^2, \quad y(t) = t^3,$$

from $t = 0$ to $t = 2$. The first step is to check that there exists a state function. Using (17.4), we see that

$$\frac{\partial M}{\partial y} = -2y = \frac{\partial N}{\partial x},$$

and thus there exists a state function. Hence, we can evaluate the integral at the endpoints of the path. Noting that

$$dx = 2t\,dt, \quad dy = 3t^2\,dt,$$

we have

$$\int_0^2 \left[\left(t^4 - t^6 \right) 2t - 6t^5 \right] dt = \int_0^2 \left(-4t^5 - 2t^7 \right) dt = \frac{-64}{3}. \qquad \diamond$$

17.1.2 First and Second Laws of Thermodynamics

The first law of thermodynamics is a restatement of the conservation of energy law in terms of the internal energy (E), work (w), and heat (q) for closed systems. Heat and work are both forms of energy flow. It has become popular in chemistry to write the first law as

$$dE = \dq + \dw, \tag{17.6}$$

where the symbol $\d$ indicates that the changes in heat (q) and work (w) depend on the path taken. The sign convention is that energy flows from the surroundings to the system are positive, and those from the system to surroundings are negative. In physics and engineering, the first law is typically written as

$$dE = \dq - \dw, \tag{17.7}$$

where work done by the system on the surroundings is regarded as positive (it is negative in the chemist's first law). Despite the outward differences between (17.6) and (17.7), the descriptions provided are the same. Consider the work $\dw = P\Delta V$ of expansion of an ideal gas at constant pressure and temperature. In the context of (17.6), we have $dE = \dq - P\Delta V$, where the minus sign arises because the surroundings do negative work on the system. On the other hand, from the point of view of (17.7), we also have $dE = \dq - P\Delta V$, since the system does positive work on the surroundings. Since the chemist's definition of the first law now dominates undergraduate teaching in the United States, we use this convention in the rest of the chapter.

The connections between heat flow and work were first investigated in the context of a simple but hypothetical engine referred to as the reversible Carnot cycle.[4]

---

[4] In honor of Nicolas Léonard Sadi Carnot (1796–1832), who first proposed it.

One cycle of this engine has the net effect of reversibly using heat provided by a hot reservoir to produce work while passing some of the heat to a cold reservoir. The quantity that is conserved in a Carnot cycle is not $đq_{rev}$, but $đq_{rev}/T$. Clausius[5] used this conserved quantity to define a new state function called the *entropy* (denoted by S), where $dS = đq_{rev}/T$.

The second law of thermodynamics states that the rate of change of entropy for a process occurring in the universe (systems plus surroundings) is

$$dS_{total} \geq \frac{đq}{T}, \tag{17.8}$$

where equality applies to a process that occurs reversibly, and inequality to one that occurs irreversibly.

To illustrate (17.8), consider an ideal gas at constant temperature and pressure for which the only work is that due to expansion. The work of expansion is

$$w = - \int_{V_1}^{V_2} P_{ext} \, dV.$$

Note that since we have evaluated $đw$ over a specified path, the result of the integration yields w. When

$$P_{ext} = \frac{nRT}{V},$$

we have

$$w = - \int_{V_1}^{V_2} \frac{nRT}{V} \, dV = -nRT \log \frac{V_2}{V_1}.$$

Since the process occurs isothermally, i.e., the total internal energy of the ideal gas is constant, it follows that $\Delta E = 0$ and

$$q_{rev} = nRT \log \frac{V_2}{V_1},$$

whence

$$\Delta S_{sys} = nR \log \frac{V_2}{V_1} > 0.$$

Since $V_2 > V_1$, it follows that ΔS_{sys} increases. However, if during a reversible process, the system absorbs q_{rev}, then it must also be true that the surroundings lose $-q_{rev}$. Consequently,

$$\Delta S_{total} = \Delta S_{sys} + \Delta S_{surr} = 0.$$

Now suppose that the expansion occurs irreversibly against zero external pressure. Since w depends only on P_{ext} and $\Delta E = 0$, we have $q = 0$. Consequently,

$$\Delta S_{total} = nR \log \frac{V_2}{V_1} + 0 > 0,$$

---

[5] Rudolf Julius Emanuel Clausius (1822–1888), German physicist and mathematician.

as required by the second law.

Example 17.3 Thermodynamics can be applied to the study of human nutritional problems, such as obesity and starvation [127, 781]. These health disorders have major medical, social, and economic impacts on human life. Although starvation is normally considered to be a problem of developing countries, symptoms of starvation arise in developed countries, at least transiently, in hospitals every time a patient is placed on an external feeding regimen. Surprisingly, even simple mathematical models for human weight change can exhibit curious properties that are not well understood [127].   ◇

17.1.3 Spontaneity

From a dynamical systems point of view, equilibrium is a consequence of the equality of the rates of two opposing reactions. In thermodynamic terms, equilibrium represents a compromise between the tendency of molecules to assume the lowest energy and the drift toward maximum disorder. When we perturb a system from equilibrium by, for example, making a sudden change in the surroundings, the system will generally react to reestablish equilibrium. In this sense, an equilibrium state exhibits stability. However, there is a proviso, which is that the system must be able to reestablish equilibrium, which is the case if and only if there are no adverse constraints. It is the existence of such constraints in the form of, for example, the flow of mass between the system and surroundings that most clearly distinguishes a system at equilibrium from a system in nonequilibrium.

The reestablishment of an equilibrium following a sudden change in the surroundings can occur reversibly or irreversibly. A reversible process requires that the state functions differ at most infinitesimally from one moment to the next and with respect to their values in the surroundings. The direction of a reversible process can be reversed at any moment in time by making an infinitesimal change in the surroundings. Hence, for example, an expansion can become a compression, and vice versa. Therefore, a completely reversible process would proceed infinitely slowly.

In contrast, an irreversible process takes place at a rate bounded away from zero, and the state variables of the system differ from those of the surroundings by amounts also bounded away from zero. Moreover, an irreversible process cannot be stopped or reversed by an infinitesimal change in the surroundings. Examples of irreversible processes are the flow of heat from hot to cold and, as we have seen, the expansion of an ideal gas into a vacuum at constant temperature and pressure. However, the common direction of reversible and irreversible processes is toward equilibrium. Thus, we must be careful not to use the term "irreversible" as a synonym for "nonequilibrium."

It follows from the above observations that heat flowing and work done during a reversible process are not the same as those taking place during an irreversible process. To illustrate this point, suppose that the only form of work done is that due to expansion. For a reversible process, P_{ext} and P_{int} are only infinitesimally different, and

$$w_{rev} = \int P_{ext}\, dV = \int (P_{int} - dP)\, dV \approx \int P_{int}\, dV\,,$$

where we have used the approximation $dP\, dV \approx 0$. In contrast, for an irreversible process, we have $P_{ext} < P_{int}$, and hence we see that

$$w_{rev} > w_{irr}\,. \tag{17.9}$$

Consequently, if we carry out the same change in state, once reversibly and once irreversibly, we see that

$$q_{rev} > q_{irr}\,.$$

This relationship provides a criterion for *spontaneity* (see below).

Since our goal is to compare thermodynamic and dynamical systems approaches to an equilibrium, it is necessary to have a thermodynamic state variable that applies only to the system in question. Fortunately, such a state function exists: the *Gibbs free energy*, denoted by G. For systems of constant entropy, the equilibrium position is in the direction of lowest energy, whereas for systems of constant energy, the equilibrium is in the direction of highest entropy. Since in most processes, neither the entropy nor the energy is held constant, Gibbs searched for a state function that incorporated both energy and entropy. The result was

$$G = H - TS\,. \tag{17.10}$$

For processes occurring under conditions of constant pressure, it is convenient to introduce a new state function called the enthalpy, denoted by H and defined as

$$H := E + PV\,.$$

When the only work is that due to expansion, we have

$$dH = dE + P\Delta V = q\,.$$

Since both H and TS are properties of the system, so is G.

Under conditions of constant temperature and pressure, we have

$$dG = dH - T\, dS = q_{irr} - q_{rev}\,,$$

where the subscript irr (for "irreversible") indicates that the path taken between the two states in an experimental realization is not necessarily a reversible path. There are three possibilities:

1. When $\Delta G = 0$, the process is reversible, and hence it is at equilibrium.
2. When $\Delta G < 0$, the process occurs irreversibly.
3. When $\Delta G > 0$, the process also occurs irreversibly, but in the opposite direction.

These observations show that an equilibrium in a closed system at constant temperature and pressure corresponds to a local minimum for the state variable G. This is qualitatively the same picture that we obtained for a locally stable fixed point of a dynamical system. In this analogy, the fixed point corresponds to the equilibrium. Whether the system is at a thermodynamic equilibrium or exactly at the fixed point dynamically, no time-dependent changes occur.

17.2 Nonequilibrium

What can be said about the thermodynamics of open dynamical systems? Matter in motion, such as the water in the water-fountain example of Chapter 3, carries internal energy as well as potential energy. Thus, the first law of thermodynamics is no longer valid. In a nonequilibrium state, the time-dependent changes in the local values of $\tilde{P}, \tilde{T}, \tilde{V}, \{\tilde{n}_i\}$ in the system give rise to local gradients that are continuously changing. These gradients depend on local interactions that are typically nonlinear.

In order to appreciate current approaches to the application of thermodynamics to nonequilibria, it is useful to consider what variables can be readily measured. For living organisms, the fundamental unit of energy for performing work is the ATP molecule. Thus, the work performed by an organism can be estimated as the number of ATP molecules produced or consumed per unit time. Modern imaging techniques, such as positron emission tomography, estimate the energy consumption of cells from the uptake of forms of glucose that cannot be metabolized [365]. However, it is a herculean task to try to estimate all the contributions due to work performed on and done by a biological system per unit time (this has recently been attempted for the human brain [470]). On the other hand, heat can be measured, for example, with a thermometer or a thermocouple. From the second law of thermodynamics, heat exchange divided by temperature gives the change in entropy. These considerations underlie the importance of the rate of entropy production Λ by a system as a variable that characterizes the nonequilibrium state.

The distinguishing feature of nonequilibrium steady states is the continuous production of entropy by the system. The total entropy production is given by

$$\left(\frac{dS}{dt}\right)_{\text{total}} \equiv \Lambda_{\text{total}} = \Lambda_{\text{sys}} - \Lambda_{\text{surr}}. \tag{17.11}$$

Since the internal energy has the same units as work, the units of Λ are joules per kelvin per second, $J \cdot K^{-1} \cdot s^{-1}$.

Entropy production has two contributions: the rate Λ_{sys} of entropy production by the system and the rate Λ_{surr} of entropy transfer from the system to the surroundings. From the second law of thermodynamics, we know that $\Lambda_{\text{sys}} \geq 0$. When $\Lambda_{\text{total}} = 0$, we have at equilibrium $\Lambda_{\text{sys}} = \Lambda_{\text{surr}} = 0$, and at steady state, $\Lambda_{\text{sys}} = \Lambda_{\text{surr}} > 0$. Using these notions, we can divide thermodynamics into three areas: equilibrium, near equilibrium, and far from equilibrium.

The first attempts to describe Λ_{sys} for dynamical systems near equilibrium were based on experimental observations suggesting that fluxes J in biological and physical systems were linearly correlated with forces X (for a more extensive discussion, see [417]). Examples include the linear dependence of heat flow on the gradient of temperature (Fourier's heat equation), the dependence of diffusion on the negative gradient of concentration (Fick's law of diffusion), and the proportionality between current and the electromotive force (Ohm's law). These observations were summarized by Onsager[6] [656, 657] in the phenomenological equations

$$\Lambda_{sys} = \sum_{i=1}^{n} J_i X_i,$$ (17.12)

where

$$J_i = \sum_{k=1}^{n} L_{ik} X_k, \quad i = 1, 2, \ldots, n,$$ (17.13)

and L_{ik} are the Onsager coefficients. In this simplified presentation of the Onsager reciprocal relations, the coefficients are proportional to T^{-1} (see Sections 17.3.1 and 17.3.2). A necessary condition for the validity of (17.12) is that the matrix of coefficients be symmetric, i.e.,

$$L_{ik} = L_{ki}.$$ (17.14)

Equation (17.12) is encouraging, since it looks very much like the nth-order system of linear differential equations we introduced in Chapter 2. Equilibrium corresponds to $J = 0$, and hence $\Lambda_{sys} = 0$. It should be noted that X is not necessarily equal to zero. It has been shown that the entropy production satisfying (17.12) is a maximum [543, 919].

Little is known about Λ_{sys} in dynamical systems that are operating far from equilibrium. Equation (17.13) is an approximation expected to hold only for small displacements from equilibrium. Indeed, this relationship is derived using linearization (Tool 7). As deviations from equilibrium become larger and larger, we can expect that neither (17.12) nor (17.13) will be true. However, it is always true for a homogeneous system that

$$\Lambda_{sys} = \frac{1}{T}\frac{dq}{dt}.$$

17.3 Dynamics and Temperature

The temperature-dependence of the rate of a chemical reaction illustrates the interplay between dynamics and thermodynamics. This connection represents an application of the steady-state approximation that we introduced in Chapter 3 to understand enzymatic reactions

---

[6] Lars Onsager (1903–1976), Norwegian-born American physical chemist and theoretical physicist.

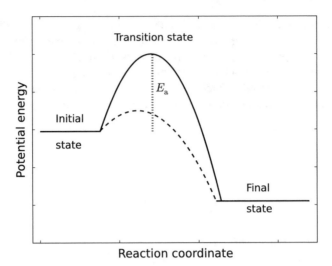

Fig. 17.2 Schematic representation of the relationship between E_a (dotted line) and the potential energy of the reactants and products of a chemical reaction solid line. The dashed line illustrates the lowering of E_a due to enzyme catalysis

$$A + B \underset{k_{-1}}{\overset{k_1}{\rightleftharpoons}} AB^{\ddagger} \overset{k_2}{\longrightarrow} P. \qquad (17.15)$$

Thus, a bimolecular chemical reaction involves the formation of an activated complex $AB^{\ddagger}$, which, in turn, breaks down to yield the product. The temperature-dependence of the rate k_2 of product formation is given by the Arrhenius[7] equation

$$k_2 = Z e^{E_a/RT}, \qquad (17.16)$$

where Z is the preexponential factor, R is the gas constant, and E_a is the activation energy of $AB^{\ddagger}$ compared to that of the reagent molecules (Figure 17.2). The concept that there is an energy barrier that must be overcome to allow a reaction to proceed is very similar to the concept, discussed in Chapter 4, of a separatrix between two attractors in a bistable dynamical system. Enzymes increase the rate of reaction by lowering the activation energy (for each 5.71 kJ/mole by which E_a is lowered, the reaction rate typically increases about tenfold [857]).

From (17.15), we have

$$\frac{d[AB^{\ddagger}]}{dt} = k_1[A][B] - (k_{-1} + k_2)[AB^{\ddagger}].$$

Making the steady-state approximation $d[AB^{\ddagger}]/dt = 0$, we obtain

$$K^{\ddagger}[A][B] = [AB^{\ddagger}]$$

---

[7] Svante August Arrhenius (1859–1927), Swedish physical chemist.

and

$$\frac{d[P]}{dt} = k_2 K^{\ddagger}[A][B].$$ (17.17)

Eyring's[8] transitional rate theory interprets $K^{\ddagger}$ in thermodynamic terms. First, it is assumed that $AB^{\ddagger}$ breaks up when a sufficiently large vibration occurs. The average energy per vibration mode is $k_B T$, where k_B is Boltzmann's constant, and the energy difference between two vibrational modes is hf, where h is Planck's constant. Thus, we can write

$$k_2 \approx f = \frac{k_B T}{h}.$$

Second, it is assumed that $K^{\ddagger}$ behaves similarly to an equilibrium constant, and thus we can write

$$K_{eq}^{\ddagger} = e^{-(\Delta G^{\circ})^{\ddagger}/RT},$$

where we have used the thermodynamic relationship for the dependence of an equilibrium constant[9]

$$\Delta G^{\ddagger} = -RT \log K^{\ddagger}.$$

Using (17.17), we have

$$\frac{d[P]}{dt} = \frac{k_B T}{h} e^{-\Delta G^{\ddagger}/RT}[A][B],$$ (17.18)

and hence the temperature-dependence of the rate constant $k(T)$ is

$$k(T) = \frac{k_B T}{h} e^{-\Delta G^{\ddagger}/RT}.$$ (17.19)

Extensions of these calculations to enzymatic catalysis leads to

$$k(T) = \gamma(T) \frac{k_B T}{h} e^{-\Delta G^{\ddagger}/RT},$$ (17.20)

where $\gamma(T)$ is the *transmission coefficient* [256]. For simple transitions, $\gamma(T) = 1$. However, in general,

$$\gamma(T) = \Gamma(T)\kappa(T)g(T),$$

where $\Gamma(T)$ is the tunneling coefficient, $\kappa(T)$ is the recrossing coefficient, and $g(T)$ describes other temperature-dependent effects. The factor $\gamma(T)$ contributes no more than a factor of 10^3 to the reaction rate, whereas the maximum effect of lowering $\Delta G^{\ddagger}$ on the reaction rate is of order 10^{19} [256]. Current interest focuses on identifying the molecular mechanisms that lower E_a.

Using (17.10) and assuming that $\gamma(T) = 1$, we have

---

[8] Henry Eyring (1901–1981), Mexican-born American theoretical chemist.

[9] More precisely, we should write $(\Delta G^{\circ})^{\ddagger}$, where the superscript o indicates that all reactants are in their standard states. We have dropped the o in the equations that follow in order to simplify the notation.

$$k(T) = \frac{k_B T}{h} e^{\Delta S^{\ddagger}/R} e^{-\Delta H^{\ddagger}/RT} . \tag{17.21}$$

Thus, there are two thermodynamic contributions that determine the rate of a reaction: the enthalpy of activation, which is related to the amount of bond formation compared to bond breaking in forming the transitional state, and the entropy of activation, which is related to the change in the number of degrees of freedom in forming the transitional state. Does the lowering of the activation energy occur because of enthalpic or entropic mechanisms?

At one time, it was thought that changes in $\Delta S^{\ddagger}$ played a dominant role in enzyme catalysis [64, 495, 664]. The argument went as follows: Consider the enzyme–substrate (ES) complex. It is expected that the formation of ES decreases the possible configurations of enzyme and substrate, and hence $\Delta S^{\ddagger}_{sys} < 0$. However, the formation of ES concurrently increases the possible configurations available to the water molecules, and hence $\Delta S^{\ddagger}_{surr} > 0$. If $\Delta S^{\ddagger}_{surr} > \Delta S^{\ddagger}_{sys}$, then the effect would be to increase the rate of reaction. However, experimental evidence suggests that the entropic changes in the enzyme and in the solution are similar [852]. Thus, it appears that in many cases, enzymatic catalysis is dominated by enthalpic considerations, namely the breaking of old bonds and the making of new ones. However, it remains possible that there may be certain situations in which entropic effects dominate [147, 495, 582, 583, 791]. Thus, the reader is cautioned to evaluate experimental observations carefully before deciding on the relative importance of entropic and enthalpic contributions for specific reactions.

17.3.1 Near Equilibrium

We anticipate that at equilibrium, dynamic and thermodynamic descriptions should be the same. Let us use the Onsager relationships to examine this expectation more closely. Consider the chemical reaction

$$A \underset{k_{-1}}{\overset{k_1}{\rightleftharpoons}} B \tag{17.22}$$

occurring in a closed dynamical system [417]. The equilibrium concentrations of A, B, respectively, $\boxed{A}$ and $\boxed{B}$, are obviously stable fixed points, since an equilibrium is the only time-independent state possible: perturbations away from the equilibrium result in irreversible flows back toward the equilibrium values. Thus, we can apply Onsager's formalism given by (17.12) and (17.13) to calculate the flux J_{kin} from a kinetic point of view and then compare it to the value J_{therm} calculated from a thermodynamic point of view.

This problem is more difficult than might appear at first glance. From a dynamics point of view, the flux is proportional to the concentration of the chemical species. However, from the thermodynamic perspective, the flux is proportional to the logarithm of the concentration of the chemical species, since

$$\Delta \mu_i := \mu_i - \mu_i^0 = RT \log[C_i] , \tag{17.23}$$

where C_i is the ith chemical species, μ_i is the chemical potential of the ith chemical species, and μ_i^0 is the chemical potential of the ith chemical species in its standard state. We remind the reader that $\Delta \mu_i$ corresponds to the force of a chemical reaction.

The *kinetic flux* J_{kin} can be obtained by applying the law of mass action (Tool 5) to (17.22) to obtain

$$\frac{d[A]}{dt} = k_{-1}[B] - k_1[A] , \quad \frac{d[B]}{dt} = k_1[A] - k_{-1}[B] .$$

Although there are two reaction flows, these are not independent, since

$$-\frac{d[A]}{dt} = \frac{d[B]}{dt} .$$

Thus, there is one distinct flux, and for convenience, we define it from left to right

$$J_{\text{kin}} = \frac{d[B]}{dt} . \tag{17.24}$$

At equilibrium, the flux vanishes, and we have

$$K = \frac{k_{-1}}{k_1} = \frac{\overline{[A]}}{\overline{[B]}} , \tag{17.25}$$

where K is the equilibrium constant.

Following along the lines of the approach to stability that we outlined in Chapter 4, it is natural to define a deviation from equilibrium u as

$$u_A = [A] - \overline{[A]} , \quad u_B = [B] - \overline{[B]} .$$

Since the system is closed, we have $u_A + u_B = 0$, and thus $[A] + [B] = \overline{[A]} + \overline{[B]}$. Consequently, we can rewrite (17.24) as

$$J_{\text{kin}} = k_1 \left(\overline{[A]} + u_A \right) - k_{-1} \left(\overline{[B]} + u_B \right) , \tag{17.26}$$

or (see Exercise 6 at the end of the chapter)

$$J_{\text{kin}} = k_1 u_A (1 + K) . \tag{17.27}$$

In order to derive J_{therm}, we need to keep in mind that the driving force for (17.22) to approach equilibrium is the difference in chemical potential

$$X = \Delta \mu_A - \Delta \mu_B ,$$

and hence at equilibrium, we have

$$\Delta \overline{\mu}_A = \Delta \overline{\mu}_B . \tag{17.28}$$

The flux J_{therm} is proportional to the force, i.e.,

$$J_{\text{therm}} = L(\Delta\mu_A - \Delta\mu_B),\tag{17.29}$$

where L is the Onsager coefficient. Using (17.23) and (17.28), we find that

$$\Delta\mu_A - \Delta\mu_B = RT\left[\log\left(1+\frac{u_A}{[A]}\right) - \log\left(1+\frac{u_B}{[B]}\right)\right].\tag{17.30}$$

Using the series expansion of $\log(1+x)$,

$$\log(1+x) = x - \frac{x^2}{2} + \frac{x^3}{3} - \cdots = \sum_{n=1}^{\infty}\frac{(-1)^{n+1}}{n}x^n,$$

we obtain for very small $u_A/[A]$ and very small $u_B/[B]$,

$$X = RT\left(\frac{u_A}{[A]} - \frac{u_B}{[B]}\right) = RT\frac{u_A}{[A]}(1+K),$$

and hence from (17.29), we conclude that

$$J_{\text{therm}} = \frac{LRT}{[A]}u_A(1+K).\tag{17.31}$$

The approximation $\log(1+x) \approx x$ is valid for only very, very small x (see Exercise 7). In other words, $J_{\text{kin}} \approx J_{\text{therm}}$ only when the concentrations of the chemical species are very close to their equilibrium values. In Section 3.1 (Tool 5), we drew attention to the potential pitfalls of applying mathematical models based on the mass action principle to biological data. Here we have shown that unless special care is taken, it is quite possible that such models may not be correctly formulated from a thermodynamic point of view.

Finally, by comparing (17.27) and (17.31), we see that

$$L = \frac{k_1[A]}{RT},$$

and hence L is not a constant, but a variable! This observation serves to remind us that in the Onsager formulation given by (17.13), the coefficients do not depend on the variables that they relate, namely the forces and fluxes. However, it is possible for them to depend on other variables.

17.3.2 *Principle of Detailed Balancing*

In applying the law of mass action (Tool 5), we implicitly assumed that at equilibrium (or steady state), the rates of the forward reactions are balanced by those in the backward direction. In the early 1900s, a great debate surrounded the study of the hypothetical cyclic chemical reaction shown in Figure 17.3. This problem, which came to be known as the *Wegscheider paradox*,[10] arose because the kinetic and thermodynamic descriptions of the situation did not agree [31, 656, 657]. The solution to this paradox was the *principle of detailed balance* [478]. It is interesting to examine this debate in more detail.

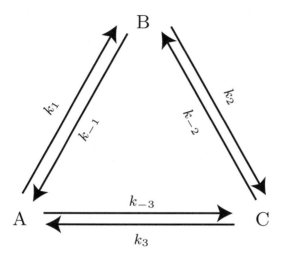

Fig. 17.3 Schematic representation of a hypothetical cyclic chemical reaction involving three chemical species.

The fluxes corresponding to Figure 17.3 are

$$J_1 = k_1[A] - k_{-1}[B], \quad J_2 = k_2[B] - k_{-2}[C], \quad J_3 = k_3[C] - k_{-3}[A].$$

Given the cyclic nature of this reaction scheme, it is clear that only two of the J's can be independent. The forces are

$$X_1 = \Delta\mu_A - \Delta\mu_B, \quad X_2 = \Delta\mu_B - \Delta\mu_C, \quad X_3 = \Delta\mu_C - \Delta\mu_A,$$

from which we see that

$$X_3 = -(X_1 + X_2).$$

Thus the total entropy production is

---

[10] After the Austrian chemist Rudolf Franz Johann Wegscheider (1859–1935).

$$\Lambda = J_1 X_1 + J_2 X_2 + J_3 X_3 = (J_1 - J_3) X_1 + (J_2 - J_3) X_2,$$

suggesting that

$$J_1 - J_3 = L_{11} X_1 + L_{12} X_2, \quad J_2 - J_3 = L_{21} X_1 + L_{22} X_2.$$

At equilibrium, $\Delta \overline{\mu}_A = \Delta \overline{\mu}_B = \Delta \overline{\mu}_C$, and hence

$$X_1 = X_2 = 0$$

and

$$J_1 = J_2 = J_3.$$

Thus, the thermodynamic condition of equilibrium does not require that all of the flows vanish, only that they all be equal. The reaction may, therefore, circulate indefinitely without producing entropy and without violating any of the classical laws of thermodynamics.

However, physical chemists argued that

$$J_1 = J_2 = J_3 = 0. \tag{17.32}$$

This conclusion is a statement of the *principle of detailed balance*, which requires not only that all independent flows vanish at equilibrium, but also that the individual flows of every reaction step become zero. Using (17.32), Onsager quickly realized that this principle provided a solution to the Wegscheider paradox. The fluxes are (see Exercise 8)

$$J_1 = \frac{k_1 \boxed{A}}{RT} X_1, \quad J_2 = \frac{k_2 \boxed{B}}{RT} X_2, \quad J_3 = -\frac{k_3 \boxed{C}}{RT} (X_1 + X_2), \tag{17.33}$$

and

$$J_1 - J_3 = L_{11} X_1 + L_{12} X_2, \quad J_2 - J_3 = L_{21} X_1 + L_{22} X_2, \tag{17.34}$$

where

$$L_{11} = \frac{k_1 \boxed{A}}{RT}, \quad L_{12} = \frac{k_3 \boxed{C}}{RT} = L_{21}, \quad L_{22} = \frac{k_2 \boxed{B} + k_3 \boxed{C}}{RT}.$$

We make two observations: (1) equilibrium implies that $J_1 = J_2 = J_3 = 0$, thus resolving the paradox; (2) we see that $L_{12} = L_{21}$ [compare with (17.14)].

It is important to keep in mind that the condition given by (17.32) is true only for chemical reactions. However, the *Wegscheider triangle* shown in Figure 17.3 has been observed in many other contexts [566, 788], including large-scale genetic, metabolic, ecological, and neural networks. Indeed, it is considered one of the basic building blocks, or network motifs, of large-scale networks. Given the widespread occurrence of this motif, it is natural to ask what its functional significance may be. We expect that mathematical interest will again focus on the Wegscheider triangle, since in these network contexts, it is not necessarily true that the principle of detailed balancing applies.

17.4 Entropy Production

One question concerns how entropy production is to be calculated. Present methods of estimation are directed toward systems that are near enough to equilibrium that the Onsager relationships are valid.

17.4.1 Minimum Entropy Production

Ilya Prigogine[11] and coworkers introduced the concept of *minimum entropy production* [272, 638, 695]. Suppose an unrestrained system is left on its own for a very long time. Then it must eventually be at equilibrium. Now suppose that we introduce one or more constraints that prevent the system from achieving equilibrium. Intuitively, one would expect the system to evolve to a state that was as close to equilibrium as possible consistent with the constraints. Hence, we expect that entropy production will be minimized.

To illustrate this concept, let us make three assumptions about a system: (1) it obeys the phenomenological equations given by (17.12); (2) the Onsager reciprocal relationships are valid (17.14); and (3) the phenomenological coefficients are independent of the forces. Now consider a system in which there are two flows and two forces. We can write

$$\Lambda_{sys} = J_1 X_1 + J_2 X_2 \tag{17.35}$$

and

$$J_1 = L_{11}X_1 + L_{12}X_2, \quad J_2 = L_{21}X_1 + L_{22}X_2. \tag{17.36}$$

Introducing (17.36) into (17.35), we obtain

$$\Lambda_{sys} = L_{11}X_1^2 + (L_{12} + L_{21})X_1 X_2 + L_{22}X_2^2.$$

Now, let us impose a constraint. Let us suppose that X_1 must remain constant. Thus, we can write

$$\frac{\partial \Lambda_{sys}}{\partial X_2} = (L_{12} + L_{21})X_1 + 2L_{22}X_2,$$

which we can simplify to

$$\frac{\partial \Lambda_{sys}}{\partial X_2} = 2(L_{21}X_1 + L_{22}X_2) = 2J_2,$$

---

[11] Ilya Romanovich Prigogine (1917–2003), Belgian physical chemist.

since we have assumed that $L_{21} = L_{12}$.

At steady state, $\Lambda_{sys} = 0$. Thus, if X_2 is unrestricted, it must be true that $J_2 = 0$, and hence

$$\frac{\partial \Lambda_{sys}}{\partial X_2} = 0.$$

This expression indicates that entropy production has an extreme value in the steady state. In a neighborhood of the value of X_2 that corresponds to the extreme value of the entropy production, we know that $\Lambda_{sys} > 0$. Thus, the value must be a minimum, and hence the stationary state of this system is characterized by minimum entropy production. It should be noted that the flow J_1 associated with fixed X_1 assumes a constant value, and hence the parameters of the system become independent of time.

The principle of minimum entropy production is equivalent to the vanishing of the flows associated with the unconstrained forces. Although the concept of minimum entropy production is very seductive [695], it holds only for stationary processes in the presence of free forces [543]. These conditions are quite restrictive, and thus it has proven difficult to demonstrate minimum entropy production for real systems [34, 616, 812]. In fact, it was subsequently shown that for systems that satisfy the Onsager relationships under less-restrictive conditions, the entropy production actually attains a maximum [543, 919].

17.4.2 Microscopic Entropy Production

In thermodynamics, the term "microscopic description" refers to interpretations based on statistical-mechanical considerations. Since statistical mechanics clarifies the meaning of the state variables used in the thermodynamic description of equilibria, can a statistical-mechanical approach also be useful in helping us understand the entropy production Λ_{sys}? Suppose we represent the nonequilibrium state by a very large number of microstates, each characterized by local, but fixed, thermodynamic variables. Thus, we can imagine the evolution of the system as a random walk between the various microstates. Along these lines, a recently emphasized approach to calculating Λ_{sys} is to employ the master-equation approach that we introduced in Section 7.2.2 for the study of random walks [745, 830]. In particular, the general form of a master equation for a system composed of a large number of microstates is [413]

$$\frac{d}{dt} P_i(t) = \sum_j [\Pi_{ij} P_j(t) - \Pi_{ji} P_i(t)], \tag{17.37}$$

where $P_i(t)$ is the probability that the ith microstate occurs at time t, and Π_{ij} describes the transition rates between the ith and jth microstates. We could imagine that this master equation describes a system maintained at nonequilibrium by placing it in contact with two different heat reservoirs. In order to apply (17.37) to a discussion of entropy production, we need to define entropy in microscopic terms and choose an appropriate expression for entropy production.

In microscopic terms, the entropy is defined as

$$S = -k_B \sum_i P_i \log P_i,$$ (17.38)

where P_i is the probability of finding the system in its ith accessible energy state. The importance of this relationship derived from statistical mechanics is that it defines a relationship between entropy and the concept of probability. The probability that the system is in a microstate whose energy is very different from the most probable energies is close to zero.

Let Ω be the set of microstates that make a significant contribution to the sum in (17.38). If we further assume that these states are indistinguishable from one another, then we have $P_i = \Omega^{-1}$ and

$$S = -k_B \Omega \left(\frac{1}{\Omega} \right) \log \left(\frac{1}{\Omega} \right) = k_B \log \Omega.$$ (17.39)

This equation is the famous Boltzmann equation for the entropy of a system. Using the natural logarithm, Boltzmann was able to reproduce all the results of classical thermodynamics that we introduced in Section 17.1. Thus, with respect to our goal, we can define

$$\Lambda_{sys}(t) = -k_B \sum_i P_i \log P_i.$$ (17.40)

Before moving on to choose an appropriate functional form for Λ_{sys}, it is useful to develop an intuitive feeling for what (17.40) means. In particular, we compare (17.40) with the Shannon entropy

$$I = -\sum_i p_i \log_2 p_i,$$ (17.41)

where the unit of entropy is a bit, and the base of the logarithm is 2. The Shannon entropy quantifies the expected information as a realization of a random variable: the higher the entropy, the less predictable the dynamics. Applying this concept to the Boltzmann entropy, we see that organization and ordering decrease the entropy (the system is more predictable), whereas mixing, disorganization, and randomization increase entropy (the system is less predictable). The following example illustrates this point.

Example 17.4 What is the information content of a coin flip?

Solution. If we assume that we have an unbiased coin, then the probability of heads and that of tails are each 0.5, and hence $p_1 = p_2 = 0.5$ and

$$I = - \left[\frac{1}{2} \log_2(0.5) + \frac{1}{2} \log_2(0.5) \right] = 1.0.$$

Thus, each coin flip contains one bit of information. In other words, for one bit of input (a flip), we get one bit of information. This is a situation of maximum uncertainly, since it is the most difficult situation for predicting the outcome. In other words, the entropy is high. On the other hand, when the probability of heads is 0.25 and tails is 0.75, we have $I = 0.81$. Since each time we flip the coin, we have a better chance of getting tails, this situation is less uncertain than that of an unbiased coin. This reduced uncertainty corresponds to a lower entropy. And so on.   ◇

Example 17.5 It has been estimated that the neural spike train generated by the H1 neuron, a motion-sensitive neuron in the visual system of the housefly, carries at least 64 bits of information per second [61, 718]. Half of this information, namely 32 bits per second, provides information concerning the stimulus. Since a precisely periodic spike train carries no information because it is completely predictable, we see that it is the variability in the timing of the spikes that carries the information. ◇

Viable candidates for the expression for entropy production Λ_{sys} are those that are nonnegative and vanish at equilibrium. A widely used expression for Λ_{sys} that meets these requirements is [745, 830]

$$\Lambda_{sys}(t) = \frac{k}{2}\sum_{ij}[\Pi_{ij}P_j(t) - \Pi_{ji}P_i(t)]\log\frac{\Pi_{ij}P_j(t)}{\Pi_{ji}P_i(t)}. \tag{17.42}$$

This expression is nonnegative for P and Π nonnegative (see Exercise 9). The reader should note the similarity between this expression for Λ_{sys} and the Lyapunov function we introduced in Section 4.6.2 for the analysis of stability in dynamical systems. Indeed, it is possible to describe Λ_{sys} in terms of a Lyapunov function and use this approach, for example, to obtain Λ_{sys} for a limit-cycle oscillation [640]. A necessary and sufficient condition for equilibrium is $\Pi_{ij}P_j = \Pi_{ji}P_i$, and for this choice, $\Lambda_{sys} = 0$ (see Exercise 10). This condition ensures microscopic reversibility and incorporates the principle of detailed balancing (see Section 17.3.2). The importance of this microscopic description for entropy production is that it can be shown to be consistent with the phenomenological description of Λ_{sys} given by (17.12) [830].

17.5 Far from Equilibrium

Many of life's vital processes, such as the heartbeat and respiratory cycle, occur far from equilibrium. This observation is reflected by the fact that the mathematical models that have been most successful in describing these dynamical systems are nonlinear and possess an unstable fixed point. However, it is evident that life is highly organized relative to its disordered surroundings [99]. Far from equilibrium, a system may still evolve to some organized steady state, but this state can no longer, in general, be characterized in terms of some suitably chosen thermodynamic potential function such as entropy production. The stability of the stationary state and its independence from initial conditions can no longer be taken for granted. It is possible, for example, that instability can arise such that instead of regressing, certain fluctuations may be amplified, invade the entire system, and compel it to evolve toward a qualitatively different regime.

An example of an organized behavior that arises far from equilibrium is synchronization in a population of oscillators [683, 811]. Another example is the complex spatial patterns that arise in reaction–diffusion systems described by partial differential equations (see, for example, discussions in [617, 883, 884]). Here we discuss another behavior, namely the possibility that complex biological systems can exhibit critical behaviors. In Chapter 5, we emphasized the importance of knowing the normal form for each type of bifurcation. This suggestion was justified by the observation that many complex dynamical systems behave as if they were organized close to stability boundaries in parameter space. Consequently, it is often possible, to a first approximation, to understand dynamics measured experimentally in terms of models based on the appropriate normal form. The question naturally arises whether it is possible for complex dynamical systems to organize themselves into a critical state, that is, whether they can achieve *self-organized criticality*.

17.5.1 The Sandpile Paradigm

The metaphor of a sandpile (Figure 17.4) drew attention to the possibility that complex dynamical systems can self-organize into a critical state [28–30] (for a very readable introduction, see [128]). Initially, as sand grains are dropped, a roughly pyramid-shaped sandpile forms whose height and base increase continuously as a function of time. However, at some point, the slope of the sides of the sandpile becomes steep enough that an avalanche of sand grains occurs.

The critical slope, or threshold, at which an avalanche occurs is determined by factors including the shape of the sand grains and the frictional contact between them. The effect of the avalanche is to redistribute mass within the sandpile and, most importantly, to decrease the slope of the sides of the sandpile. However, since

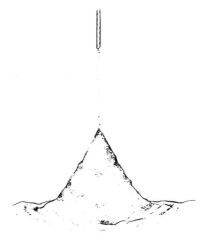

Fig. 17.4 Sandpile metaphor for self-organized criticality. In this representation, the slope of the sandpile is already critical. Thus the addition of grains of sand causes the formation of avalanches of arbitrary size. Figure prepared by Rachel Lee (Pomona College).

sand grains are continually being added, the slope of the sides of the sandpile increases again, eventually exceeding the critical threshold. A new avalanche occurs, and the cycle repeats. At some point, the sandpile reaches a quasistationary state in which avalanches of varying size and duration occur intermittently as sand grains are added. It is observed that there are many small avalanches, fewer medium-sized avalanches, and very few large avalanches.

A surprising prediction by Per Bak[12] and colleagues is that a dynamical system that is slowly driven toward a threshold releases energy in the form of avalanches intermittently rather than continuously [28–30, 128]. Moreover, the size and duration of the avalanches obey well-defined statistical laws that span several orders of magnitude. Here the phrase "slowly driven" means that the time between avalanches is much larger than their average duration. Their key mathematical insight was that the pdf $p(s)$ for the size of an avalanche in a given system obeys a power law of the form

$$p(s) \approx s^{\beta} ,$$

where $\beta = -3/2$.

Dynamical systems that exhibit intermittent behaviors that obey a $-3/2$ power law are said to exhibit self-organized criticality (SOC). Scientists soon discovered that intermittent behaviors generated by many physical and biological systems exhibit $-3/2$ power laws (for reviews, see [29, 389]). Examples include size distributions of raindrops [128], the occurrence of earthquakes and epileptic seizures [662, 663], and the propagation of activity in neural populations [38, 569]. Perhaps the most careful investigation into the universality of SOC is the study of the occurrence of avalanches in rice piles as rice grains are slowly fed into the gap between two vertical parallel glass plates [244]. The advantage of this experimental paradigm is that a number of key parameters can be readily controlled, including the orientation of the rice grains, their shape and frictional coefficients, and the maximum size of the avalanches, referred to as the dimension of the system. It has been observed that the nature of avalanches and their statistical properties depend greatly on the shape of the rice grains. In particular, a power law with $\beta = -3/2$ occurs only when the aspect ratio of the rice grain, namely the ratio of its length to its diameter, is sufficiently large. Thus, SOC is but one of a wide range of possible intermittent phenomena that occur when a dynamical system is slowly driven toward a threshold [391, 425, 526, 531].

---

[12] Per Bak (1948–2002), Danish theoretical physicist.

17.5.2 Power Laws

Power-law behaviors are very common in nature. Sometimes, power laws simply reflect geometric constraints [744]. For example, the volume of a cube depends on ℓ^3, where ℓ is the side length; the circumference of a circle of radius r is πr^2, and so on. In other situations, an explanation for a power-law behavior is unknown, such as the power-law behavior exhibited by the power spectrum $W(f)$ of certain types of noise, as discussed in Section 14.7. Finally, in the case of anomalous diffusion, power-law behavior is exhibited by the associated pdf, namely

$$p(x) = ax^\beta, \tag{17.43}$$

where $p(x)$ is the pdf, x is referred to as the laminar phase, and β is the power-law exponent. Typically, the time series exhibits a bursting quality, as shown schematically in Figure 17.5. In these situations, the laminar phases x are determined by setting an arbitrary threshold and then measuring the times δt between successive threshold crossings, say in the upward direction, and/or the duration δD of the events associated with each consecutive pair of threshold crossings (e.g., the first in the upward direction, the next downward). Figure 17.5a shows how the laminar phases are determined when activity is measured using a single recording electrode.

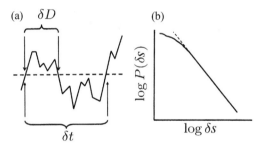

Fig. 17.5 (a) Two definitions of the laminar phase: (1) the interevent interval dt and (2) the event duration dD. The horizontal dashed line is an arbitrary threshold. (b) Normalized laminar phase probability distribution $P(\delta s)$, where δs is either δt or δD. The power-law exponent β is determined from the linear slope of the tail of the distribution.

An important property of a power law is that it is *scale-invariant*. A more popular term is *scale-free*. Scale-invariance means that if we scale the argument x by a constant factor c, then this causes only a proportional scaling of the function itself, i.e.,

$$p(cx) = a(cx)^\beta = c^\beta p(x) \propto p(x).$$

In other words, scaling by a constant simply multiplies the original power-law relation by the constant c^β. Thus, it follows that all power laws with a particular scaling exponent are equivalent up to a constant factor, since each is simply a scaled version of the others. An example of scale-invariance is the observation that the frictional coefficient is independent of the contact area between two surfaces [255].

In general, the pdf for a power law has the general form

$$p(x) \propto L(x)x^{-\alpha'},$$

where $\alpha' > 1$ is the power-law exponent related to the tail of $p(x)$, and $L(x)$ is a *slowly varying function*, that is, a function that satisfies

$$\lim_{x \to \infty} \frac{L(tx)}{L(x)} = 1$$

for some constant t. The necessity of this property follows directly from the requirement that $p(x)$ be asymptotically scale-invariant: thus, the form of $L(x)$ controls only the shape and finite extent of the lower tail. For example, suppose $L(x)$ is the constant function. Then we have a power law that holds for all values of x.

Typically, it is not possible to write down in closed form an expression for $p(x)$ for a process that exhibits a power law. This means that we cannot, in general, estimate the moments (e.g., mean and variance) using the pdf as described in Section 14.2.3. One way to circumvent this problem is to use the characteristic function $\varphi(f)$, which often can be obtained in closed form and hence can be used to estimate the moments (see Section 14.4.5).

An alternative approach is to assume a lower bound x_{min} beyond which the power law holds. By restricting attention to the "tail of the distribution," we can give the pdf the form

$$p(x) = \frac{\alpha' - 1}{x_{min}} \left(\frac{x}{x_{min}} \right)^{-\alpha'}, \qquad (17.44)$$

where the constant is necessary to ensure that $p(x)$ is normalized. In general, $\beta \neq \alpha'$.

Since $p(x)$ is a probability density function, we know that the moments are given by

$$\langle x^m \rangle = \int_{x_{min}}^{\infty} x^m p(x) dx = \frac{\alpha' - 1}{\alpha' - 1 - m} x_{min}^m, \qquad (17.45)$$

which is well defined for $m < \alpha' - 1$. In other words, all moments diverge when $m \geq \alpha' - 1$: when $\alpha' < 2$, the mean and all higher-order moments are infinite. It is of interest to note that when $2 < \alpha' < 3$, the mean exists, but the variance and higher-order moments are infinite. For finite-size samples drawn from such distributions, this behavior implies that the central moment estimates (such as the mean and the variance) for divergent moments will never converge, i.e., as more data are accumulated, they continue to grow.

17.5.3 Measuring Power Laws

At first glance, it appears straightforward to estimate the power-law exponent β: a plot of $\log p(x)$ versus $\log x$ should be linear with slope β (Figure 17.5b). In practice, the plot is not linear over the whole range of x but only in the region for x greater than a certain value. This means that the power-law exponent is most typically evaluated from the tail of the distribution. There are no criteria for objectively determining how to choose the portion of the tail that gives the most reliable estimate of β. Moreover, estimates of β depend of the length of time series, the digitization rate used to collect the time series, and the choice of threshold. Finally, care must be taken to make sure that the power law is not attributable to the measuring device itself.

Although the linearity of a log–log plot is considered the signature of a power law, linearity is not sufficient to prove the existence of a power law. A frequently cited rule of thumb for the existence of a power law is that it must extend over at least three decades [128]. However, since data sets are necessarily of finite length, it can be difficult to satisfy this criterion, and thus it is possible that other types of statistical distributions also fit the observations. Consequently, increasing attention is being focused on the development of more rigorous statistical approaches.

One approach combines maximum likelihood estimates with Kolmogorov–Smirnov[13] statistics to obtain a robust method for testing the significance of the estimate of a power-law exponent [133]. Relevant computer programs are freely downloadable,[14] and thus the use of this method is readily accessible to experimentalists. An alternative strategy is not to rely on a single test for the presence of a power law, but rather to apply a battery of tests for detecting the presence of one [663]: each test that confirms the presence of a power law increases the likelihood that a significant power law exists.

Discussions based solely on the results of statistical tests can spark lively debate (e.g., [187]). Arguments become muted when plausible models become available that predict not only the presence of a power law, but also the correct magnitude of the power-law exponent [94, 673, 716, 717]. Scientific interest in power-law relationships stems from the ease with which certain general classes of mechanisms generate them [94, 334, 635, 673, 687, 780]. Thus, the demonstration that two apparently different phenomena, say earthquakes and epileptic seizures, share the same power laws [663] necessitates an evaluation of the phenomena in light of the mechanisms known to generate them. Research on the origins of power-law relations and efforts to observe and validate them in the real world is an active topic of research in many fields of science, including physics, computer science, linguistics, geophysics, sociology, economics, neuroscience, physiology, and more.

---

[13] Andrey Nikolaevich Kolmogorov (1903–1987), Soviet mathematician; Nikolai Vasilyevich Smirnov (1900–1966), Soviet mathematician.

[14] For example at http://tuvalu.santfe.edu/~aarmc/powerlaws/.

17.5.4 Epileptic Quakes

Figure 17.6 shows that a $-5/3$ power law is observed for seizure energies in patients with medically intractable epilepsy [662, 663]. This power law was measured from 16,032 electrographic seizures arising from the mesial temporal and frontal lobes of 60 patients undergoing surgical evaluation. Curiously, the same $-5/3$ power law is observed for the seismic moments determined from 81,977 earthquakes recorded between 1984 and 2000, in southern California. This similarity in power-law behaviors is thought to reflect that both dynamical systems can be thought of as being composed of populations of coupled nonlinear relaxation oscillators [663].

There are two implications of the observations in Figure 17.6. First, they imply that statistical methods developed to predict the occurrence of earthquakes might be applicable to the prediction of epileptic seizures. The ability to predict when the next seizure will occur would greatly benefit a patient with epilepsy. For example, if a patient with epilepsy was driving an automobile and realized that a seizure was likely to occur within the next minute, this would give the patient enough time to pull off the road.

It is important to realize that the existence of a power law does not in itself imply predictability [820]. The presence of a power law only indicates that very rare events, such as the birth of a black swan [820], occur more frequently than would be predicted from, for example, the moments of a Gaussian pdf. However, predictability is increased when departures occur in the power law over certain length scales (see, for example, $\downarrow$ in Figure 17.7). Common causes for these depar-

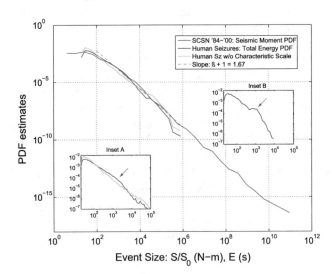

Fig. 17.6 Log–log plot of the pdf estimate versus seizure energy for data from 60 human subjects [663]. Although for some subjects, the log–log plot was linear over the whole range of energies (lower inset), in others, there were clear-cut departures from linearity over a finite range of seizure energies (see upper inset).

tures, referred to as "dragon kings" [785, 786], include the critical phenomena we discussed in Section 5.4. Since critical phenomena are associated with dynamical systems operating close to an edge of stability, it is hoped that the detection of dragon kings may lead to practical methods to predict the occurrence of seizures.

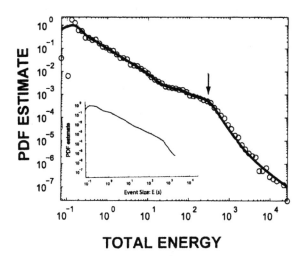

TOTAL ENERGY

Fig. 17.7 Log–log plot of the pdf versus seizure energy for rats treated with 3-mercaptoproprionic acid (3-MPA), a compound that induces seizures [663]. The inset shows that the characteristic size regime indicated by ↓ in the full figure is not seen when the 3-MPA dose is lowered.

The second implication is based on the universality of the mechanism that likely produces the power-law behavior in epileptic seizures and earthquakes. Figure 17.7 shows the pdf of seizure energies for seizure recurrences in rats treated with 3-MPA, a compound that inhibits the synthesis of GABA, the main inhibitory neurotransmitter in the brain. When the dose of 3-MPA is low (inset), a linear power-law behavior is observed. However, when the dose of 3-MPA is increased, the log–log plot develops a region of nonlinearity (↓). One of the known effects of 3-MPA is to reduce membrane noise [198]. Gil and Sornette[15] [268] showed that in a Landau–Ginzburg[16] model for self-organized criticality, changing the noise intensity produces qualitatively the same changes as seen in Figure 17.7. Taken together, these observations suggest that it might be possible to evaluate the hypothesis concerning self-organized criticality experimentally by examining the dynamics seen in animal models of epilepsy.

---

[15] Didier Sornette, Swiss mathematical physicist

[16] Lev Davidovich Landau (1908–1968), Soviet physicist; Vitaly Lazarevich Ginzburg (1916–2009), Soviet physicist.

17.6 What Have We Learned?

1. What is a state function and how is it defined mathematically?

   A state function is a thermodynamic function whose net change in going from state A to state B and then from state B to state A is zero. This result is independent of the path taken. It is defined mathematically as a function whose cyclic integral, given by (17.5), is zero. Equivalently, we can infer that the cyclic integral is zero from the condition for an exact (Pfaffian) differential equation given by (17.4).

2. Why are work and heat not state functions?

   In general, the amount of heat transferred to a system or work done on a system in going from state A to state B, and vice versa, depends on the path taken. In other words, work and heat are not state functions. A useful analogy concerns the location of two points on the surface of the earth. The net change in altitude that one experiences in traveling between the two points is a fixed constant that depends only on the altitude of each of the two points. However, the distance traveled in going between the two points depends on the route taken.

3. What is an equation of state?

   An equation of state is a relationship between the intensive and extensive variables that describe an equilibrium state. For example, the ideal gas law given by (17.1) relates V to a function of n, P, and T. Equations of state do not follow directly from the laws of thermodynamics but must be determined empirically from experimental observations. A reversible process is defined as one in which the changes between state A and state B occur by a sequence of steps, each of which in never more than infinitesimally removed from a state of equilibrium. Thus the equation of state provides a valid description of these changes.

4. What are the similarities between a thermodynamic equilibrium and a stable fixed point in a dynamical system?

   They are both time-independent states, and they can be characterized by the minimum of an appropriately defined variable.

5. To what extent do dynamic and thermodynamic descriptions of an equilibrium state agree?

   Even in the simplest case, these descriptions agree only when the deviations in the concentrations of the chemical species from their equilibrium values are very, very small. It should be noted that this conclusion is based in part on the assumption that the law of mass action (Chapter 3) is valid.

6. Is an equilibrium state the same as a steady state?

   An equilibrium state is not the same as a steady state. For example, in the water-fountain experiment discussed in Chapter 3, the steady state corresponds to a fixed water level in the cup when the faucet is open. The equilibrium state is obtained by turning the water off at the faucet and corresponds to no water in the cup. The basic issue is that equilibrium thermodynamics do not apply to situations in which mass flows through the system. In this chapter, we showed that the entropy production for an equilibrium state is zero, but for a steady state, the rate is positive.

7. What is the principle of detailed balancing and why is it important?

> The principle of detailed balancing arises in the description of dynamical systems that can be decomposed into elementary processes (see Chapter 3). It requires that at equilibrium, each elementary process must be precisely balanced by its reverse. This principle does not arise from thermodynamic considerations but must be included to ensure the equivalence of dynamic and thermodynamic descriptions of a system's behavior.

8. What are examples of biological systems that operate far from equilibrium?

> All processes that occur in a living organisms are examples of such systems.

9. What is the thermodynamic variable that is currently thought to best describe nonequilibria?

> The entropy production of a system is the variable. Provided that a system is near equilibrium, the entropy production is related to the linear product of a force and its associated flux. In theory, situations far from equilibrium would be characterized by nonlinear relationships between forces and fluxes.

10. How does the master-equation approach for random walks arise in the description of nonequilibria?

> This approach arises in the description of systems operating near equilibrium. It assumes that one can define a microstate characterized by local values of the extensive and intensive thermodynamic variables. The presence of a given microstate and switches between existing microstates as the system evolves is modeled as a random walk.

11. What is self-organized criticality and what experimental measurements suggest that such phenomena may be present?

> Self-organized criticality is a property of nonlinear dynamic systems that have a critical state as their attractor. It occurs in slowly driven nonlinear dynamical systems operating under conditions of nonequilibrium. It is characterized by the presence of a power law with a $-3/2$ exponent and hence scale-invariant properties (see Section 17.5.2).

12. What is a power law? What is meant by scale-invariance?

> Power laws have the form $f(x) = ax^\beta$, and thus a log–log plot of $f(x)$ versus f is linear with slope β. Scale-invariance means that if we scale the argument x by a constant factor c, then this causes only a proportional scaling of the function itself. For example, the area of a circle is always related in the same way to r^2 no matter what unit we use to measure the radius.

13. How do we distinguish between a power-law distribution and other types of statistical distributions?

> Power-law exponents are typically measured from the tails of a distribution. It is difficult to know whether a power law measured in this way is statistically significant. A variety of statistical approaches are becoming available to help answer this question [133, 663]. The strongest evidence for a power law is the existence of a model for the phenomenon that predicts both the occurrence of a power law and the measured value of the power-law exponent.

17.7 Exercises for Practice and Insight

1. Calculate the work of expansion of a van der Waals gas under conditions of constant pressure using (17.2). How does this compare to that determined for an ideal gas?

2. Using the definitions of an exact differential and the Gibbs free energy G, show that

$$\left(\frac{\partial S}{\partial P}\right)_T = -\left(\frac{\partial V}{\partial T}\right)_P.$$

   Similarly, if we have $dA = -S\,dT - P\,dV$, where A is the Helmholtz free energy, show that

$$\left(\frac{\partial S}{\partial V}\right)_T = \left(\frac{\partial P}{\partial T}\right)_V.$$

   These equations are referred to as Maxwell's equations. If $(\partial V/\partial T)_P$ and $(\partial P/\partial T)_V$ can be determined from an appropriate equation of state, then Maxwell's relationships can be used to calculate changes in entropy.

3. Evaluate the line integral along the curve whose parametric equations are given by

   a. $\int_C (2x - xy)\,dx + x^2\,dy$, where C is defined by $x(t) = 2t^2$ and $y(t) = t^3 + 3t$, from $t = -1$ to $t = 3$.

   b. $\int_C \sqrt{y}\,dx + (2x - y)\,dy$, where C is $x(t) = 4t^3$ and $y(t) = t$, from $t = 0$ to $t = 1$.

4. Show that the value of the given line integral is independent of the path and then evaluate it

   a. $\int_{(1,3)}^{(-1,2)} xy^2\,dx + x^2 y\,dy.$

   b. $\int_{(-3,1)}^{(3,3)} \log y^2\,dx + 2xy^{-1}\,dy.$

5. Show that (17.26) can be rewritten as (17.27).

6. For small x, we have $\log(1+x) \approx x$. How large must x be so that the difference between x and $\log(1+x)$ is 1%? 10%?

7. Derive equations (17.33) and (17.34).

8. Show that (17.42) is nonnegative. (Hint: Observe that equation (17.42) has the form $x - y\log(x/y)$.)

9. Show that (17.42) vanishes at equilibrium, namely, when the detailed balance condition is obeyed.

10. On-off inrmittency can be produced by a quadratic map with parametric noise, i.e.

$$x_{t+1} = r(\xi_t)(1 - x_t) \tag{17.46}$$

   where $r(\xi_t) = kU_t$ where U_t is a uniformly distributed discrete varable on the interval $[0,1]$ and $k > 1$ [334]. XPPAUT provides a convenient way to demonstrate the $-3/2$-power law.

- Write an XPPAUT program to integrate (17.46).
- Once you have loaded the program into XPPAUT, go to NUMERICS then POINCARE MAP then choose PERIOD. You will see a menu: take the following choices:

  – Variable is X
  – Section is 0.01. This is how you choose the threshold for the laminar phases. The choice 0.01 is a good starting point.
  – Direction is +1. This means that you are defining a threshold crossing in the upward direction
  – Stop on Section is N

- Now integrate the program. Once done click on 'Data'. You will see two columns. The first column is the time difference between successive threshold crossings in the upward direction. This is our definition of a laminar phase.
- Now go to STOCHASTICS, then HISTOGRAM. The white bar at the top of the window will ask you

  – Number of bins: type in your guess, say 100
  – Low: smallest bin
  – High: largest bin
  – Variable: X
  – condition: Press 'ENTER'

- Now click on 'DATA'. You will see two columns: the first column is called TIME and the position of the bin. The second column is X and represents the number of laminar phases that had length equal to the size of this bin. You can see what your histogram looks like by using Viewaxes as usual.
- What we want to do is to plot log X versus log T? Do you have any zeros in your data? Likely, yes: what are you going to do about this?
- You can use the Data Menu command REPLACE to change the numbers of log10 numbers.
- Now when you use Viewaxes you will have a loglog plot. Try to estimate the slope of the linear part as best as you can.
- We used 10000000 iterations and tried to concentrate the bindings for the smaller laminar phases, e.g., 500 bins with the lowest being 0 and the highest being 500.

  a. Show that in order to obtain a $-3/2$ power that $k \geq 2.71828\cdots$.
  b. Prove that the onset of on-off intermittency requires that $k \geq 2.71828\cdots$¿ The reader can check their proof by examining the discussion in [334].

Chapter 18
Concluding Remarks

In the not too distant past, the relationship between biology and mathematics closely resembled that of the *two solitudes* described by the Canadian writer Hugh MacLennan in his famous book *Barometer Rising* [525]: "together, but separate, alone, but together." However, the last few decades have seen a dramatic increase in the role of mathematical and computer modeling in biological research. Indeed, mathematical modeling presently plays an integral role in projects such as the industrial use of genetically modified organisms, the human genome and virtual human projects, the development of new therapeutic strategies and drug delivery systems, the control of epidemics, and assessments of the impact of climate change on the biosphere. Despite two seminal reports on the need for educational reform in science at the undergraduate level [626, 798], it is clear that the undergraduate preparation of biologists, mathematicians, and computer scientists remains far below what is demanded by society's current requirements.

The recent success of mathematical biology owes much to a handful of gifted scientists whose training and interests led to a mastery of both biology and mathematics. Their contributions form the basis of the rather forbidding requirements for postgraduate training in mathematical and theoretical biology: one must have a solid grounding in mathematics, biology, computer science, and appropriate laboratory skills.[1] However, such an educational model is likely to be feasible for only some aspiring biologists, since there is simply too much to learn.

Future successes in mathematical biology will increasingly be accomplished by interdisciplinary teams of researchers each of whose members possesses at least one of the requisite complementary skill sets. The success of team research depends less on individual skills than on the ability to work together effectively to deliver implementable solutions to problems in a timely manner [551]. The important skill sets for productive teamwork include critical thinking, problem solving, project management, and effective communication [588, 590]. Sadly, these are the very aspects of undergraduate biology training that are the most neglected. Thus, instead of trying

---

[1] Consult http://life.biology.mcmaster.ca/~brian/biomath/careers.theo.biol.html for more on this.

© Springer Nature Switzerland AG 2021
J. Milton and T. Ohira, *Mathematics as a Laboratory Tool*,
https://doi.org/10.1007/978-3-030-69579-8_18

to train a "superscientist" who has mastered all the relevant disciplines, it is likely to be more effective for departments of biology to have students specialize in a particular area of interest supplemented with practical experiences gained from work on interdisciplinary teams.

In this book, we have specifically addressed the issue of developing communication skills between biologists and mathematicians. We focused on the use of mathematics in the collection of data in a laboratory setting simply because good data is the cornerstone of both experimental and modeling research. We used a variety of scenarios to introduce biologists to a variety of useful mathematical techniques and to introduce mathematicians to how these tools are used in practical applications.

A fundamental limitation of courses in biomathematics is that the number of biology students who are qualified to take such courses is typically less than five percent of the population of registered biology students. One way to overcome this problem is to incorporate appropriate computer/mathematical modeling modules into existing laboratory-based biology courses. Of course, to be useful, the module should discuss a computational or mathematical skill that is relevant to the subject matter of the course. For example, one could include a module on filters and spectral analysis in a course on physiology or kinesiology and a module on feedback control in most biology courses, including those that discuss climate change.

The Internet is having an increasing impact on undergraduate education. Indeed, it is now possible for students to listen to prerecorded lectures given by experts at major universities at any time of the day with just a few keystrokes. Insufficient attention has been given to preparing students to use such resources effectively. Thus, moving mathematics into the biology laboratory is likely to be just one of the changes that will fundamentally reshape education.

References

1. M. Abeles, E. Vaadia, H. Bergman, Firing patterns of single units in the prefrontal cortex and neural network models. Network **1**, 13–25 (1990)
2. R.H. Abraham, C.D. Shaw, *Dynamics: The Geometry of Behavior. Part 4: Bifurcation Behavior* (Aerial Press, Santa Cruz, CA, 1998)
3. K. Abu-Hassan, J.D. Taylor, P.G. Morris, E. Donati, Z.A. Bortolotto, G. Indiveri, J.F.R. Paton, A. Nogaret, Optimal solid state neuron. Nature Comm. **10**, 5309 (2019)
4. D. Acheson, A pendulum theorem. Proc. R. Soc. London A **443**, 239–245 (1993)
5. D. Acheson, *From Calculus to Chaos: An Introduction to Dynamics* (Oxford University Press, New York, 1997)
6. R.K. Adair, *The Physics of Baseball* (Harper-Collins, New York, 2002)
7. W.G. Aiello, H.I. Freedman, J. Wu, Analysis of a model representing stage-structured population growth with state-dependent time delay. SIM J. Appl. Math. **52**, 855–869 (1992)
8. K. Aihara, G. Matsumoto, Two stable steady states in the Hodgkin-Huxley axons. Biophys. J. **41**, 87–89 (1983)
9. R.M. Alexander, *Principles of Animal Locomotion* (Princeton University Press, Princeton, New Jersey, 2003)
10. U. Alon, Network motifs: theory and experimental approaches. Nature Rev. Genet. **8**, 450–461 (2007)
11. U. Alon, M.C. Surette, N. Barkai, S. Leibler, Robustness in bacterial chemotaxis. Nature **397**, 168–171 (1999)
12. S.A. Aoki, G. Lillaci, A. Gupta, A. Baumschlager, D. Schweingruber, M. Khammash. A universal biomolecular integral feedback controller for robust perfect adaptation. Nature **570**, 533–537 (2019)

© Springer Nature Switzerland AG 2021

J. Milton and T. Ohira, *Mathematics as a Laboratory Tool*,

https://doi.org/10.1007/978-3-030-69579-8

13. Y. Asai, Y. Tasaka, K. Nomura, T. Nomura, M. Casidio, P. Morasso, A model of postural control in quiet standing: robust compensation of delay-induced instability using intermittent activation of feedback control. PLoS ONE **4**, e6169 (2009)

14. A. Atay, Balancing the inverted pendulum with position feedback. Appl. Math. Lett. **12**, 51–56 (1999)

15. U. an der Heiden, Delays in physiological systems. J. Math. Biol. **8**, 345–364 (1979)

16. U. an der Heiden, Flexible modeling, mathematical analysis, and applications, in *Neurodynamics: Proceedings of the 9th Summer Workshop on Mathematical Physics* (World Scientific, Singapore, 1991), pp. 49–95

17. U. an der Heiden, M.C. Mackey, The dynamics of production-destruction: Analytic insight into complex behavior. J. Math. Biol. **16**, 75–101 (1982)

18. U. an der Heiden, M.C. Mackey, Mixed feedback: a paradigm for regular and irregular oscillations, in *Temporal Disorders in Human Oscillatory Systems*, ed. by L. Rensing, U. an der Heiden, M.C. Mackey (Springer, New York, 1987), pp. 30–46

19. R.L. Anderson, Distribution of the serial correlation coefficient. Ann. Math. Stat. **13**, 1–13 (1942)

20. R.M. Anderson, R.M. May, *Infectious Disease of Humans: Dynamics and Control* (Oxford University Press, London, 1991)

21. A.P. Arkin, J. Ross, Computational functions in biochemical reaction networks. Biophys. J. **67**, 560–578 (1994)

22. T. Asai, T. Fukai, S. Tanaka, A subthreshold MOS circuit for the Lotka-Volterra neural network producing the winners-take-all solution. Neural Netw. **12**, 211–216 (1999)

23. Y. Asai, S. Tateyama, T. Nomura, Learning an intermittent control strategy for postural balancing using an EMG-based human computer interface. PLoS ONE **8**, e62956 (2013)

24. K.J. Astrom, R.M. Murray, *Feedback Systems: An Introduction for Scientists and Engineers* (Princeton University Press, Princeton, N.J., 2008)

25. S.M. Baer, T. Erneux, J. Rinzel, The slow passage through a Hopf bifurcation: delay, memory effects, and resonance. SIAM J. Appl. Math. **49**, 55–71 (1989)

26. S. Bahar, A.B. Neiman, P. Jung, J. Kurths, L. Schimansky-Geier, K. Showalter, Introduction to focus issue: nonlinear and stochastic physics in biology. Chaos **21**, 047501 (2011)

27. N.T.J. Bailey, *The Elements of Stochastic Processes* (Wiley, New York, 1990)

28. P. Bak, *How Nature Works: The Science of Self-Organized Criticality* (Copernicus, New York, 1996)

29. P. Bak, C. Tang, K. Wiesenfeld, Self-organized criticality: an explanation of $1/f$ noise. Phys. Rev. Lett. **59**, 381–384 (1987)

30. P. Bak, C. Tang, K. Wiesenfeld, Self-organized criticality. Phys. Rev. A **38**, 364–375 (1988)

31. T.A. Bak, *Contributions to the Theory of Chemical Equilibria* (W. A. Benjamin, New York, 1963)

32. G.L. Baker, J.A. Blackburn, *The Pendulum: A Case Study in Physics* (Oxford University Press, New York, 2005)

33. P. Ball, The physical modelling of human social systems. Complexus **1**, 190–206 (2004)

34. E. Barbera, On the principle of minimal entropy production for Navier-Stokes-Fourier fluids. Continuum Mech. Thermodynam. **11**, 327–330 (1999)

35. J.C. Bastos de Figueiredo, L. Diambra, L. Glass, C.P. Malta, Chaos in two-loop negative feedback system. Phys. Rev. E **65**, 051905 (2002)

36. C. Beaulieu, Z. Kisvarday, P. Somoygi, M. Cynader, A. Cowey, Quantitative distribution of GABA-immunopositive and -immunonegative neurons and synapses in the monkey striate cortex (Area 17). Cerebral Cortex **2**, 295–309 (1992)

37. J. Bechhoefer, Feedback for physicists: a tutorial essay on control. Rev. Mod. Phys. **77**, 783–836 (2005)

38. J.M. Beggs, D. Plenz, Neuronal avalanches in neocortical circuits. J. Neurosci. **23**, 11167–11177 (2003)

39. N. Bekiaris-Liberis, M. Jankovic, M. Krstic, Compensation of state-dependent state delay for nonlinear systems. Syst. Control Lett. **61**, 849–856 (2012)

40. J. Bélair, Periodic pulsatile stimulation of a nonlinear oscillator. J. Math. Biol. **24**, 217–232 (1986)

41. J. Bélair, Population models with state-dependent delays, in *Lecture Notes in Pure and Applied Mathematics*, vol. 131, ed. by O. Arino, D.E. Axelrod, M. Kimmel (Marcel Dekker, New York, 1991), pp. 165–176

42. J. Bélair, S.A. Campbell, Stability and bifurcations of equilibria in a multiple-delayed differential equation. SIAM J. Appl. Math. **54**, 1402–1424 (1994)

43. J. Bélair, L. Glass, Universality and self-similarity in the bifurcations of circle maps. Physica D **16**, 143–154 (1985)

44. J. Bélair, L. Glass, U. an der Heiden, J. Milton, *Dynamical Disease: Mathematical Analysis of Human Illness* (American Institute of Physics, Woodbury, New York, 1995)

45. J. Bélair, M.C. Mackey, J. Mahaffey, Age-structured and two delay models for erythropoiesis. Math. Biosci. **128**, 317–346 (1995)

46. J. Bélair, J.G. Milton, Itinerary of a discontinuous map from the continued fraction expansion. Appl. Math. Lett. **1**, 339–342 (1988)

47. A. Bellen, M. Zennaro, *Numerical Methods for Delay Differential Equations* (Oxford Science Publications, Oxford, 2013)

48. R. Bellman, *Adaptive Control Processes: A Guided Tour* (Princeton University Press, Princeton, 1961)

49. J.S. Bendat, A.G. Piersol, *Random Data: Analysis and Measurement Procedures*, 2nd edn. (Wiley, New York, 1986)

50. B.W. Bequette, *Process Control: Modeling, Design and Simulation* (Prentice Hall, New York, 2003)

51. H.C. Berg, *Random Walks in Biology* (Princeton University Press, Princeton, New Jersey, 1993)

52. H.C. Berg, *E. coli in Motion* (Springer, New York, 2004)

53. H.C. Berg, E.M. Purcell, Physics of chemoreception. Biophys. J. **20**, 193–219 (1977)

54. H.C. Berg, P.M. Tedesco, Transient response to chemotactic stimuli in Escherichia coli. Proc. Natl. Acad. Sci. USA **72**, 3235–3239 (1975)

55. O.G. Berg, R.B. Winter, P.H. von Hippel, Diffusion-driven mechanisms of protein interactions on nucleic acids. 1. Model and theory. Biochemistry **20**, 6929–6948 (1981)

56. R.D. Berger, J.P. Saul, R.J. Cohen, Transfer function analysis of autonomic regulation. I. Canine atrial rate function. Am. J. Physiol. **256**, H142–H152 (1989)

57. C. Bernard, *Leçons Sur Les Phénomènes de La Vie Commun Aux Animaux et Aux Végétaux* (Ballière, Paris, 1878)

58. D.S. Bernstein, Robust static and dynamic output-feedback stabilization. Deterministic and stochastic perspectives. IEEE Trans. Autom. Control **AC-32**, 1076–1084 (1987)

59. M.J. Berry, M. Meister, Refractoriness and neural precision. J. Neurosci. **18**, 2200–2210 (1998)

60. H.S. Bhat, N. Kumar, Spectral solution of delayed random walks. Phys. Rev. E **86**, 045701 (2012)

61. W. Bialek, F. Rieke, R.R. de Ruyter, van Steveninck, D. Warland, Reading a neural code. Science **252**, 1854–1857 (1991)

62. J.Y. Bigot, A. Daunois, P. Mandel, Slowing down far from the limit points in optical bistability. Phys. Lett. A **123**, 123–127 (1987)

63. E. Bloch-Salisbury, P. Indic, F. Bednarek, D. Paydarfar, Stabilizing immature breathing patterns of preterm infants using stochastic mechanosensory stimulation. J. Appl. Physiol. **107**, 1017–1027 (2009)

64. D. Blow, So do we understand how enzymes work? Structure **8**, R77–R81 (2000)

65. J. Boeckh, K.-E. Kaissling, D.A. Schneider, Insect olfactory receptors. Cold Spring Harbor Symp. Quant. Biol. **30**, 1263–1280 (1965)

66. J.L. Bogdanoff, S.J. Citron, Experiments with an inverted pendulum subject to rapid parametric excitation. J. Acoust. Soc. Amer. **38**, 447–452 (1964)

67. K.F. Bonhoeffer, Activation of passive iron as a model for the excitation of nerve. J. Gen. Physiol. **32**, 441–486 (1948)

68. A. Borsellino, A. De Marco, A. Allazetta, S. Rinesi, B. Bartolini, Reversal time distribution in the perception of visual ambiguous stimuli. Kybernetik **10**, 139–144 (1972)

69. A. Bottaro, M. Casidio, P.G. Morasso, V. Sanguineti, Body sway during standing. Is it the residual chattering of an intermittent stabilization process? Hum. Mov. Sci. **24**, 588–615 (2005)

70. A. Bottaro, Y. Yasutake, T. Nomura, M. Casidio, P. Morasso, Bounded stability of the quiet standing posture: an intermittent control model. Hum. Mov. Sci. **27**, 473–495 (2008)

71. S. Bowles, S.N. Durlauf, K. Hoft, *Poverty Traps* (Princeton University Press, Princeton, New Jersey, 2006)

72. W.E. Boyce, R.C. DiPrima, *Elementary Differential Equations and Boundary Value Problems*, 9th edn. (Wiley, New York, 2005)

73. R.N. Bracewell, *The Fourier Transform and Its Applications*, 2nd edn., revised (McGraw-Hill, San Francisco, 1986)

74. F. Brauer, C. Castillo-Chavez, *Mathematical Models in Population Biology and Epidemiology* (Springer, New York, 2001)

75. W. Braune, O. Fischer, *On the Center of Gravity of the Human Body*, Translated by P.G.J. Maquet, R. Furong (Springer, Berlin, 1984)

76. D. Bray, Bacterial chemotaxis and the question of gain. Proc. Natl. Acad. Sci. USA **99**, 7–9 (2002)

77. H. Bremer, P.P. Denis, Modulation of chemical composition and other parameters of the cell by growth rate, in *Escherichia coli and Salmonella: Cellular and Molecular Biology.* ed. by F.C. Neidhardt (A. S. M. Press, Washington, D.C., 1996), pp. 1553–1569

78. P.C. Bressloff, C.V. Wood, Spontaneous oscillation in a nonlinear delayed-feedback shunting model of the pupil light reflex. Phys. Rev. E **58**, 3597–3606 (1997)

79. P.C. Bressloff, C.V. Wood, P.A. Howarth, Nonlinear shunting model of the pupil light reflex. Proc. R. Soc. Ser. B **263**, 953–960 (1996)

80. C. Briart, A. Gupta, M. Khammash, Antithetic integral feedback ensures robust perfect adaptation in noisy biomolecular networks. Cell Syst. **2**, 15–16 (2016)

81. P.W. Bridgman, *Dimensional Analysis* (Yale University Press, New Haven, 1922)

82. E.O. Brigham, *The Fast Fourier Transform and Its Applications* (Prentice Hall, Englewood, N. J., 1988)

83. D. Brockman, T. Giesel, The ecology of gaze shifts. Neurocomputing **32–33**, 643–650 (2000)

84. R. Brown, A brief account of microscopical observations made in the months of June, July, and August, 1827, on the particles contained in the pollen of plants; and on the general existence of active molecules in organic and inorganic bodies. Edinburgh New Philos. J. **5**, 358–371 (1828). An electronic copy can be downloaded from sciweb.nybg.org/science2/pdfs/dws/Brownian.pdf

85. L.M. Browning, K.J. Lee, T. Huang, P.D. Nallathamby, J.E. Lowman, X.-H.N. Xu, Random walk of single gold nanoparticles in zebrafish embryos leading to stochastic toxic effects on embryonic developments. Nanoscale **1**, 138–152 (2009)

86. G.A. Bryant, Animal signals and emotion in music: coordinating affect across group. Front. Psychol. **4**, 1–12 (2013)

87. J. Buck, Synchronization of rhythmic flashing of fireflies II. Quart. Rev. Biol. **63**, 265–285 (1988)

88. M.I. Budyko, The effect of solar radiation variations on the climate of the earth. Tellus **21**, 611–619 (1969)

89. M.A. Buice, J.D. Cowan, Statistical mechanics of the neocortex. Prog. Biophys. Mol. Biol. **99**, 53–86 (2009)

90. E. Burdet, T.E. Milner, Quantization of human motions and learning of accurate movements. Biol. Cybern. **78**, 307–318 (1998)

91. F.L. Burton, S.M. Cobbe, Dispersion of ventricular repolarization and refractory period. Cardiovasc. Res. **50**, 10–23 (2001)

92. O. Buse, R. Pérez, A. Kuznetsov, Dynamical properties of the repressilator model. Phys. Rev. E **81**, 066206 (2010)

93. J.L. Cabrera, R. Bormann, C. Eurich, T. Ohira, J. Milton, State-dependent noise and human balance control. Fluctuations Noise Lett. **4**, L107–L118 (2004)

94. J.L. Cabrera, J.G. Milton, On-off intermittency in a human balancing task. Phys. Rev. Lett. **89**, 158702 (2002)

95. J.L. Cabrera, J.G. Milton, Human stick balancing: tuning Lévy flights to improve balance control. CHAOS **14**, 691–698 (2004)

96. J.L. Cabrera, J.G. Milton, On-off intermittency and survival times. Nonlinear Stud. **11**, 305–317 (2004)

97. J.L. Cabrera, J.G. Milton, Stick balancing, falls and Dragon Kings. Eur. J. Phys. Spec. Top. **205**, 231–241 (2012)

98. B. Calancie, B. Needham-Shropshire, P. Jacobs, K. Willer, G. Zynch, B.A. Green, Involuntary stepping after chronic spinal cord injury: evidence for a central pattern generator for locomotion in man. Brain **117**, 1143–1159 (1994)

99. S. Camazine, J.-L. Deneubourg, N.R. Franks, J. Sneyd, G. Theraulaz, E. Bonabeau, *Self-Organization in Biological Systems* (Princeton University Press, Princeton, NJ, 2001)

100. S.A. Campbell, Stability and bifurcation of a simple neural network with multiple time delays, in *Fields Institute Communications*, vol. 21, ed. by S. Ruan, G.S.K. Wolkowicz, J. Wu (1999), pp. 65–79

101. S.A. Campbell, J. Bélair, T. Ohira, J. Milton, Limit cycles, tori and complex dynamics in a second-order differential equation with delayed negative feedback. J. Dyn. Diff. Eqns. **7**, 13–236 (1995)

102. C.J. Canavier, R.J. Butera, Functional phase response curves: a method for understanding synchronization of adapting neurons. J. Neurophysiol. **102**, 387–398 (2009)

103. W.B. Cannon, Organization for physiological homeostasis. Physiol. Rev. **36**, 399–431 (1929)

104. D.W. Carley, D.C. Shannon, A minimal mathematical model of human periodic breathing. J. Appl. Physiol. **65**, 1389–1399 (1988)

105. C.E. Carr, Processing of temporal information in the brain. Ann. Rev. Neurosci. **16**, 223–243 (1993)
106. I.R. Cassar, N.D. Titus, W.M. Grill, An improved genetic algorithm for designing optimal temporal patterns of neural stimulation. J. Neurol. Eng. **14**, 066013 (2017)
107. B. Castaing, G. Gunaratne, F. Heslot, L. Kadahoff, A. Libschaber, S. Thomae, X.-Z. Wu, S. Zaleski, G. Zanetti, Scaling of hard turbulence in Rayleigh-Bénard convection. J. Fluid Mech. **204**, 1–30 (1989)
108. H. Caswell, M.G. Neubert, Reactivity and transient dynamics of discrete-time ecological systems. J. Diff. Eq. Appl. **11**, 295–310 (2005)
109. S.R. y Cajal, *Histology of the Nervous System of Man and Vertebrates, Volumes 1 & 2* (Oxford University Press, New York, 1995)
110. R.D. Cazé, M. Humphries, B. Gutkin, Passive dendrites enable single neurons to compute linearly non-separable functions. PLoS Comp. Biol. **9**, e1002867 (2013)
111. B. Chance, The kinetics of the enzyme-substrate compound of peroxidase. J. Biol. Chem. **151**, 553–755 (1943)
112. J. Chang, D. Paydarfar, Switching neuronal state: optimal stimuli revealed using a stochastically-seeded gradient algorithm. J. Comp. Neurosci. **37**, 569–582 (2014)
113. J. Chang, D. Paydarfar, Evolution of extrema features reveals optimal stimuli for biological state transitions. Sci. Rep. **8**, 3403 (2018)
114. J. Chang, D. Paydarfar, Optimizing stimulus waveforms for electroceuticals. Biol. Cybern. **113**, 191–199 (2019)
115. R.A. Chapman, Repetitive responses in squid giant axons and their premature annihilation by additional brief depolarizing currents. Quart. J. Exp. Physiol. **65**, 1–7 (1980)
116. K.M. Chapman, R.S. Smith, A linear transfer function underlying impulse frequency modulation in a cockroach mechanoreceptor. Nature **197**, 699–700 (1963)
117. Z. Chen, P.C. Ivanov, K. Hu, H.E. Stanley, Effect of nonstationarities on detrended fluctuation analysis. Phys. Rev. E **65**, 041107 (2002)
118. Y. Chen, J. Wu, Slowly oscillating periodic solutions for a delayed frustrated network of two neurons. J. Math. Anal. Appl. **259**, 188–201 (2001)
119. N.S. Cherniack, G.S. Longobardo, Cheyne-Stokes breathing, an instability in physiologic control. New Engl. J. Med. **288**, 952–957 (1973)
120. N.S. Cherniack, G.S. Longobardo, Mathematical models of periodic breathing and their usefulness in understanding cardiovascular and respiratory disorders. Exp. Physiol. **91**(2), 295–305 (2006)
121. L. Chi, A.C.G. Uprichard, B.R. Lucchesi, Profibrillatory action of pinacidil in a conscious canine model of sudden coronary death. J. Cardiovasculat Pharm. **15**, 452–464 (1990)
122. H.J. Chiel, R.D. Beer, The brain has a body: adaptive behavior emerges from interactions of nervous system, body and environment. TINS **20**, 553–557 (1997)

123. R.A. Chisholm, E. Filotas, Critical slowing down as an indicator of transitions in two-species models. J. Theoret. Biol. **257**, 142–149 (2009)

124. A.V. Chizhov, A.V. Zefirov, D.V. Amakhin, E.Y. Smirnova, A.V. Zaitsev, Minimal model of interictal and ictal discharges "Epileptor-2." PLOS Comp. Biol. **14**, e1006186 (2018)

125. P.J. Choi, L. Cai, K. Fieda, X.S. Xie, A stochastic single-molecule event triggers phenotype switching of a bacterial cell. Science **322**, 442–446 (2008)

126. C.C. Chow, J.J. Collins, A pinned polymer model of postural control. Phys. Rev. E **52**, 907–912 (1995)

127. C.C. Chow, K.D. Hall, The dynamics of human body weight change. PLoS Comp. Biol. **4**, e1000045 (2008)

128. K. Christensen, N.R. Moloney, *Complexity and Criticality* (Imperial College Press, London, 2005)

129. P.H. Chu, J.G. Milton, J.D. Cowan, Connectivity and the dynamics of integrate-and-fire neural networks. Int. J. Bifurc. Chaos **4**, 237–243 (1994)

130. A.N. Churilov, J. Milton, E.R. Salakhova, An integrate-and-fire model for pulsativity in the neuroendocrine system. Chaos **30**, 083132 (2020)

131. F.H. Clarke, Y.S. Ledyaev, E.D. Sontag, A.I. Subbotin, Asymptotic controllability implies feedback stabilization. IEEE Trans. Autom. Control **20**, 1–13 (1999)

132. W.L. Clarke, S. Anderson, M. Breton, S. Patek, L. Kashmer, B. Kovatchev, Closed-loop artificial pancreas using subcutaneous glucose sensing and insulin delivery and a model predictive control algorithms: the Virginia experience. J. Diabetes Sci. Tech. **3**, 1031–1038 (2009)

133. A. Clauset, C.R. Shalizi, M.E.J. Newman, Power-law distributions in empirical data. SIAM Rev. **51**, 661–703 (2009)

134. J.R. Clay, Excitability of the squid giant axon revisited. J. Neurophysiol. **80**(903–913), 1998 (1998)

135. J.R. Clay, D.B. Folger, D. Paydarfar, Ionic mechanism underlying optimal stimuli for neuronal excitation: role of Na^+ channel inactivation. PLoS ONE **7**, e45983 (2012)

136. J.R. Clay, D. Paydarfar, D.A. Forger, A simple modification of the Hodgkin-Huxley equation explains Type 3 excitability in squid giant axons. J. R. Soc. Interface **5**(1421–1428), 2008 (2008)

137. T. Cluff, R. Balasubramanian, Motor learning characterized by changing Lévy distributions. PLoS ONE **4**, e5998 (2009)

138. B.A. Cohen, A. Sances, Stationarity of the human electroencephalogram. Med. Biol. Eng. Comput. **15**, 513–518 (1977)

139. K.S. Cole, *Membranes, Ions, and Impulses* (University of California Press, Berkeley, 1968)

140. B.J. Cole, Is animal behavior chaotic? Evidence from the activity of ants. Proc. R. Sovc. Lond. B **244**, 253–259 (1991)

141. M.J. Coleman, A. Ruina, An uncontrolled walking toy that cannot stand still. Phys. Rev. Lett. **80**, 3658–3661 (1998)

142. J.J. Collins, C.J. de Luca, Random walking during quiet standing. Phys. Rev. Lett. **73**, 764–767 (1994)

143. R.F. Constantino, R.A. Desharnais, J.M. Cushing, B. Dennis, Chaotic dynamics in an insect population. Science **275**, 389–391 (1997)

144. J.W. Cooley, J.W. Tukey, An algorithm for machine calculation of complex Fourier series. Math. Comput. **19**, 297–301 (1965)

145. R.M. Corless, G.H. Gonnet, D.E.G. Hare, D.J. Jeffrey, D.E. Knuth, On the Lambert W function. Ad. Comp. Math. **5**, 329–359 (1996)

146. G. Cornelissen, Cosinor-based rhythmometry. Theoret. Biol. Medical Mod. **11**, 16 (2014)

147. A. Cornish-Bowden, Enthalpy-entropy compensation: a phantom phenomenon. J. Biosci. **27**, 121–126 (2002)

148. R.F. Costantino, J.M. Cushing, B. Dennis, R.A. Desharnais, Experimentally induced transitions in the dynamic behavior of insect populations. Nature **375**, 227–230 (1995)

149. K. Cranmer, Kernel estimation in high-energy physics. Comput. Phys. Commun. **136**, 198–207 (2001)

150. E. Cruck, R. Moitie, N. Seube, Estimation of basins of attraction for uncertain systems with affine and Lipschitz dynamics. Dyn. Control **11**, 211–227 (2001)

151. H. Cruse, J. Storer, Open loop analysis of a feedback mechanism controlling the leg position in the stick insect Carausius morosus: comparison between experiment and simulation. Biol. Cybern. **25**, 143–153 (1977)

152. G. Csernak, G. Stepan, Life expectancy of transient microchaotic behavior. J. Nonlinear Sci. **15**, 63–91 (2005)

153. G. Csernak, G. Stepan, Digital control as a source of chaotic behavior. Int. J. Bifurc. Chaos **20**, 1365–1378 (2010)

154. J. Cui, C.C. Canavier, R.J. Butera, Functional phase response curves: a method for understanding synchronization of adapting neurons. J. Neurophysiol. **102**, 387–398 (2009)

155. C.A. Czeisler, J.S. Allan, Acute circadian phase reversal in man via bright light exposure: application to jet-lag. Sleep Res. **16**, 605 (1987)

156. C.A. Czeisler, M.P. Johnson, J.F. Duffy, E.N. Brown, J.M. Ronda, Exposure to bright light and darkness to treat physiologic maladaptation to night work. New. Engl. J. Med. **322**, 1253–1259 (1990)

157. M.A. Dahlem, S. Rode, A. May, N. Fujiwara, Y. Hirata, K. Aihara, J. Kürths, Towards dynamical network biomarkers in neuromodulators of episodic migraine. Transl. Neurosci. **4**, 282–294 (2013)

158. F. Lopes da Silva, W. Blanes, S.N. Kalitizin, J. Parra, P. Suffczynski, D.N. Velis, Epilepsies as dynamical diseases of brain systems: basic model of the transition between normal and epileptic activity. Epilepsia **44**(Suppl. 12), 72–83 (2003)

159. T.R. Darlington, J.M. Beck, S.G. Lisberger, Neural implementation of Bayesian inference in a sensorimotor behavior. Nature Neurosi. **21**, 1442–1451 (2018)

160. T. Dashevskiy, G. Cymbalyuk, Propensity for bistability of bursting and silence in the leech heart interneuron. Front. Comp. Neurosci. **12**, 5 (2018)

161. W.B. Davenport, W.L. Root, *An Introduction to the Theory of Random Signals and Noise* (IEEE Press, New York, 1987)

162. J. Day, J.E. Rubin, C.C. Chow, Competition between transients in the rate of approach to a fixed point. SIAM J. Appl. Dyn. Syst **8**, 1523–1563 (2009)

163. D.A. Dean II., D.F. Forger, E.B. Klerman, Taking the lag out of jet lag through model-based schedule design. PLoS Comp. Biol. **5**, e1000418 (2009)

164. G. de Vries, T. Hillen, M. Lewis, J. Müller, B. Schönfisch, *A Course in Mathematical Biology: Quantitative Modeling with Mathematical and Computational Methods* (SIAM, Philadelphia, 2006)

165. J.E. Deffeyes, R.T. Harbourne, A. Kyvelidou, W.A. Stuberg, N. Stergiou, Nonlinear analysis of sitting postural sway indicates developmental delay in infants. Clin. Biomech. **24**, 564–570 (2009)

166. H. Degn, A.V. Holden, L.F. Olsen, *Chaos in Biological Systems* (Plenum Press, New York, 1987)

167. H. Degn, I.F. Olsen, J.W. Perram, Bistability, oscillations and chaos in an enzyme reaction. Ann. N. Y. Acad. Sci. **316**, 623–637 (1979)

168. M.C. DeSantis, J.-L. Li, Y.M. Wang, Protein sliding and hopping kinetics on DNA. Phys. Rev. E **83**, 021907 (2011)

169. R.A, Desharnais, Population dynamics of Tribolium, in *Structured Population Models in Marine, Terrestrial, and Freshwater Systems*, ed. by S. Tuljapurkar, H. Caswell (Chapman & Hall, New York, 1997), pp. 303–328

170. J.M. Deutsch, Probability distributions for one component equations with multiplicative noise. Physica A **246**, 430–444 (1994)

171. R.L. Devaney, *An Introduction to Chaotic Dynamical Systems* (Benjamin Cummings, Menlo Park, CA, 1986)

172. A. Diniz, M.L. Wijnants, K. Torre, J. Barreiros, N. Crato, A.M.T. Bosman, F. Hasselman, R.E.A. Cox, G.C. van Order, D. Delognières, Contemporary theories of $1/f$ noise in motor control. Hum. Mov. Sci. **30**, 895–905 (2011)

173. P.S. Dodds, D.J. Watts, A generalized model of social and biological contagion. J. Theoret. Biol. **232**, 587–604 (2005)

174. R. Dodla, G. Svirskis, J. Rinzel, Well-timed, brief inhibition can promote spiking: postinhibitory facilitation. J. Neurophysiol. **95**, 2664–2677 (2006)

175. E. Doedel, indy.cs.concordia.ca/auto

176. R.J. Douglass, L. Wilkens, E. Pantazelou, F. Moss, Noise enhancement of information transfer in crayfish mechanoreceptors by stochastic resonance. Nature **365**, 337–340 (1989)

177. E.H. Dowell, M. Ilgamova, *Studies in Nonlinear Aeroelasticity* (Springer, New York, 1988)

178. P.G. Drazin, W.H. Reid, *Hydrodynamic Stability* (Cambridge University Press, Cambridge, England, 1981)

179. T. Drengstig, H.R. Ueda, P. Ruoff, Predicting perfect adaptation motifs in reaction kinetic networks. J. Phys. Chem. B **112**, 16752–16758 (2008)

180. J.C. Dessing, C.M. Craig, Bending it like Beckham: how to visually fool the goalkeeper. PLoS One **5**, e13161 (2010)

181. S.F. Donkers, M. Roerdink, A.J. Greven, P.J. Beck, Regularity of center-of-pressure trajectories depends on the amount of attention invested in postural control. Exp. Brain Res. **181**, 1–11 (2007)

182. R.D. Driver, D.W. Susser, M.L. Slater, The equation $\dot{x}(t) = ax(t) + bx(t - \tau)$ with "small delay.". Am. Math. Mon. **80**, 990–995 (1973)

183. W.N. Dudley, R. Wickham, N. Coombs, An introduction to survival statistics: Kaplan-Meier analysis. J. Adv. Pract. Oncol. **7**, 9–100 (2016)

184. S.E. Duenwald, R. Vanderby, R.S. Lakes, Constitutive equations for ligaments and other soft tissue: evaluation by experiment. Acta Mechanica **205**, 23–33 (2009)

185. G. Earhart, A. Bastian, Form switching during human locomotion: traversing wedges in a single step. J. Neurophysiol. **84**, 605–615 (2000)

186. L. Edelstein-Keshet, *Mathematical Models in Biology* (Random House, New York, 1988)

187. A.M. Edwards, R.A. Phillips, N.W. Watkins, M.P. Freeman, E.J. Murphy, V. Afanasyev, S.V. Buldyrev, M.G.E. da Luz, E.P. Raposo, H.E. Stanley, Revisiting Lévy flight search patterns of wandering albatrosses, bumblebees and deer. Nature **449**, 1044–1049 (2007)

188. P. Ehrenfest, T. Ehrenfest, Über zwei bekannte Einwände gegen der Boltzmannschen H-Theorem. Phys. Zeit. **8**, 311–314 (1907)

189. W. Einhäuser, K.A. Martin, P. König, Are switches in perception of the Necker cube related to eye position? Eur. J. Neurosci. **20**, 2811–2818 (2004)

190. A. Einstein, On the movement of small particles suspended in stationary liquids required by the molecular-kinetic theory of heat. Ann. Physik **17**, 549–560 (1905)

191. K. El Houssaini, A.I. Ivanov, C. Bernard, V.K. Jirsa, Seizures, refractory status epilepticus, and depolarization block as endogenous brain states. Phys. Rev. E **91**, 010701 (2015)

192. A.M. Ellison, Bayesian inference in ecology. Ecol. Lett. **7**, 509–520 (2004)

193. S.P. Ellner, J. Guckenheimer, *Dynamic Models in Biology* (Princeton University Press, Princeton, New Jersey, 2006)

194. M.B. Elowitz, S. Leibler, A synthetic oscillatory network of transcriptional regulators. Nature **403**, 335–338 (2000)

195. M.B. Elowitz, A.J. Levine, E.D. Siggia, P.S. Swain, Stochastic gene expression in a single cell. Science **297**, 1183–1186 (2002)

196. N. El-Samad, J.P. Goff, M. Khammash, Calcium homeostasis and parturient hypercalcemia: an integral feedback perspective. J. Theoret. Biol. **214**, 17–29 (2002)

197. A.K. Engel, W. Singer, Temporal binding and the neural correlates of sensory awareness. Trends Cog. Sci. **5**, 16–25 (2001)

198. D. Engel, I. Pahner, K. Schulze, C. Frahm, H. Jarry, G. Ahnert-Hilger, A. Draguhn, Plasticity of rat central inhibitory synapses through GABA metabolism. J. Physiol. **535**(2), 473–482 (2001)

199. K. Engelborghs, T. Luzyanina, D. Roose, Numerical bifurcation analysis of delay differential equations using DDE-BIFTOOL. ACM Trans. Math. Softw. (TOMS) **28**, 1–21 (2002)

200. E. Enikov, G. Stepan, Micro-chaotic motion of digitally controlled machines. J. Vib. Control **4**, 427–443 (1998)

201. C.E. Epstein, J. Shotland, The bad truth about Laplace's transform. SIAM Rev. **50**, 504–520 (2008)

202. B. Ermentrout, *Simulating, Analyzing, and Animating Dynamical Systems: A Guide to XPPAUT for Researchers and Students* (SIAM, Philadelphia, 2002)

203. G.B. Ermentrout, An adaptive model for synchrony in the firefly Pteroptyx malaccae. J. Math. Biol. **29**, 571–585 (1991)

204. G.B. Ermentrout, Type I membranes, phase resetting curves and synchrony. Neural Comput. **8**, 979–1001 (1996)

205. G.B. Ermentrout, L. Glass, B.E. Oldeman, The shape of phase-resetting curves in oscillators with a saddle node on an invariant circle bifurcation. Neural Comp. **24**, 3111–3125 (2012)

206. G.B. Ermentrout, N. Kopell, Oscillator death in systems of coupled neural oscillators. SIAM J. Appl. Math. **50**, 125–146 (1990)

207. G.B. Ermentrout, D.H. Terman, *Mathematical Foundations of Neuroscience* (Springer, New York, 2010)

208. T. Erneux, *Applied Delay Differential Equations* (Springer, New York, 2009)

209. C.W. Eurich, J.G. Milton, Noise-induced transitions in human postural sway. Phys. Rev. E **54**, 6681–6684 (1996)

210. C.W. Eurich, M.C. Mackey, H. Schwegler, Recurrent inhibitory dynamics: the role of state-dependent distribution of conduction delay times. J. Theoret. Biol. **216**, 31–50 (2002)

211. C.W. Eurich, K. Pawelzik, J.D. Cowan, J.G. Milton, Dynamics of self-organized delay adaptation. Phys. Rev. Lett. **82**, 1594–1597 (1999)

212. C.W. Eurich, A. Thiel, L. Fahse, Distributed delays stabilize ecologial feedback systems. Phys. Rev. Lett. **94**, 158104 (2005)

213. M. Evans, N. Hastings, B. Peacock, *Statistical Distributions*, 2nd edn. (Wiley-Interscience, New York, 1993)

214. M. Fairweather, Skill learning principles: implications for coaching practice, in *Coaching Process: Principles and Practice for Sport*. ed. by N. Cross, J. Lyle (Butterworth-Heineman, New York, 2003), pp. 113–129

215. A.A. Faisal, L.P. Selen, D.M. Wolpert, Noise in the nervous system. Nature Rev. **9**, 292–303 (2008)

216. G. Fan, S.A. Campbell, G.S.K. Wolkowicz, H. Zhu, The bifurcation study of 1 : 2 resonance in a delayed system of two coupled neurons. J. Dyn. Diff. Eqns. **25**, 193–216 (2013)

217. J.D. Farmer, Chaotic attractors of an infinite-dimensional dynamical system. Physica D **4**, 366–393 (1982)

218. P. Fatt, B. Katz, Some observations on biological noise. Nature **166**, 597–598 (1950)

219. S. Fauve, F. Heslot, Stochastic resonance in a bistable system. Phys. Lett. A **97**, 5–7 (1983)

220. O. Feely, A tutorial introduction to non-linear dynamics and chaos and their applications to sigma-delta modulators. Int. J. Circ. Theor. Appl. **25**, 347–367 (1997)

221. M.J. Feigenbaum, Quantitative universality for a class of non-linear transformations. J. Stat. Phys. **19**, 25–52 (1978)

222. J.L. Feldman, J.D. Cowan, Large-scale activity in neural nets I: Theory with application to motoneuron pool responses. Biol. Cybern. **17**, 29–38 (1975)

223. W. Feller, *An Introduction to Probability Theory and Its Applications*, vol. I, 3rd edn. (Wiley, New York, 1968)

224. R.D. Fields, Myelination: an overlooked mechanism of synaptic plasticity? Neuroscientist **11**, 528–531 (2005)

225. R. Fitzhugh, Mathematical models of threshold phenomena in the nerve membrane. Bull. Math. Biophys. **17**, 257–278 (1955)

226. R. Fitzhugh, Impulses and physiological states in theoretical models of nerve membranes. Biophys. J. **1**, 445–466 (1961)

227. R. Fitzhugh, *Mathematical Models for Excitation and Propagation in Nerve, in Mathematical Engineering* (McGraw Hill, New York, 1969), pp. 1–85

228. R. Fitzpatrick, D.K. Rogers, D.I. McCloskey, Proprioceptive, visual and vestibular thresholds for the perception of sway during standing in humans. J. Physiol. **478**, 173–186 (1994)

229. E.H. Flach, S. Schnell, Use and abuse of the quasi-steady-state approximation. Syst. Biol. **153**, 187–191 (2006)

230. E.H. Flach, S. Schnell, Stability of open pathways. Math. Biosci. **228**, 147–152 (2010)

231. I. Flügge-Lotz, *Discontinuous and Optimal Control* (McGraw-Hill, New York, 1968)

232. C. Foley, M.C. Mackey, Mathematical model for G-CSF administration after chemotherapy. J. Theoret. Biol. **19**, 25–52 (2009)

233. D.B. Forger, D. Paydarfar, J.R. Clay, Optimal stimulus shapes for neuronal excitation. PLoS Comp. Biol. **7**, e1002089 (2011)

234. J. Fort, D. Jana, J. Humet, Multidelayed random walk: theory and applications to the neolithic transition in Europe. Phys. Rev. E **70**, 031913 (2004)

235. E. Fontich, J. Sardanyés, General scaling law in the saddle-node bifurcation: a complex phase space study. J. Physics A: Math. Theoret. **41**, 015102 (2008)

236. J. Foss, A. Longtin, B. Mensour, J. Milton, Multistability and delayed recurrent loops. Phys. Rev. Lett. **76**, 708–711 (1996)

237. J. Foss, F. Moss, J. Milton, Noise, multistability and delayed recurrent loops. Phys. Rev. E **55**, 4536–4543 (1997)

238. J. Foss, J. Milton, Multistability in recurrent neural loops arising from delay. J. Neurophysiol. **84**, 975–985 (2000)

239. W.J. Freeman, *Neurodynamics: An Exploration of Mesoscopic Brain Dynamics* (Springer, New York, 2000)

240. W.J. Freeman, M.D. Holmes, Metastability, instability, and state transitions in neocortex. Neural Net. **18**, 497–504 (2005)

241. T. Frei, C.H., Chang, M., Filo, M., Khammash. Genetically engineered integral feedback controllers for robust perfect adaptation in mammalian cells. bioRxiv https://doi.org/10.1101/2020.12.06.412304 (2020)

242. A.S. French, A.V. Holden, Alias-free sampling of neuronal spike trains. Kybernetik **5**, 165–171 (1971)

243. A.S. French, P.H. Torkkeli, The power law of sensory adaptation: simulation by a model of excitability in spider mechanoreceptor neurons. Ann. Biomed. Eng. **36**, 153–161 (2008)

244. V. Frette, K. Christensen, A. Malthe-Sorenssen, J. Feder, T. Jossang, P. Meakin, Avalanche dynamics in a pile of rice. Nature **379**, 49–52 (1996)

245. A.A. Frost, R.G. Pearson, *Kinetics and Mechanism*, 2nd edn. (Wiley, New York, 1961)

246. T.D. Frank, Delay Fokker-Planck equations, Novikov's theorem, and Boltzmann distributions as small delau approximations. Phys. Rev. E **72**, 01112 (2005)

247. V. Fugère, R. Krahe, Electric signals and species recognition in the wave-type gymnotiform fish Apteronotus leptorhynchus. J. Exp. Biol. **213**, 225–236 (2010)

248. R.P. Futrelle, How molecules get to their detectors: the physics of diffusion of insect pheromones. Trends Neurosci. **7**, 116–120 (1984)

249. C.P. Gabbard, *Lifelong Motor Development*, 4th edn. (Pearson-Benjamin Cummings, New York, 2004)

250. R.F. Galan, G.B. Ermentrout, N.N. Urban, Predicting synchronized neural assemblies from experimentally estimated phase-resetting curves. Neurocomputing **69**, 1112–1115 (2006)

251. R. Gambell, *Birds and Mammals-Antarctica, in Antarctica* (Pergamon Press, New York, 1985), pp. 223–241

252. L. Gammaitoni, Stochastic resonance and the dithering effect in threshold physical systems. Phys. Rev. E **52**, 4691–4698 (1995)
253. L. Gammaitoni, P. Hänggi, P. Jung, F. Marchesoni, Stochastic resonance. Rev. Mod. Phys. **70**, 223–288 (1998)
254. A. Gandhi, S. Levin, S. Orszag, "Critical slowing down" in time-to-extinction: an example of critical phenomena in ecology. J. Theoret. Biol. **192**, 363–376 (1998)
255. J. Gao, W.D. Luedtke, D. Gourdon, M. Ruths, J.N. Israelachville, U. Landman, Frictional forces and Amonton's law: from the molecular to the macroscopic scale. J. Phys. Chem. B **108**, 3410–3425 (2008)
256. J. Gao, S. Ma, D.T. Major, K. Nam, J. Pu, D.G. Truhlar, Mechanisms and free energies of enzymatic reactions. Chem Rev. **106**, 3188–3209 (2006)
257. C.W. Gardiner, *Handbook of Stochastic Methods for Physics, Chemistry and the Natural Sciences* (Springer, New York, 1990)
258. T.S. Gardner, C.R. Cantor, J.J. Collins, Construction of a genetic toggle switch in Escherichia coli. Nature **403**, 339–342 (2000)
259. L. Garrido, A. Juste, The determination of probability density functions by using neural networks. Comp. Phys. Commun. **115**, 25–31 (1998)
260. J.-P. Gauthier, B. Berret, F. Jean, A biomechanical inactivation principle. Proc. Steklov Inst. Math. **268**, 93–116 (2010)
261. P.J. Gawthrop, L. Wang, Intermittent model prediction control. Proc. Int. Mech. Eng. part I. Mech. Eng. I-J Syst. **211**, 1007–1018 (2007)
262. P. Gawthrop, I. Loram, K. Lakie, H. Gollee, Intermittent control: a computational theory of human control. Biol. Cybern. **104**, 31–51 (2007)
263. P. Gawthrop, I. Loram, H. Gollee, M. Lakie, Intermittent control models of human standing: similarities and differences. Biol. Cybern. **108**, 159–168 (2014)
264. W.B. Gearhart, H.S. Shultz, The function $\sin x/x$. Coll. Math. J. **21**, 90–99 (1990)
265. H.L. Gerber, R.P. Joshi, C.C. Tseng, Using Bode plots to access intracellular coupling. IEEE Trans. Plasma Sci. **36**, 1659–1664 (2008)
266. P. Giesel, H. Wagner, Lyapunov functions and their basins of attraction for a single-point muscle-skeletal model. J. Math. Biol. **54**, 453–464 (2007)
267. L. Giuggioli, Z. Neu, Fokker-Planck representation of non-Markov representations of non-Markov equations: application to delayed systems. Phil. Trans. R. Soc. A. **377**, 20180131 (2018)
268. L. Gil, D. Sornette, Landau-Ginzburg theory of self-organized criticality. Phys. Rev. Lett. **76**, 3991–3994 (1996)
269. D.T. Gillespie, Exact stochastic simulation of coupled chemical reactions. J. Phys. Chem. **81**, 2340–2361 (1977)
270. M. Gitterman, Simple treatment of correlated multiplicative and additive noise. J. Phys. A: Math. Gen. **32**, L293–L297 (1999)

271. L. Giuggioli, Z. Neu, Fokker-Planck representations of non-Markov Langevin equations: applications to delayed systems. Phil. Trans. R. Soc. A **377**, 20180131 (2019)

272. P. Glansdorff, I. Prigogine, *Thermodynamic Theory of Structure, Stability and Fluctuations* (Wiley, New York, 1971)

273. L. Glass, Resetting and entraining biological rhythms, in *Nonlinear Dynamics in Physiology and Medicine*. ed. by A. Beuter, L. Glass, M.C. Mackey, M.S. Titcombe (Springer, New York, 2003), pp. 124–148

274. L. Glass, C. Graves, G.A. Petrillo, M.C. Mackey, Unstable dynamics of a periodically driven oscillator in the presence of noise. J. Theoret. Biol. **85**, 455–473 (1980)

275. L. Glass, M.C. Mackey, Pathological conditions resulting from instabilities in physiological control systems. Ann. N. Y. Acad. Sci. **316**, 214–235 (1979)

276. L. Glass, M.C. Mackey, *From Clocks to Chaos: The Rhythms of Life* (Princeton University Press, Princeton, New Jersey, 1988)

277. L. Glass, C.P. Malta, Chaos in multi-looped negative feedback systems. J. Theor. Biol. **145**, 217–223 (1990)

278. L. Glass, P. Hunter, A. McCulloch, *Theory of Heart: Biophysics, Biomechanics and Nonlinear Dynamics of Cardiac Function* (Springer, New York, 1991)

279. A.L. Goldberg, Nonlinear dynamics, fractals and chaos: application to cardiac electrophysiology. Ann. Biomed. Eng. **18**, 195–198 (1990)

280. A.L. Goldberger, V. Bharagava, B.J. West, A.J. Mandell, On the mechanism of cardiac electrical stability: the fractal hypothesis. Biophys. J. **48**, 525–528 (1985)

281. M. Gómez-Schiavon, L.F. Chen, A.E. West, N.G. Buchler, BayFish: Bayesian inference of transcription dynamics from population snapshots of single-molecule RNA-FISH in single cells. Genome Biol. **18**, 164 (2017)

282. M. Goodfellow, K. Schindler, G. Baier, Self-organized transients in a neural mass model of epileptogenic tissue dynamics. NeuroImage **55**, 2644–2660 (2012)

283. B.C. Goodwin, *Temporal Organization in Cells* (Academic Press, New York, 1963)

284. J. Gotman, Measurement of small time differences between EEG channels method and its application to epileptic seizure propagation. Electroenceph. Clin. Neurophysiol. **56**, 501–514 (1983)

285. P.S. Grassia, Delay, feedback and quenching in financial markets. Eur. Phys. J. B **17**, 347–362 (2000)

286. J.S. Griffith, Mathematics of cellular control processes. I. Negative feedback to one gene. J. Theor. Biol. **20**, 202–216 (1968)

287. Y.Y. Grinberg, J.G. Milton, R.P. Kraig, Spreading depression sends microglia on Lévy flights. PLoS ONE **6**, e19294 (2011)

288. C. Grotta-Ragazzo, K. Pakdaman, C.P. Malta, Metastability for delayed differential equations. Phys. Rev. E. **60**, 6230–6233 (1999)

289. H.L. Grubin, T.R. Govindan, Quantum contributions and violations of the classical law of mass action. VLSI Design **6**, 61–64 (1998)

290. J. Guckenheimer, A robust hybrid stabilization strategy for equilibria. IEEE Trans. Autom. Control **40**, 321–326 (1995)

291. J. Guckenheimer, P. Holmes, *Nonlinear Oscillations, Dynamical Systems, and Bifurcations of Vector Fields* (Springer, New York, 1983)

292. J. Guckenheimer, K. Hoffman, W. Weckesser, Numerical computation of canards. Int. J. Bifurc. Chaos **10**, 2669–2687 (2000)

293. J. Guckenheimer, R.A. Oliva, Chaos in the Hodgkin-Huxley model. SIAM J. Appl. Dyn. Sys. **1**, 105–114 (2002)

294. M.R. Guevara, Bifurcations involving fixed points and limit cycles in biological systems, in *Nonlinear Dynamics in Physiology and Medicine*, ed. by A. Beuter, L. Glass, M.C. Mackey, M.S. Titcombe (Springer, 2003), pp. 41–85

295. M.R. Guevara, Dynamics of excitable cells, in *Nonlinear Dynamics in Physiology and Medicine*, ed. by A. Beuter, L. Glass, M.C. Mackey, M.S. Titcombe (Springer, 2003), pp. 87–121

296. M.R. Guevara, L. Glass, A. Shrier, Phase locking, period-doubling bifurcations, and irregular dynamics in periodically stimulated cardiac cells. Science **214**, 1350–1353 (1981)

297. M.R. Guevara, A. Shrier, L. Glass, Phase resetting in a model of sinoatrial nodal membrane: Ionic and topological aspects. Amer. J. Physiol. **251**, H1298–H1305 (1986)

298. S. Guillouzic, I. L'Heureux, A. Longtin, Small delay approximation of stochastic delay differential equations. Phys. Rev. E **59**, 3970–3982 (1999)

299. H.M. Gupta, J.R. Campanha, The gradually truncated Lévy flight for systems with power-law distributions. Physica A **268**, 231–239 (1999)

300. H.M. Gupta, J.R. Campanha, The gradually truncated Lévy flight: stochastic process for complex systems. Physica A **275**, 531–543 (2000)

301. P. Guptasarma, Does the replication-induced transcription regulate synthesis of the myriad low number proteins in Escherichia coli? Bioassays **17**, 987–997 (1995)

302. T. Gulati, L. Guo, D.R. Ramanathan, A. Bodepudi, K. Ganguly, Neural reactivation drug sleep determines network credit assignment. Nature Neurosci. **20**, 1277–1284 (2017)

303. R. Guttman, S. Lewis, J. Rinzel, Control of repetitive firing in squid axon membrane as a model for a neuron oscillator. J. Physiol. (London) **305**, 377–395 (1980)

304. A.C. Guyton, J.W. Crowell, J.W. Moore, Basic oscillating mechanism of Cheyne-Stokes breathing. Am. J. Physiol. **187**, 395–398 (1956)

305. A. Haidar, L. Legault, M. Dallaire et al., Glucose-responsive insulin and glucagon delivery (dual-hormone artificial pancreas) in adults with type 1 diabetes: a randomized crossover trial. Can. Med. Assoc. J. **185**, 297–305 (2013)

306. H. Haken, *Brain Dynamics: Synchronization and Activity Patterns in Pulse-Coupled Neural Nets with Delays and Noise* (Springer, New York, 2002)

307. J.R.S. Haldane, *Enzymes* (Longmans, Green and Company, London, 1930)

308. J. Hale, H. Koçak, *Dynamics and Bifurcations* (Springer, New York, 1991)

309. J.K. Hale, *Theory of Functional Differential Equations* (Springer, New York, 1977)

310. S.E. Halford, J.F. Marko, How do site-specific DNA-binding proteins find their targets? Nucleic Acids Res. **32**, 3040–3052 (2004)

311. G. Haller, G. Stepan, Micro-chaos in digital control. J. Nonlin. Sci. **6**, 415–448 (1996)

312. O.P. Hamill, A. Marty, B. Sakmann, F.J. Sigworth, Improved patch-clamp techniques for high-resolution recording from cells and cell-free membrane patches. Pflugers Arch. **391**, 85–100 (1981)

313. C. Hammond, H. Bergman, P. Brown, Pathological synchronization in Parkinson's disease: networks, models and treatments. Trends Neurosci. **30**, 357–364 (2007)

314. C.H. Hansen, R.G. Endres, N.S. Wingreen, Chemotaxis in Escherichia coli: a molecular model for robust precise adaptation. PLoS Comput. Biol **4**, 14–27 (2008)

315. A.V. Harcourt, W. Esson, On the laws of connexion between the conditions of a chemical change and its amount. Phil. Trans. Roy. Soc. (London) **156**, 193–221 (1866)

316. A.L. Harris, Darwin's legacy: an evolutionary view of women's reproductive and sexual functioning. J. Sex Res. **50**, 207–246 (2013)

317. C.M. Harris, D.M. Wolpert, Signal-dependent noise determines motor planning. Nature **394**, 780–784 (1998)

318. S. Harrison, B. Backus, Disambiguating Necker cube rotation using a location cue: what types of spatial location signal can the visual system learn? J. Vision **10**, 1–15 (2010)

319. A. Hastings, Transients: the key to long-term ecological understanding? Trends Ecol. Evol. **19**, 39–45 (2004)

320. S.P. Hastings, J.J. Tyson, D. Webster, Existence of periodic solutions for negative feedback control systems. J. Diff. Equ. **25**, 39–64 (1977)

321. N.G. Hatsopoulos, J.P. Donoghue, The science of brain-machine interface technology. Ann. Rev. Neurosci. **32**, 229–266 (2009)

322. J.M. Hausdorff, Gait dynamics, fractals and falls: finding meaning in the stride-to-stride fluctuations of human walking. Hum. Mov. Sci. **26**, 555–589 (2007)

323. J. Hausdorff, C. Peng, Multiscaled randomness: a possible source of $1/f$ noise in biology. Phys. Rev. E **54**, 2154–2157 (1996)

324. J.M. Hausdorff, M.E. Cudkowicz, R. Firtion, J.Y. Wei, A.L. Goldberger, Gait variability and basal ganglia disorders: stride-to-stride variation of gait cycle timing in Parkinson's disease and Huntington's disease. Mov. Disord. **13**, 428–437 (1998)

325. J.M. Hausdorff, H.K. Edelberg, S.L. Mitchell, A.L. Goldberger, J.Y. Wei, Increased gait unsteadiness in community dwelling elderly fallers. Arch. Phys. Med. Rehab. **78**, 278–283 (1997)

326. J.M. Hausdorff, S.L. Mitchell, R. Firtion, C.K. Peng, M.E. Cudkowicz, J.Y. Wei, A.L. Goldenberg, Altered fractal dynamics of gait: reduced stride interval correlations with aging and Huntington's disease. J. Appl. Physiol. **82**, 262–269 (1996)

327. J.M. Hausdorff, C.-K. Peng, Z. Ladin, J.Y. Wei, A.L. Goldberger, Is walking a random walk? evidence for long-range correlations in stride interval of human gait. J. Appl. Physiol. **78**, 349–358 (1995)

328. W.A. Hauser, J.F. Annegers, L.T. Kurland, The incidence of epilepsy in Rochester, Minnesota, 1935–1984. Epilepsia **34**, 453–468 (1993)

329. H. Hayashi, S. Ishizuka, Chaotic nature of bursting discharges in the Onchidium pacemaker neuron. J. Theor. Biol. **156**, 269–291 (1992)

330. H. Hayashi, S. Ishizuka, Chaotic responses of the hippocampal CA3 region to a mossy fiber stimulation in vitro. Brain Res. **686**, 194–206 (1995)

331. H. Hayashi, M. Nakao, K. Nirakawa, Entrained, harmonic, quasiperiodic and chaotic responses of the self-sustained oscillation of Nitella to sinusoidal stimulation. J. Physiol. Soc. Japan **52**, 344–351 (1983)

332. N.D. Hayes, Roots of the transcendental equation associated with a certain difference-differential equation. J. Lond. Math. Soc. **25**, 226–232 (1950)

333. Q. He, Y. Lin, Molecular mechanism of light responses in Neurospora: from light-induced transcription to photoadaptation. Genes Dev. **19**, 2888–2899 (2005)

334. J.F. Heagy, N. Platt, S.M. Hammel, Characterization of on-off intermittency. Phys. Rev. E **49**, 1140–1150 (1994)

335. C.N. Heck, D. King-Stephens, A.D. Massey, et al., Two-year seizure reduction in adults with medically intractable partial-onset epilepsy treated with responsive neurostimulation: final results of the RNS System Pivotal trial. Epilepsia **55**, 432–441

336. D. Helbing, P. Molnár, I.J. Farkas, K. Bolay, Self-organizing pedestrian movement. Environ. Plan. B: Plan. Design **28**, 361–383 (2001)

337. F.A. Hellebrandt, R.H. Tepper, G.L. Braun, Location of the cardinal anatomical orientation planes passing through the center of weight in young adult women. Am. J. Physiol. **121**, 465–470 (1938)

338. L.F. Henderson, The statistics of crowd fluids. Nature **229**, 381–383 (1971)

339. P. Hersen, M.N. McClean, L. Mahadevan, S. Ramanathan, Signal processing by the HOG MAP kinase pathway. Proc. Natl. Acad. Sci. USA **105**, 7165–7170 (2008)

340. M.H. Higgs, C.J. Wilson, Measurement of phase resetting curves using optogenetic barrage stimuli. J. Neurosci. Meth. **289**, 23–30 (2017)

341. A.V. Hill, The possible effects of the aggregation of the molecules of haemoglobin on its dissociation curves. J. Physiol. **40**(Suppl), iv–vii (1910)

342. W.E. Hick, On the rate of gain of information. Quart. J. Exp. Psych. **4**, 11–26 (1952)

343. B. Hille, *Ion Channels of Excitable Membranes*, 3rd edn. (Sinnauer Associated, Sunderland, MA, 2001)

344. A.L. Hodgkin, The local electric changes associated with repetitive action in a non-modulated axon. J. Physiol. **107**, 165–181 (1948)

345. A.L. Hodgkin, The Cronian lecture: Ionic movements and electrical activity in giant nerve fibers. Proc. R. Soc. London, Ser. B **148**, 1–37 (1958)

346. A.L. Hodgkin, *The Conduction of the Nervous Impulse* (Liverpool University Press, Liverpool, 1964)

347. A.I. Hodgkin, A.F. Huxley, A quantitative description of membrane current and its application to conduction and excitation in nerve. J. Physiol. (London) **117**, 500–544 (1951)

348. A.L. Hodgkin, R.D. Keynes, Experiments on the injection of substances into squid giant axons by means of a microsyringe. J. Physiol. (London) **131**, 592–616 (1956)

349. J.J. Hopfield, Neural networks and physical systems with emergent collective computational abilities. Proc. Natl. Acad. Sci. USA **79**, 2554–2558 (1982)

350. F. Horn, R. Jackson, *General Mass Action Kinetics* (Springer, New York, 1972)

351. W. Horsthemenke, R. Lefever, *Noise-Induced Transitions: Theory and Applications in Physics, Chemistry and Biology* (Springer, New York, 1984)

352. T. Huang, L.M. Browning, X.-H.N. Xu, Far-field photostable optical nanoscopy (photon) for real-time super-resolution single-molecular imaging of signaling pathways of single live cells. Nanoscale **4**, 2797–2812 (2012)

353. R. Huang, I. Chavey, K.M. Taule, B. Lukic, S. Jeney, M.G. Raizer, E.-L. Florin, Direct observation of the full transition from ballistic to diffusive Brownian motion in a liquid. Nature Phys. **7**, 576–580 (2011)

354. M. Hubbard, The flight of sports projectiles, in *Biomechanics in Sport: Performance Enhancement and Injury Prevention (Volume IX of the Encyclopedia of Sports Medicine)*. ed. by V. Zatsiorsky (Blackwell Science, London, 2000), pp. 381–400

355. D.H. Huebel, T.N. Wiesel, M.P. Stryker, Orientation columns in macaque monkey visual cortex demonstrated by the 2-deoxyglucose autoradiographic technique. Nature **269**, 328–330 (1977)

356. J.P. Huelsenbeck, F. Ronquist, R. Nielsen, J.P. Bollback, Bayesian inference of phylogeny and its impact on evolutionary biology. Science **294**, 2310–2314 (2001)

357. N.E. Humphries, N. Queiroz, J.R.M. Dyer, N.G. Pade, M.K. Musyl, K.M. Schaefer, D.W. Fuller, J.M. Brunnschweiler, T.K. Doyle, J.D.R. Houghton, G.C. Hays, C.S. Jones, L.R. Noble, V.J. Wearmouth, E.J. Southall1, D.W. Sims, Environmental context explains Lévy and Brownian movement patterns of marine predators. Nature **465**, 1066–1069 (2010)

358. J.D. Hunter, J. Milton, Using inhibitory interneurons to control neural synchrony, in *Epilepsy as Dynamic Disease*. ed. by J. Milton, P. Jung (Springer, New York, 2003), pp. 115–130

359. J.D. Hunter, J.G. Milton, Amplitude and frequency dependence of spike timing: implications for dynamic regulation. J. Neurophysiol. **90**, 387–394 (2003)

360. J.D. Hunter, J.G. Milton, H. Lüdtke, B. Wilhelm, H. Wilhelm, Spontaneous fluctuations in pupil size are not triggered by lens accommodation. Vis. Res. **40**, 567–573 (2000)

361. J.D. Hunter, J.G. Milton, P.J. Thomas, J.D. Cowan, Resonance effect for neural spike time reliability. J. Neurophysiol. **80**, 1427–1438 (1998)

362. G.E. Hutchinson, Circular cause systems in ecology. Ann. New York Acad. Sci. **50**, 221–246 (1948)

363. W.D. Hutchinson, R.J. Allan, H. Opitz, R. Levy, J.O. Dostrovsky, A.G. Lang, A.M. Lozano, Neurophysiological identification of subthalamic nucleus in surgery for Parkinson's disease. Ann. Neurol. **44**, 622–628 (1998)

364. A. Huxley, From overshoot to voltage clamp. Trends Neurosci. **25**, 553–558 (2002)

365. T. Ido, C-N. Wan, V. Casella, J.S. Fowler, A.D. Wold, M. Reivich, D.E. Kuhl, Labeled 2-deoxy-D-glucose analogs. ^{18}F-labeled 2-deoxy-2-fluoro-D-glucose, 2-deoxy-2-fluoro-D-mannose, and ^{14}C-2-deoxy-2-fluoro-D-glucose. J. Lab. Compounds Radiopharm. **14**, 175–182 (1978)

366. E.L. Ince, *Ordinary Differential Equations* (Dover Publications, New York, 1956)

367. T. Insperger, G. Stepan, Semi-discretization method for delayed systems. Int. J. Numer. Meth. Eng. **55**, 503–518 (2002)

368. T. Insperger, Act-and-wait concept for continuous-time control systems with feedback delay. IEEE Trans. Control Sys. Technol. **14**, 974–977 (2007)

369. T. Insperger, On the approximation of delayed systems by Taylor series expansion. J. Comp. Nonlin. Dyn. **10**, 024503 (2015)

370. T. Insperger, J. Milton, Sensory uncertainty and stick balancing at the fingertip. Biol. Cybern. **108**, 85–101 (2014)

371. T. Insperger, J. Milton, G. Stepan, Acceleration feedback improves balancing against reflex delay. J. R. Soc. Interface **10**, 20120763 (2013)

372. T. Insperger, G. Stepan, Stability of the damped Mathieu equation with time delay. J. Dyn. Syst. Meas. Control **125**, 166–171 (2003)

373. T. Insperger, G. Stepan, *Semidiscretization for Time-Delay Systems* (Springer, New York, 2011)

374. T. Insperger, J. Milton, G. Stepan, Semidiscretization for time-delayed neural balance control. SIAM J. Appl. Dyn. Sys. **14**, 1258–1277 (2015)

375. T. Insperger, G. Stepan, Act-and-wait control concept for discrete-time systems with feedback delay. IET Control Theory A **1**, 553–557 (2007)

376. T. Insperger, G. Stépán, J. Turi, State-dependent delay in regenerative turning processes. Nonlin. Dyn. **47**, 275–283 (2007)

377. S. Ishizuka, H. Hayashi, Chaotic and phase-locked responses of the somatosensory cortex to a periodic medial lemniscus stimulation in the anesthetized rat. Brain Res. **723**, 46–60 (1996)

378. E.M. Izhikevich, Resonate-and-fire neurons. Neural Netw. **14**, 883–894 (2001)

379. E.M. Izhikevich, Neural excitability, spiking and bursting. Int. J. Bifurc. Chaos **10**, 1171–1266 (2000)

380. E.M. Izhikevich, Simple models of spiking neurons. IEEE Trans. Neural Net. **14**, 1569–1572 (2003)

381. E.M. Izhikevich, Which model to use for cortical spiking neurons? IEEE Trans. Neural Net. **15**, 1063–1070 (2004)

382. E.M. Izhikevich, *Dynamical Systems in Neuroscience: The Geometry of Excitability and Bursting* (MIT Press, MIT, 2007)

383. E.W. Izhikevich, G.M. Edelman, Large-scale model of mammalian thalamocortical system. Proc. Natl. Acad. Sci. USA **105**, 3593–3598 (2008)

384. E. Jakobsson, R. Guttman, The standard Hodgkin-Huxley model and squid axons in reduced external C^{++} fail to accommodate to slowly rising currents. Biophys. J. **31**, 293–298 (1980)

385. J. Jalife, Mathematical approaches to cardiac arrhythmias. Ann. N. Y. Acad. Sci. **591**, 1–417 (1990)

386. J. Jalife, G.K. Moe, Effect of electrotonic potential on pacemaker activity of canine Purkinje fibers in relation to parasystole. Circ. Res. **39**, 801–808 (1976)

387. C. Janse, A neurophysiological study of the peripheral tactile system of the pond snail Lymnaea stagnalis. Neth. J. Zool. **24**, 93–161 (1973)

388. G.M. Jenkins, D.G. Watts, *Spectral Analysis and Its Applications* (Emerson-Adams Press Inc, Boca Raton, Florida, 1969)

389. H.J. Jensen, *Self-Organized Criticality: Emergent Complex Behavior in Physical and Biological Systems* (Cambridge University Press, New York, 1998)

390. K. Jerger, S.J. Schiff, Periodic pacing and in vitro epileptic focus. J. Neurophysiol. **73**, 876–879 (1995)
391. L.C. Jia, P.-Y. Lai, C.K. Chan, Scaling properties of avalanches from a collapsing granular pile. Phys. A **281**, 404–412 (2000)
392. D.L. Jindrich, R.J. Full, Dynamic stabilization of rapid hexapedal locomotion. J. Exp. Biol. **205**, 2803–2823 (2002)
393. V.K. Jirsa, P. Fink, P. Foo, J.A.S. Kelso, Parametric stabilization of biological coordination: a theoretical model. J. Biol. Phys. **26**, 85–112 (2000)
394. V.K. Jirsa, W.C. Stacey, P.P. Quilichini, A.I. Ivanov, C. Bernard, On the nature of seizure dynamics. Brain **137**, 2210–2230 (2014)
395. V.K. Jirsa, T. Proix, D. Perdikis, M.M. Woodman, H. Wang, J. Gonzalex-Martinez, C. Bernard, C. Bénar, M. Guye, P. Chauvel, F. Bartolomei, The virtual epileptic patient: individualized whole-brain models of epilepsy spread. NeuroImage **145**, 377–388 (2017)
396. B.R. Johnson, R.A. Wyttenbach, R. Wayne, R.R. Hong, Action potentials in a giant algal cell: a comparative approach to mechanisms and evolution of excitability. J. Undergrad. Neurosci. Edu. **1**, A23–A27 (2002)
397. T.P. Jorgensen, *The Physics of Golf*, 2nd edn. (Springer, New York, 1999)
398. K.T. Judd, K. Aihara, Pulse propagation networks: a neural network model that uses temporal coding by action potentials. Neural Nr. **6**, 203–215 (1993)
399. M. Kac, Random walk and the theory of Brownian motion. Amer. Math. Monthly **54**, 369–391 (1947)
400. M. Kaern, T.C. Elston, W.J. Blake, J.J. Collins, Stochasticity in gene expression: from theories to phenotypes. Nat. Rev. Genet. **6**, 451–464 (2005)
401. A. Kamimura, T. Ohira, Group chase and escape. New J. Phys. **12**, 053013 (2010)
402. A. Kamimura, T. Ohira, *Group Chase and Escape: Fusion of Pursuits-Escapes and Collective Motions* (Springer, New York, 2019)
403. T. Kiemel, K.S. Oie, J.J. Jeka, Slow dynamics of postural sway are in the feedback loop. J. Neurophysiol. **95**, 1410–1418 (2006)
404. A.F. Kohn, A. Friestada Rocha, J.P. Segundo, Presynaptic irregularity and pacemaker inhibition. Biol. Cybern. **41**, 5–18 (1981)
405. B.A. Kovacs, J. Milton, T. Insperger, Virtual stick balancing: sensory uncertainties in angular displacement and velocity. Roy. Soc. Open Sci. **6**, 191006 (2019)
406. P. Krauss, A. Zankl, A. Schelley, H. Schulz, V. Metzner, Analysis of structure and dynamics of three-neuron motifs. Front. Com. Neurosci. **13**, 5 (2019)

407. H.I. Krebs, M.L. Aisen, B.T. Volpe, N. Hogan, Quantization of continuous arm movements in humans with brain injury. Proc. Natl. Acad. Sci. USA **96**, 4645–4639 (1999)
408. T. Krogh-Madsen, L. Glass, E.J. Doedel, M.R. Guevara, Apparent discontinuities in the phase-resetting response of cardiac pacemakers. J. Theoret. Biol. **230**, 499–517 (2004)
409. T. Krogh-Madsen, R. Butera, G.B. Ermentrout, L. Glass, Phase resetting neural oscillators: topological theory versus real world, in *Phase Response Curves in Neuroscience: Theory, Experiment, and Analysis*, ed. by N.W. Schultheiss, et al. Springer Series in Computational Neuroscience 6
410. A.M. Kuncel, W.M. Grill, Selection of stimulus parameters for deep brain stimulation. Clin. Neurophysiol. **115**, 2431–2441 (2004)
411. A.M. Kunsyz, A. Shrier, L. Glass, Bursting behavior during fixed-delay stimulation of spontaneously beating chick heart cell aggregates. Amer. J. Physiol. **273**, C331–C346 (1997)
412. I.L. Kurtzer, Long-latency reflexes account for limb biomechanics through several supraspinal pathways. Front. Integr. Neurosci. **8**, 99 (2015)
413. N.G. Van Kampen, *Stochastic Processes in Physics and Chemistry*, 3rd edn. (Elsevier, New York, 2007)
414. D. Kaplan, L. Glass, *Understanding Nonlinear Dynamics* (Springer, New York, 1995)
415. S. Kar, W.T. Baumann, W.R. Paul, J.J. Tyson, Exploring the roles of noise in the eukaryotic cell cycle. Proc. Natl. Acad. Sci. USA **106**, 6471–6476 (2009)
416. S. Karlin, J. McGregor, Ehrenfest urn models. J. Appl. Prob. **2**, 352–376 (1965)
417. A. Katchalsky, P.F. Curran, *Nonequilibrium Thermodynamics in Biophysics* (Harvard University Press, Cambridge, Massachusetts, 1967)
418. M. Kawato, Transient and steady state response curves of limit cycle oscillators. J. Math. Bio. **12**, 13–30 (1981)
419. M. Kawato, Internal model for motor control. Curr. Opin. Neurobiol. **9**, 718–727 (1999)
420. R.E. Kearney, I.W. Hunter, System identification of human joint dynamics. CRC Crit. Rev. Biomed. Eng. **18**, 55–87 (1990)
421. J. Keener, A. Panfilov, A biophysical model for defibrillation of cardiac tissue. Biophys. J. **71**, 1335–1345 (1996)
422. J. Keener, J. Sneyd, *Mathematical Physiology* (Springer, New York, 1998)
423. J.A.S. Kelso, *Dynamic Patterns: The Self-Organization of Brain and Behavior* (MIT Press, Boston, 1995)
424. J. Kempe, Quantum random walks: an introductory overview. Cont. Phys. **44**, 307–327 (2003)

425. M. Khfifi, M. Loulidi, Scaling properties of a rice-pile model: inertia and friction effects. Phys. Rev. E **78**, 051117 (2008)
426. M. Khosroyamiand, G.K. Hung, A dual-mode dynamic model of the human accommodative system. Bull. Math. Biol. **64**, 285–299 (2002)
427. K.A. Kleopa, Autoimmune channelopathies of the nervous system. Curr. Neuropharmacol. **9**, 458–467 (2011)
428. M.D. Klimstra, E. Thomas, R.H. Stoloff et al., Neuromechanical considerations for incorporating rhythmic arm movement in the rehabilitation of walking. Chaos **19**, 026102 (2009)
429. D. Kleinfeld, F. Raccula-Behling, H.J. Chiel, Circuits constructed from identified Aplysia neurons exhibit multiple patterns of persistent activity. Biophys. J. **57**, 697–715 (1990)
430. D.L. Kleinman, Optimal control of linear systems with time-delay and observational noise. IEEE Trans. Automat. Control **14**, 524–527 (1969)
431. F.M. Klis, A. Boorsma, P.W.J. DeGroot, Cell wall construction in Saccharomycetes cerevisiae. Yeast **23**, 185–202 (2006)
432. P.E. Kloeden, E. Platen, *Numerical Solution of Stochastic Differential Equations* (Springer, New York, 1992)
433. B.K. Knight, Dynamics of encoding in a population of neurons. J. Gen. Physiol. **59**, 734–766 (1972)
434. J. Kofránek, J. Rusz, Restoration of Guyton's diagram for regulation of the circulation as a basis for quantitative physiological model development. Physiol. Rev. **59**, 897–908 (2010)
435. S. Kogan, D. Williams, Characteristic function based estimation of stable parameters, in *A Practical Guide to Heavy Tailed Data*. ed. by R. Adler, R. Feldman, M. Tagga (Birkhäuser, Boston, 1998), pp. 311–338
436. V.B. Kolmanovski, V.R. Nosov, *Stability of Functional Differential Equations* (Academic Press, London, 1986)
437. Y. Kondoh, M. Hisada, Neuroanatomy of the terminal (sixth abdominal) ganglion of the crayfish, Procambarus clarkii (Girard). Cell Tissue Res. **243**, 273–288 (1986)
438. I. Koponen, Analytic approach to the problem of convergence of truncated Lévy flights towards the Gaussian stochastic process. Phys. Rev. E **52**, 1197–1199 (1995)
439. K.P. Körding, D.M. Wolpert, Bayesian integration in sensorimotor learning. Nature **427**, 244–247 (2004)
440. K.P. Körding, D.M. Wolpert, Bayesian decision theory in sensorimotor control. Trends Cog. Sci. **10**, 319–326 (2006)
441. T. Kostova, R. Ravindran, M. Schonbek, Fitzhugh-Nagumo revisted: types of bifurcations, periodic forcing and stability regions by a Lyapunov functional. Int. J. Bifurc. Chaos **14**, 913–925 (2004)
442. P. Kowalcyzk, G. Glendinning, M. Brown, G. Medrano-Cerda, H. Dallali, J. Shapiro, Modelling human balance using switched systems with linear feedback control. J. R. Soc. Interface **9**, 234–245 (2012)

443. D. Krapf, E. Marinari, R. Metzler, G. Oshanin, X. Xu, A. Squarcini, Power spectral density of a single Brownian trajectory: what one can and cannot learn from it. New. J. Phys. **20**, 023029 (2018)

444. D. Krapf, N. Kukat, E. Marinari, R. Metzler, G. Oshanin, C. Selhunter-Unkel, A. Squarcini, L. Stadler, M. Weiss, X. Xu, Spectral content of a single non-Brownan trajectory. Phys. Rev. X **9**, 0110219 (2019)

445. M. Krstic, *Delay Compensation for Nonlinear, Adaptive, and PDE Systems* (Birkäuser, Boston, MA, 2009)

446. J.M. Kruger, D.M. Rector, S. Roy, H.P.A. Van Dingen, G. Belelnk, J. Panksepp, Sleep as a fundamental property of neuronal assemblies. Nat. Rev. Neurosci. **9**(12), 910–919 (2008)

447. P. Kruse, M. Stadler, *Ambiguity in Mind and Nature* (Springer, New York, 1995)

448. Y. Kuang, *Delay Differential Equations with Application in Population Dynamics* (Academic Press, San Diego, 1993)

449. U. Küchler, B. Mensch, Langevin's stochastic differential equation extended by a time-delayed term. Stoch. Stoch. Rep. **40**, 23–42 (1992)

450. P. Kügler, Early afterdischarges with growing amplitudes via delayed subcritical Hopf bifurcations and unstable manifolds of saddle foci in cardiac action potential dynamics. PLoS One **11**, e0151178 (2016)

451. T.A. Kuiken, L.A. Miller, R.D. Lipschutz, B.A. Lock, K. Stubble-field, P.D. Marasso, P. Zhou, G. Dumanian, Targeted reinnervation for enhanced prosthetic arm function in a woman with a proximal amputation: a case study. Lancet **369**, 371–380 (2007)

452. A.D. Kuo, The six determinants of gait and the inverted pendulum analogy: a dynamic walking perspective. Hum. Mov. Sci. **26**, 617–656 (2007)

453. K. Kuriyama, R. Stickgold, M.P. Walker, Sleep-dependent learning and motor-skill complexity. Learn. Mem. **11**, 705–713 (2004)

454. A. Kusumi, Y. Sako, M. Yamamoto, Confined lateral diffusion of membrane receptors as studied by single particle tracking (nanovid microscopy). Effects of calcium-induced differentiation in cultured epithelial cells. Biophys. J. **65**, 2021–2040 (1993)

455. Y.A. Kuznetsov, *Elements of Applied Bifurcation Theory* (Springer, New York, 1998)

456. A. Kuznetsov, V. Afraimovich, Heteroclinic cycles in the repressilator model. Chaos, Solitons Fractals **45**, 660–665 (2012)

457. K.A. Kwon, R.J. Shipley, M. Edirisingle, D.G. Ezra, G. Rose, S.M. Best, R.E. Cameron, High-speed camera characterization of voluntary eye blinking kinematics. J. R. Soc. Interface **10**, 20130227 (2013)

458. S.J. Lade, T. Gross, Early warning signals for critical transitions: a generalized modeling approach. PLoS Comp. Biol. **8**, e1002360 (2012)

459. Y.-C. Lai, T. Tél, *Transient Chaos: Complex Dynamics on Finite Time Scales* (Springer, New York, 2011)

460. M. Landry, S.A. Campbell, K. Morrisand, A.O. Aguilar, Dynamics of an inverted pendulum with delay feedback control. SIAM J. Appl. Dyn. Sys. **4**, 333–351 (2005)

461. P. Langevin, Sur la théorie du mouvement brownien. C. R. Acad. Sci. (Paris) **146**, 530–533 (1908)

462. A. Lasota, M.C. Mackey, *Chaos, Fractals, and Noise: Stochastic Aspects of Dynamics* (Springer, New York, 1994)

463. G.W. Lee, A. Ishihara, K.A. Jacobsen, Direct observation of Brownian motion of lipids in a membrane. Proc. Natl. Acad. Sci. USA **88**, 6274–6278 (1991)

464. E.T. Lee, J.W. Wang, *Statistical Methods for Survival Data Analysis*, 4th edn. (Wiley, Hobaken, NJ, 2013)

465. D. Lehotzky, Numerical methods for the stability and stabilizability analysis of delayed dynamical systems. Ph.D. thesis. Budapest University of Technology and Economics (2017)

466. J. Lehtonen, The Lambert W function in ecological and evoltionary models. Meth. Ecology Evol. **7**, 1110–1118 (2016)

467. E.M. Leise, Modular construction of nervous system: a basic principle of design in invertebrates and vertebrates. Brain Res. Brain Res. Rev. **15**, 1–23 (1990)

468. I. Lengyel, I.R. Epstein, Modeling of the Turing structures in the chlorine-iodide-malonic acid-starch reaction. Science **251**, 650–652 (1991)

469. I. Lengyel, G. Rabai, I.R. Epstein, Experimental and modeling study of oscillations in the chlorine dioxide-malonic acid reaction. J. Am. Chem. Soc. **112**, 9104–9110 (1990)

470. P. Lennie, The cost of cortical computation. Current Biol. **13**, 493–497 (2003)

471. I. Lestas, J. Paulsson, N.E. Ross, G. Vinnicomb, Noise in gene regulatory networks. IEEE Trans. Autom. Control (Special Issue) **53**, 189–200 (2008)

472. H.K. Leung, Stochastic transient of noisy Lotka-Volterra model. Chin. J. Phys. **29**, 637–652 (1991)

473. A. Levchenko, P.A. Iglesias, Models of eukaryotic gradient sensing: application to chemotaxis and neutrophils. Biophys. J. **82**(1 Pt. 1), 50–63 (2002)

474. M. Levi, W. Weckesson, Stabilization of the inverted linearized pendulum by high frequency vibration. SIAM Rev. **37**, 219–223 (1995)

475. H. Levine, Pattern formation in the microbial world: Dictyostelium discoideum, in *Epilepsy as a Dynamic Disease.* ed. by J. Milton, P. Jung (Springer, New York, 2003), pp. 189–211

476. J. Lewis, M. Bachoo, L. Glass, C. Polosa, Complex dynamics resulting from repeated stimulation on the respiratory rhythm: phase resetting and aftereffects. Phys. Lett. A **125**, 119–122 (1987)

477. S.V.F. Levy, C. Tsallis, A.M.C. Souza, R. Maynard, Statistical-mechanical foundation of the ubiquity of Lévy distributions in nature. Phys. Rev. Lett. **75**, 3589–3593 (1995)

478. G.W. Lewis, A new principle of equilibrium. Proc. Natl. Acad. Sci. **11**, 179–183 (1925)

479. T.J. Lewis, M.R. Guevara, $1/f$ power spectrum of the QRS complex does not imply fractal activation of the ventricles. Biophys. J. **60**, 1297–1300 (1991)

480. S.P. Leys, G.O. Mackie, R.W. Meech, Impulse conduction is a sponge. J. Exp. Biol. **202**, 1139–1150 (1999)

481. T. Li, S. Kheifets, D. Medellin, M.G. Raizen, Measurement of the instantaneous velocity of a Brownian particle. Science **328**, 1673–1675 (2010)

482. Y. Li, W.S. Levine, G.E. Loeb, A two-joint human postural control model with realistic time delays. IEEE Trans. Neural Syst. Rehab. Eng. **20**, 738–748 (2012)

483. T.Y. Li, J.A. Yorke, Period three implies chaos. Amer. Math. Monthly **82**, 985–982 (1975)

484. E. Libby, P.B. Rainey, Exclusion rules, bottlenecks and the evolution of stochastic phenotype switching. Proc. R. Soc. B **278**, 3574–3583 (2011)

485. A. Likas, Probability density estimation using artificial neural networks. Comp. Phys. Commun. **135**, 167–175 (2001)

486. J.C. Lilly, G.M. Austin, W.W. Chambers, Threshold movements produced by excitation of cerebral cortex and efferent fibes with some parametric regions of rectangular current pulses (cats and monkeys). J. Neurophysiol. **15**, 319–341 (1952)

487. C.J. Limb, J.T. Robinson, Current research on music perception in cochlear implant users. Otolaryngol. Clin. North Am. **45**, 129–140 (2012)

488. O.C.J. Lippold, Oscillation in the stretch reflex arc and the origin of the rhythmical 8–12 c/s component of essential tremor. J. Physiol. **206**, 359–382 (1970)

489. L.A. Lipsitz, M. Lough, J. Nieme, T. Travison, H. Howlett, B. Manor, A shoe insole delivering subsensory vibratory noise improves balance and gait in healthy elderly people. Ach. Phys. Med. Rehabil. **96**, 432–439 (2015)

490. A. Lit, The magnitude of the Pulfrich stereo-phenomenon as a function of target velocity. J. Exp. Psychol. **59**, 165–175 (1960)

491. Y. Liu, J. Milton, S.A. Campbell, Outgrowing seizures in childhood absence epilepsy: time delays and bistability. J. Comp. Neurosci. **46**, 197–209 (2019)

492. D.B. Lockhart, L.H. Ting, Optimal sensorimotor transformations for balance. Nature Neurosci. **10**, 1329–1336 (2007)

493. T. Loddenkemper, A. Pan, Deep brain stimulation in epilepsy. J. Clin. Neurophysiol. **116**, 217–234 (2001)

494. I.E. Loewenfeld, *The Pupil: Anatomy, Physiology, and Clinical Applications*, vol. I (Iowa State University Press, Ames, Iowa, 1993)

495. R.B. Loftfield, E.A. Eigner, A. Pastuszyn, T.N. Lövgren, H. Jakubowski, Conformational changes during enzyme catalysis: role of water in the transition state. Proc. Natl. Acad. Sci. USA **77**, 3374–3378 (1980)

496. A. Longtin, Noise-induced transitions at a Hopf bifurcation in a first-order delay-differential equation. Phys. Rev. A **44**, 4801–4813 (1991)

497. A. Longtin, A. Bulsara, F. Moss, Time interval sequences in bistable systems and the noise induced transmission of information by sensory neurons. Phys. Rev. Lett. **67**, 656–659 (1991)

498. A. Longtin, J.G. Milton, Complex oscillations in the human pupil light reflex with "mixed" and delayed feedback. Math. Biosci. **90**, 183–199 (1988)

499. A. Longtin, J.G. Milton, Insight into the transfer function, gain and oscillation onset for the pupil light reflex using delay-differential equations. Biol. Cybern. **61**, 51–58 (1989)

500. A. Longtin, J.G. Milton, Modelling autonomous oscillations in the human pupil light reflex using nonlinear delay-differential equations. Bull. Math. Biol. **51**, 605–624 (1989)

501. A. Longtin, J.G. Milton, J.E. Bos, M.C. Mackey, Noise and critical behavior of the pupil light reflex at oscillation onset. Phys. Rev. A **41**, 6992–7005 (1990)

502. I.D. Loram, C.N. Maganaris, M. Lakie, Human postural sway results from frequent, ballistic bias impulses by soleus and gastrocnemius. J. Physiol. **56**, 295–311 (2005)

503. I.D. Loram, M. Lakie, I. Di Giulo, C.N. Maganaris, The consequences of short-range stiffness and fluctuating muscle activity for proprioception of postural joint rotations: the relevance to human standing. J. Neurophysiol. **102**, 460–474 (2009)

504. I.D. Loram, H. Gollee, M. Lakie, P.J. Gawthrop, Human control of an inverted pendulum: is continuous control necessary? Is intermittent control effective? Is intermittent control physiological? J. Physiol. **589**(2), 307–324 (2011)

505. S.R. Lord, C. Sherrington, H.B. Menz, *Falls in Older People: Risk Factors and Strategies for Prevention* (Cambridge University Press, New York, 2001)

506. E.N. Lorenz, Deterministic nonperiodic flow. J. Atmos. Sci. **20**, 282–293 (1963)

507. A.J. Lotka, *Elements of Mathematical Biology* (Dover Publications, New York, 1956)

508. D.W. Loving, R. Kapur, K.J. Meadot, M.J. Morrell, Differential neuropsychological outcomes following targeted responsive neurostimulation for partial-onset epilepsy. Epilepsia **56**, 1836–1844 (2015)

509. Y. Liu, J. Milton, S.A. Campbell, Outgrowing seizures in childhood absence epilepsy: time delays and bistability. J. Comp. Neurosci. **46**, 197–209 (2019)

510. K. Liu, H. Wang, J. Xiao, Z. Taha, Analysis of human standing balance by largest lyapunov exponent. Comput. Intell. Neurosci. **2015**, 158478 (2015)

511. E.W. Lund, Guldberg and Waage and the law of mass action. J. Chem, Ed. **42**, 548–550 (1965)

512. J. Luo, J. Wang, T.M. Ma, Z. Sun, Reverse engineering of bacterial chemotaxis pathway via frequency domain analysis. PLoS ONE **5**, e9182 (2010)

513. A.M. Lyapunov, *The General Problem of the Stability of Motion* (Taylor & Francis, Washington, DC, 1992)

514. W.W. Lytton, Computer modeling of epilepsy. Nature Rev. **9**, 626–637 (2008)

515. J. Ma, J. Wu, Multistability in spiking neuron models of delayed recurrent inhibitory loops. Neural Comp. **19**, 2124–2148 (2007)

516. J. Ma, J. Wu, Patterns, memory and periodicity in two-neuron recurrent inhibitory loops. Math. Model. Nat. Phenom. **5**, 67–99 (2010)

517. D.K.C. MacDonald, The statistical analysis of electrical noise. Phil. Mag. **41**, 814–818 (1950)

518. D.K.C. MacDonald, *Noise and Fluctuations: An Introduction* (Wiley, New York, 1962)

519. N. MacDonald, *Biological Delay Systems: Linear Stability Theory* (Cambridge University Press, New York, 1989)

520. M.C. Mackey, Periodic auto-immune hemolytic anemia: an induced dynamical disease. Bull. Math. Biol. **41**, 829–834 (1979)

521. M.C. Mackey, U. an der Heiden, The dynamics of recurrent inhibition. J. Math. Biol. **19**, 211–225 (1984)

522. M.C. Mackey, L. Glass, Oscillation and chaos in physiological control systems. Science **197**, 287–289 (1977)

523. M.C. Mackey, J.G. Milton, Dynamical diseases. Ann. N. Y. Acad. Sci. **504**, 16–32 (1987)

524. M.C. Mackey, J.G. Milton, A deterministic approach to survival statistics. J. Math. Biol. **28**, 33–48 (1990)

525. H. Maclennan, *Barometer Rising* (William Collins, Toronto, 1941)

526. C.F.M. Magalhães, J.G. Moreira, Catastrophic regime in the discharge of a sandpile. Phys. Rev. E **82**, 051303 (2010)

527. J.M. Mahaffy, J. Bélair, M.C. Mackey, Hematopoietic model with moving boundary condition and state dependent delay. J. Theoret. Biol. **190**, 135–146 (1998)

528. J.M. Mahaffy, C.V. Pao, Models of genetic control by repression with time delays and spatial effects. J. Math. Biol. **20**, 39–57 (1984)

529. J.M. Mahaffy, E.S. Savev, Stability analysis for mathematical models of the lac operon. Quart. Appl. Math. **57**, 37–53 (1999)

530. B.G. Malkiel, *A Random Walk Down Wall Street* (W. W. Norton & Company, New York, 2003)

531. A. Malthe-Sorenssen, J. Feder, K. Christensen, V. Frette, T. Jossang, Surface fluctuations and correlations in a pile of rice. Phys. Rev. Lett. **83**, 764–767 (1999)

532. T.R. Malthus, *An Essay on the Principle of Population* (Penguin, Baltimore, MD, 1970)

533. J.G. Mancilla, T.J. Lewis, D.J. Pinto, J. Rinzel, B.W. Connors, Synchronization of electrically coupled pairs of inhibitory interneurons in neocortex. J. Neurosci. **27**, 2058–2078 (2007)

534. S. Mandal, R.R. Sarkar, S. Sinha, Mathematical models of malaria: a review. Malaria J. **10**, 202 (2011)

535. B.B. Mandlebrot, J.W. Van Ness, Fractional Brownian motions, fractional noises, and applications. SIAM Rev. **10**, 422–437 (1968)

536. A.Z. Manitis, A.W. Olbrot, Finite spectrum assignment problem for systems with delays. IEEE Trans. Autom. Control **AC-24**, 541–553 (1979)

537. R.N. Mantegna, H.E. Stanley, Scaling behavior in the dynamics of an economic index. Nature (London) **376**, 46–49 (1995)

538. M.W. Marcellin, M.A. Lepley, A. Bilgin, T.J. Flohr, T.T. Chinen, J.H. Kasner, An overview of quantization in JPEG. Signal Proc. Image Commun. **17**(73–84), 2002 (2000)

539. H. Margenau, G.M. Murphy, *The Mathematics of Physics and Chemistry* (Van Nostrand, Toronto, 1968)

540. P.J. Marin, A.J. Herrero, J.G. Milton, T.J. Hazell, D. Garcia-Lopez, Whole-body vibration applied during upper body exercise improves performance. J. Strength Conditioning Res. **27**, 1807–1812 (2013)

541. J.B. Marion, *Classical Dynamics of Particles and Systems*, 2nd edn. (Academic Press, New York, 1970)

542. O.D. Martinez, J.P. Segundo, Behavior of a single neuron in a recurrent excitatory loop. Biol. Cybern. **47**, 33–41 (1983)

543. L.M. Martyushev, V.D. Seleznev, Maximum entropy production principle in physics, chemistry and biology. Phys. Rep. **426**, 1–45 (2006)

544. G. Matsumoto, K. Aihara, Y. Hanyu, N. Takahashi, S. Yoshizawa, J. Nagumo, Chaos and phase locking in normal squid axons. Phys. Lett. A **123**, 162–166 (1987)

545. G. Matsumoto, K. Aihara, M. Ichikawa, A. Tasaki, Periodic and non-periodic responses of membrane potentials in squid giant axons during sinusoidal current stimulation. J. Theoret. Neurobiol. **3**, 1–14 (1984)

546. G. Matsumoto, T. Kunisawa, Critical slowing-down near the transition region from the resting to time-ordered states in squid giant axons. J. Phys. Soc. Japan **44**, 1047–1048 (1978)

547. R.M. May, Simple mathematical models with very complicated dynamics. Nature **261**, 459–467 (1976)

548. C. Maurer, R. Peterka, A new interpretation of spontaneous sway measures based on a simple model of human postural control. J. Neurophysiol. **93**, 189–200 (2005)

549. R.M. Mazo, *Brownian Motion: Fluctuations, Dynamics and Applications* (Oxford Science Publications, New York, 2002)

550. M.K. McClintock, Menstrual synchrony and suppression. Nature **229**, 244–245 (1971)

551. S. McKay, *The Secret Life of Bletchley Park: The WWII Codebreaking Center and the Men and Women Who Worked There* (Aurum Press, London, 2010)

552. R.W. Meech, G.O. Mackie, *Evolution of Excitability in Lower Metazoans, in Invertebrate Neurobiology* (Cold Spring Harbor Laboratory Press, New York, 2007), pp. 581–616

553. B. Mehta, S. Schaal, Forward models in visuomotor control. J. Neurophysiol. **88**, 942–953 (2002)

554. B.A. Mello, Y. Tu, Perfect and near-perfect adaptation in a model of bacterial chemotaxis. Biophys. J. **84**, 2943–2956 (2003)

555. B.A. Mello, Y. Tu, Effects of adaptation in maintaining high sensitivity over a wide range of backgrounds for Escherichia coli chemotaxis. Biophys. J. **92**, 2329–2337 (2007)

556. J.W. Mellor, *Higher Mathematics for Students of Chemistry and Physics (Green, and Co* (Longmans, New York, 1902)

557. K. Mergenthaler, R. Engbert, Modeling the control of fixational eye movements with neurophysiological delays. Phys. Rev. Lett. **98**, 138104 (2007)

558. N. Metropolis, M.L. Stein, P.R. Stein, On finite limit sets for transformations on the unit interval. J. Comb. Theory **15**, 25–44 (1973)

559. J.T. Mettetal, D. Murray, C. Gómez-Uribe, A. van Oudenaarden, The frequency dependence of osmo-regulation in Saccharomyces cerevisiae. Science **319**, 482–484 (2008)

560. U. Meyer, J. Shao, S. Chakrabarty, S.E. Brandt, H. Luksch, R. Wessel, Distributed delays stabilize neural feedback. Biol. Cybern. **99**, 79–87 (2008)

561. G. Meyer-Kress, I. Choi, N. Weber, R. Bargas, A. Hübler, Musical signals from Chua's circuit. IEEE Trans. Circuits Syst. II Analog Digital Signal Process. **40**, 688–695 (1993)

562. L. Michaelis, M.L. Menten, Die Kinetik der Invertinwirkung. Biochem. Zts. **49**, 333 (1913)

563. M.J. Miller, S.H. Wei, M.D. Chandler, I. Parker, Autonomous T cell trafficking examined in vivo with intravital two-photon microscopy. Proc. Natl. Acad. Sci. USA **100**, 2604–2609 (2003)

564. M.J. Miller, S.H. Wei, I. Parker, M.D. Cahalan, Two-photon imaging of lymphocyte motility and antigen response in intact lymph node. Science **296**, 1869–1873 (2002)

565. R. Miller, What is the contribution of axonal conduction delay to temporal structure of brain dynamics?, in *by C.* ed. by O.E.-R.Brain. Dynamics (New York, Pantex (Academic Press, 1994), pp. 53–57

566. R. Milo, S. Shen-Orr, S. Itzkovitz, N. Kashtan, D. Chklovskii, U. Alon, Network motifs: simple building blocks of complex networks. Science **298**, 824–827 (2002)

567. J. Milton, *Dynamics of Small Neural Populations* (American Mathematical Society, Providence, Rhode Island, 1996)

568. J. Milton, Pupil light reflex: delays and oscillations, in *Nonlinear Dynamics in Physiology and Medicine*. ed. by A. Beuter, L. Glass, M.C. Mackey, M.S. Titcombe (Springer, New York, 2003), pp. 271–302

569. J. Milton, Neuronal avalanches, epileptic quakes and other transient forms of neurodynamics. Eur. J. Neurosci. **35**, 2156–2163 (2012)

570. J.G. Milton, J. Bélair, Chaos, noise and extinction in models of population growth. Theor. Pop. Biol. **37**, 273–290 (1990)

571. J. Milton, D. Black, Dynamic diseases in neurology and psychiatry. Chaos 8–13 (1995)

572. J.G. Milton, A. Fuerte, C. Bélair, J. Lippai, A. Kamimura, T. Ohira, Delayed pursuit-escape as a model for virtual stick balancing. Nonlinear Theory Appl. IEICE **4**, 129–137 (2013)

573. J. Milton, J. Gyorffy, J.L. Cabrera, T. Ohira, Amplitude control of human postural sway using Achilles tendon vibration. USNCTAN2010 **2010**, 791 (2010)

574. J. Milton, P. Jung, *Epilepsy as a Dynamic Disease* (Springer, New York, 2003)

575. J. Milton, P. Naik, C. Chan, S.A. Campbell, Indecision in neural decision making models. Math. Model. Nat. Phenom. **5**, 125–145 (2010)

576. J.G. Milton, Epilepsy: multistability in a dynamic disease, in *Self-Organized Biological Dynamics & Nonlinear Control*. ed. by J. Walleczek (Cambridge University Press, New York, 2000), pp. 374–386

577. J.G. Milton, Epilepsy as a dynamic disease: a tutorial of the past with an eye to the future. Epil. Behav. **18**, 33–44 (2010)

578. J.G. Milton, The delayed and noisy nervous system: implications for neural control. J. Neural Eng. **8**, 065005 (2011)

579. J.G. Milton, Intermittent motor control: the "drift-and-act" hypothesis, in *Progress in Motor Control: Neural, Computational and Dynamic Approaches*. ed. by M.J. Richardson, M. Riley, K. Shockley (Springer, New York, 2013), pp. 169–193

580. J.G. Milton, J.L. Cabrera, T. Ohira, Unstable dynamical systems: delays, noise and control. Europhys. Lett. **83**, 48001 (2008)

581. J.G. Milton, J. Foss, Oscillations and multistability in delayed feedback control, in *The Art of Mathematical Modeling: Case Studies in Ecology, Physiology and Cell Biology*. ed. by H.G. Othmer, F.R. Adler, M.A. Lewis, J.C. Dallon (Prentice Hall, New York, 1997), pp. 179–198

582. J.G. Milton, W.C. Galley, Evidence for heterogeneity in DNA-associated solvent mobility from acridine phosphorescence spectra. Biopolymers **25**, 1673–1684 (1986)

583. J.G. Milton, W.C. Galley, Rate constants and activation parameters for the mobility of bulk and DNA-associated glycol-water solvents. Biopolymers **25**, 1685–1695 (1986)

584. J.G. Milton, A. Longtin, Evaluation of pupil constriction and dilation from cycling measurements. Vis. Res. **30**, 515–525 (1990)

585. J.G. Milton, A. Longtin, A. Beuter, M.C. Mackey, L. Glass, Complex dynamics and bifurcations in neurology. J. Theoret. Biol. **138**, 129–147 (1989)

586. J.G. Milton, A. Longtin, H.H. Kirkham, G.S. Francis, Irregular pupil cycling as a characteristic abnormality in patients with demyelinative optic neuropathy. Amer. J. Opthalmology **105**, 402–407 (1988)

587. J.G. Milton, A.R. Quan, I. Osorio, Nocturnal frontal lobe epilepsy: metastability in a dynamic disease?, in *Epilepsy: Intersection of Neuroscience, Biology, Mathematics, Engineering and Physics.* ed. by I. Osorio, N.P. Zaveri, M.G. Frei, S. Arthurs (CRC Press, New York, 2011), pp. 445–450

588. J.G. Milton, A.E. Radunskaya, A.H. Lee, L.G. de Pillis, D.F. Bartlett, Team research at the biology-mathematics interface: project management perspectives. CBE Life Sci. Educ. **9**, 316–322 (2010)

589. J. Milton, T. Insperger, G. Stepan, Human balance control: dead zones, intermittency and micro-chaos, in *Mathematical approaches to Biological Systems: Networks, Oscillations, and Collective Motions.* ed. by T. Ohira, T. Uzawa (Springer, New York, 2015), pp. 1–28

590. J. Milton, A. Radunskaya, W. Ou, T. Ohira, A thematic approach to undergraduate research in biomathematics: balance control. Math. Model. Nat. Phenom. **6**, 260–277 (2011)

591. J. Milton, R. Meyer, M. Zhvanetsky, S. Ridge, T. Insperger, Control at stability's edge minimizes energetic costs: expert stick balancing. J. R. Soc. Interface **13**, 20160212 (2016)

592. J.G. Milton, T. Insperger, W. Cook, D.M. Harris, G. Stepan, Microchaos in human postural balance: sensory dead zones and sampled time-delayed feedback. Phys. Rev. E **98**, 022223 (2018)

593. J. Milton, T. Insperger, Actng together destabilizing influences can stabilize human balance. Phil. Trans. R. Soc. A **377**, 20180126 (2019)

594. J.G. Milton, S.S. Small, A. Solodkin, On the road to automatic: dynamic aspects in the development of expertise. J. Clin. Neurophysiol. **21**, 134–143 (2004)

595. J.G. Milton, T. Ohira, J.L. Cabrera, R.M. Fraiser, J.B. Gyorffy, F.K. Ruiz, M.A. Strauss, E.C. Balch, P.J. Marin, J.L. Alexander, Balancing with vibration: a prelude for "drift and act" balance control. PLoS ONE **4**, e7427 (2009)

596. J.G. Milton, J.L. Townsend, M.A. King, T. Ohira, Balancing with positive feedback: the case for discontinuous control. Phil. Trans. R. Soc. A **367**, 1181–1193 (2009)

597. J. Milton, J. Wu, S.A. Campbell, J. Bélair, Outgrowing neurological disease: microcircuits, conduction delay and childhood absence epilepsy, in *Computational Neurology - Computational Psychiatry: Why and How?*, ed. by P. Erdi, S. Bhattacharya, A. Cochran (Springer, New York), pp. 11–47 (2017)

598. R.E. Mirollo, S.H. Strogatz, Synchronization of pulse-coupled biological oscillators. SIAM J. Appl. Math. **50**, 1645–1662 (1990)

599. J. Milton, J.L. Cabrera, T. Ohira, S. Tajima, Y. Tonosaki, C.W. Eurich, S.A. Campbell, The time-delayed inverted pendulum: implications for human balance control. Chaos **19**, 026110 (2009)

600. L. Mirny, M. Slutsky, Z. Wunderlich, A. Tativigi, J. Leigh, A. Kosmrlj, How a protein searches for its site on DNA: the mechanism of facilitated diffusion. J. Phys. A. Math. Theor. **42**, 434013 (2009)

601. P. Moller, *Electric Fishes: History and Behavior* (Chapman & Hall, London, 1995)

602. J. Monod, F. Jacob, General conclusions: teleonomic mechanisms in cellular metabolism, growth and differentiation. Cold Spring Harbor Symp. Quant. Biol. **26**, 389–401 (1961)

603. L.H.A. Monteiro, A. Pellizari Filho, J.G. Chaui-Berlinck, J.R.C. Piquiera, Oscillation death in a two-neuron network with delay in a self-connection. J. Biol. Sci. **15**, 49–61 (2007)

604. H.J. Morowitz, *Energy Flows in Biology* (Ox Bow Press, Woodbridge, CT, 1970)

605. H.J. Morowitz, *Entropy for Biologists: An Introduction to Thermodynamics* (Academic Press, New York, 1970)

606. K. Morris, *An Introduction to Feedback Controller Design* (Harcourt/Academic Press, New York, 2001)

607. C. Morris, H. Lecar, Voltage oscillations in the barnacle giant muscle fiber. Biophys. J. **35**, 193–213 (1981)

608. R.P. Morse, E.F. Evans, Enhancement of vowel coding for cochlear implants by addition of noise. Nature Med. **2**, 928–932 (1996)

609. F. Moss, Stochastic resonance: from the ice ages to the monkey's ear, in *Some Problems in Statistical Physics*. ed. by G.H. Weiss (SIAM Frontiers in Applied Mathematics, Philadelphia, 1994), pp. 205–253

610. F. Moss, J.G. Milton, Balancing the unbalanced. Nature **425**, 911–912 (2003)

611. G.K. Motamedi, R.P. Lesser, D.L. Miglioretti, Y. Mizuon-Matsumoto, B. Gordon, W.R.S. Webber, D.C. Jackson, J.P. Sepkuty, N.E. Crone, Optimizing parameters for terminating cortical after discharges with pulse stimulation. Epilepsia **43**, 836–846 (2002)

612. V.B. Mountcastle, The columnar organization of the neocortex. Brain **20**, 701–712 (1997)

613. M. Moussaid, D. Helbing, G. Theraulaz, How simple rules determine pedestrian behavior and crowd disasters. Proc. Natl. Acad. Sci. USA **108**, 6884–6888 (2011)

614. M.K. Muezzinoglu, I. Tristan, R. Huerta, V.S. Afraimovich, M.I. Rabinovich, Transients versus attractors in complex networks. Int. J. Bifurc. Chaos **20**, 1653–1675 (2010)

615. S. Mukherji, A. van Oudenaarden, Synthetic biology: understanding biological design from synthetic circuits. Nature Rev. Gen. **10**, 859–871 (2009)

616. I. Müller, Entropy in nonequilibrium, in *Entropy*. ed. by A. Greven, G. Keller, G. Warnecke (Princeton University Press, Princeton, NJ, 2003), pp. 79–105

617. J.D. Murray, *Mathematical Biology* (Springer, New York, 1989)

618. J.D. Murray, *Mathematical Biology: I: An Introduction*, 3rd edn. (Springer, New York, 2002)

619. R.M. Murray, Z. Li, S.S. Sastry, *A Mathematical Introduction to Robotic Manipulation* (CRC Press, Boca Raton, 1994)

620. T. Musha, S. Sato, M. Yamamoto, *Noise in Physical Systems and $1/F$ Fluctuations* (IOS Press, Kyoto, Japan, 1992)

621. J.S. Nagumo, S. Arimoto, S. Yoshizawa, An active pulse transmission line simulating a nerve axon. Proc. IRE **50**, 2061–2070 (1962)

622. K.I. Naka, W.A. Rushton, S-potentials from color units in the retina of fish. J. Physiol. **185**, 584–599 (1966)

623. H. Nakao, Asymptotic power law of moments in a random multiplicative process with weak additive noise. Phys. Rev. E **58**, 1591–1601 (1998)

624. A.M. Nakashima, M.J. Borland, S.M. Abel, Measurement of noise and vibration in Canadian forces armored vehicles. Ind. Health **45**, 318–327 (2007)

625. P.D. Nallathamby, K.J. Lee, X.-H.N. Xu, Design of stable and uniform single photonics for in vivo dynamics imaging of nanoenvironments of zebrafish embryonic fluids. ACS Nano **2**, 1371–1380 (2008)

626. National Research Council, *Transforming Undergraduate Education for Future Research Biologists* (National Academy Press, 2003)

627. L.A. Necker, Observations on some remarkable optical phenomena seen in Switzerland and on an optical phenomenon which occurs on viewing of a crystal or geometrical solid. London Edinburgh Phil. Mag. J. Sci. **1**, 329–337 (1832)

628. P.D. Neilson, M.D. Neilson, N.J. O'Dwyer, Internal models and intermittency: a theoretical account of human tracking behavior. Biol. Cybern. **58**, 101–112 (1988)

629. S.R. Nelson, M.Y. Ali, K.M. Trybus, D.M. Warshaw, Random walk of processive, quantum dot-labeled myosin Va molecules within the actin cortex of COS-7 cells. Biophys. J. **97**, 509–518 (2009)

630. J.A. Nessler, C.J. de Leone, S. Gilliland, Nonlinear time series analysis of knee and ankle kinematics during side by side treadmill walking. Chaos **19**, 026104 (2009)

631. J.A. Nessler, S. Gilliland, Interpersonal synchronization during side by side treadmill walking is influenced by leg length differential and altered sensory feedback. Hum. Mov. Sci. **28**, 772–788 (2009)

632. J.A. Nessler, T. Spargo, A. Craig-Jiones, J.G. Milton, Phase resetting behavior in human gait is influenced by treadmill walking speed. Gait Posture **43**, 187–191 (2016)

633. J.A. Nessler, S. Heredia, J. Bélair, J. Milton, Walking on a vertically oscillating treadmill: phase synchronization and gait kinematics. PLoS ONE **12**, e0169924 (2017)

634. T.L. Netoff, C.D. Acker, J.C. Bettencourt, J.A. White, Beyond two-cell networks: experimental measurement of neuronal responses to multiple synaptic inputs. J. Comp. Neurosci. **18**, 287–295 (2005)

635. M.E.J. Newman, Power laws, Pareto distributions and Zipf's law. Cont. Phys. **46**, 323–351 (2005)

636. K.M. Newell, S.M. Slobounov, E.S. Slobounov, P.C.M. Molenaar, Stochastic processes in postural center-of-pressure profiles. Exp. Brain Res. **113**, 158–164 (1997)

637. L.G. Ngo, M.R. Rousweel, A new class of biochemical oscillator models based on competitive binding. Eur. J. Biochem. **245**, 182–190 (1997)

638. G. Nicolis, I. Prigogine, *Self-Organization in Non-Equilibrium Systems* (Wiley, New York, 1977)

639. E. Niedermeyer, F. Lopes da Silva, *Electro-Encephalography: Basic Principles, Clinical Applications and Related Fields* (Urban & Schwarzenberg, Baltimore, 1987)

640. J.M. Nieto-Villar, R. Quintana, J. Rieumont, Entropy production rate as a Lyapunov function in chemical systems: proof. Physica Scripta **68**, 163–165 (2003)

641. R. Nijhawan, The flash-lag phenomenon: object motion and eye movements. Perception **3**, 263–282 (2001)

642. R. Nijhawan, S. Wu, Compensating, time delays with neural predictions: are predictions sensory or motor?? Phil. Trans. T. Soc. **367**, 1063–1078 (2009)

643. A. Nimmerjahn, F. Kirchhoff, F. Helmchen, Resting microglial cells are highly dynamic surveillants of brain parenchyma in vivo. Science **308**, 1314–1318 (2005)

644. L. Noethen, S. Walcher, Quasi-steady state in the Michaelis–Menten system. Nonlinear Anal.: Real World Appl. **8**, 1512–1535 (2007)

645. T. Nomura, K. Kawa, Y. Suzuki, M. Nakanishi, T. Yamasaki, Dynamic stability and phase resetting during biped gait. Chaos **19**, 026103 (2009)

646. T. Nomura, S. Oshikawa, Y. Suzuki, K. Kiyone, P. Morasso, Modeling human postural sway on intermittent control and hemodynamic perturbations. Math. Biosci. **245**, 86–95 (2013)

647. A. Novich, M. Wiener, Enzyme induction as an all-or-none phenomenon. Proc. Natl. Acad. Sci. USA **43**, 553–566 (1957)

648. P.L. Nunez, *Electric Fields of the Brain: The Neurophysics of EEG* (Oxford University Press, New York, 1981)

649. Y. Nutku, Hamiltonian structure of the Lotka-Volterra equations. Phys. Lett. A **145**, 27–28 (1990)

650. E.P. Odum, *Fundamentals of Ecology* (W. B. Saunders, Philadephia, 1953)

651. T. Ohira, J. Milton, Delayed random walks: investigating the interplay between noise and delay, in *Delay Differential Equations: Recent Advances and New Directions*. ed. by B. Balachandran, T. Kalmár-Nagy, D.E. Gilsinn (Springer, New York, 2009), pp. 305–335

652. T. Ohira, J.G. Milton, Delayed random walks. Phys. Rev. E **52**, 3277–3280 (1995)

653. T. Ohira, N. Sazuka, K. Marumo, T. Shimizu, M. Takayasu, H. Takayasu, Preditability of currency market exchange. Physica A **308**, 368–374 (2002)

654. T. Ohira, T. Yamane, Delayed stochastic systems. Phys. Rev. E **61**, 1247–1257 (2000)

655. K. Omori, H. Kojima, R. Kakani, D.H. Stavit, S.M. Blaugrund, Acoustic characteristics of rough voice: subharmonics. J. Voice **11**, 40–47 (1997)

656. L. Onsager, Reciprocal relation in irreversible processes. I. Phys. Rev. **37**, 405–426 (1931)

657. L. Onsager, Reciprocal relation in irreversible processes. II. Phys. Rev. **38**, 2265–2779 (1931)

658. S. Oprisan, C. Canavier, Phase resetting and phase locking in hybrid circuits of one model and one biological neuron. Biophys. J. **87**, 2282–2298 (2004)

659. G. Orosz, R.E. Wilson, G. Stepan, Traffic jams: dynamics and control. Philos. Trans. R. Soc. A **368**, 4455–4479 (2010)

660. J.S. Orr, J. Kirk, K.G. Gary, J.R. Anderson, A study of the interdependence of red cell and bone marrow stem cell populations. Brit. J. Haematol. **15**, 23–34 (1968)

661. I. Osorio, M.G. Frei, Seizure abatement with single DC pulses: is phase resetting at play? Int. J. Neural Sys. **19**, 1–8 (2009)

662. I. Osorio, M.G. Frei, D. Sornette, J. Milton, Pharmaco-resistant seizures: self-triggering capacity, scale-free properties and predictability? Eur. J. Neurosci. **30**, 1554–1558 (2009)

663. I. Osorio, M.G. Frei, D. Sornette, J. Milton, Y.-V.C. Lai, Epileptic seizures: quakes of the brain? Phys. Rev. E **82**, 021919 (2010)

664. M.I. Page, W.P. Jencks, Entropic contributions to rate acceleration in enzymic and intramolecular reactions and the chelate effect. Proc. Natl. Acad. Sci. USA **68**, 1678–1683 (1971)

665. K. Pakdaman, C. Grotta-Ragazzo, C.P. Malta, Transient regime duration in continuous-time neural networks with delay. Phys. Rev. E **58**, 3623–3627 (1998)

666. K. Pakdaman, C. Grotta-Ragazzo, C.P. Malta, O. Arino, J.-F. Vibert, Effect of delay on the boundary of the basin of attraction in a system of two neurons. Neural Netw. **11**, 509–519 (1998)

667. C.E. Palmer, Studies of the center of gravity in the human body. Child Dev. **15**, 99–180 (1944)

668. A. Papoulis, *Probability, Random Variables and Stochastic Processes* (McGraw-Hill, New York, 1965)

669. E. Parzen, On the estimation of a probability density function and mode. Ann. Math. Stat. **33**, 1065–1076 (1962)

670. E. Parzen, *Stochastic Processes* (SIAM, Philadelphia, 1999)

671. J.H. Pasma, T.A. Boonstra, J. van Kordekin, V.V. Spyropoulou, A.C. Shouten, A sensitivity analysis of an inverted pendulum balance control model. Front. Comp. Neurosci. **11**, 59 (2017)

672. J. Pastor, B. Peckham, S. Bridgham, J. Weltzin, J. Chen, Plant community dynamics, nutrient cycling, and alternative stable equilibria in peatlands. Amer. Nat. **160**, 553–568 (2002)

673. F. Patzelt, K. Pawelzik, Criticality of adaptive control dynamics. Phys. Rev. Lett. **107**, 238103 (2011)

674. A.R. Peacocke, J.N.H. Skerrett, The interaction of aminoacridines with nucleic acids. Trans. Faraday Soc. **52**, 261–279 (1956)

675. P. Pearle, B. Collett, K. Bart, What Brown saw and you can too. Am. J. Phys. **78**, 1278–1289 (2010)

676. D. Pelz, *Dave Pelz's Putting Bible* (Doubleday, New York, 2002), pp. 89–94

677. C.-K. Peng, S.V. Buldyrev, A.L. Goldberger, S. Havlin, F. Sciortino, M. Simons, H.E. Stanley, Long-range correlations in nucleotide sequences. Nature **356**, 168–170 (1992)

678. C.-K. Peng, S.V. Buldyrev, A.L. Goldberger, S. Havlin, M. Simons, H.E. Stanley, Finite-size effects on long-range correlations: implications for analyzing DNA sequences. Phys. Rev. E **47**, 3730–3733 (1993)

679. D.H. Perkel, J.H. Schulman, T.H. Bullock, G.P. Moore, J.P. Segundo, Pacemaker neurons: effects of regularly spaced synaptic inputs. Science **145**, 61–63 (1964)

680. J. Perrin, *Brownian Movement and Molecular Reality* (Dover Publications, New York, 2005)

681. R.J. Peterka, A.C. Sanderson, D.P. O'Leary, Practical considerations in the implementation of the French-Holden algorithm for sampling neuronal spike trains. IEEE Trans. Biomed. Engng. **25**, 192–195 (1978)

682. A. Peters, The effects of normal aging on myelinated nerve fibers in monkey central nervous system. Front. Neuroanatomy **3**, 11 (2009)

683. A. Pikovsky, M. Rosenblum, J. Kurths, *Synchronization: A Universal Concept in Nonlinear Sciences* (Cambridge University Press, New York, 2001)

684. H.M. Pinsker, *Aplysia* bursting neurons as endogenous oscillators. I. Phase–response curves for pulsed inhibitory input. J. Neurophysiol. **40**, 527–543 (1977)

685. M. Planck, Hamiltonian structures for the n-dimensional Lotka-Volterra equation. J. Math. Phys. **36**, 3520–3524 (1995)

686. R.E. Plant, A Fitzhugh differential-difference equation modeling recurrent neural feedback. SIAM J. Appl. Math. **40**, 150–162 (1981)

687. N. Platt, E.A. Spiegel, C. Tresser, On-off intermittency: a mechanism for bursting. Phys. Rev. Lett. **70**, 279–282 (1993)

688. R.J. Polge, E.M. Mitchell, Impulse response determination by cross correlation. IEEE Trans. Aerospace Elec. Sys. **6**, 91–97 (1970)

689. Y. Pomeau, P. Manneville, Intermittent transition to turbulence in dissipative dynamical systems. Commun. Math. Phys. **74**, 189–198 (1980)

690. N.P. Poolos, D. Johnston, Dendritic ion channelopathy in acquired epilepsy. Epilepsia **53**(Suppl 9), 32–40 (2012)

691. S.A. Prescott, Y. De Koninck, T.J. Sejnowski, Biophysical basis for three distinct dynamical mechanisms of action potential initiation. PLoS Comp. Biol. **4**, e1000198 (2008)

692. W.H. Press, S.A. Teukolsky, W.T. Vetterling, B.P. Flannery, *Numerical Recipes: The Art of Scientific Computing*, 3rd edn. (Cambridge University Press, New York, 2007)

693. A.J. Preyer, R.J. Butera, Neuronal oscillators in aplysia californica that demonstrate weak coupling in vivo. Phys. Rev. Lett. **95**, 138103 (2005)

694. G.A. Prieto, R.L. Parker, F.L. Vernon III., A Fortran 90 library for multitaper spectrum analysis. Comput. Geosci. **35**, 1701–1710 (2009)

695. I. Prigogine, I. Stengers, *Order Out of Chaos: Man's Dialogue with Nature* (Bantam Books, New York, 1984)

696. J.W.S. Pringle, V.J. Wilson, The response of a sense organ to harmonic stimulus. J. Exp. Bio **29**, 220–235 (1952)

697. A. Priplata, J. Niemi, M. Salen, J. Harry, L.A. Lipsitz, Noise-enhanced human balance control. Phys. Rev. Lett. **89**, 238101 (2002)

698. A.A. Priplata, J.B. Niemi, J.D. Harry, L.A. Lipsitz, J.J. Collins, Vibratory insoles and balance control in elderly people. Lancet **362**, 1123–1124 (2003)

699. R.M. Pritchard, Visual illusions viewed on stabilized retinal images. Quart. J. Exp. Psychol. **10**, 77–81 (1958)

700. P.R. Protachevicz, F.S. Borges, E.L. Lameu, P. Ji, K.C. Iarosz, A.H. Kihara, I.L. Caldas, J.D. Szezech Jr., M.S. Baptista, E.E.N. Macau, V.G. Antonpoulos, A.M. Batista, J. Kurths, Bistable firing patterns in a neural network model. Front. Comp. Neurosci. **13**, 19 (2019)

701. R. Pušenjak, Application of Lambert function in the control of production systems with delay. Int. J. Eng. Sci. **6**, 28–38 (2017)

702. P.N. Pusey, Brownian motion goes ballistic. Science **332**, 802–803 (2011)

703. K. Pyragus, Continuous control of chaos by elf-controlling feedback. Phys. Lett. A **170**, 421–428 (1992)

704. A. Quan, I. Osorio, T. Ohira, J. Milton, Vulnerability to paroxysmal oscillations in delayed neural networks: a basis for nocturnal frontal lobe epilepsy? Chaos **21**, 047512 (2011)

705. M.I. Rabinovich, R. Heurta, G. Laurent, Transient dynamics for neural processing. Science **321**, 48–50 (2008)

706. M.I. Rabinovich, P. Varona, A.I. Selveston, H.D.I. Abarbanel, Dynamical principles in neuroscience. Rev. Mod. Phys. **78**, 1213–1265 (2006)

707. G. Raivich, Like cops on the beat: the active role of resting microglia. Trends Neurosci. **28**, 671–573 (2005)

708. W. Rall, Distinguishing theoretical synaptic potentials computed for different soma-dendritic distributions of synaptic inputs. J. Neurophysiol. **30**, 1138–1168 (1967)

709. C.V. Rao, D.M. Wolf, A.P. Arkin, Control, exploitation and tolerance of intracellular noise. Nature **420**, 231–237 (2002)

710. A. Rapoport, "Ignition" phenomena in random nets. Bull. Math. Biophys. **14**, 35–44 (1952)

711. J.M. Raser, E.K. O'Shea, Control of stochasticity in eukaryotic gene expression. Science **304**, 1811–1814 (2004)

712. F. Ratliff, H.K. Hartline, W.H. Miller, Spatial and temporal aspects of retinal inhibitory interaction. J. Opt. Soc. Am. **53**, 110–120 (1963)

713. F. Ratliff, B.W. Knight, N. Graham, On tuning and amplification by lateral inhibition. Proc. Natl. Acad. Sci. USA **62**, 733–740 (1868)

714. J.P.H. Reulen, J.T. Marcus, M.J. van Gilst, D. Koops, J.E. Bos, G. Tiesinga, F.R. de Vries, K. Boshuizen, Stimulation and recording of dynamic pupillary reflex: the iris technique. Med. Bio. Eng. Comp. **26**, 27–32 (1988)

715. A.D. Reyes, E.E. Fetz, Effects of transient depolarizing potentials on firing rate of cat neocortical neurons. J. Neurophysiol. **69**, 1673–1683 (1993)

716. A.M. Reynolds, Bridging the gap between correlated random walks and Lévy walks: autocorrelation as a source Lévy walk movement patterns. J. R. Soc. Interface **7**, 1753–1758 (2010)

717. A.M. Reynolds, Can spontaneous cell movements be modeled as Lévy walks? Physica A **389**, 273–277 (2010)

718. F. Rieke, D. Warland, W. Cialek, Coding efficiency and information rates in sensory neurons. Europhys. Lett. **22**, 151–156 (1993)

719. A.D. Riggs, S. Bourgeois, M. Cohn, The lac repressor-operator interaction. 3. Kinetic studies. J. Mol. Biol. **53**, 401–417 (1970)

720. J. Rinzel, S.M. Baer, Threshold for repetitive activity for a slow stimulus ramp: a memory effect and its dependence on fluctuations. Biophys. J. **54**, 551–555 (1988)

721. J. Rinzel, G.B. Ermentrout, Analysis of neural excitability and oscillations, in *Method in Neuronal Modeling: From Synapses to Networks*. ed. by C. Koch, I. Segev (MIT Press, Cambridge, 1989), pp. 135–169

722. H. Risken, *The Fokker-Planck Equation* (Springer, New York, 1989)

723. M.A.J. Roberts, E. August, A. Hamadeh, P.K. Maini, P.E. McSharry, J.P. Armitage, A. Papchritodoulou, A model invalidation-based approach for elucidating biological signaling pathways, applied to the chemotaxis pathway in *R. sphaeroides*. BMC Syst. Biol. **3**, 105 (2009)

724. P.A. Rock, *Chemical Thermodynamics: Principles and Applications* (Collier-Macmillan Ltd, London, 1969)

725. X. Rodet, Sound and music from Chua's circuit. J. Circuits Syst. Comput. **3**, 49–61 (1993)

726. E. Ronco, T. Arsan, P.J. Gawthrop, Open-loop intermittent feedback control: practical continuous GPC. IEE Proc. Part D: Control. Theor. Appl. **146**, 426–436 (1999)

727. J.E. Rose, J.F. Bruggee, D.J. Anderson, J.E. Hind, Phase-locked response to low-frequency tones in a single auditory nerve fiber of the squirrel monkey. J. Neurophysiol. **30**, 769–793 (1967)

728. R. Rosen, *Dynamical System Theory in Biology*, vol. 1 (Wiley-Interscience, New York, 1970)

729. J. Rosenhouse, *The Monty Hall Problem: The Remarkable Story of Math's Most Contentious Brain Teases* (Oxford University Press, New York, 2009)

730. R. Ross, Some a priori pathometric equations. Br. Med. J. **1**, 546–547 (1911)

731. L.L. Rubchinsky, C. Park, R.M. Worth, Intermittent neural synchronization in Parkinson's disease. Nonlinear Dyn. **68**, 329–346 (2012)

732. J. Rubin, D. Terman, High frequency stimulation of the subthalamic nucleus eliminates pathological thalamic rhythmicity in a computational model. J. Comput. Neurosci. **16**, 211–235 (2004)

733. J. Rudnick, G. Gaspari, *Elements of the Random Walk: An Introduction for Advanced Students and Researchers* (Cambridge University Press, New York, 2004)

734. U. Ryde-Pettersson, Oscillation in the photosyntheic Calvin cycle: examination of a mathematical model. Acta Chem. Scand. **46**, 406–408 (1992)

735. Y. Sako, A. Kusumi, Compartmentalized structure of the plasma membrane for receptor movements as revealed by a nanometer-level motion analysis. J. Cell Bio. **125**, 1251–1264 (1994)

736. M. Santillan, M.C. Mackey, Dynamic regulation of the tryptophan operon: a modeling study and comparison with experimental data. Proc. Natl. Acad. Sci. USA **98**, 1364–1369 (2001)

737. C.B. Saper, P.M. Fuller, N.P. Pederson, J. Lui, T.E. Scammell, Sleep state switching. Neuron **68**, 1023–1042 (2010)

738. A.-H. Sato, H. Takayasu, Y. Sawada, Invariant power laws distribution of Langevin systems with colored multiplicative noise. Phys. Rev. E **61**, 1081–1087 (2000)

739. A.-H. Sato, H. Takayasu, Y. Sawada, Power law fluctuation generator based on analog electrical circuit. Fractals **8**, 219–225 (2009)

740. M.J. Saxton, Single-particle tracking: the distribution of diffusion coefficients. Biophys. J. **72**, 1744–1753 (1997)

741. M. Scheffer, *Critical Transitions in Nature and Society* (Princeton University Press, Princeton, NJ, 2009)

742. A.M. Schillings, B.M.H. Van Wezel, T.H. Mülder, J. Duysens, Muscular responses and movement strategies during stumbling over obstacles. J. Neurophysiol. **83**, 2093–2102 (2000)

743. K.A. Schindler, C.A. Bernasconi, R. Stoop, P.H. Goodman, R.J. Douglas, Chaotic spike patterns evoked by periodic inhibition of rat cortical neurons. Z. Naturforsch **25a**, 509–512 (1997)

744. Schmidt-Nielsen, *Scaling: Why Is Animal Size so Important?* (Cambridge University Press, Cambridge, MA, 1985)

745. J. Schnakenberg, Network theory of microscopic and macroscopic behavior of master equation systems. Rev. Mod. Phys. **48**, 571–585 (1976)

746. J.P. Scholz, J.A.S. Kelso, G. Schöner, Nonequilibrium phase transitions in coordinated biological motion: critical slowing down and switching time. Phys. Lett. A **123**, 390–394 (1987)

747. H. Schulman, R. Duvivier, R. Kapral, The uterine contractility index. Am. J. Obstet. Gynecol. **145**, 1049–1058 (1983)

748. J. Schwarzenbach, K.F. Gill, *System Modeling and Control*, 3rd edn. (Halsted Press, New York, 1992)

749. L.A. Segel, *Modeling Dynamic Phenomena in Molecular and Cellular Biology* (Cambridge University Press, New York, 1984)

750. L.A. Segel, On the validity of the steady state assumption of chemical kinetics. Bull. Math. Biol. **50**, 579–593 (1988)

751. L.A. Segel, A. Jäger, D. Elias, I.R. Cohen, A quantitative model of autoimmune disease and T-cell vaccination: does more mean less? Immunol. Today **16**, 80–84 (1995)

752. L.A. Segel, M. Slemrod, The quasi-steady-state assumption: a case study in perturbation. SIAM Rev. **31**, 446–477 (1989)

753. D. Selmeczi, L. Li, L.I.I. Pedersen, S.F. Nrrelykke, P.H. Hagedorn, S. Mosler, N.B. Larsen, E.C. Cox, H. Flyvbjerg, Cell motility as random motion: a review. Eur. Phys. J. Spec. Top. **157**, 1–15 (2008)

754. R. Shadmehr, M.A. Smith, J.W. Krakauer, Error correction, sensory prediction, and adaptation in motor control. Annu. Rev. Neurosci. **33**, 89–108 (2010)

755. L.F. Shampine, S. Thompson, Solving DDE's in MATLAB. Appl. Num. Math. **37**, 441–458 (2001)

756. R. Shankar, *Basic Training in Mathematics: A Fitness Program for Science Students* (Springer, New York, 2006)

757. A.A. Sharp, L.F. Abbott, E. Marder, Artificial electrical synapses in oscillatory networks. J. Neurophysiol. **67**, 1691–1694 (1992)

758. A.A. Sharp, M.B. O'Neil, L. Abbott, E. Marder, Dynamic clamp: Computer-generated conductances in real neurons. J. Neurophysiol. **69**, 992–995 (1993)

759. A.A. Sharp, M.B. O'Neil, L. Abbott, E. Marder, The dynamic clamp: artificial conductances in biological neurons. Trends Neurosci. **16**, 389–394 (1993)

760. L.P. Shayer, S.A. Campbell, Stability, bifurcation, and multistability in a system of two coupled neurons with multiple time delays. SIAM J. Appl. Math. **61**, 673–700 (2000)

761. M.J. Shelley, L. Tao, Efficient and accurate time-stepping schemes for integrate-and-fire neuronal networks. J. Comp. Neurosci. **11**, 111–119 (2001)

762. G.M. Shepherd, *The Synaptic Organization of the Brain* (Oxford University Press, New York, 1990)

763. R.E. Sheridan, H.A. Lester, Functional stoichiometry at the nicotinic receptor: the photon cross section for phase 1 corresponds to two bis-Q molecules per channel. J. Gen. Phys. **80**, 499–515 (1982)

764. L.P. Shilnokov, A contribution to the problem of the structure of an extended neighborhood of a rough state of saddle-focus type. Math. USSR Sbornik **10**, 91–102 (1970)

765. A.I. Shilnikov, G.S. Cymbalyuk, Homoclinic saddle-node orbit bifurcations on a route between tonic spiking and bursting in neuron models. Regul. Chaotic Dyn. **9**, 281–297 (2004)

766. Y. Shimmer, E. Yokata, Cytoplasmic streaming in plants. Curr. Opin. Cell Biol. **16**, 68–72 (2004)

767. Y.-J. Shin, B. Hencey, S.M. Liplan, X. Shen, Frequency domain analysis reveals external periodic fluctuations can generate sustained p53 oscillations. PLoS ONE **6**, e22852 (2011)

768. H. Shinozaki, T. Mori, Robust stability analysis of linear time delay system by Lambert W function. Automatica **42**, 1791–1799 (2006)

769. A. Shumway-Cook, M.H. Woollacott, *Motor Control: Theory and Practical Applications* (Lippincott Williams & Wilkins, New York, 2001)

770. J. Sieber, B. Krauskopf, Extending the permissible control loop latency for the controlled inverted pendulum. Dyn. Sys. **20**, 189–199 (2005)

771. F.H. Sieling, C.C. Canavier, A.A. Prinz, Predictions of phase-locking in excitatory hybrid networks: Excitation does not promote phase-locking in pattern-generating networks as reliably as inhibition. J. Neurophysiol. **102**, 69–84 (2009)

772. J.A. Simmons, Resolution of target range by echolocating bats. J. Acosut. Soc. Amer. **54**, 157–173 (1973)

773. L. Simon, *Drug-Delivery Systems for Chemical, Biomedical and Pharmaceutical Engineering* (Wiley, New York, 2012)

774. W. Simon, *Mathematical Techniques for Biology and Medicine* (Academic Press, New York, 1972)

775. H. Smith, *An Introduction to Delay Differential Equations with Applications to the Life Sciences* (Springer, New York, 2010)
776. J. Maynard Smith, *Mathematical Ideas in Biology* (Cambridge University Press, Princeton, New Jersey, 1968)
777. L.K. Smith, E.L. Weiss, L.D. Lehmkuhl, *Brunnstrom's Clinical Kinesiology*, 5th edn. (F. A. Davis, Philadelphia, 1983)
778. D. Sobel, *Longitude: The True Story of a Genius Who Solved the Greatest Scientific Problem of His Time* (First Estate, London, 1996)
779. S. Sokol, The Pulfrich stero-illusion as an index of optic nerve dysfunction. Surv. Opthalmol. **20**, 432–434 (1976)
780. S. Solomon, M. Levy, Spontaneous scaling emergence in generic stochastic systems. Int. J. Mod. Phys. **7**, 745–751 (1996)
781. B. Song, D.M. Thomas, Dynamics of starvation in humans. J. Math. Biol. **54**, 27–43 (2007)
782. E.D. Sontag, A universal construction of Artstein's theorem in nonlinear stabilization. Syst. Control **13**, 117–123 (1989)
783. D. Sornette, *Why Stock Markets Crash: Critical Events in Complex Financial Systems* (Princeton University Press, Princeton, NJ, 2003)
784. D. Sornette, *Critical Phenomena in Natural Sciences: Chaos, Fractals, Selforganization and Disorder: Concepts and Tools* (Springer, New York, 2004)
785. D. Sornette, Dragon-Kings, black swans, and the prediction of crisis. Int. J. Terraspace Sci. Eng. **2**, 1–18 (2009)
786. D. Sornette, G. Ouillon, Dragon kings: mechanism, statistical methods, and empirical evidence. European Phys. J Spec. Top. **205**, 1–26 (2012)
787. B. Spagnolo, A. Fiasconaro, D. Valenti, Noise-induced phenomena in Lotka-Volterra systems. Fluctuation Noise Lett. **3**, L177–L185 (2003)
788. O. Sporns, R. Kötter, Motifs in brain networks. PLoS Biol. **2**, e369 (2004)
789. L.R. Stanford, Conduction velocity variations minimize conduction time differences among retinal ganglion cell axons. Science **238**, 358–360 (1987)
790. H.E. Stanley, *Phase Transitions and Critical Phenomena* (Oxford University Press, London, 1971)
791. E.B. Starikov, B. Nordén, Enthalpy-entropy compensation: a phantom or something useful? J. Phys. Chem. **111**, 14431–14435 (2007)
792. L. Stark, Environmental clamping of biological systems: pupil servomechanism. J. Opt. Soc. Amer. A **52**, 925–930 (1962)
793. L. Stark, *Neurological Control Systems: Studies in Bioengineering* (Plenum Press, New York, 1968)
794. L. Stark, The pupil as a paradigm example of a neurological control systems. IEEE Trans. Biomed. Eng. **31**, 919–930 (1984)
795. L. Stark, F.W. Campbell, J. Atwood, Pupillary unrest: an example of noise in a biological servo-mechanism. Nature **182**, 857–858 (1958)

796. L. Stark, T.N. Cornsweet, Testing a servoanalytic hypothesis for pupil oscillations. Science **127**, 588 (1958)

797. L. Stark, P.M. Sherman, A servoanalytic study of consensual pupil reflex to light. J. Neurophysiol. **20**, 17–26 (1957)

798. L.A. Steen, *Math & Bio 2010: Linking Undergraduate Disciplines* (The Mathematical Association of America, Washington, D.C., 2005)

799. R.B. Stein, R.G. Lee, T.R. Nichols, Modifications of ongoing tremors and locomotion by sensory feedback. Electroencephalogr. Clin. Neurophysiol. (suppl) **34**, 511–519 (1978)

800. G. Stepan, *Retarded Dynamical Systems: Stability and Characteristic Functions* (Wiley, New York, 1989)

801. G. Stepan, Delay effects in the human sensory system during balancing. Phil. Trans. R. Soc. A **367**, 1195–1212 (2009)

802. G. Stépán, T. Insperger, Stability of time-periodic and delayed systems: a route to act-and-wait control. Ann. Rev. Control **30**, 159–168 (2006)

803. G. Stepan, J.G. Milton, T. Insperger, Quantization improves stabilization of dynamical systems with delayed feedback. Chaos **27**, 114306 (2017)

804. H.J. Stern, A simple method for the early diagnosis of abnormality of the pupillary reaction. Brit. J. Opthalmol. **28**, 275–276 (1944)

805. A. Stephenson, On a new type of stability. Manchester Memoirs **52**, 1–10 (1908)

806. C.F. Stevens, How cortical interconnectedness varies with network size. Neural Comp. **1**, 473–479 (1965)

807. B. Stevens, S. Tanner, R.D. Fields, Control of myelination by specific patterns of nerve impulses. J. Neurosci. **1**, 9303–9311 (1998)

808. J.R. Stirling, M.S. Zakynthinaki, Stability and the maintenance of balance following a perturbation from quiet stance. Chaos **14**, 96–105 (2004)

809. A.D. Straw, B. Branson, T.R. Neumann, M.H. Dickinson, Multi-camera real-time three-dimensional tracking of multiple flying animals. J. Roy. Soc. Interface **8**, 6900–6914 (2010)

810. S.H. Strogatz, *Nonlinear Dynamics and Chaos* (Addison-Wesley, New York, 1994)

811. S.H. Strogatz, *SYNC: The Emerging Science of Spontaneous Order* (Hyperion, New York, 2003)

812. H. Struchtrup, W. Weiss, Maximum of the local entropy production becomes minimal in stationary process. Phys. Rev. Lett. **80**, 5048–5051 (1998)

813. Y. Sugiyama, M. Fukui, M. Kikuchi, K. Hasebe, A. Nakayama, Traffic jams without bottle necks: experimental evidence for the physical mechanism of the formation of a jam. New J. Phys. **10**, 033001 (2008)

814. D. Sul, P.C.B. Phillips, C.-Y. Choi, Prewhitening bias in HAC estimation. Oxford Bull. Econ. Stat. **67**, 517–546 (2005)

815. F.C. Sun, L. Stark, Switching control of accommodation: experimental and simulation responses to ramp inputs. IEEE Trans. Biomed. Eng. **37**, 73–79 (1990)

816. P.S. Swain, M.B. Elowitz, E.D. Siggia, Intrinsic and extrinsic contributions to stochasticity in gene expression. Proc. Natl. Acad. Sci. USA **99**, 12795–12800 (2002)

817. M. Swat, A. Kel, H. Herzel, Bifurcation analysis of the regulatory modules of the mammalian G1/S transition. Bioinformatics **20**, 1506–1511 (2004)

818. V.V. Sychev, *The Differential Equations of Thermodynamics*, 2nd edn. (Hemisphere Pub. Co., New York, 1991)

819. H. Takayasu, A.-H. Sato, M. Takayasu, Stable infinite variance fluctuation in randomly amplified Langevin systems. Phys. Rev. Lett. **79**, 966–969 (1997)

820. N.N. Taleb, *The Black Swan: The Impact of the Highly Improbable* (Random House, New York, 2007)

821. I. Tasaki, Demonstration of two stable states of the nerve membrane in potassium-rich media. J. Physiol. (Lond.) **148**, 306–331 (1959)

822. P.A. Tass, *Phase Resetting in Medicine and Biology* (Springer, New York, 1999)

823. A. Tempestini, C. Monico, L. Gardini, F. Vanzi, F.S. Pavine and M. Capitanio, Sliding of a single lac repressor protein along DNA is tuned by DNA-sequence and molecular switching. Nucleic Acids Res. **46**, 5001–5011 (2018)

824. M.H. Thaut, G.C. McIntosh, R.R. Rice, R.A. Miller, J. Rathburn, J.M. Brault, Rhythmic auditory stimulation in gait training for Parkinson's disease patients. Mov. Disord. **11**, 193–200 (1996)

825. R. Thom, *Structural Stability and Morphogenesis* (W. A. Benjamin Inc, Reading, MA, 1975)

826. S.T. Thornton, J.B. Marion, *Classical Dynamics of Particles and Systems*, 4th edn. (Holt, Rinehardt & Winston, Austin, TX, 1995)

827. M. Timme, F. Wolf, The simplest problem in the collective dynamics of neural networks: is synchrony stable? Nonlinearity **21**, 1579–1599 (2008)

828. J.E. Toettcher, D. Gong, W.A. Lim, O.D. Weiner, Light-based feedback for controlling intracellular signaling dynamics. Nature Meth. **8**, 837–839 (2011)

829. J.E. Toettcher, C.A. Voigt, O.D. Weiner, W.A. Lim, The promise of optogenetics in cell biology: interrogating molecular circuits in space and time. Nature Meth. **8**, 35–38 (2011)

830. T. Tomé, M.J. de Oliveira, Entropy production in nonequilibrium systems at stationary states. Phys. Rev. Lett. **108**, 020601 (2012)

831. G.M. Tondel, T.R. Candy, Human infants' accommodation responses to dynamic stimuli. Invest. Ophthalmol. Vis. Sci. **48**, 949–956 (2007)

832. J.R. Tredicce, G.L. Lippi, P. Mandel, B. Charasse, A. Chevalier, B. Picqué, Critical slowing down at a bifurcation. Am. J. Phys. **72**, 799–809 (2004)

833. E. Trucco, The smallest value of the axon density for which "ignition" can occur in a random net. Bull. Math. Biophys. **14**, 365–374 (1952)

834. W.M. Tsang, A.L. Stone, Z.N. Aldworth, J.G. Hildebrand, T.L. Daniel, A.I. Akinwande, J. Voldman, Flexible split-ring electrode for insect flight biasing using multisite neural stimulation. IEEE Trans. Biomed. Eng. **57**, 1757–1764 (2010)

835. J.J. Tyson, H.G. Othmer, *The Dynamics of Feedback Control Circuits in Biochemical Pathways, in Progress in Biophysics*, vol. 5 (Academic Press, New York, 1978), pp. 1–62

836. S.M. Ulam, J. von Neumann, On combination of stochastic and deterministic processes. Bull. Am. Math. Soc. **53**, 1120 (1947)

837. M. Ursino, C. Cuppini, E. Magosso, Multisensory Bayesian inference depends on synapse maturation during training: theoretical analysis and neural modeling implementation. Neural Comp. **29**, 735–782 (2017)

838. M. Ursino, A. Crisafullim, G. de Pellegrino, E. Magosso, V. Cuppini, Development of a Bayesian estimator for audio-visual integration: a neurocomputational study. Front. Comp. Neurosci. **11**, 89 (2017)

839. T. Ushio, K. Hirai, Chaotic behavior in piecewise linear sampled data control systems. Int. J. Non-linear Mech. **20**, 493–506 (1985)

840. T. Ushio, H. Hirai, H. Hirayama, Bifurcations and chaos in sampled data systems with dead zone. Electron. Comm. Japan **66**, 36–45 (1984)

841. C. van de Kemp, P.J. Gawthrop, H. Gollee, I.D. Loram, Refractoriness in sustained visuo-manual control: is the refractory duration intrinsic or does it depend on external system properties? PLoS ONE **9**, e1002843 (2013)

842. H. van der Krooj, E. van Asselonk, F.C.T. van der Helm, Comparison of different methods to identify and quantify balance contol. J. Biomech. **145**, 175–203 (2005)

843. P.W. van der Pes, The discovery of the Brownian motion. Scientiarum Historia **13**, 27–35 (1971)

844. B. van der Pol, Forced oscillations in a circuit with non-linear resistance (Reception with reactive triode). Phil. Mag. (London, Edinburgh, and Dublin) **3**, 65–80 (1927)

845. B. van der Pol, J. van der Mark, The heart beat considered as a relaxation oscillator, and an electrical model of the heart. Phil. Mag. (7th ser.) **6**, 763–775 (1928)

846. W. van Drongelen, *Signal Processing for Neuroscientists: Introduction to the Analysis of Physiological Signals* (Academic Press, New York, 2007)

847. G.C. van Orden, J.G. Holden, M.T. Turvey, Human cognition and $1/f$ scaling. J. Exp. Psychol. Gen. **134**, 117–122 (2005)

848. M. Venkadesan, J. Guckenheimer, F.J. Valero-Cuevas, Manipulating the edge of stability. J. Biomech. **40**, 1653–1661 (2007)

849. S.C. Venkataramani, T.M. Antonsen, E. Ott, J.C. Sommerer, Power spectrum and fractal properties of time series. Physica D **96**, 66–99 (1996)

850. A.A. Verveen, L.J. DeFelice, Membrane noise. Prog. Biophys. Mol. Biol. **28**, 189–265 (1974)

851. J.F. Vibert, M. Davis, J.P. Segundo, Recurrent inhibition: its influence upon tranduction and afferen discharges in slowly-adapting stretch receptor organs. Biol. Cybern. **33**, 167–178 (1979)

852. J. Villa, M. Strajbl, T.M. Glennon, Y.Y. Sham, Z.T. Chu, A. Warshel, How important are entropic contributions to enzyme catalysis? Proc. Natl. Acad. Sci. USA **97**, 11899–11904 (2000)

853. M. Vince, The intermittency of control movements and the psychological refractory period. Brit. J. Psychol. Gen. Sect. **38**, 149–157 (1948)

854. S. Visser, H.G.E. Meijer, H.C. Lee, W. van Drongelen, M.J.A.M. Putten, S.A. van Gils, Comparing epileptiform behavior of mesoscale detailed models and population models of neocortex. J. Clin. Neurophysiol. **27**, 471–478 (2010)

855. G.M. Viswanathan, V. Afanazyev, S.V. Buldyrev, E.J. Murph, H.E. Stanley, Lévy flight search patterns of wandering albatrosses. Nature **381**, 413–415 (1996)

856. G.M. Viswanathan, S.V. Buldyrev, S. Havlin, M.G.E. da Luz, E.P. Raposo, H.E. Stanley, Optimizing the success of random searches. Nature **401**, 911–914 (1999)

857. D. Voet, J.G. Voet, *Biochemistry*, vol. I (Biomolecular, Mechanism of Enzyme Action and Metabolism (Wiley, New York, 2003)

858. S. Vogel, *Comparative Biomechanics: Life's Physical World* (Princeton University Press, Princeton, NJ, 2003)

859. Y.M. Wang, R.H. Austin, E.C. Cox, Single molecule measurements of repressor protein 1D diffusion on DNA. Phys. Rev. Lett. **97**, 048302 (2006)

860. Y. Wang, S.S. Toon, R. Gautam, D.B. Henson, Blink frequency and duration during perimetry and their relationship to test-retest threshold variability. Inv. Ophthal. Visual Sci. **52**, 4546–4550 (2011)

861. S. Wang, M.M. Musharoff, C.C. Canavier, S. Gasparini, Hippocampal CA1 pyramidal neurons type 1 phase-response curves and type 1 excitability. J. Neurophysiol. **109**, 2757–2766 (2013)

862. R.M. Warner, *Spectral Analysis of Time-Series Data* (Guilford Press, New York, 1998)

863. R. Warwick, P.L. Williams, *Gray's Anatomy*, 35th edn. (Longman, Edinburgh, 1973)

864. R. Wayne, Excitability in plant cells. Am. Sci. **81**, 140–151 (1993)

865. S.G. Waxman, M.V.L. Bennett, Relative conduction velocities of small myelinated and non-myelinated fibers in the central nervous system. Nature New Biol. **238**, 217–219 (1972)

866. J. Wei, Axiomatic treatment of chemical reaction systems. J. Chem. Phys. **36**, 1578–1584 (1962)

867. J. Wei, C.D. Prater, The structure and analysis of complex reaction schemes. Adv. Catal. **13**, 203–392 (1962)

868. E.W. Weinstein, *Encyclopedia of Mathematics* (Chapman & Hill/ CRC, New York, 2003)

869. G.H. Weiss, *Aspects and Applications of Random Walks* (North-Holland, New York, 1994)

870. J.N. Weiss, The Hill equation revisited: uses and misuses. FASEB J. **11**, 835–841 (1997)

871. R. Werman, Stoichiometry of GABA-receptor interactions: GABA modulates the glycine receptor. Adv. Exp. Med. Biol. **123**, 287–301 (1979)

872. C.F. Westburg, Bayes' rule for clinicians: an introduction. Front. Psychol. **1**, 192 (2010)

873. R. Weron, Lévy-stable distributions revisited: tail index > 2 does not exclude the Lévy-stable regime. Int. J. Mod. Phys. C **12**, 209–223 (2001)

874. D.T. Westwick, R.E. Kearney, *Identification of Nonlinear Physiological Systems* (Wiley-Interscience, New York, 2003)

875. E.L. White, *Cortical Circuits: Synaptic Organization of the Cerebral Cortex Structure, Function and Theory* (Birkhäuser, Boston, 1989)

876. I.K. Wiesmeier, D. Dalin, C. Maurer, Elderly use proprioception rather than visual and vestibular cues for postural motor control. Front. Aging Neurosci. **7**, 97 (2015)

877. S. Wiggins, *Chaotic Transport in Dynamical Systems* (Springer, New York, 1992)

878. K. Willson, D.P. Francis, R. Wensel, A.J.S. Coats, K.H. Parker, Relationship between detrended fluctuation analysis and spectral analysis of heart-rate variability. Physiol. Meas. **23**, 385–401 (2002)

879. H.R. Wilson, Simplified dynamics of human and mammalian neocortical neurons. J. Theoret. Biol. **200**, 375–388 (1999)

880. H.R. Wilson, *Spikes, Decisions and Actions: Dynamical Foundations of Neurosciences* (Oxford University Press, New York, 1999)

881. H.R. Wilson, J.D. Cowan, Excitatory and inhibitory interactions in localized populations of model neurons. Biophys. J. **12**, 1–224 (1972)

882. A.T. Winfree, *The Geometry of Time* (Springer, New York, 1972)

883. A.T. Winfree, *When Time Breaks Down: The Three-Dimensional Dynamics of Electrochemical Waves and Cardiac Arrhythmias* (Princeton University Press, Princeton, NJ, 1987)

884. A.T. Winfree, Are cardiac waves relevant to epileptic wave propagation?, in *Epilepsy as a Dynamic Disease*. ed. by J. Milton, P. Jung (Springer, New York, 2003), pp. 165–188

885. D.A. Winter, *Biomechanics and Motor Control of Human Movement*, 3rd edn. (Wiley, Toronto, 2005)

886. D.A. Winter, A.E. Patla, E. Prince, M. Ishac, K. Gielo-Perczak, Stiffness control of balance during quiet standing. J. Neurophysiol. **80**, 1211–1221 (1998)

887. J.R. Wolpaw, D.J. McFarland, Control of a two-dimensional movement signal by a noninvasive brain-computer interface in humans. Proc. Natl. Acad. Sci. USA **101**, 17849–17854 (2004)

888. G.F. Woodman, A brief introduction to the use of event-related potentials (ERPs) in studies of perception and memory. Attn. Disorders Psychophys. **72**(2031–2064), 2010 (2010)

889. M. Woollacott, A. Shumway-Cook, Changes in posture control across the life span: a systems approach. Phys. Ther. **70**, 799–807 (1990)

890. E. Wright. A nonlinear difference-differential equation. J. Reine Angew. Math. **494**: 66–87 (1955)

891. J. Wu, H. Zivari-Piran, J.D. Hunter, J.G. Milton, Projective clustering using neural networks with adaptive delay and transmission loss. Neural Comp. **23**, 1568–1604 (2011)

892. H. Wu, P.A. Robinson, Modeling and investigation of neural activity in the thalamus. J. Theoret. Biol. **244**, 1–14 (2007)

893. C. Wunsch, The spectral description of climate change including the 100 ky energy. Clim. Dyn. **20**, 353–363 (2003)

894. N. Yamada, Chaotic swaying of the upright posture. Hum. Mov. Sci. **14**, 711–716 (1995)

895. K. Yamamoto, K. Shimada, K. Ito, S. Hamada, A. Ishijima, M. Tazawa, Chara myosin and the energy of cytoplasmic streaming. Plant Cell. Physiol. **47**, 1427–1431 (2006)

896. T. Yamamoto, C.E. Smith, Y. Suzuki, K. Kiyono, T. Tanahashi, S. Sakuda, P. Morasso, Y. Nomura, Universal and individual characteristics of postural sway during quiet standing in healthy young adults. Phys. Rep. **3**, e12329 (2015)

897. S. Yanchuk, M. Wolfrum, P. Hövel, E. Schöll, Control of steady states by long delay feedback. Phys. Rev. E **74**, 026201 (2006)

898. Y. Yarom, Rhythmogenesis in a hybrid system interconnecting an olivary neuron to an analog network of coupled oscillators. Neuroscience **44**, 263–275 (1991)

899. C. Yau, K. Campbell, Bayesian statistical learning of big data biology. Biophys. Rev. **11**, 95–102 (2019)

900. Y. Yarom, Rhythmogenesis in a hybrid system interconnecting an olivary neuron to an analog network of coupled oscillators. Neurosci. **44**, 263–275 (1991)

901. J.I. Yellot, Spectral consequences of photoreceptor sampling in the rhesus retina. Science **221**, 382–385 (1983)

902. S. Yi, S. Duan, P.W. Nelson, A.G. Ulsoy, The Lambert W function approach to time delay system on the LambertW$_{DDE}$ toolbox, in *Proceedings of the 10th IFAC Workshop on Time Delay Systems*, Boston (2012)

903. T.-M. Yi, Y. Huang, M.I. Simon, J. Doyle, Robust perfect adaptation in bacterial chemotaxis through integral feedback control. Proc. Natl. Acad. Sci. USA **97**, 4649–4653 (2000)

904. Z. Yi, K.K. Tan, Dynamic stability conditions for Lotka-Volterra recurrent neural networks with delays. Phys. Rev. E **66**, 011910 (2002)

905. G.-S. Yi, J. Wang, K.-M. Tsang, X.-L. Wei, B. Ding, Biophysical insights into how spike threshold depends on the rate of membrane depolarization in Type I and Type II neurons. PLoS ONE **10**, e01300250 (2015)

906. N. Yildirim, M.C. Mackey, Feedback regulation in the lactose operon: a mathematical modeling study and comparison with experimental data. Biophys. J. **84**, 2841–2851 (2003)

907. N. Yildrim, M. Santilan, D. Horike, M.C. Mackey, Dynamics and stability in a reduced model of the lac operon. Chaos **14**, 1279–292 (2004)

908. J. Yu, Empirical characteristic function estimation and its applications. Econometric Rev. **23**, 93–123 (2004)

909. X. Yu, E.R. Lewis, Studies with spike initiators: linearization by noise allows continuous signal modulation in neural networks. IEEE Trans. Biomed. Eng. **36**, 36–43 (1989)

910. H. Zakon, J. Oestreich, S. Tallarovic, F. Triefenbach, EOD modulations of brown ghost electric fish: JARs, chirps, rises, and dips. J. Physiol. Paris **96**, 451–458 (2002)

911. M.S. Zakynthinaki, J.R. Stirling, C.A. Cordent Martinez, A. López Diíaz de Durana, M.S. Quintana, G.R. Romo, J.S. Molinueve, Modeling the basin of attraction as a two-dimensional manifold from experimental data: applications to balance in humans. Chaos **20**, 013119 (2010)

912. B. Zalc, R.D. Fields, Do action potentials regulate myelination? Neuroscientist **6**, 1–12 (2000)

913. V.M. Zatsiorsky, *Kinetics of Human Motion* (Human Kinetics, Champaign, Ill., 2002)

914. V.M. Zatsiorsky, M. Duarte, Rambling and trembling in quiet standing. Motor Control **4**, 185–200 (2000)

915. J.P. Zbilut, *Unstable Singularities and Randomness: Their Importance in the Complexity of Physical, Biomedical and Social Systems* (Elsevier, New York, 2004)

916. Ya..B.. Zedovitch, *Higher Mathematics for Beginners and Its Applications to Physics* (MIR Publishers, Moscow, 1973)

917. F-G. Zeng, Q-J.Fu, R.P. Morse, Human hearing enhanced by noise. Brain Res. **869**, 251–255 (2000)

918. A. Zgonnikov, I. Lubashevsky, S. Kanemoto, T. Miyazawa, T. Suzuki, To act or not to react? Intrinsic stochasticity of human control in virtual stick balancing. J. R. Soc. Interface **11**, 20140636 (2015)

919. H. Ziegler, Some extremum principles in irreversible thermodynamics, in *Progress in Solid Mechanics*, vol. 4, ed. by I.N. Snedden, R. Hill (North Holland, Amsterdam, 1963), pp. 91–193

920. V.M. Zolotarev, *One-Dimensional Stable Distributions* (American Mathematical Society, Providence, RI, 1986)
921. W. Zou, D.V. Senthilkumar, A. Koseska, J. Kurths, Generalizing the transition from amplitude to oscillator death in coupled oscillators. Phys. Rev. E **88**, 050901 (2013)
922. W. Zou, Y. Tang, L. Li, J. Kurths, Oscillator death in asymmetrically delay-coupled oscillators. Phys. Rev. E **85**, 046206 (2012)

Index

© Springer Nature Switzerland AG 2021
J. Milton and T. Ohira, *Mathematics as a Laboratory Tool*,
https://doi.org/10.1007/978-3-030-69579-8

Printed in the United States
by Baker & Taylor Publisher Services